Age
and
Growth
of
Fish

Age and Growth of Fish

ROBERT C. SUMMERFELT
MANAGING EDITOR

GORDON E. HALL
TECHNICAL EDITOR

 IOWA STATE UNIVERSITY PRESS / AMES

Library of Congress Cataloging-in-Publication Data

Age and growth of fish.

"Papers presented at the International Symposium on Age and Growth of Fish, Des Moines, Iowa, June 9–12, 1985"—Pref.
Sponsored by Iowa State University, College of Agriculture, and others.
Includes bibliographies and index.
1. Fishes—Growth—Congresses. 2 Fishes—Age—Congresses. I. Summerfelt, Robert C. II. Hall, Gordon E. III. International Symposium on Age and Growth of Fish (1985 : Des Moines, Iowa) IV. Iowa State University. College of Agriculture.

QL639.15.A35   1987   597'.03'1   86–21508
ISBN 0-8138-0733-6

Contents

Kenneth D. Carlander

AN APPRECIATION
BY **ROBERT J. MUNCY**
MISSISSIPPI COOPERATIVE FISH AND WILDLIFE RESEARCH UNIT
MISSISSIPPI STATE, MISSISSIPPI 39762

These proceedings of the 1985 International Symposium on Age and Growth of Fish are dedicated to Dr. Kenneth Dixon Carlander, who has devoted much of his professional career since 1938 to documenting and providing a better understanding of growth and aging processes in fishes. His interest and knowledge has been always graciously extended to anyone expressing interest and requesting assistance. His quest for better documentation and more readily accessible information to accommodate or facilitate developing widespread professional interest in this subject area resulted in the publication of his Handbook of Freshwater Biology (Carlander 1953) and complementary volumes (Carlander 1969, 1977). He offered suggestions for standardizing and evaluating age and growth studies, based on his own efforts to process diverse records of biological data on North American freshwater fishes.

A brief history of work on age and growth in fish is revealed in Carlander's publications and those of his students as well as in his paper (Carlander 1986) in "Historical Perspective" in Session I of this symposium. Detailed studies of growth and age structure in populations of major sport fishes in Iowa lakes appeared in publication soon after Carlander's appointment in 1946 as Assistant Professor of Zoology at Iowa State University and Leader of the Iowa Cooperative Fisheries Research Unit. In the ensuing years, more than 50 scientific papers were published by Carlander and his graduate students, who used age and growth techniques in evaluating the performance of fish populations in Iowa streams, natural lakes, reservoirs, and farm ponds. Carlander evaluated new techniques for aging fish in 10 other papers, and sought to explain fish growth and aging to the general public in 10 popular articles. He also wrote 12 scientific critiques addressed to professional colleagues.

Long-term studies of fish population in Clear Lake, Iowa, in the 1950s and 1960s revealed the problems posed by missing scale annuli. Other studies showed that scale analyses did not adequately demonstrate the suspected impacts of flooding on fish populations in Iowa streams. Carlander attempted through his

graduate students' studies to evaluate index marks on fish scales, as well as to use RNA-DNA ratio techniques on wild fish subjected to various environmental stressors.

Carlander's professional career, which has already spanned 47 years, has enriched the fisheries field far beyond his personal research contributions. Since he began teaching at Iowa State University in 1946, he has directed programs of 34 Ph.D. graduates and 59 M.S. students (of whom 22 also completed Ph.D. programs--12 at Iowa State). In addition, he offered enthusiastic encouragement and opportunities for personal involvement in aquatic studies to developing undergraduate and graduate students pursuing studies in other fields, thereby expanding understanding and appreciation of the aquatic sciences.

Carlander has provided many opportunities for foreign students to study at Iowa State University to enable them to assume prominent professional roles in their own countries. While serving as a visiting professor in Egypt (1965-66) and Indonesia (1977-78), he substantially increased his personal knowledge of the environmental and social problems that confront foreign fisheries students. His expanded views not only enriched his teaching and advisory roles, but also improved and influenced educational opportunities of students in the United States. Professionals trained by him fill many major positions in foreign countries, as well as in universities, private businesses, and state and federal governments in the United States.

Kenneth D. Carlander's accomplishments in fisheries research and education, as well as his contributions to Iowa, have been recognized in many ways: selection to four scholastic honor societies; selection as a Fellow by the American Association for the Advancement of Science, the American Institute of Fishery Research Biologists, the Iowa Academy of Science, and the International Academy of Fishery Scientists; appointment in 1974 by Iowa State University as Charles F. Curtiss Distinguished Professor; appointment by Iowa governors to various councils and boards; and invitations to lecture at more than 30 universities and scientific laboratories. Carlander has been a member of more than 30 professional societies, serving on committees and boards of 11 and being elected President of the American Fisheries Society (1960-61), of Sigmi Xi, Iowa Chapter (1963-64), and of the Iowa Academy of Science (1968-69). He was presented the Award of Excellence by the American Fisheries Society in 1979 and the Distinguished Fellow Award by the Iowa Academy of Science in 1980.

Probably his greatest reward has been the continued professional recognition by, and active involvement in the growth of, the more than a century old fisheries profession in the United States. His colleagues--and especially those who studied under and worked with him--take pleasure in this special opportunity to further recognize his many valued contributions.

ACKNOWLEDGMENTS

Suggestions by Robert Summerfelt, Paul Eschmeyer, Joseph Kutkuhn, James McCann, Richard Noble, and Michael Van Den Avyle were most helpful and appreciated.

REFERENCES

CARLANDER, K.D. 1953. Handbook of freshwater fishery biology with the first supplement. Wm. C. Brown Company, Dubuque, Iowa, USA.

CARLANDER, K.D. 1969. Handbook of freshwater fishery biology. Volume 1, Iowa State University Press, Ames, Iowa, USA.

CARLANDER, K.D. 1977. Handbook of freshwater fishery biology. Volume 2, Iowa State University Press, Ames, Iowa, USA.

CARLANDER, K.D. 1986. A history of scale age and growth studies of North American freshwater fishes. In R.C. Summerfelt and G.E. Hall, editors. Age and growth of fish. Iowa State University Press, Ames, Iowa, USA.

Preface

BY **ROBERT C. SUMMERFELT**

This book contains papers presented at the International Symposium on Age and Growth of Fish, Des Moines, Iowa, June 9-12, 1985. This symposium was arranged to honor Dr. Kenneth D. Carlander at the time of his retirement from Iowa State University and to provide a forum for all who share interest in the subject.

Sustained yield management of fish, crustaceans, and shellfish resources, many of which are heavily exploited, requires a substantial body of scientific information. One important part of this information base is knowledge of age and growth. The age of fish tells their longevity, age at maturity, age at migration to sea and return to freshwater, and age when they are recruited to the fishery. Age composition data from commercial fishery catches are used to develop catch curves from which annual mortality rates are calculated. Carefully determined fish age data are essential for the calculation of growth. Growth and mortality rates, along with a measure of recruitment, provide the three most important population rate functions that are essential to the proper assessment and management of fisheries.

It is necessary and productive in all branches of science to periodically evaluate basic precepts and methodology and to assess problems and progress. There is need to evaluate aging techniques and procedures used in calculating and describing fish growth. Furthermore, there must be conceptual and practical ways to apply this information to management of fish stocks, which must be evaluated too.

The first comprehensive examination of age and growth of fish by a cosmopolitan group of scientists took place at the International Symposium on the Ageing of Fish, held in Reading, England, July 1973. Twenty-one research reports and reviews presented at that symposium were published in Ageing of Fish, edited by T. B. Bagenal. In the 12 years between the Reading and Des Moines symposia research on daily growth of larval and juvenile fish has increased greatly.

Interest in aging fish goes back many years. Scales were first used for aging fish in 1898. In contrast, the occurrence of daily growth rings in fish otoliths has been known only since Giorgio Pannella's 1971 publication in Science, "Fish Otoliths: Daily Growth Layers and Periodical Patterns."

Pannella's discovery has been considered the most important advancement in age determination in recent times.

Aging fish has suffered, however, from becoming commonplace, and too many aspects of the process have been taken for granted. Methods have sometimes been misunderstood or misapplied and validation of age determination has often been lacking. Recent findings have uncovered significant errors in aging, usually underaging, with serious consequences. Underestimates of age have contributed to overexploitation of fish stocks that have a much lower turnover rate than assumed. Therefore, in the present symposium, fundamental concepts were critically examined, and matters of variability, error, bias, and validation were strongly emphasized.

Aging of fish continues to be a tedious, labor-intensive activity. Thus, in both the Reading and the present symposia, we find a continuing search for faster and more efficient ways—such as use of microcomputers and visual analysis systems—to improve the efficiency of collecting and processing age and growth data.

Between the Reading and the Des Moines symposia, the Proceedings of the International Workshop on Age Determination of Oceanic Pelagic Fishes: Tunas, Billfishes, and Sharks was published. The proceedings contain papers presented at the meeting in Miami, Florida, February 15-18, 1982. That workshop concerned age assessment and growth of a taxonomically diverse group of pelagic marine fishes that traverse temperate and tropical oceans. Therefore, in planning for the 1985 meeting in Des Moines, we attempted both to update the information since the 1973 Reading Symposium and go beyond the scope of the 1982 Miami meeting.

The 1985 symposium in Des Moines was attended by a group of scientists largely different from those attending the 1973 symposium in Reading; only Carlander, Casselman, Hirschhorn, and Mathews presented papers at both meetings. A total of 164 persons registered for the Des Moines meeting; most were from the United States and Canada, but there were representatives from 11 other nations as well. The program included 58 oral papers presented in 12 sessions, a poster session of 15 papers and 2 panel discussions, each with a convener and 4 panelists. The subject matter was taxonomically diverse, including clams and lobsters, and sharks and bony fishes; the habitats included freshwater and marine, tropical marine, mountain streams, and man-made lakes thermally enriched from power plant effluents. Topics ranged from study of length-frequency distributions, enumeration of marks on hard parts—including scales, otoliths, and other bones—validation methods, biochemical procedures for study of short-term growth, and mathematical-statistical considerations for back-calculating and describing fish growth.

The present book of 40 papers is an outgrowth of the 1985 symposium, but not a complete proceedings. This is so because not all speakers submitted manuscripts and not all manuscripts survived the rigorous peer review process that was necessary to

obtain a quality product. Unfortunately, we were unable to
include the introductory remarks of the conveners and the
panelists. Nevertheless, this book includes 37 of 73 oral and
poster papers presented at the symposium. It contains several
substantive reviews, a good representation of contemporary
research, and an extensive bibliography.

Three papers not presented at the meeting were added to serve
special needs. The papers by Buckley and Bulow, and Busacker and
Adelman were invited to provide basic biochemical methodology for
the study of current growth rate. The paper by E. B. Brothers
("Methodological Approaches to the Examination of Otoliths in
Aging Studies") was solicited to obtain insight for study of
otoliths from his considerable experience with this topic, which
is now the most intensively investigated area in fish aging
studies. The report by the Glossary Committee, chaired by C. A.
Wilson, adds to the standardization and a better understanding of
basic terminology.

In addition to serving as a forum for those studying age and
growth of fish, the symposium and publication of this book was to
honor the professional life of Kenneth D. Carlander on the
occasion of his retirement from the faculty of Iowa State
University. R. J. Muncy was invited to prepare the dedication.
Carlander devoted much of his professional career since 1938 to
research on age and growth of fish. He is internationally known
to fishery professionals for his Handbook of Freshwater Fishery
Biology. This book was first published in 1950; three years
later it was republished in revised form with a supplement added.
Carlander then expanded it to two volumes, published in 1969 and
1977. He is currently writing the third volume. As a faculty
member at Iowa State University for 39 years, he directed many
graduate students. A substantial number of these were from
outside of the United States and they have made significant
contributions to fishery science throughout the world. It is a
pleasure to dedicate this book to Kenneth D. Carlander.

Although there is an extensive list of acknowledgements, I
wish to express appreciation to the many people who assisted in
the symposia: the speakers, panelists, and conveners who were
participants in the symposium; and the students, secretaries,
colleagues at Iowa State, and my wife, Deanne, for helping with
many activities before, during, and after the meeting. The work
of many persons who reviewed the manuscripts for this book deserve
special mention. All manuscripts were examined by at least 3
reviewers, some as many as 7. The 119 technical reviewers listed
in the acknowledgements conducted about 180 reviews. The
reviewer's comments substantially improved the quality of the
completed manuscripts. I am also grateful for the financial
assistance of the sponsors and the cooperation of the Iowa State
University Press. The congenial personalities of my editorial
colleagues, Gordon E. Hall and Carol Sanderson, made the whole
process easier. Their red and blue pencils covered every line of
manuscript at least twice.

Acknowledgments

INTERNATIONAL SYMPOSIUM ON AGE AND GROWTH OF FISH

Program: Kenneth D. Carlander, Bruce W. Menzel and Robert C.
   Summerfelt, Chairman

Program Preparation: Marie W. Anderson

Registration: Marie W. Anderson, Janice E. Berhow and Linda J.
   Ritland

Poster Session: Robert B. Moorman

Audio-Visuals and Preview Room: Gary Atchison, Chr., Gary Marty,
   Mark Sandheinrich, Roger Vancil and James Wang.

Hatchery Tour: Mike Mason, Hatchery Manager

Glossary Committee: Charles A. Wilson, Chr., Richard J. Beamish,
   Edward B. Brothers, Kenneth D. Carlander, John M. Casselman,
   John M. Dean, Ambrose Jearld, Jr., Eric D. Prince, and Alex
   Wild.

Session Conveners: Joseph H. Kutkuhn, James C. Schmulbach,
   Michael Van Den Avyle, Edward B. Brothers, Jerome V.
   Shireman, Charles W. Caillouet, Jr., Robert O'Gorman, James
   G. Wiener, Robert B. Moorman, Richard C. Hennemuth, Ambrose
   Jearld, Jr., Gene R. Huntsman, Richard L. Noble, Clarence A.
   Carlson, and James B. Reynolds.

Panel Members: Incorporating age and growth studies in policy
   development and decision making - Donald A. Mc Caughran,
   Daniel Pauly, Ronal W. Smith, and Keith Sainsbury

Panel Members: Methodology - New Directions - Steven E. Campana,
   R. J. Beamish, John M. Casselman, and Ira R. Adelman.

Invited Speakers: Ira R. Adelman, Richard J. Beamish, Frank J.
   Bulow, Kenneth D. Carlander, John M. Casselman, Peter D. M.
   MacDonald, Jacques Moreau, and Daniel Pauly

Symposium Banquet: Master of Ceremonies, Bruce W. Menzel;
    Dedication, Robert J. Muncy; Speaker, Gilbert C. Radonski;
    Entertainer, Dan Hunter

Sponsors: Iowa State University, College of Agriculture and the
    Graduate College, Iowa Conservation Commission Fisheries
    Section, and the Sport Fishing Institute

PUBLICATION OF AGE AND GROWTH OF FISH

Managing Editor: Robert C. Summerfelt

Technical Editor: Gordon E. Hall

Production Editor: Carol Sanderson

Technical Reviewers: Gary J. Atchison, Willard E. Barber, Robert
    C. Barkman, Mark D. Barnes, Richard J. Beamish, David H.
    Bennett, Roger A. Bergstedt, Jeff Boxrucker, Michael J.
    Bradford, Edward B. Brothers, Bradford E. Brown, Craig
    Brown, C. Fred Bryan, Lawrence J. Buckley, Frank J. Bulow,
    Gregor M. Cailliet, Charles W. Caillouet, Jr., Steven E.
    Campana, Kenneth D. Carlander, Daniel W. Coble, Richard E.
    Condrey, Rodney Cook, Charles C. Coutant, David Cox, Victor
    Crecco, P. J. Dare, John M. Dean, William P. Dwyer, D. M.
    Ellis, G. J. Farmer, David A. Fournier, R. A. Fredine,
    Richard V. Frie, Audrey J. Geffen, Christopher T. Gledhill,
    Robert J. Graham, Samuel H. Gruber, Terry A. Haines, Roy C.
    Heidinger, William T. Helm, Richard C. Hennemuth, George
    Hirschhorn, John M. Hoenig, N. A. Holme, William H. Horns,
    Wayne A. Hubert, Robert L. Hune, Gene R. Huntsman, C. Phillip
    Goodyear, John A. Gulland, Steven J. Gutreuter, Arne Johan
    Jensen, Barry Johnson, Cynthia Jones, Harold E. Klaassen,
    Eugene L. Lange, R. Weldon Larimore, Peter A. Larkin, Dennis
    W. Lee, K. P. Lone, R. Gregory Lough, Scott L. Marshall,
    Donald A. Mc Caughran, Gordon A. McFarlane, Bruce W. Menzel,
    Richard Methot, Steven J. Miller, Larry Mitzner, Timothy
    Modde, John R. Moring, Robert J. Muncy, John D. Neilson,
    Bryce R. Nielson, Richard L. Noble, Donald J. Orth, Steven S.
    Parker, James D. Parrish, Daniel Pauly, T. J. Pitcher, Harold
    L. Pratt, Eric D. Prince, Richard L. Radtke, Stephen Ralston,
    P. F. Randerson, James B. Reynolds, Lawrence M. Riley,
    Douglas S. Robson, James A. Rice, William E. Ricker, William
    F. Royce, Saul B. Saila, Keith Sainsbury, Mark B.
    Sandheinrich, Dennis L. Scarnecchia, James C. Schneider, Jon
    T. Schnute, Frank J. Schwartz, David A. Secor, William L.
    Shelton, Gary R. Shepherd, Jerome V. Shireman, Robert C.
    Siegfried, Michael P. Sissenwine, Gregg J. Small, William W.
    Taylor, Dale W. Toetz, John E. Thorpe, David W. Townsend,
    James H. Uchiyama, Eric C. Volk, Gordon T. Waring, Stanley M.
    Warlen, Sanford Weisberg, Bobby Gene Whiteside, Alexander
    Wild, Ron Williams, David A. Wright, and Dale L. Zimmerman

I
History, problems, and current status

A history of scale age and growth studies of North American freshwater fish[1]

KENNETH D. CARLANDER
DEPARTMENT OF ANIMAL ECOLOGY
IOWA STATE UNIVERSITY, AMES, IA 50011

ABSTRACT
    The first verified demonstration of the use of scales in
aging fish was on carp Cyprinus carpio in 1898; for the next 30
years, most studies were on marine fishes. The scale method has
been widely used on North American freshwater fish only since
1930, but much data have accumulated since then. Van Oosten's
1929 review and critique established the pattern for most
studies. Plastic impressions and computer programs were major
changes in techniques. Computers resulted in greater use of the
Fraser-Lee method or of curvilinear regressions in back calculation. They also resulted in widespread use of the regression
method, which does not correct for variance in scale size of
fish at a given length as do the traditional methods. Difficulties in interpreting scale markings were often recognized, but
in general, were given too little attention.
    A new paradigm is evolving based upon a more thorough under-
standing of factors involved in scale and body growth. Methods
of growth study are now under renewed scrutiny. We should strive
for 1) standardization of terminology, 2) better criteria for
annulus recognition, 3) improvement of back-calculation methods,
4) improved statistical interpretation of growth, 5) more validation of age determination, and 6) specification of objectives
for each study.

    Scales have been used to estimate the age and growth of fish
for almost a century, but most of the studies on freshwater fish
have been in the last 50 years. In the last few years, we have

---

[1] Journal Paper No. J-11935 of the Iowa Agriculture and Home
Economics Experiment Station, Ames, Iowa, Project No. 2378.

---

begun to question the general application of scale studies, and
we recognize a need for a new paradigm, in the sense of Kuhn
(1970). We are in a scientific revolution, perhaps not a revolution but at least a skirmish.

Van Oosten (1929), in his historical review of the scale
method, indicated that, although Hintze demonstrated in 1888 that
the age of carp could be determined from examination of scales,
Hoffbauer in 1898 usually is recognized as the first to accurately age fish from scales and to validate the results (through the
first 3 years), using carp reared in ponds. The method was
quickly adopted by Einar Lea, Knut Dahl, and others working on
the North Sea fisheries. By 1912, the methods were applied to
Pacific salmon in Canada by C. M. Fraser and C. H Gilbert.

Freshwater fishery management was quite limited at that
time, and growth studies did not develop until the 1920's. In
general, freshwater and marine fisheries investigations have been
independent of each other, and communication between the two has
not been as extensive as would be desirable. I believe that one
reason for the lack of communication is that most marine fishery
investigations are on a few major fisheries, with extensive
annual collection of data to monitor the fishery and to establish
models and plans for optimum yields, whereas freshwater investigations are of necessity less intensive and involve a greater
diversity of species and of waters. The emphasis in freshwater
management has been upon propagation and habitat maintenance
rather than maximum sustained yields. At this symposium, we
probably have the best mixture of persons related to age and
growth of marine and of freshwater fish yet attained.

Bennett (1971) credited Borodin (1924) and Barney (1924)
with bringing the scale method to the attention of fishery
workers in America. Their papers were the first on the subject
in the Transactions of the American Fisheries Society, but
earlier studies had appeared elsewhere. Barney and Anson (1920,
1923) published on growth of pygmy sunfish Elassoma zonatum and
orangespotted sunfish Lepomis humilis. The Ontario Fisheries
Research Laboratory did a number of early studies (Adamstone
1922; Clemens 1922; Couch 1922; Harkness 1922; and others),
probably stimulated by Huntsman (1918). Van Oosten (1923) reported on scales of whitefish Coregonus clupeaformis collected
from Lake Huron in 1917.

I believe that Creaser's (1926) study of scale structure of
pumpkinseed Lepomis gibbosus and Van Oosten's (1929) life history
of lake herring Leucichthys artedi were major stimuli in introducing the techniques to freshwater fishery biologists. Van
Oosten's critique was particularly valuable and still should be
required reading. These two papers were published about the same
time as the need for more information for fishery management in
inland waters was being recognized. The New York Conservation
Commission started biological surveys of the major river drain-
ages in that state in 1926, but these did not include growth
studies until 1932. The surveys were conducted in the summer by

college professors and students, because there were few
technically-trained fishery biologists at that time.
   Extensive surveys of fish growth were started in Wisconsin
by Juday and Schneberger in 1930 (Bennett 1937), Ohio (Grimm and
Bangham 1930), Michigan in 1931 (Beckman 1948), and Minnesota in
1935 (Eddy and Carlander 1942).   I started in fish-growth
studies in 1936 at the University of Minnesota under Dr. Samuel
Eddy.   I was assisted by WPA (Works Progress Administration) and
NYA (National Youth Administration) personnel; the scales were
collected by CCC (Civilian Conservation Corps) crews as part of
their lake surveys.   Similar assistance was used in other states.
These agencies, started by President Roosevelt to provide employ-
ment during the depression of the 1930's, were major stimuli to
conservation programs.   The work of Van Oosten and of Hile (1936,
1941) in the Great Lakes Fisheries Investigations at Ann Arbor
provided much of the leadership during the 1930-50's.

DEVELOPMENT OF METHODS
   The methodology described by Van Oosten (1929) has been used
in most studies, but some modifications have been made over the
years.   Tedious cleaning of scales before examination has largely
been eliminated, because examination of a few scales usually gave
at least one that was clean enough to read.   Some laboratories
use ultrasonic methods to clean scales when necessary.   Van
Oosten mounted the scales on glass slides in glycerin-gelatin,
but other mounting media, or even scales placed between slides,
were often more satisfactory.   Impressions of the outer surface
of the scales often were found to include the features needed for
recognition of annuli more clearly than did the entire scale.
The first impressions were on celluloid softened with acetone
(Nesbit 1934).   Plastics, developed in the 1940's and later improved, provided a more satisfactory material for impressions
(e.g., Dery 1983), and a number of presses were developed with
which several scales at a time could be impressed into the
plastic, which was sometimes softened by heating.
   Van Oosten (1923, et al. 1934) described an apparatus to
project an enlarged image of a scale on ground glass, which was
copied in many laboratories and was made available commercially.
Several other types of projectors have been used, and many
studies now use the readily available microfilm projectors.
Microfilm readers rarely permit an image as large as the one
Van Oosten used, and thus the measurements may not be as precise.
Recent use of sensitized screens connected to microcomputors
increase the precision of measurement and simplify calculations
(Frie 1982).

RECOGNITION OF ANNULI
   The criteria used by Van Oosten (1929) for recognizing
annuli were generally accepted, but many studies did not have

adequate samples to evaluate the interpretations, as was done in
some studies which described the features used for recognizing
annuli and gave evidence of validity on the basis of consistency
of year-class abundance and of growth rates. These rarely indicated much difference from Van Oosten.

Difficulties in interpretation were often recognized, and
the reported results were often just best guesses. In some instances, the fish with scales that were difficult to interpret
were simply eliminated from the sample. This practice biases
the results, particularly with respect to longevity and age-class
composition, because most of the difficult scales are from the
older fish, which are thus underrepresented in the final evaluation. Some workers read the scales two or more times independently to check the reproducibility of their results, but many
did not. Some believed they got better results if they knew the
size of the fish when they examined the scales, but many thought
such knowledge tended to bias the result and thus always read
the scales without that knowledge.

Special difficulty was experienced with fish that failed to
form a detectable annulus the first year when growth was minimal,
but this was believed to be of limited occurrence in most populations and could readily be recognized when it did occur. Other
examples of scales failing to register annuli were detected occasionally, but were thought to occur so infrequently that they
would not interfere with most growth studies (Carlander 1974).

BACK CALCULATION

Most of the early studies back-calculated lengths by the
direct proportion (Dahl-Lea) method because the calculations were
by hand, slide rules, or simple crank calculators. Some more
detailed studies used the Fraser-Lee method or even a curvilinear
body-scale regression. Nomographs, introduced in the 1940's,
simplified calculations (Carlander and Smith 1944, Hile 1970).

Since the 1960's the availability of computers for data
analysis and specialized programs for back-calculation made it
relatively simple to calculate body-scale regressions from each
study and make the calculations on this basis. Carlander (1981,
1982) pointed out the dangers in this practice when the samples
are not adequate for body-scale regression analysis, as when the
size range of the fish in the sample is limited, particularly
when the first few age groups are missing. Some computer programs calculate the lengths by substituting the scale measurement
into the body-scale regression based upon the sample. This
method, often referred to as the regression method, will give
very similar averages to the use of the Fraser-Lee method with
the same intercept, but the variances will be greater, and the
means for separate year classes may differ from those with the
Fraser-Lee method. The Fraser-Lee method takes into account the
fact that scale size may differ on fish of the same length and
also within the site specified for scale collection.

ACCUMULATION AND USE OF DATA

Data on many species of freshwater fishes, particularly those of interest to fishermen, accumulated rapidly from the mid-30's on. Some concept of the amount of data can be secured from the tabulations in my Handbooks (Table 1) first appearing in 1950. Much additional unpublished data are in files and reports of many laboratories. Complications in tabulating the data were the lack of uniformity in measurement and in method of back calculation, and the failure to adequately indicate the techniques used. In most early studies, standard length was used, but as biologists made more contacts with fishermen, fork or total length was used, and Hile (1948) recommended the general use of total length. In the first edition and first supplement of my Handbook, I attempted to tabulate the data independently by the type of measurement. In the later editions, I converted standard and fork lengths to total lengths even though such conversion involves some approximation. Errors also come from differences in the way individuals measure the lengths. It also was necessary to combine data from many studies in later volumes.

The main purpose of scale studies, at first, was to contribute general knowledge of the life history of the various species, to get some idea of the average growth rate, age at first maturity, and average life span. For management purposes, we wanted to know the time it took for fish to reach catchable size or legal size limit, age-class composition of the catch, and whether the growth in a given population was above or below average for the species. It was soon recognized that some lakes

Table 1. Numbers of references and species included in various editions of "Handbook of freshwater fishery biology" by K. D. Carlander.

Date of Publication	Number of references		Number of species	
	New	Previously Listed	New	Previously Listed
1950	1118		155	
1953	458		30	
1969	1884	936	157	126
1977	582	437	7	22
MSS	352	162	83	15
Totals	4394		432	
Adjusted	2985		308	

Note: The 1950 and 1953 editions included only growth data but the later ones also included other life-history data. The adjusted totals are on the basis that only half of the additional papers included growth data which is the approximate number as estimated from sample counts. The 1969 volume includes the fish other than Perciformes, the 1977 volume, the Centrarchidae, and MSS represents the manuscripts that I have prepared on the Percichthyidae and the Etheostomini (darters). Yellow perch Perca flavescens, walleye Stizostedion vitreum, sauger S. canadense, freshwater drum Aplodinotus grunniens, and a few minor species remain to be tabulated.

had slow-growing populations of yellow perch Perca flavescens, sunfish Lepomis spp., or other species due to overcrowding and that elimination of that population and restocking was often the most effective method of improving the fishing. Comparison of growth rates in various waters gave evidence of some of the environmental factors affecting growth rate, such as temperature, length of growing season (and these two factors as integrated in latitude or altitude), and water chemistry. These studies also provided evidence of differences in the relative abundance of year classes, and from this, evidence of factors related to recruitment. Year-class abundance was used to demonstrate the failure of fry planting to affect future catches of fish in several populations (Van Oosten 1937, Hile 1937, Carlander 1945, Rose 1955, Smith and Krefting 1954) and the success of such planting in other situations (Carlander et al. 1960).

Age and growth data are vital to many of the models of maximum sustained yield that have been widely used in major marine fisheries, particularly since Beverton and Holt (1957) and Ricker (1958). These models are less applicable in freshwater fisheries because they assume fairly long-term stability in the environment, which is seldom characteristic of lakes or streams. Furthermore, few freshwater fisheries justify the research effort needed to apply these models. Freshwater fishery biologists usually are responsible for many populations and for fisheries involving a variety of species; thus, they utilize the age and growth data only for clues as to the condition of the population and types of management required. Models recently have been applied to several inland fisheries despite the difficulties and they may help in guiding management.

RECOGNITION OF FAILURES OF SCALE INTERPRETATION

Some marine investigators early recognized that scale data were not as reliable as data collected from otoliths or spines. Missing annulis were also acknowledged in some inland studies, but the irregularities were not considered to be very common, nor to cause much general error. By the late 1950's and 1960's, however, some of us were beginning to realize that ages of the older fish were frequently underestimated. These errors did not greatly affect the estimates of growth or mortality rates during the early years of life, although they do affect the von Bertalannfy growth curves and the models for estimating yields, which were not extensively used in freshwater investigations.

I was rather slow to recognize the need to examine otoliths or other hard structures to check on the scale interpretations because I believed that the factors causing growth checks on scales would be the same as on these other structures and that interpretation difficulties would be similar. Simkiss (1974) showed that otolith and bone growth might be more valid because they have a higher priority in utilization of calcium, which might actually be resorbed from scales. Aass' 1972 comparison of scale and otolith readings for Coregonus albula (the same genus as in Van Osten's 1929 classic study) was also convincing evidence that we needed to be more critical of scale results. Otoliths had long been recognized as superior for age determinations of several marine fishes but usually were neglected in freshwater fishes until the 1970's when the significance of microstructure and daily rings was recognized.

Beamish and McFarlane (1983) have recently called attention to the common failure to validate age determinations in each study, a need that had been mentioned in many earlier papers, but often neglected. The need for more critical analysis was a major theme of the International Workshop on Age Determination of Oceanic Pelagic Fishes in Miami, Florida, February 1982 (see particularly Casselman 1983, Brothers 1983, and Smith 1983). Most of the emphasis at that workshop was on age determination, but Casselman said, "If we think we have problems in age assessment, we haven't seen anything until we start looking at back calculation" (Smith 1983).

WHAT NOW?

Where should we go from here in the use of scales for age and growth studies? The approach to scale studies has changed rather drastically in the last 15 years. To complete this revolution and to develop a new model for growth studies, we need a better understanding of the growth of scales and other hard structures in relation to body growth, and of the formation of annuli and other marks on the scales. This will involve physiological analysis of the growth processes. Much of this information may already be available, but I do not know of a good summary relating such knowledge to the determination of age and growth from scales.

We have been doing age and growth studies for fishery management purposes without much more knowledge of the biological bases for our method than did the early investigators. Many of us have less understanding than they did because we have assumed that they established an adequate base. Now questions are coming to the front.

Should scale studies be abandoned in favor of otoliths and bones? I think that we will continue to rely primarily on scales in many studies because of the ease of collection, preparation, and reading, and also because scales can be collected without injury to the fish. Although the interpretations of age of older fish may be subject to more error than with otoliths or bones, the data derived on growth in the earlier years probably are fairly valid and adequate for many management decisions. Nevertheless we need to be more critical of our methods and to standardize the techniques to a greater extent.

I hope that, at this Symposium, we can (1) recommend studies needed in the near future to provide better understanding and better technology and (2) provide instructions (which may need periodic revision) for fish growth studies. As a starter I will make some suggestions:

1. Standardization of terminology. A good start on this has been made by Jearld (1983) and by the glossary in Prince and Pulos (1983).
2. Improvement of recognition of annuli. This should involve (a) research on the physiological and environmental processes in formation of annuli and other checks, (b) development of criteria for recognition of annuli for various species with clearly marked photographs, and (c) development of electronic scanning devices for recognition of the annuli. It is likely that the electronic scanners will have to be programmed for each species and, possibly, for individual populations. Exchange of data for comparisons of readings from different laboratories should also be encouraged (Casselman 1983).
3. Improvement of back-calculation methods. This involves determination of (a) the relationship between the growth of body length and scale measurement (physiology and synchrony or asynchrony of the two), (b) the best area for scale collection, (c) the best method of scale measurement, (d) the body-scale relationship, and (e) the best method of calculation. The study of body-scale relationships should include (a) determination of whether the curvilinearity is significant to back calculation; (b) differences between body-scale relationships of different year classes within the same population, and the effect of these differences on body-scale relationships based upon one-time collections; and (c) seasonal changes in body-scale relationships, if the body and scale growth are not synchronous.
4. Development of better statistical procedures for interpretation and comparisons of growth (Dapson 1980).
5. Validation techniques should be used wherever possible, though it is recognized that these may not be feasible in many investigations.
6. For many species, we no longer need studies just to add to the general description of their growth. The reasons for each scale study should be clearly stated before undertaking the work. The questions to be answered should be specified and the sampling and analysis be planned to answer these questions most efficiently.

Scale studies are entering a new era with new challenges. It will be interesting to see how in 1990 we view the tremendous mass of growth data that have been accumulated so far.

REFERENCES

AASS, P. 1972. Age determination and year-class fluctuations of cisco, Coregonus albula L., in the Mjosa hydroelectric reservoir, Norway. Institute of Freshwater Research, Drottningholm (Sweden) Report 52:5-22.

ADAMSTONE, F. B. 1922. Rates of growth of the blue and yellow pikeperch. Publications of Ontario Fisheries Research Laboratory 5:77-86.

BARNEY, R. L. 1924. A confirmation of Borodin's scale method of age-determination of Connecticut River shad. Transactions of the American Fisheries Society 54:168-177.

BARNEY, R. L., and B. J. ANSON. 1920. Life history and ecology of the pigmy sunfish, Elassoma zonatum. Ecology 1(4):241-256.

BARNEY, R. L., and B. J. ANSON. 1923. Life history and ecology of the orange-spotted sunfish (Lepomis humilis). Report of the U.S. Fish Commission, 1922, Appendix 15:1-16.

BEAMISH, R. J., and G. A. McFARLANE. 1983. Validation of age determination estimates: the forgotten requirement. Transactions of the American Fisheries Society 12:735-743.

BECKMAN, W. C. 1948. The length-weight relationship, factors for conversions between standard and total lengths, and coefficients of condition for seven Michigan fishes. Transactions of the American Fisheries Society 75:237-256.

BENNETT, G. W. 1937. The growth of the large mouthed black bass, Huro salmoides (Lacepede), in the waters of Wisconsin. Copeia 1937(2):104-118.

BENNETT, G. W. 1971. Management of lakes and ponds. 2nd Edition. Van Nostrand Reinhold Company, New York, USA.

BEVERTON, R. J. H., and S. J. HOLT. 1957. On the dynamics of exploited fish populations. United Kingdom Ministry of Agriculture and Fisheries, Fishery Investigations, Series II, Volume XIX. London, England.

BORODIN, N. 1924. Age of shad (Alosa sapidissima Wilson) as determined by the scales. Transactions of the American Fisheries Society 54:178-184.

BROTHERS, E. B. 1983. Summary of round table discussions on back calculations. Pages 35-44 in L. D. Prince and L. M. Palos, editors. Proceedings of the International Workshop on Age Determination of Oceanic Pelagic Fishes: Tunas, Billfishes, and sharks. United States National Marine Fisheries Service, NOAA Technical Report NMFS 8, Miami, Florida, USA.

CARLANDER, K. D. 1945. Age, growth, sexual maturity and population fluctuations of the yellow pike-perch, Stizostedion vitreum vitreum (Mitchill), with reference to the commercial fisheries, Lake of the Woods, Minnesota. Transactions of the American Fisheries Society 73:90-107.

CARLANDER, K. D. 1950. Handbook of freshwater fishery biology. Wm. C. Brown Company, Dubuque, Iowa, USA.

CARLANDER, K. D. 1953. First supplement to Handbook of fishery biology. Wm. C. Brown Company, Dubuque, Iowa, USA.

CARLANDER, K. D. 1969. Handbook of freshwater fishery biology.
    Volume one. Iowa State University Press, Ames, Iowa, USA.
Carlander, K. D. 1974. Difficulties in ageing fish in relation
    to inland fishery management. pages 200-205 in T. B. Bagenal,
    Editor, Ageing of fish, Unwin Brothers Limited, Old Woking,
    Surrey, England.
CARLANDER, K. D. 1977. Handbook of freshwater fishery biology.
    Volume two. Iowa State University Press, Ames, Iowa, USA.
CARLANDER, K. D. 1981. Caution on the use of the regression
    method of back-calculating length from scale measurements.
    Fisheries (Bethesda) 6(1):2-4. (see also Corrections. ibid.
    8(5):25).
CARLANDER, K. D. 1982. Standard intercepts for calculating
    lengths from scale measurements for some centrarchid and
    percid fishes. Transactions of the American Fisheries
    Society 111:332-336.
CARLANDER, K. D., and L. L. SMITH, Jr. 1944. Some uses of nomo-
    graphs in fish growth studies. Copeia 1944(3):157-162.
CARLANDER, K. D., R. R. WHITNEY, E. B. SPEAKER, and K. MADDEN.
    1960. Evaluation of walleye fry stocking in Clear Lake,
    Iowa, by alternate-year planting. Transactions of the Amer-
    ican Fisheries Society 89:249-254.
CASSELMAN, J. M. 1983. Age and growth assessment of fish from
    their calcified structures -- techniques and tools. Pages
    1-17 in L. D. Prince and L. M. Palos, editors. Proceedings
    of the International Workshop on Age Determination of
    Oceanic Pelagic Fishes: Tunas, Billfishes, and sharks.
    United States National Marine Fisheries Service, NOAA Tech-
    nical Report NMFS 8, Miami, Florida, USA.
CLEMENS, W. A. 1922. A study of the ciscoes of Lake Erie.
    Contributions to Canadian Biology 4:75-85.
COUCH, J. H. 1922. The rate of growth of the white fish
    (Coregonus albus) in Lake Erie. Publications of Ontario
    Fisheries Research Laboratory 7:97-107.
CREASER, C. W. 1926. The structure and growth of the scales of
    fishes in relation to the interpretation of their life
    history, with special reference to the sunfish, Eupomotis
    gibbosus. University of Michigan, Museum of Zoology, Mis-
    cellaneous Publications 17:1-82.
DAPSON, R. W. 1980. Guidelines for statistical usage in age
    estimation techniques. Journal of Wildlife Management 44:541
    -548.
DERY, L. M. 1983. Use of laminated plastic to impress fish
    scales. Progressive Fish-Culturist 45:88-89.
EDDY, S., and K. D. CARLANDER. 1942. Growth rate studies of
    Minnesota fishes. Minnesota Department of Conservation,
    Fisheries Research Investigational Report 28, St. Paul,
    Minnesota, USA.
FRIE, R. V. 1982. Measurement of fish scales and back-calcula-
    tion of body lengths using a digitizing pad and microcompu-
    ter. Fisheries (Bethesda) 7(6):5-8.

GRIMM, W. V., and R. V. BANGHAM. 1930. Growth of Buckeye Lake fishes in 1930-- 3 common species compared. Ohio Division of Conservation Bulletin 71:1-11.

HARKNESS, W. J. K. 1922. The rate of growth of the yellow perch (Perca flavescens) in Lake Erie. Publications of Ontario Fisheries Research Laboratory 6:23-31.

HILE, R. 1936. Age and growth of the cisco, Leucichthys artedi (LeSueur), in the lakes of the northeastern highlands, Wisconsin. U.S. Bureau of Fisheries Bulletin 48(19):211-317.

HILE, R. 1937. The increase in abundance of yellow pikeperch, Stizostedion vitreum (Mitchill), in Lakes Huron and Michigan, in relation to the artificial propagation of the species. Transactions of the American Fisheries Society 66: 143-159.

HILE, R. 1941. Growth of the rock bass, Ambloplites rupestris (Rafinesque), in Nebish Lake, Wisconsin. Transactions of the Wisconsin Academy of Science, Arts and Letters 33:189-337.

HILE, R. 1948. Standardization of methods of expressing lengths and weights of fish. Transactions of the American Fisheries Society 75:157-164.

HILE, R. 1970. Body-scale relation and calculation of growth in fishes. Transactions of the American Fisheries Society 99: 468-474.

HUNTSMAN, A. G. 1918. The scale method of calculating the rate of growth in fishes. Transactions of Royal Society of Canada, Series III, 12(4):47-52.

JEARLD, A., Jr. 1983. Age determination. Pages 301-324 in L. A. Nielsen and D. L. Johnson, editors. Fishery techniques. The American Fisheries Society, Bethesda, Maryland, USA.

KUHN, T. S. 1970. The structure of scientific revolutions. 2nd edition enlarged. International encyclopedia of unified science, Vol. 2, No. 2. University of Chicago Press, Chicago, Illinois, USA.

NESBIT, R. 1934. A convenient method for preparing celluloid impressions of fish scales. Journal du Conseil 9:373-376.

PRINCE, L. D., and L. M. PALOS, editors. 1983 Proceedings of the International Workshop on Age Determination of Oceanic Pelagic Fishes: Tunas, Billfishes, and sharks. United States National Marine Fisheries Service, NOAA Technical Report NMFS 8, Miami, Florida, USA.

RICKER, W. E. 1958. Handbook of computations for biological statistics of fish populations. Fisheries Research Board of Canada Bulletin 119. Ottawa, Canada.

ROSE, E. T. 1955. The fluctuations in abundance of walleyes in Spirit Lake, Iowa. Proceedings of the Iowa Academy of Science 62:567-575.

SIMKISS, K. 1974. Calcium metabolism of fish in relation to ageing. Pages 1-12 in Ageing of fish, edited by T. B. Bagenal. Unwin Brothers Limited, Surrey, England.

SMITH, C. L. 1983. Summary of round table discussions on back calculation. Pages 45-47 in L. D. Prince and L. M. Palos,

editors. Proceedings of the International Workshop on Age
Determination of Oceanic Pelagic Fishes: Tunas, Billfishes,
and sharks. United States National Marine Fisheries Service,
NOAA Technical Report NMFS 8. Miami, Florida, USA.

SMITH, L. L. Jr., and L. W. KREFTING. 1954. Fluctuations in
production and abundance of commercial species in the Red
Lakes, Minnesota, with special reference to changes in the
walleye population. Transactions of the American Fisheries
Society 83:131-160.

VAN OOSTEN, J. 1923. The whitefishes (Coregonus clupeaformis).
A study of the scales of whitefishes of known ages. Zoolog-
ica 2:380-412.

VAN OOSTEN, J. 1929. Life history of the lake herring,
(Leucichthys artedi Le Sueur) of Lake Huron as revealed by
its scales, with a critique of the scale method. U.S.
Bureau of Fisheries Bulletin 44:265-428.

VAN OOSTEN, J. 1937. Artificial propagation of commercial fish
of the Great Lakes. Transactions of the North American
Wildlife Conference 2:605-612.

VAN OOSTEN, J., H. J. DEASON, and F. W. JOBES. 1934. A micro-
projection machine designed for the study of fish scales.
Journal du Conseil 9:241-248.

Current trends in age determination methodology

RICHARD J. BEAMISH AND
GORDON A. McFARLANE
DEPARTMENT OF FISHERIES AND OCEANS
FISHERIES RESEARCH BRANCH
PACIFIC BIOLOGICAL STATION
NANAIMO, BRITISH COLUMBIA V9R 5K6
CANADA

ABSTRACT
   The most important advancement in age-determination studies
from 1970 until the present has been the discovery of daily
growth rings. To date, this discovery has been limited in its
application to very young or short-lived fish.
   Recent studies have demonstrated that scale ages for some
species can result in serious underestimates of age. These
underestimates occur for freshwater and marine species but
appear to be more serious for marine fishes. The awareness that
scale ages were not always reliable, renewed the interest in
developing accurate methods of age determination and evaluating
the implications of ageing error in stock assessment. The
realization that fish live for long periods, sometimes with
minimal growth, has stimulated interest in the importance of
longevity.

   Summarizing the current trends in age determination
methodology requires a rather subjective decision about when to
begin the review. We considered 1970 to be an appropriate
starting point because the discovery of daily growth increments
occurred at this time. Our review covers methodology for both
freshwater and saltwater species. It is obvious that we can not
review all areas relating to age determination methodology in
such a short paper and therefore have omitted topics that are
less familiar to us such as automated processing of ageing
structures and some of the mathematical attemps to estimate
age. We have summarized major contributions into 6 categories
beginning with the recognition of daily growth increments.

RECOGNITION OF DAILY GROWTH INCREMENTS
   The discovery of daily growth zones (Pannella 1971) is the

most significant, recent advancement in age-determination
methodology. Our review is based on 42 papers published after
Pannella's original work. Although the discovery that daily
growth zones form in otoliths was a milestone, it has been of
limited use in resolving the traditional problems of precision
and accuracy. It is useful for identifying the position of the
first annulus (Fig. 1) (Wild and Foreman 1980; Powell 1982;
Beamish et al. 1983a), for ageing species with brief life spans
(Pannella 1974; Brothers et al. 1976) and for validating the
younger ages of a few species (Brothers et al. 1976; Taubert and
Tranquilli 1982). However, the problems of separating checks
from annuli and determining if all zones counted on a structure
equal the age of the animal, still remain.

Mugiya et al. (1981) described the development of daily
growth increments in relation to the calcium cycle in fish.
Aggregates of needle-like crystals of calcium carbonate formed
on the outer margin of the otolith and produced an incremental
zone. A second zone, called the discontinuous zone, formed
between the incremental zones. The incremental zones were
predominantly calcium and the discontinuous zone predominantly
organic matrix (Fig. 2). Both zones formed in one day.

There were conflicting reports on how environmental factors
affected the formation of these zones. There was agreement that
an endogenous rhythm was responsible for initiating development
of the increments (Taubert and Coble 1977; Tanaka et al. 1981;
Campana and Neilson 1982; Neilson and Geen 1982; Radtke and Dean
1982). Various studies concluded that feeding frequency,
photoperiod and temperature either did or did not have an effect
on the number and width of the increments (Taubert and Coble
1977; Brothers 1981; Campana and Neilson 1982; Neilson and Geen
1982; Campana 1983; Marshall and Parker 1982).

While it was clear that increments formed during periods of
starvation (Taubert and Coble 1977; Marshall and Parker 1982;
Campana 1983) and under constant photoperiod (Campana and
Neilson 1982), there was some evidence that the endogenous
rhythm could be synchronized with environmental photoperiod
(Mugiya et al. 1981). Although detailed analysis of the otolith
growth process, and factors that influence it, remains to be
documented, there is general agreement that increments could be
correlated to daily cycles.

Figure 1. Daily growth increments from the center, middle
and edge of an otolith from a 26 cm sablefish Anoplopoma
fimbria. A. Nucleus area showing the zones thought to
form immediately after hatching (1-19), and after
absorption of yolk (19-32). B. Center of otolith
showing evenly spaced wide zones. C. Edge of otolith
showing the compression of zones in the area thought to
be the annulus. (Reproduced from Beamish et al. 1983a
with permission of Alaska Sea Grant, publisher).

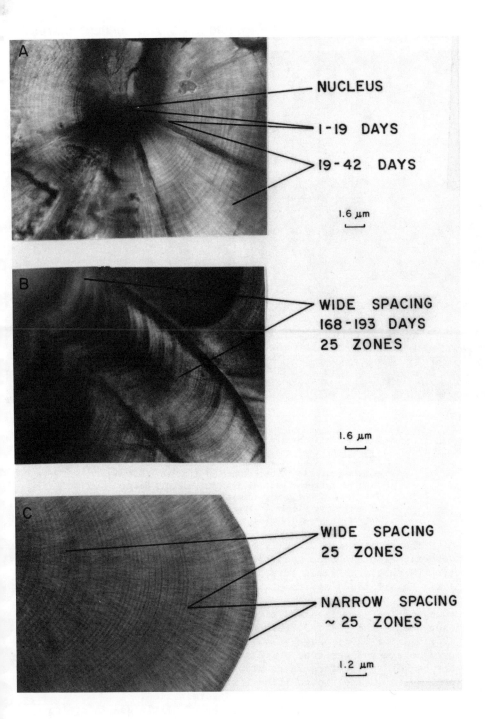

A — NUCLEUS
1-19 DAYS
19-42 DAYS
1.6 μm

B — WIDE SPACING
168-193 DAYS
25 ZONES
1.6 μm

C — WIDE SPACING
25 ZONES
NARROW SPACING
~ 25 ZONES
1.2 μm

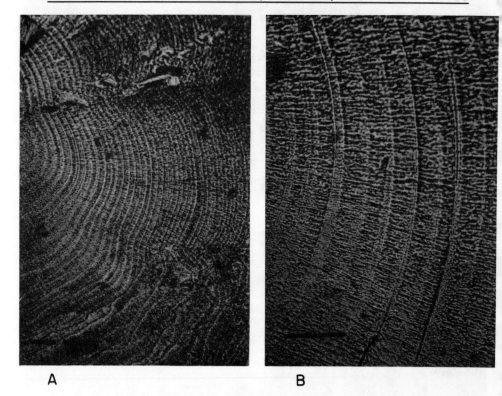

A   B

Figure 2. Daily growth increments in an otolith from
Cynoscion jamaicensis (reproduced from Pannella 1974,
with permission of the author and Unwin Brothers Ltd.,
publisher). A. Central part of the otolith. Very
regular patterns made of 2-4 increments. Scale:
100 µm. B. High magnification of A, showing structure
of zones, particularly aragonitic needles, at right
angles to growth surfaces. Scale 100 µm.

The application of daily growth increments to age
determination has been used primarily to age larval and juvenile
fish (Methot and Kramer 1979; Steffensen 1980; Barkman et al.
1981; Powell 1982) and tropical species (Pannella 1974; Ralston
1976; Struhsaker and Uchiyama 1976; Victor 1982). Growth of
young-of-the-year fish and timing of behavioural changes have
been documented (Fig. 3) (Brothers and McFarland 1981; Methot
1981; Victor 1982; Lough et al. 1982). The technique has been
used to age younger tuna (Wild and Foreman 1980; Uchiyama and
Struhsaker 1981; Brothers et al. 1983; Radtke 1983). The
ability to locate the position of the first annulus has been a
major contribution to routine age determination.

Daily growth increments have been of little use in ageing
older fish or in solving interpretation problems in adults. It
is possible that future studies of the factors influencing daily

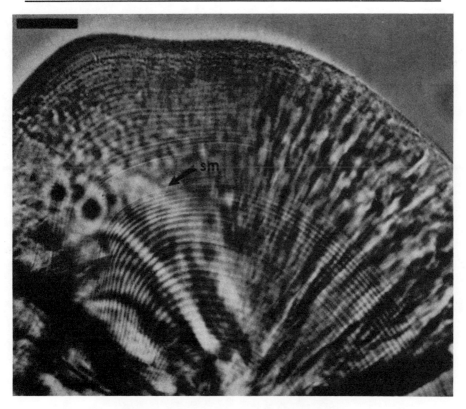

Figure 3.   Daily growth increments on sagitta of a 14.3 mm
bluehead wrasse Thalassoma bifasciatum, showing
settlement mark (sm).   Scale 40 μm. (Reproduced from
Victor 1982, with permission of the author and the
editors of Marine Biology, publisher).

increment formation will provide important information about the
fish ageing process. This will result in the development of
methodologies that measure the ageing process more directly.
Thus, the major contribution of the discovery of daily growth
zones to age determination may be in the future when the process
of ageing is better understood.

DECREASED CONFIDENCE IN THE SCALE METHOD OF AGE DETERMINATION
     In the past, the scale method of age determination was
generally accepted as a routine and accurate method for ageing
all fish in a population. Unfortunately, the method was seldom
validated (Beamish and McFarlane 1983) and investigators did not
consider the possibility that fish could live for many years

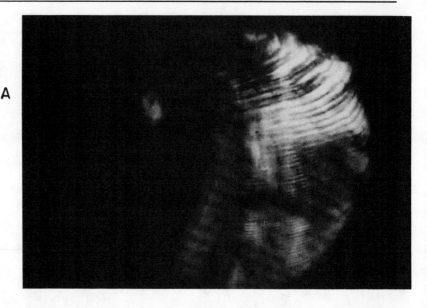

A

B

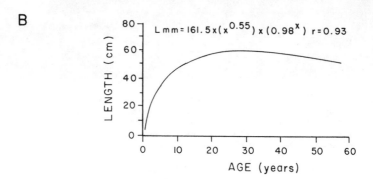

$L\,mm = 161.5 \times (x^{0.55}) \times (0.98^{x})\ r = 0.93$

Figure 4.   A.   Broken and burnt section of an otolith from a 57 year-old lake whitefish <u>Coregonus</u> <u>clupeaformis</u>. Note the thickness of the otolith.

    B.   Growth curve of lake whitefish from Lake Minto (57°13'N, 74°53'W) derived from otolith section ages. (Both figures reproduced from Power 1978, with permission of the author and the editors of the Journal of the Fisheries Research Board, publisher).

with little or no growth (Figs. 4, 5). To understand the problems associated with the scale method of age determination it is necessary to understand the development and function of the scale.

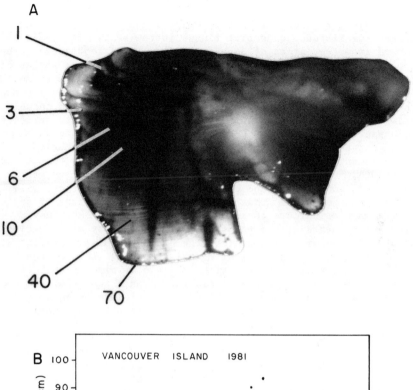

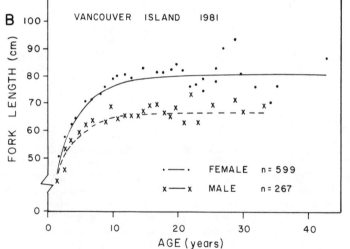

Figure 5. A. Broken and burnt section of an otolith from a 70-year-old sablefish _Anoplopoma_ _fimbria_. Note that otolith growth was confined to the ventral surface after approximately age 5+ to 6+, and close spacing of annuli after age 10+. B. Growth curve for male and female sablefish showing rapid growth to approximately age 10+ and little or no growth thereafter.

Scales of teleost fishes develop as bony plates embedded in pockets of fibrous connective tissue in the skin. The scale is formed from two layers of bone-forming cells (osteoblasts). The upper (outer) layer of the scale is characterized by a series of ridges called circuli and the lower (inner) layer forms a smooth surface. Active scale formation by osteoblasts proceeds around the periphery of the scale. In many fishes most of the annual growth occurs in a relatively short time. Scale growth reflects this pattern of fish growth resulting in periods during which little or no scale growth occurs. The length of this period varies among species, but, in general, is longer than the period of growth. Therefore, if the period of no or reduced growth becomes 12 months long and continues for many years there will be little or no scale growth (Fig. 6). Thus the nature of scale growth should indicate to investigators that scale ages may become inaccurate at the point in the life history of a species when growth becomes asymptotic. There are a number of recent studies (Table 1; Fig. 7) that clearly show that this is the case and that scales underestimate the age of older fish. If incorrect ages are assigned to older fish as a result of the use of the scale method then there is an accumulation of ages at the point where the method breaks down. This results in a serious overestimate of production.

There is sufficient evidence to show that the scale method should not be used to age some species and the older age groups of many species (Table 1). When it is applicable and has been validated for all age groups in the population it is still necessary to validate its application to other populations. Because the consequences of ageing errors can be important (Beamish and McFarlane 1983; Leaman and Beamish 1984) the general applicability of the scale method or any other method should never be assumed.

RECOGNITION THAT SOME SPECIES ARE MUCH OLDER THAN PREVIOUSLY THOUGHT

Once it was realized by some investigators that scales underestimated the age of older fish, it was found that some species were considerably older than previously thought. For example, lake whitefish Coregonus clupeaformis and lake trout Salvelinus namaycush were reported to exceed 60 and 50 yr, respectively (Power 1978); common white sucker Catostomus commersoni 23 yr (Beamish and McFarlane 1983); brook trout Salvelinus fontinalis 24 yr (Reimers 1979); and cisco Coregonus artedii 21 yr (Stelfox, Alberta Fish and Wildlife, pers. comm.). For marine fishes, a number of species have been found to be older than reported (Chilton and Beamish 1982; Berkeley and Houde 1983; Cailliet et al. 1983; Lee et al. 1983). One of the most striking examples of the change in our understanding of maximum ages has occurred for groundfish species off the west coast of Canada (Chilton and Beamish 1982), in which most of the

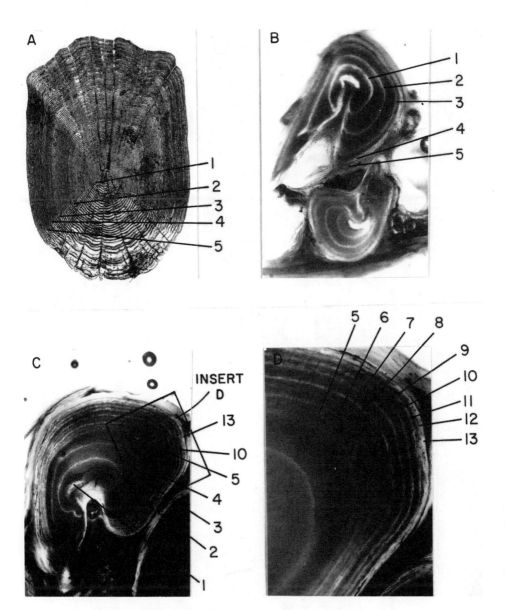

Figure 6.   A.   Scale of a white sucker <u>Catostomus</u>
<u>commersoni</u>, tagged at age 5 and recaptured 8 years
later. Using the criteria outlined by Spoor (1938) the
fish would have been aged 5 at the time of recapture.
B.   Pectoral fin ray section of the same fish
aged 5+ years at time of tagging (May 1973) and 13 years
at the time of recapture (June 1981) (C). D. Insert
shows reduction in spacing between the annuli with age.
(Sections are from different fin rays, hence their
morphologies differ).

Table 1. Examples of ageing error using the scale method. Species were selected that were relevant to this symposium and published since 1970.

Species	Reference	Comment
Salvelinus namaycush	Simard and Magnin 1972	Scale growth slows down at maturity while otoliths continue to grow in relation to length increases.
Salvelinus fontinalis	Dutil and Power 1977	Scale age was generally less than otolith age and could underestimate ages by 40%.
Coregonus albula	Aass 1972	Scale ages are mostly lower than otolith ages with disagreements at all age levels and differences as great as 7 yr.
Prosopium cylindraceum	Jessop 1972	Scales underestimated the age of oldest fish.
Coregonus clupeaformis	Power 1978	Otolith cross-section ages up to 35 yr older than previously recorded maximum age using scales.
Coregonus clupeaformis	Mills and Beamish 1980	Fin-ray ages were older than scale ages for slower growing fish.
Coregonus clupeaformis	Barnes and Power 1984	Scale ages underestimate otolith ages of western Labrador lake whitefish. Discrepancies occur from age 4 or 5 and can be considerable.
Thymallus arcticus	Craig and Poulin 1975	Scale ages underestimated the ages of old fish.
Thymallus arcticus	Sikstrom 1983	Ages determined from scales may underestimate the true age and should be used cautiously unless validated.
Esox masquinongy	Harrison and Hadley 1979	Fish older than 9 could not be aged from scales while ages of 16 were obtained using cleithra.
Rutilus rutilus	Hansen 1978	Scale ages underestimate older ages and are less suitable than opercular bones for age determination.

24

Table 1 (continued)

Species	Reference	Comment
Catostomus commersoni	Beamish 1973	Scale ages are unreliable beyond the age of maturity.
Catostomus commersoni	Beamish and McFarlane 1983	Tagging studies confirmed that scale age substantially underestimates true age and that fin-ray ages are more reliable.
Micropterus salmoides	Maraldo and MacCrimmon 1979	Scales are likely to underestimate true ages of older fish.
Perca flavescens	Schmitt and Hubert 1981	For older perch cleithra provides more reliable ages than scales.
Stizostedion vitreum vitreum	Campbell and Babaluk 1979	Scale method underestimates age of older fish in unexploited populations. Dorsal fin spines were recommended for age determination.
Stizostedion vitreum vitreum	Belanger and Hogler 1982	Otolith or pectoral fin-ray ageing methods should be used as an occasional check for scale ages.
Aplodinotus grunniens	Goeman et al. 1984	Oldest fish underestimated by scale method, scales only 61% reliable for ageing this species.
Gadus macrocephalus	Foucher et al. 1984	Length-frequency analysis indicated more older Pacific cod, slower growth, and greater longevity than the scale method in all regions of the Northeast Pacific.
Anoplopoma fimbria	Beamish and Chilton 1982	Ages from otolith sections from sablefish in the commercial fishery range from 8-45 yr compared to estimates of 5-10 yr by the scale method.
Ophiodon elongatus	Beamish and Chilton 1977	Scale ages are unreliable for fish shortly after the age of maturity. Scale ages can be only 1/2 of true age.

25

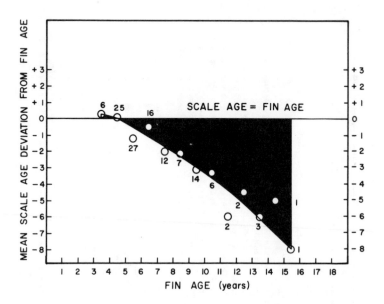

Figure 7. Deviation of mean age determined by the scale
method from age determined by the fin ray method for
lingcod <u>Ophiodon</u> <u>elongatus</u> captured in the Gulf Islands,
Strait of Georgia. Curve fitted by eye. Sample size is
indicated for each point.

commercially important species are now believed to be much
older. Nine percent of the species have estimated maximum ages
of less that 20 yr while 48% range from 21 to 49 yr and the
remainder have estimated ages from 50 to greater than 100 yr
(Table 2). For example, a rougheye rockfish <u>Sebastes</u> <u>aleutianus</u>
was estimated to be 140 yr, one of the oldest recorded ages for
any fish.

Some of the oldest estimated ages have been obtained from
examinations of otolith sections. It has been found that unlike
scales, otoliths continue to grow as the fish ages. However,
otolith growth becomes allometric because deposition occurs
predominantly on the inner surface (Figs. 4, 5). Otoliths from
older fish are obviously much thicker than from younger fish;
and when examined in cross section the thickened area contains a
prominent pattern of growth zones. If these zones are
interpreted to form annually then the fish is quite old. Some
investigators reject this interpretation although no empirical
data have been provided to support such a rejection. For some
species the validity of these older ages has been proven, and

Table 2. Maximum ages for commercially important groundfish species off
the west coast of Canada; revised from Chilton and Beamish (1982).

Family and common name	Scientific name	Maximum age(yr)
Anoplopomatidae		
Sablefish	Anoplopoma fimbria	70
Gadidae		
Pacific cod	Gadus macrocephalus	8
Pacific hake	Merluccius productus	23
Walleye pollock	Theragra chalcogramma	12
Hexagrammidae		
Lingcod	Ophiodon elongatus	21
Pleuronectidae		
Arrowtooth flounder	Atheresthes stomias	22
Rock sole	Lepidopsetta bilineata	25
Dover sole	Microstomus pacificus	45
English sole	Parophrys vetulus	22
Scorpaenidae		
Rougheye rockfish	Sebastes aleutianus	140
Pacific ocean perch	Sebastes alutus	90
Shortraker rockfish	Sebastes borealis	120
Silvergray rockfish	Sebastes brevispinis	80
Darkblotched rockfish	Sebastes crameri	47
Widow rockfish	Sebastes entomelas	58
Yellowtail rockfish	Sebastes flavidus	64
Bocaccio	Sebastes paucispinis	36
Canary rockfish	Sebastes pinniger	75
Redstripe rockfish	Sebastes proriger	41
Yellowmouth rockfish	Sebastes reedi	71
Harlequin rockfish	Sebastes varigatus	43
Sharpchin rockfish	Sebastes zacentrus	45
Squalidae		
Spiny dogfish	Squalus acanthias	80

this proof provides circumstantial evidence that interpretations
for other species may be correct.

An example of a species that has recently been found to be
quite old, is the sablefish Anoplopoma fimbria, one of the most
commercially important groundfish species off the west coast of
North America. Prior to 1981, sablefish were aged exclusively
by scales. Scale ages indicated fish in the commercial fishery
were primarily 3 to 8 yr (Low et al. 1976). Using these ages,
Low et al. developed management strategies that assumed
sablefish were relatively fast-growing, short-lived, and very
productive.

In 1981, Beamish and Chilton described a method for ageing
sablefish using sections of otoliths. Using this method the
majority of fish in the commercial fishery ranged from 4 to
40 yrs. Ages determined from otoliths indicated that males and
females matured at about the same age (5+) and a size of about
55 cm; however, male growth rates were slower after maturity and
no males grew larger than 70 cm. Females continued to grow and
reach a size greater than 100 cm, with an average size of 70
cm. Because large fish were preferred, the catches contained

more females than males. While natural mortality was shown to be similar (McFarlane and Beamish 1983), fishing mortality was higher for females. More precise ageing indicated that there was an extended period where fish continue to spawn but showed little or no increases in length (Fig. 5). This change in interpretation in the ages of sablefish, as might be expected, was met by considerable skepticism by other workers.

Ages were validated by injecting fish with 50 mg/kg of oxytetracycline (OTC) (Beamish et al. 1983a) and by examining growth of tagged fish (Beamish et al. 1983b; McFarlane et al. 1984). When these fish were recovered, and the otoliths examined, the number of zones that formed beyond the OTC mark equalled the number of years at liberty in most cases. The tagging studies confirmed that very little growth occurred and that fish were older than indicated from scale ages (McFarlane et al. 1984). Of particular interest was the large number of fish that were recovered which showed no increase in length or were shorter than when released (Table 3). The absence of growth occurred more frequently in males than females and the decreases in length could not be explained by measurement error (Beamish et al. 1983a).

Table 3. Growth of tagged sablefish at liberty for four years. Growth for males and females is recorded separately.

Release year Recapture year	1977 1981				1978 1982				1979 1983			
	≤60 cm (%)		>60 cm (%)		≤60 cm (%)		>60 cm (%)		≤60 cm (%)		>60 cm (%)	
	♂	♀	♂	♀	♂	♀	♂	♀	♂	♀	♂	♀
No growth	3(21.4)	3(50.0)	4(22.2)	7(12.5)	–	–	3(30.0)	4(20.0)	6(31.6)	1(33.3)	5(16.1)	6(16.7)
–ve growth	10(71.4)	2(33.3)	10(55.6)	31(55.4)	1(33.3)	–	4(40.0)	6(30.0)	6(31.6)	–	17(54.8)	10(27.8)
+ve growth	1(7.1)	1(16.7)	4(22.2)	18(32.1)	2(66.7)	–	3(30.0)	10(50.0)	7(36.8)	2(66.7)	9(29.1)	20(55.5)

Using this new age and growth information it was estimated that exploitation of sablefish in the northeast Pacific Ocean should have been 20 to 30% of exploitation rates determined from scale ages. Management strategies based on the new age information have resulted in a stable fishery (McFarlane et al. 1985), whereas management based on scale ages appears to have resulted in overfishing.

A number of other commercially important groundfishes, such as rockfishes, are considerably older than previously reported. Similar results undoubtably will be obtained when validation is undertaken for other species.

IMPORTANCE OF AGEING ERROR

The assessment of experimental error is fundamental in science. However, few studies discuss the error associated with age determination (Beamish and McFarlane 1983).

The realization that significant errors do occur has stimulated research into their implications. One area of investigation is the quantification and reporting of precision and accuracy. While a variety of statistical procedures were available, it was common practice to compare the precision of ages by percent agreement. This method of reporting error was insensitive to the age distribution in the fishery. For example, if young Pacific herring Clupea harengus pallasi are aged incorrectly by one year it can incorrectly indicate good or poor recruitment. Similarly, incorrectly ageing Pacific salmon Oncorhynchus sp. by one year has major implications on harvest and management strategy. In contrast, an error of up to 5 years for a long-lived species, such as an old sablefish or lake trout, may not be critical for its management. Beamish and Fournier (1981) developed a method for assessing the relative importance of error, and Chang (1982) subsequently incorporated statistical procedures to evaluate the reproducibility of age estimates. These procedures allow the precision of estimates to be compared objectively for all species.

Another area of investigation is the development of statistically based models for cohort analysis and other stock assessment techniques. The incorporation of statistics in these models emphasizes the need to assess errors. For example, the cohort analysis model of Fournier and Archibald (1982) examines the abundance of age-groups in response to natural and fishing mortalities and derives the most appropriate sequences of cohort decay. Within this process it also allows observed abundance to reflect errors in ageing and corrects these observations using the more complete information derived from the decay series, within limits specified by the user. The application of this model to Pacific ocean perch Sebastes alutus (Archibald et al. 1983) in Queen Charlotte Sound, Canada, indicated that overfishing was more serious than previously thought.

More detailed discussion of the consequences of ageing errors on stock dynamics are presented in Powers (1983).

RENEWED INTEREST IN AGE VALIDATION

The recognition that ageing error is important has stimulated an interest in the validation of age determination methods. Validation means proving a technique is accurate. Accuracy can be proven or estimated; estimates of accuracy are less valuable, but in some cases (Archibald et al. 1983) only an estimate is possible.

In a previous review (Beamish and McFarlane 1983) it was indicated that validation of ages of older fish requires either a mark-recapture study or the identification of known-age fish in the population. It may be possible to find some "natural" mark on a structure that identifies a particular year class (Thomas, 1983), but such marks appear to be very rare. For most species, the only choice is to undertake a mark-recapture study for validation.

At the time of tagging, scales or fin rays can be removed for comparison with similar structures upon recapture of the fish. It is also possible to use chemicals to mark structures. Fluorochrome labels are most commonly used, especially when a structure cannot be removed at time of marking. Chemicals are readily available, easily administered, and provide a distinct fluorescent mark on a variety of structures when exposed to ultraviolet light (Fig. 8). Appropriate dosages for oxytetracycline are discussed by Beamish et al. (1983a). The use of radionuclide ratios can be used to confirm an age range (Bennett et al. 1982).

We recognize that proper validation can only be accomplished over a number of years. While the method is being validated, age estimates should be compared using several structures. Even for a validated method we strongly recommend that comparisons among structures be a routine procedure for any laboratory providing age-determination estimates for management.

In general, the age-determination method selected for many studies probably provides accurate estimates of age over the range of more rapid growth (Beamish and McFarlane 1983). However, when growth is reduced because of the sex of an individual, maturation, food limitations, changes in behavior, environmental conditions, or other causes, it is likely that there will be a change in the appearance of an annulus and that the method of age determination will have to be modified. Only by validating the method can it be proven that fish are not older than estimated and that older fish are not an important

Figure 8. Structures used for age determination that were marked by oxytetracycline (OTC) injections. Arrow indicates OTC mark. A. Broken section of an otolith from a sablefish Anoplopoma fimbria. B. Fin ray section from a lingcod Ophiodon elongatus (reproduced from Cass and Beamish, 1983, with permission of the authors and the editors of the North American Journal of Fisheries Management, publisher). C. Broken section of an otolith from a black rockfish Sebastes malanops, (photo courtesy of B. Culver and D. Ayres, Washington Department of Fisheries, Montesano, Washington). D. Broken section of an otolith from a rock sole Lepidopsetta bilineata (photo courtesy of J. Fargo, Pacific Biological Station, Nanaimo). E. Second dorsal spine from a spiny dogfish Squalus acanthias. F. Centrum face of a vertebra from a leopard shark Triakis semifasciata, (reproduced from Smith 1984, with permission of the author and editors of Transactions of the American Fisheries Society, publisher).

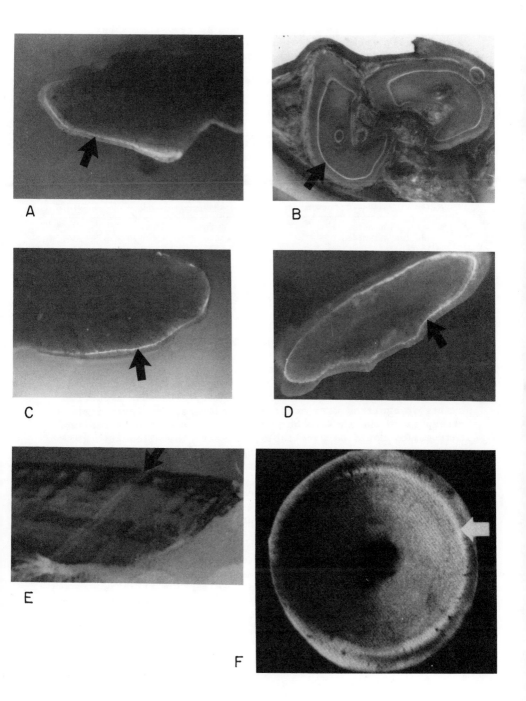

component of the population. Because so few studies have
successfully validated age-determination methods or even
attempted to do so (Table 4) (Beamish and McFarlane 1983), and
because ages are so fundamental to our understanding of biology
and stock dynamics, there is no alternative other than to
determine the accuracy of any method.

Table 4. Number of attempts to validate methods of age determinations
    from 500 publications that included age estimates for fish, from
    Beamish and McFarlane (1983).

Degree of validation	Pre-1940	1940-1969	1970-1980	Total
No validation attempted	41	80	49	170
Validation attempted	44	156	130	330
Validation of first few age groups only	42	125	109	276
Validation of all age groups	2	31	21	54
Validation successful	1	11	5	17

INCREASED USE OF NON-TRADITIONAL STRUCTURES
    Traditionally, ages were estimated using scales and the
surfaces of otoliths. Other structures such as fin rays,
vertebrae, spines, opercular bones, cleithra and length-
frequency analysis were used occasionally. Because ages from
scales or otoliths were considered accurate, few investigators
attempted to use these other methods. As researchers realized
that scales could be unreliable, and that the allometric growth
of otoliths resulted in a crowding of annuli on the edge, more
emphasis was placed on non-traditional structures and
interpretations.
    Sections of fin rays have been found to be useful where
scale ages underestimated the age of older fish (Beamish 1981),
or for stocks where other structures could not be used, e.g.,
walleye pollock Theragra chalcogramma (Beamish 1981) or tuna
(Antoine et al. 1983; Cayr and Diouf 1983). Fin rays, like
scales, may be removed without sacrificing the animal. The
clarity of the growth pattern of a ray may vary among rays on
the same fish, thus it is necessary to section all fins and all
rays before selecting a structure. Proper preparation and
viewing of fin-ray sections are essential. Sectioning at
incorrect angles and thicknesses and viewing with insufficient
light will obscure annuli. The annulus can be difficult to
interpret for the first 1 or 2 years, but it is usually quite
clear for older fish.
    Whole vertebrae have been used to age some fish. However,
the use of vertebral sections produced older ages; in
particular, examination of sections of the centrum has been

useful in ageing tunas and sharks (Gruber and Stout 1983; Smith 1984; Prince et al. 1985). As with otoliths, all parts of the vertebrae did not grow proportionally. Because the interpretations from these sections have been validated in some cases, there can be little doubt that growth zones that appear in only some parts of the vertebrae are indeed annuli (Fig. 9).

Cleithra, opercular and other bones in the head have been used by some investigators (Casselman 1974), however, they remain less popular than other structures.

Length-frequency analysis is used because age is interpreted directly from length, making this method rapid and inexpensive. It is a useful technique if a particular analysis requires information only about the fastest-growing phase in the life history of the species (Foucher and Fournier 1982). However, a main difficulty is the interpretation of ages in the distributions representing the largest fish. If it can be established by tagging, or through the use of other structures, that the frequency distribution of the largest fish contains only a few year-classes, then length-frequency analysis may be the simplest age-determination method. If this cannot be determined then length-frequency analysis is unreliable, once the distributions merge or overlap.

An example of the error that can result using length-frequency analysis is illustrated in the age determination of lamprey. It was believed from this analysis that the life cycle of lamprey lasted from 5 to 9 years (Hardisty 1961, 1969). However, Purvis (1980) proved, through the use of known age fish, that the length of the life cycle could be about 20 years. While the use of length frequencies is tempting in some situations, we stress that this method, like other techniques, must also be validated.

Traditionally, investigators have attempted to use only one method to age all fish in the population. If it is possible to do this then there are obvious advantages to using only one structure. If a particular structure is most useful during one phase of the life cycle and not in another, we see no reason why a combination of structures cannot be used to estimate age. For example, scales may be the most appropriate structures for accurately ageing young fish and sections of fin rays may be used for obtaining reliable ages for older fish in the population. Changing from one structure to another for some species, such as lingcod (Chilton and Beamish, 1982) or chinook salmon Onchorhynchus tshawytscha (Chilton and Bilton 1986) may introduce a small error, but this error is smaller than the error that results from using only one structure.

CONCLUSION

We believe that there will continue to be advances which will improve or refine the accuracy and precision of existing techniques or successfully enable new species to be aged.

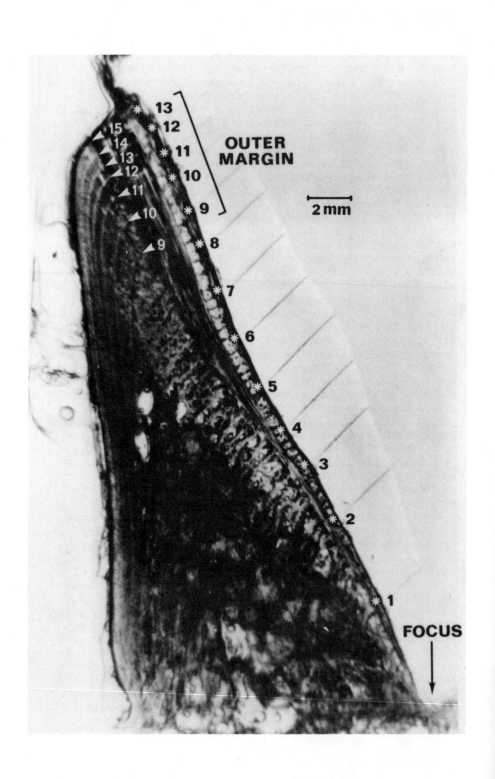

Figure 9. A sectioned vertebra from a 227.3 kg (254 cmFL)
   giant Atlantic bluefin tuna Thunnus thynnus, viewed
   under transmitted light. Internal bands (white numbers
   and arrows), external bands (black numbers and white
   asterisks), focus, and outer margin are shown as they
   relate to the whole vertebra (13 yr) and vertebral
   section (15 yr) methods of assigning ages. The
   vertebral section method involves adopting a counting
   procedure which enumerates distal internal bands in
   sections, as well as proximal external bands on the cone
   surface. (Reproduced from Prince et al. 1985, with
   permission of the authors and the editors of the
   Canadian Journal of Fisheries and Aquatic Science,
   publisher).

However, entirely new methods of age determination have to be
found that will monitor the ageing process of tissues directly
in some manner other than counting growth zones.

REFERENCES
AASS, P. 1972. Age determination and year-class fluctuations
   of cisco, (Coregonus albula) L., in the Mjosa hydroelectric
   reservoir. Institute of Freshwater Research,
   Drottningholm, 52:5-22.
ANTOINE, M. L., J. MENDOZA, AND P. M. CAYR . 1983. Progress of
   age and growth assessment of Atlantic skipjack tuna,
   (Euthynnus pelamis), from dorsal fin spines. Pages 91-97
   in E. D. Prince and L. M. Pulos, editors. Proceedings of
   the international workshop on age determination of oceanic
   pelagic fishes: tunas, billfishes, and sharks. United
   States National Marine Fisheries Service Technical Report
   8.
ARCHIBALD, C. P., D. FOURNIER, AND B. M LEAMAN. 1983.
   Reconstruction of stock history and development of
   rehabilitation strategies for Pacific ocean perch in Queen
   Charlotte Sound, Canada. North American Journal of
   Fisheries Management 3:283-294.
BARKMAN, R. C., D. A. BENGSTON AND A. D. BECK. 1981. Daily
   growth of the juvenile fish (Menidia menidia) in the
   natural habitat compared with juveniles reared in the
   laboratory. Rapports et Proces-Verbaux des Reunions
   Commission International pour l'Exploration Scientific de
   la Mer 178:324-326.
BARNES, M. A. AND G. POWER. 1984. A comparison of otolith and
   scale ages for western Labrador lake whitefish (Coregonus
   clupeaformis). Environmental Biology of Fishes 10:297-299.
BEAMISH, R. J. 1973. Determination of age and growth of
   populations of the white sucker (Catostomus commersoni)
   exhibiting a wide range in size at maturity. Journal of
   the Fisheries Research Board of Canada 30:607-616.

BEAMISH, R. J. 1981. Use of fin-ray sections to age walleye
    pollock, Pacific cod, and albacore, and the importance of
    this method. Transactions of the American Fisheries
    Society 110:287-299.
BEAMISH, R. J. AND D. E. CHILTON. 1977. Age determination of
    lingcod (Ophiodon elongatus) using dorsal fin rays and
    scales. Journal of the Fisheries Research Board of Canada
    34:1305-1313.
BEAMISH, R. J. AND D. E. CHILTON. 1981. Preliminary evaluation
    of a method to determine the age of sablefish (Anoplopoma
    fimbria). Canadian Journal of Fisheries and Aquatic
    Sciences 39:277-287.
BEAMISH, R. J. AND D. A. FOURNIER. 1981. A method for
    comparing the precision of a set of age determinations.
    Canadian Journal of Fisheries and Aquatic Sciences
    38:982-983.
BEAMISH, R. J. AND G. A. MCFARLANE. 1983. The forgotten
    requirement for age validation in fisheries biology.
    Transactions of the American Fisheries Society 112:735-743.
BEAMISH, R. J., G. A. MCFARLANE, AND D. E. CHILTON. 1983a. Use
    of oxytetracycline and other methods to validate a method
    of age determination for sablefish (Anoplopoma fimbria).
    Pages 95-116 in Proceedings of the international sablefish
    symposium. Alaska Sea Grant Report 83-3.
BEAMISH, R. J., G.A. MCFARLANE, R. SCARSBROOK, D. CHILTON, I.
    BARBER, K. BEST, A. CASS, AND W. SHAW. 1983b. A summary
    of sablefish tagging and biological studies conducted
    during 1980 and 1981 by the Pacific Biological Station.
    Canadian Manuscript Report of Fisheries and Aquatic
    Sciences 1732:135 p.
BENNETT, J. T., G. W. BOEHLERT, AND K. K. TUREKIAN. 1982.
    Confirmation of longevity in Sebastes diploproa (Pisces:
    Scorpaenidae) from $^{210}Pb/^{226}Ra$ measurements in otoliths.
    Marine Biology 71:209-215.
BELANGER, S. E. AND S. R. HOGLER. 1982. Comparison of five
    ageing methodologies applied to walleye (Stizostedion
    vitreum) in Burt Lake, Michigan. Journal of Great Lakes
    Research 8:666-671.
BERKELEY, S. A. AND E. D. HOUDE. 1983. Age determination of
    broadbill swordfish, Xiphias gladius, from the Straits of
    Florida, using anal fin spine sections. Pages 137-143 in
    E. D. Prince and L. M. Pulos, editors. Proceedings of the
    international workshop an age determination of oceanic
    pelagic fishes: tunas, billfishes, and sharks. United
    States Department of Commerce, NOAA Technical Report,
    NMFS 8.
BROTHERS, E. B. 1981. What can otolith microstructures tell us
    about daily and subdaily events in the early life history
    of fish? Rapports et Proces-Verbaux des Reunions
    Commission Internationale pour l'Exploration Scientific de
    la Mer 178:393-394.

BROTHERS, E. B. AND W. N. MCFARLAND. 1981. Correlations between otolith microstructure, growth, and life history transitions in newly recruited French grunts [Haemulon flavolivcatum (Desmarest), Haemulidae]. Rapports et Proces-Verbaux des Reunions Commission Internationale pour l'Exploration Scientific de la Mer 178:369-374.

BROTHERS, E. B., C. P. MATHEWS, AND R. LASKER. 1976. Daily growth increments in otoliths from larval and adult fishes. Fishery Bulletin 74:1-8.

BROTHERS, E. B., E. D. PRINCE, AND D. W. LEE. 1983. Age and growth of young-of-the-year bluefin tuna, Thunnus thynnus, from otolith microstructure. Pages 49-59 in E. D. Prince and L. M. Pulos, editors. Proceedings of the international workshop on age determination of oceanic pelagic fishes: tunas, billfishes, and sharks. United States Department of Commerce, NOAA Technical Report, NMFS 8.

CAILLIET, G. M., L. K. MARTIN, J. T. HARVEY, D. KUSHER, AND B. A. WELDEN. 1983. Preliminary studies on the age and growth of the blue, Prionace glauca, common thresher, Alopias vulpinus, and shortfin mako, Isurus oxyrinchus, sharks from California waters. Pages 179-188 in E. D. Prince and L. M. Pulos, editors. Proceedings of the international workshop on age determination of oceanic pelagic fishes: tunas, billfishes, and sharks. United States Department of Commerce, NOAA Technical Report, NMFS 8.

CAMPANA, S.E. 1983. Feeding periodicity and the production of daily growth increments in otoliths of steelhead trout (Salmo gairdneri) and starry flounder (Platichthys stellatus). Canadian Journal of Zoology 61:1591-1597.

CAMPANA, S.E. AND J. D. NEILSON. 1982. Daily growth increments in otoliths of starry flounder (Platichthys stellatus) and the influence of some environmetal variable in their production. Canadian Journal of Fisheries and Aquatic Sciences 39:937-942.

CAMPBELL, J. S. AND J. A. BABALUK. 1979. Age determination of walleye, (Stizostedion vitreum vitreum) (Mitchell), based on the examination of eight different structures. Canadian Fisheries and Marine Service Technical Report 849.

CASS, A. J. AND R. J. BEAMISH. 1983. First evidence of validity of the fin-ray method of age determination for marine fishes. North American Journal of Fisheries Management 3:182-188.

CASSELMAN, J. M. 1974. Analysis of hard tissue of pike (Esox lucius L.) with special reference to age and growth. Pages 13-27 in T.B. Bagenal, editor. The ageing of fish. Unwin Brothers, Ltd., England.

CAYRE, P. M. AND T. DIOUF. 1983. Estimating age and growth of little tunny, (Euthynnus alletteratus), off the coast of Senegal, using dorsal fin spine sections. Pages 105-110 in E. D. Prince and L. M. Pulos, editors. Proceedings of the international workshop on age determination of oceanic pelagic fishes: tunas, billfishes, and sharks. United States Department of Commerce, NOAA Technical Report, NMFS 8.

CHANG, W. Y. B. 1982. A statistical method for evaluating the reproducibility of age determination. Canadian Journal of Fisheries and Aquatic Sciences 39:1208-1210.

CHILTON, D. E. AND R.J. BEAMISH. 1982. Age determination methods for fishes studied by the Groundfish Program at the Pacific Biological Station. Canadian Special Publication of Fisheries and Aquatic Sciences 60.

CHILTON, D. E. AND H. T. BILTON. 1986. A new method using dorsal fin rays for ageing chinook salmon (Oncorhynchus tschawytscha) and evidence of its validity. Canadian Journal of Fisheries and Aquatic Sciences (In press).

CRAIG, P. C. AND V. A. POULIN. 1975. Movements and growth of Arctic grayling (Thymallus arcticus) and juvenile arctic char (Salvelinus aplinus) in a small arctic stream, Alaska. Journal of the Fisheries Research Board of Canada, 32:689-697.

DUTIL, J. D. AND G. POWER. 1977. Validité de la lecture des otolithes comparee a celle de la lecture des ecailles pour la determination de l'age de l'omble de fontaine. (Salvelinus fontinalis). Naturaliste Canada 104:361-367.

FOUCHER, R. P. AND D. FOURNIER. 1982. Derivation of Pacific cod age composition using length-frequency analysis. North American Journal of Fisheries Management 2:276-284.

FOUCHER, R. P., R. G. BAKKALA, AND D. FOURNIER. 1984. Comparison of age frequency derived by length-frequency analysis and scale reading for Pacific cod in the north Pacific Ocean. International North Pacific Fisheries Commission Bulletin 42:232-242.

FOURNIER, D. AND C. P. ARCHIBALD. 1982. A general theory for analyzing catch-at-age data. Canadian Journal of Fisheries and Aquatic Sciences 39:1195-1207.

GOEMAN, T. J., D. R. HELMS, AND R. C. HEIDINGER. 1984. Comparison of otolith and scale age determinations for freshwater drum from the Mississippi River. Proceedings of the Iowa Academy of Science 91:49-51.

GRUBER, S. H. AND R. G. STOUT. 1983. Biological materials for the study of age and growth in a tropical marine elasmobranch, the lemon shark, (Negaprion brevirostris) (Poey). Pages 193-205 in E. D. Prince and L. M. Pulos, editors. Proceedings of the international workshop on age determination of oceanic pelagic fishes: tunas, billfishes, and sharks. United States Department of Commerce, NOAA Technical Report NMFS 8.

HANSEN, L. P. 1978. Age determination of roach, Rutilus rutilus (L.) from scales and opercular bones. Archiv fur Fischereiwissenshaft 29(1/2):93-98.

HARDISTY, N. W. 1961. The growth of larval lampreys. Journal of Animal Ecology 30:357-371.

HARDISTY, N. W. 1969. Information on the growth of the ammocoete larva of the anadromous sea lamprey, Petromyzon marinus in British Rivers. Journal of Zoology 159:139-144.

HARRISON, E. J. AND W. F. HADLEY. 1979. A comparison of the use of cleithra to the use of scales for age and growth studies. Transactions of the American Fisheries Society 108:452-456.

JESSOP, B. M. 1972. Ageing round whitefish (Prosopium cylindraceum) of the Leaf River, Ungava, Quebec, by otoliths. Journal of the Fisheries Research Board of Canada 29:452-454.

LEAMAN, B. M. AND R. J. BEAMISH. 1984. Ecological and management implications of longevity in some northeast Pacific groundfishes. International North Pacific Fisheries Commission Bulletin 42:85-97.

LEE, D. W., E. D. PRINCE, AND M. E. CROW. 1983. Interpretation of growth bands on vertebrae and otoliths of Atlantic bluefin tuna, Thunnus thynnus. Pages 61-69 in E. D. Prince and L. M. Pulos, editors. Proceedings of the international workshop on age determination of oceanic pelagic fishes: tunas, billfishes, and sharks. United States Department of Commerce, NOAA Technical Report, NMFS 8.

LOUGH, R. G., M. PENNINGTON, G. R. BOLZ, AND A. A. ROSENBERG. 1982. Age and growth of larval Atlantic herring (Clupea harengus) L., in the Gulf of Maine-Georges Bank region based on otolith growth increments. United States National Marine Fisheries Service Fishery Bulletin 80: 187-199.

LOW, L. L., G. TANONAKA, AND H. H. SHIPPEN. 1976. Sablefish of the northeastern Pacific Ocean and Bering Sea. Northwest and Alaska Fishery Center processed report, NMFS, Seattle, Washington, U.S.A.

MCFARLANE, G. A. AND R. J. BEAMISH. 1983. Biology of adult sablefish (Anoplopoma fimbria) in waters off western Canada. Pages 59-80 in Proceedings of the international sablefish symposium. Alaska Sea Grant Report 83-3.

MCFARLANE, G. A., R. J. BEAMISH, AND R. DEMORY. 1984. Additional information on the validity of the ageing technique for sablefish developed by Beamish and Chilton (1982): Unpublished report submitted to the International North Pacific Fisheries Commission, Anchorage, Alaska, U.S.A.

MCFARLANE, G. A., W. SHAW, AND A. V. TYLER. 1985. Sablefish: coastwide stock assessments. Pages 163-186 in A. V. Tyler and G. A. McFarlane, editors. Groundfish stock assessments off the west coast of Canada in 1984 and recommended yield options for 1985. Canadian Manuscript Report of Fisheries and Aquatic Sciences 1813.

MARALDO, D. C. AND H. R. MACCRIMMON. 1979. Comparison of ageing methods and growth rates for largemouth bass, (Micropterus salmoides) Lacepede, from northern latitudes. Environmental Biology of Fishes 4:263-271.

MARSHALL, S. L. AND S. S. PARKER. 1982. Pattern identification

in the microstructure of sockeye salmon (<u>Oncorhynchus</u>
<u>nerka</u>) otoliths. Canadian Journal of Fisheries and Aquatic
Sciences 39:542-547.

METHOT, R. D. JR. 1981. Spatial covariation of daily growth
rates of larval northern anchovy, (<u>Engraulis mordax</u>), and
northern lampfish, (<u>Stenobrachius leucopsarus</u>). Rapports
et Proces-Verbaux des Reunions Commission Internationale
pour l'Exploration Scientific de la Mer 178:424-431.

METHOT, R. D. JR., AND D. KRAMER. 1979. Growth of northern
anchovy, (<u>Engraulis mordax</u>), larvae in the sea. United
States National Maine Fisheries Service Fishery Bulletin
77:413-423.

MILLS, K. H. AND R. J. BEAMISH. 1980. Comparison of fin-ray
and scale age determinations for lake whitefish (<u>Coregonus</u>
<u>clupeaformis</u>) and their implications for estimates of
growth and annual survival. Canadian Journal of Fisheries
and Aquatic Sciences 37:534-544.

MUGIYA, Y., N. WATABE, J. YAMADA, J. M. DEAN,
D. G. DUNKELBERGER, AND M. SHIMIZU. 1981. Diurnal rhythm
in otolith formation in the goldfish (<u>Carassius auratus</u>).
Comparative Biochemistry and Physiology 68A:659-662.

NEILSON, J. D. AND G. H. GEEN. 1982. Otoliths of chinook
salmon (<u>Oncorhynchus tshawytscha</u>): daily growth increments
and factors influencing their production. Canadian Journal
of Fisheries and Aquatic Sciences 39:1340-1347.

PANNELLA, G. 1971. Fish otoliths:daily growth layers and
periodical patterns. Science, New York 173:1124-1127.

PANNELLA, G. 1974. Otolith growth patterns: an aid in age
determination in temperate and tropical fishes. Pages
28-39 <u>in</u> T. B. Bagenal, editor. The ageing of fish. Unwin
Brothers Ltd., England.

POWER, G. 1978. Fish population structure in Arctic lakes.
Journal of the Fisheries Research Board of Canada 35:53-59.

POWERS, J. E. 1983. Some statistical characteristics of ageing
data and their ramifications in population analysis of
oceanic pelagic fishes. Pages 19-27 <u>in</u> E. D. Prince and
L. M. Pulos, editors. Proceedings of the international
workshop on age determination of oceanic pelagic fishes:
tunas, billfishes, and sharks. United States Department of
Commerce, NOAA Technical Report, NMFS 8.

POWELL, A. B. 1982. Annulus formation on otoliths and growth
of young summer flounder from Pamlico Sound, North
Carolina. Transactions of the American Fisheries Society
111:688-693.

PRINCE, E. D., D. W. LEE AND J. C. JAVECH. 1985. Internal
zonations in sections of vertebrae from Atlantic bluefin
tuna, <u>Thunnus thynnus</u>, and their potential use in age
determination. Canadian Journal of Fisheries and Aquatic
Sciences 42:938-946.

PURVIS, H. A. 1980. Effects of temperature on metamorphosis

and the age and length at metamorphosis in sea lamprey,
(Petromyzoan marinus) in the Great Lakes. Canadian Journal
of Fisheries and Aquatic Sciences 37:1827-1834.

RADTKE, R. L. 1983. Otolith formation and increment deposition
in laboratory-reared skipjack tuna (Euthynnus pelamis),
larvae. Page 99-103 in E. D. Prince and L. M. Pulos,
editors. Proceedings of the international workshop on age
determination of oceanic pelagic fishes: tunas,
billfishes, and sharks. United States Department of
Commerce, NOAA Technical Report, NMFS 8.

RADTKE, R. L. AND J. M. DEAN. 1982. Increment formation in the
otoliths of embryos, larvae, and juveniles of the
mummichog, (Fundulus heteroclitus). United States National
Marine Fisheries Service Fishery Bulletin 80:201-215.

RALSTON, S. 1976. Age determination of a tropical reef
butterflyfish utilizing daily growth rings of otoliths.
Fishery Bulletin 74:990-994.

REIMERS, N. 1979. A history of a stunted brook trout
population in an alpine lake: A lifespan of 24 years.
California Fish and Game 69:196-215.

SCHMITT, D. N. AND W. A. HUBERT. 1982. Comparison of cleithra
and scales for age and growth analysis of yellow perch.
Progressive Fish Culturist 44:87-88.

SIKSTROM, C. B. 1983. Otolith, pectoral fin ray, and scale age
determinations for Arctic grayling. Progressive Fish
Culturist 45:220-223.

SIMARD, A. AND E. MAGNIN. 1972. Method of determination of the
age and growth of lake trout, (Salvelinus namaycush)
Walbaum, of Lake L'Assomption and of Lake Tremblant,
Quebec. Nature Canada (Quebec) 99:561-579.

SMITH, S. 1984. Timing of vertebral-band deposition in
tetracycline-injected leopard sharks. Transactions of the
American Fisheries Society 113:308-313.

SPOOR, H. 1938. Age and growth of the sucker, Catostomus
commersoni (Lacepede), in Muskellunge Lake, Vilas County,
Wisconsin. Transactions of the Wisconsin Academy of
Science 31:457-505.

STEFFENSEN, E. 1980. Daily growth increments observed in
otoliths from juvenile East Baltic cod. DANA 1:29-37.

STRUHSAKER, P. AND J. H. UCHIYAMA. 1976. Age and growth of the
nehu, (Stolephorus purporeus) (Pisces: Engraulidae), from
the Hawaiian Islands as indicated by daily growth
increments of sagittae. United States National Marine
Fisheries Service Fishery Bulletin 74:9-17.

TANAKA, K., Y. MUGIYA, AND J. YAMADA. 1981. Effects of
photoperiod and feeding on daily growth patterns in
otoliths of juvenile Tilapia nilotica. United States
National Marine Fisheries Service Fishery Bulletin
79:459-466.

TAUBERT, B. D. AND D. W. COBLE. 1977. Daily rings in otoliths of three species of Lepomis and Tilapia mossambica. Journal of the Fisheries Research Board of Canada 34:332-340.

TAUBERT, B. D. AND J. A. TRANQUILLI. 1982. Verification of the formation of annuli in otoliths of largemouth bass. Transactions of the American Fisheries Society 111:531-534.

THOMAS, R. M. 1983. Back-calculation and time of hyaline ring formation in the otoliths of the Pilchard off South West Africa. South African Journal of Marine Science 1:3-18.

UCHIYAMA, J. H. AND P. STRUHSAKER. 1981. Age and growth of skipjack tuna, (Katsuwonus pelamis), and yellowfin tuna, Thunnus albacores, as indicated by daily growth increments of sagittae. United States National Marine Fisheries Service Fishery Bulletin 79:151-162.

VICTOR, B. C. 1982. Daily otolith increments and recruitment in two coral-reef wrasses, Thalassoma bifasciatum and Halichoeres bivittatus. Marine Biology 71:203-208.

WILD A. AND T. J. FOREMAN. 1980. The relationship between otolith increments and time for yellowfin and skipjack tuna marked with tetracycline. Inter-American Tropical Tuna Commission Bulletin 17:509-560.

II
Characterization of growth

RNA-DNA ratios as indicators of growth in fish: A review

FRANK J. BULOW
DEPARTMENT OF BIOLOGY
TENNESSEE TECHNOLOGICAL UNIVERSITY
COOKEVILLE, TN 38505

ABSTRACT
This paper presents a review of developments and applications in the use of RNA and DNA in fish age and growth studies during the last 15 years. RNA-DNA ratios of whole fish and of various tissues have been used as indicators of short-term or current growth rates. RNA-DNA ratios, RNA content per individual, and RNA concentrations in tissues have also been utilized as indicators and predictors of rate of protein synthesis, level of metabolic activity, relative cell size, and relative condition. Protein-DNA ratio has also been employed as an index to relative cell mass, and DNA content and DNA concentration of whole fish and of various tissues used as indicators of cell number. These indicators have been applied in a variety of studies of experimental and natural populations. In applying these indicators, it is recommended that white muscle tissue be used for adult fish and whole fish for larval fish studies. Growth rate comparisons based on RNA-DNA ratios should be limited to the same species and to restricted size and life history stages. Factors such as temperature and maturation may have significant effects on levels and activity of the nucleic acids.

In normal cells, deoxyribonucleic acid (DNA) is found primarily in the nucleus in association with chromosomal material. The total quantity of DNA per cell is constant in normal somatic tissue within a given species, and this amount is apparently not

The Age and Growth of Fish, edited by Robert C. Summerfelt and Gordon E. Hall © 1987 The Iowa State University Press, Ames, Iowa 50010.

altered by starvation or other stress. The constant
DNA concentration has become a standard reference for
determining the total number of cells in a given
tissue.

Ribonucleic acid (RNA), however, is present in
variable quantity in the nucleus and cytoplasm. It is
concerned with transfer of the genetic code of the
nuclear DNA into the cytoplasm and with the actual
synthesis of new protein. The quantity of RNA varies
directly with the activity of protein synthesis;
therefore, it is expected to be more concentrated in
tissues undergoing faster growth or protein synthesis.
Since the amount of DNA per cell is constant within a
species, the ratio of RNA to DNA (RNA per unit DNA) is
indicative of the amount of RNA per cell. This ratio
is usually considered a more accurate index of protein
synthetic activity than RNA concentration alone,
because the ratio is not affected by differences in
cell numbers.

While these facts have long been known, only
recently have RNA-DNA ratios been used in the study of
rate of protein synthesis and hence growth in fish
(Love 1980). I proposed use of the RNA-DNA ratio as an
indicator of the recent growth rate of a fish based on
observation that RNA-DNA ratios of whole golden shiners
Notemigonus crysoleucas were higher for those
undergoing faster growth and protein synthesis in
response to higher feeding rates (Bulow 1970). DNA
remained fairly constant during 14 days of starvation
of shiners followed by six days of feeding, but RNA
declined during starvation and rapidly increased with
resumed feeding. Shiners starved for 45 days showed
decreased RNA, but increased DNA concentrations. The
increased DNA was attributed to reduction in cell
volume with food deprivation and, consequently, a
greater number of cells were present in a given weight
of tissue sample.

In feeding experiments with bluegills Lepomis
macrochirus, RNA-DNA ratios of liver tissue were most
indicative of changes in body length and weight,
followed by muscle tissue, stomach, and intestine,
respectively (Bulow 1971).

The early literature (Circa 1950-1973) concerning
the relationship between nucleic acids and growth rates
of fish has previously been reviewed (Bulow 1974). The
purpose of this paper is to review developments and
applications of the RNA-DNA ratio technique that have
occurred since this 1974 review, along with some
relevant papers missed earlier. A series of nucleic
acid papers by Dr. Yoshihiro Satomi has recently come
to the author's attention. Because these articles

appear in Japanese with German abstracts, lack of
interpretation has apparently delayed their
incorporation into western literature on this subject.

EXTRACTION AND QUANTIFICATION
     There is presently no standardization of
techniques for the extraction and quantification of
fish nucleic acids. The methods of Webb and Levy
(1955) have been used for nucleic acid extraction
followed by RNA quantification by the orcinol method of
Schneider (1957) and DNA quantification by the
diphenylamine method of Burton (1956). These methods
were described by Bulow (1970) and variously modified
in subsequent studies (Bulow 1971; Haines 1973; Mustafa
1977b; Bulow et al. 1981; Thorpe et al. 1982). Similar
techniques, but with the substitution of the DNA method
of Ashwell (1957), were used by Mustafa (1977a, 1978).
     Zeitoun et al. (1977) described use of perchloric
acid (PCA) extraction (Munro and Fleck 1966) with RNA
being quantified by the orcinol method of Brown (1946)
and DNA by the diphenylamine method of Giles and Myers
(1965). Lone and Matty (1980a) described use of
modifications of the methods of Wannemacher et al.
(1965), Shibko et al. (1967), and Munro and Fleck
(1969) for extraction and quantification of the nucleic
acids.
     Various modifications of the Schmidt-Thannhauser
(1945) method, as modified by Munro and Fleck (1966),
have been used with the substitution of several
alternative methods for RNA and DNA quantification
(Cowey et al. 1974; Emmersen and Emmersen 1976). A
description of the Schmidt-Thannhauser method, as
adapted for use with relatively large quantities of
tissue, was provided by Kayes (1978). Wilder and
Stanley (1983) described additional modifications of
the Schmidt-Thannhauser method (Munro and Fleck 1966)
for use with relatively large tissue samples.
     Buckley (1979) outlined an adaptation of the
Schmidt-Thannhauser method (Munro and Fleck 1966) to
extract and quantify the microquantities of RNA and DNA
present in larval fish and eggs. However, Buckley's
flow chart should be modified to include the addition
of 1.4 ml of 0.6 N PCA just prior to the RNA reading
(Buckley, personal communication). Barron and Adelman
(1984) also provided a detailed description of similar
techniques adapted for use with larval fish.
     Fluorometric methods have also been used for DNA
quantification in fish eggs and larvae (Buckley 1980)
and other planktonic marine organisms (Holm-Hansen et
al. 1968), and for RNA and DNA quantification in

planktonic marine organisms (Dortch et al. 1983).
Fluorometric methods may be especially useful in
studies dealing with small quantities of nucleic acids.
Rosenlund et al. (1983) described the extraction of
RNA from ribosomes of fish tissue and the use of
fluorometric methods for DNA quantification.

In summary, variations of the Schmidt-Thannhauser
method, as modified by Munro and Fleck (1966), are now
most commonly used for extraction and quantification of
RNA and DNA. Fluorometric techniques are being used
for quantification when RNA and DNA concentrations are
low. There is a need for standardization of
techniques for RNA-DNA studies of fish to facilitate
use of these techniques by fisheries biologists.

TISSUE COMPARISONS AND GENETIC ASPECTS

Mustafa (1977b) compared red and white muscle of
the carp Catla catla, Labeo rohita, and Labeo bata, and
found higher RNA and DNA concentrations in red muscle.
He reasoned that red muscle had higher RNA because it
was more metabolically active than white muscle. The
higher DNA concentration of red muscle indicated
smaller cell size and, consequently, a greater number
of cells per unit weight of tissue.

Mustafa (1978) observed higher DNA concentration
in the tail musculature of the catfish Heteropneustes
fossilis than the trunk region, which indicated a
smaller cell size in tail muscle. Because DNA
concentration appeared to decline with linear growth of
the adult fish, Mustafa proposed that growth proceeded
by increase in size of cells rather than their number.
However, studies reviewed by Love (1980) indicated that
growth of muscle and liver included both increase in
cell size and increase in number of cells. Zeitoun et
al. (1977) used whole egg and whole fry DNA
concentration to estimate multiplication of cells, and
RNA concentration to estimate cell enlargement in the
ontogenesis of rainbow trout Salmo gairdneri. They
found an increase in both RNA and DNA as development
proceeded from prehatching through the postyolk
absorption stage, indicating both cell multiplication
and cell enlargement. Changes in DNA tended to precede
changes in RNA by one sampling period.

Kayes (1978, 1979) used DNA concentration to
present evidence that growth of young black bullheads
Ictalurus melas was primarily a function of cell
multiplication rather than cell enlargement. Barron
and Adelman (1985) found that early larval growth of
fathead minnows Pimephales promelas occurred by cell
multiplication rather than by cell enlargement.

Weatherley (1972) reviewed other studies which
indicated that both cell multiplication and cell
enlargement are involved in fish growth at various
growth stages. It appears that the degree to which
each process is involved in fish growth varies with the
life history stage. Lone and Ince (1983) presented
evidence that fish muscle cell multiplication continues
throughout life whereas cell size increases up to a
stage beyond which no further increase occurs. These
trends, along with the usual decline in growth rate
with increase in age and size, suggest that growth rate
comparisons based on RNA-DNA ratios should be limited
to specific size ranges and life history stages.

Rosenlund et al. (1983) cautioned that growth rate
studies based on RNA-DNA ratios should be limited to
comparisons within one species, because the
ultrastructural organization of white muscle cells
varies with the species (Love 1970; Nag 1972).
Hinegardner and Rosen (1972) proposed that cellular DNA
content is related to evolutionary history of the fish.
Highly specialized fish tend to have less DNA per cell
than the more generalized and less evolved fish of the
same phyletic grouping.

MATURATION, SPAWNING, AND EARLY LIFE

Emmersen and Emmersen (1976) found increases in
liver DNA and RNA associated with initiation of ovarian
growth in the flounder Platichthys flesus, with RNA
reaching maximum levels in December when ovarian growth
was rapid and liver weights maximal. The liver
produces the protein vitellogenin which functions in
egg development. They also observed that estrogen
treatment caused increased liver protein synthetic
activity, expressed by increased liver RNA, and that
the amount of DNA per liver increased in relation to
liver weight. Both liver RNA and DNA declined after
spawning.

Satomi and Ishida (1976) used the RNA-DNA ratio as
an indicator of changes in biological activity of
liver, red muscle, and white muscle tissues during the
maturation process in the Ayu-fish Plecoglossus
altivelis. In females the ratio in red and white
muscle decreased during maturation, but the ratio in
the liver increased during early stages of maturation.
Their results and those of Emmersen and Emmersen (1976)
reveal that high liver RNA-DNA ratios may not
necessarily indicate rapid body growth. In both sexes,
the RNA-DNA ratio in the liver was correlated with
changes in liver size, and the ratio in red and white
muscle was correlated with changes in the condition

factor. During maturation, ratios in white muscle
declined as liver ratios increased in the female and as
red muscle ratios increased in the male.
        Mustafa and Jafri (1977) used RNA and protein
concentrations of epaxial white muscle to follow the
growth dynamics of Channa punctatus. Both parameters
were high in age 0 fish, declined with onset of sexual
maturity, and then increased until they leveled off as
growth slowed at advanced ages. Mustafa (1977a)
noted that maturation of Clarias batrachus was
accompanied by a steady increase in DNA concentration and
decrease in RNA concentration of epaxial white muscle.
He reasoned that the higher DNA concentration reflected
a decrease in cell size (more cells required to make up
a given weight of tissue) as a result of biochemical
transfer toward gonad production, coupled with a
reduction in food consumption. The same depletion was
thought to contribute to the decline in white muscle
RNA concentration.
        In summary, liver and red muscle nucleic acid
levels can change during maturation and reflect
metabolic activity unrelated to somatic growth of the
fish. The maturation process can also influence changes
occurring in white muscle nucleic acid levels in
response to the demands of gonad production. The RNA-
DNA ratio of white muscle tissue seems to be the best
indicator of somatic growth rate.

POPULATION ANALYSIS AND SEASONAL VARIATIONS
        Satomi (1966) found a relationship between the
RNA-DNA ratio of liver and muscle tissue and growth
rate in the silver carp Hypophthalmichthys molitrix,
crucian carp Carassius carassius, and common carp
Cyprinus carpio. Haines (1973) suggested that the RNA-
DNA ratio could be used as an indicator of long-term
population growth of smallmouth bass Micropterus
dolomieui and common carp when population age structure
was known and recruitment was controlled. He found
that RNA-DNA ratios were higher in 1-year-old fish than
in 2-year-old fish at a comparable growth stage.
        Satomi (1972) noted a positive relationship
between the RNA-DNA ratio and growth rate, but a
decrease in DNA concentration with increase in fish
weight in fish collected over the interval May to
October. The latter probably reflected an increase in
cell size as alluded to earlier. Nasiri (1972) found
that muscle RNA-DNA ratios of three species of river-
dwelling carpsuckers Carpiodes spp. in Iowa increased
from April to August and declined through October.
Sable (1974) observed a similar pattern with river-

dwelling channel catfish Ictalurus punctatus from the same area.

Satomi and Tanaka (1978a) found that RNA-DNA ratios of whole cultured eels Anguilla japonica showed significant positive correlations with body weight and total body length. They also studied seasonal changes from August to April in nucleic acid content of liver, red muscle, and white muscle of the cultured eel (Satomi and Tanaka 1978b). During the growing season RNA-DNA ratios in the three tissues were not affected by different diets, but ratios in liver tissue were higher in fish with higher liver-somatic indexes. While liver RNA remained fairly constant, liver DNA was higher during the winter for eels on commercial diets. Liver RNA-DNA ratios, therefore, decreased during the winter. The condition factor of eels fed frozen mackerel showed a correlation with ratios in white muscle during all seasons examined, and in red muscle only in the winter season.

Bulow et al. (1978) found seasonal variations in liver RNA-DNA ratios of bluegills associated with gonad maturation, spawning, temperature, and oxygen concentrations. The ratio indicated a longer summer growth depression in a smaller, shallower lake and faster growth in a larger, deeper lake. Scale analysis verified these conclusions.

Satomi (1978) found that the RNA-DNA ratio in red muscle of three cyprinids provided a useful index to pond water fertility. Analyzing epaxial white muscle filets from Channa punctatus, Mustafa (1979) and Mustafa and Zofair (1983) found that RNA and protein concentrations increased with "filet condition factor" (weight of filet x 1000/length of filet). Earlier, Spigarelli and Smith (1976) found that RNA-DNA ratios were not related to the traditional K-factor (coefficient of condition); they concluded that the ratio reflected short-term growth while the K-factor reflected long-term growth histories. Other studies, however, have revealed correlations between condition and muscle RNA-DNA ratios (Satomi and Ishida 1976; Satomi and Tanaka 1978b).

Haines (1980) found the RNA-DNA ratio in white muscle of black crappie Pomoxis nigromaculatus declined during the spawning season, reached a low in mid-May, increased after spawning, declined slightly in mid-summer, and rose again during late summer and early autumn. RNA-DNA ratio was significantly correlated with body weight and with water temperature. The study indicated that the ratio may be a useful predictor of skeletal muscle growth over short time intervals. Bulow et al. (1981) found that white muscle RNA

concentrations and RNA-DNA ratios of bluegills from a
central Tennessee lake were maximum in the spring,
declined during maturation and spawning, were low
during the summer, increased again in the fall, and
were low during the winter. Liver RNA, RNA-DNA ratios,
and liver-somatic indexes of females peaked in May,
when gonosomatic indexes also peaked, and then declined
as spawning occurred. Emmersen and Emmersen (1976) had
previously observed a similar pattern and discussed the
role of the liver and ovary in production of yolk
protein. Bulow et al. (1981) reasoned that summer
depressions in the ratios may have been associated with
thermal stratification and dissolved oxygen stress.

Thorpe et al. (1982) found that white muscle RNA-
DNA ratio of juvenile Atlantic salmon <u>Salmo salar</u>
provided a valid short-term index of growth. They
concluded that the ratio could be used to determine the
time of divergence of two growth groups more precisely
than was possible using size frequency distributions.

In a comparative study of juvenile striped bass
<u>Morone saxatilis</u> from rivers and hatcheries of the
eastern United States, Buckley et al. (1985) found that
hatchery fish had higher relative liver weights and
lower liver RNA, DNA, and protein concentrations than
wild fish. The hatchery-reared fish had a high fat
diet and highly vacuolated livers. Among wild
populations, Hudson River fish were singled out for
having high liver weight-body weight ratios, low liver
DNA, poorer bone quality, poor swimming stamina, and
low muscle protein levels. The Hudson River fish also
had high liver RNA-DNA ratios compared with their
average muscle RNA-DNA ratios. While the large liver
and high liver RNA-DNA ratio could be indicative of
healthy, well-fed, rapidly growing fish, other
indicators revealed that these fish were actually
stressed and had suffered liver damage. The high RNA-
DNA ratios were due to low DNA, not high RNA levels.
Low protein concentration in the muscle also indicated
severe depletion. Histopathological observations
indicated that the enlarged livers of Hudson River fish
were not due to increased vacuolation of liver cells as
occurred with hatchery-reared fish; rather a high
incidence of necrosis of liver cells suggested that the
liver enlargement was due to parasitism. Hudson River
fish had the poorest bone structural integrity and were
heavily infested with larval cestodes, accompanied by
extensive loss of muscle bundles. The latter was
consistent with their poor swimming stamina and low
muscle protein concentrations. This study was
especially valuable as a caution in the interpretation
of RNA-DNA ratio data.

STARVATION, DIET, AND DRUG FEEDING
     Satomi (1969) observed decreased RNA concentration
with starvation followed by an increase after
refeeding. DNA concentration changes were limited, but
RNA-DNA ratios decreased with starvation and increased
with refeeding. Acute starvation caused a decrease in
the turnover time of total nuclear and ribosomal RNA in
carp liver (Bouche et al. 1977). Buckley (1979) found
the RNA-DNA ratio to be a sensitive indicator of the
nutritional state and growth rate of Atlantic cod
Gadus morhua larvae. Starved larvae had lower RNA and
higher DNA concentrations than fed larvae. Larvae held
at higher densities had higher RNA-DNA ratios and
faster growth rates than those held at lower plankton
densities, and the ratios were significantly correlated
with protein growth rates. The ratio has subsequently
been used to assess nutritional condition and growth
rate in several other larval fish studies (Buckley
1980, 1981; Buckley and Lough 1985; Martin et al.
1985).
     Mustafa and Mittal (1982) found that 5 months of
food deprivation caused a decrease in RNA-DNA ratio and
weight of the liver, while the ratio and weight of the
brain maintained stability. Refeeding caused an
increase in liver ratio and weight while the brain
again remained stable. Lied et al. (1983) observed
that the DNA concentration in white trunk muscle of
Atlantic cod remained essentially unchanged through 8
days of starvation followed by 12 hours of refeeding.
Ribosomal RNA concentration and RNA-DNA ratios were
stable after 3 days of starvation, but decreased after
5 and 8 days of starvation. During the initial 4 hours
of refeeding, the muscle RNA increased to 88% of that
of normal-fed fish.
     Love (1980) reviewed a series of studies by Bouche
and associates concerning nucleic acid responses of
carp to longer periods (11 months) of starvation
followed by refeeding. Although muscle DNA typically
remained stable and could be used as a "tissue
constant", it did diminish after 11 months of
starvation, indicating lysis of the nuclei. Similarly,
liver DNA diminished between the 8th and 10th months of
starvation. Additionally, the response to refeeding
was slower following such a severe starvation.
     Satomi and Tanaka (1973) found that different
protein diets had no effect on DNA in the liver of
rainbow trout fingerlings, but did produce differences
in liver RNA-DNA ratios. Cowey et al. (1974) found
that liver DNA of similar-sized cultured and wild
plaice Pleuronectes platessa did not differ
significantly, irrespective of dietary history. Liver

RNA, meanwhile, was highest in cultured fish fed high
protein diet and lowest in the wild fish.   Lied and
Rosenlund (1984) studied the effect of the ratio of
protein energy to total energy (PE/TE) in feed on the
content of ribosomal RNA, DNA, and RNA-DNA ratio in
white trunk muscle of Atlantic cod. Protein
concentrations of less than 47.4% PE/TE reduced
ribosomal RNA and RNA-DNA ratios.

Kayes (1978, 1979) found that hypophysectomy
decreased the RNA concentration and the RNA-DNA ratio
of black bullhead Ictalurus melas epaxial muscle.
Beef growth hormone replacement therapy had the
opposite effect, and the RNA-DNA ratio was positively
correlated with changes in body size. Numerous other
studies have utilized the RNA-DNA ratio to define the
role of hormones in the growth of fish (Lone and Matty
1980a,b,c; 1982a,b; 1983; 1984; Matty et al. 1982;
Lone and Ince 1983; Sower et al. 1983).

TEMPERATURE EFFECTS

Spigarelli and Smith (1976) used the RNA-DNA ratio
of epaxial muscle to compare relative growth rates of
fish in a thermal plume to growth of fish from two
ambient temperature areas of Lake Michigan.   They found
that the mean ratio of plume brown trout Salmo trutta
was significantly higher than that of combined control
fish, but no such differences occurred with rainbow
trout and chinook salmon Oncorhynchus tshawytscha.
Buckley (1982) observed positive linear relationships
between RNA-DNA ratio and growth rate of whole larval
winter flounder at 5,7, and 10 degrees C. The
increased growth rate observed at higher temperatures
was not accomplished by an increase in the RNA-DNA
ratio at the higher temperatures, but rather by an
increased growth rate at a given ratio. He concluded
that the utility of the RNA-DNA growth relation for
estimation of growth rate can be improved by
introducing temperature as a second independent
variable.

Buckley et al. (1984), in a study of larval sand
lance Ammodytes americanus, found that growth rate and
RNA-DNA ratio increased with higher feeding level. A
direct linear relation observed between the ratio and
growth rate was improved by adding temperature as a
second independent variable. Buckley (1984) developed
a general model for estimating larval fish growth at
sea based on water temperature, RNA-DNA ratio, and
growth of eight species of temperate marine fish larvae
reared in the laboratory. The relationship was not
affected by size or age over most of the larval period.

Goolish et al. (1984) found that RNA concentration in carp white muscle tissue increased with higher growth rate for all acclimation temperatures, while DNA, protein content, and tissue hydration were unaffected. For a given growth rate, RNA concentration and RNA-DNA ratio were higher in colder acclimated fish as a probable compensation for decreased RNA activity. Maximum growth resulted in an approximate doubling of the RNA-DNA ratio while, over the range of 12-30 degrees C, cold acclimation caused an increase of approximately 50%. In view of these acclimation effects, it would be of value to reexamine studies, such as those of Uchiyama et al. (1969) and Bulow et al. (1981), which have reported seasonal changes in RNA-DNA ratios in fish. Goolish et al. (1984) also observed an increase in cell mass with warmer acclimation temperatures and a high correlation between muscle RNA-DNA ratios and the rate of in vitro glycine uptake by scales.

CHEMICAL EXPOSURE, pH, AND TOXICANT STRESS

Mudge et al. (1977) found reduction in cell size and decreased cytoplasmic RNA content in interrenal tissue of brook trout Salvelinus fontinalis exposed to low pH (pH 4.0). Stroganov (1977) found that a 24-day exposure to zinc at 10 mg/L or higher reduced RNA concentrations in the liver and intestine of carp but did not affect DNA concentrations. Cadmium exposure, however, had no affect on RNA or DNA concentrations. Kearns and Atchison (1979) used the RNA-DNA ratio of young-of-the-year yellow perch Perca flavescens to study the effects of cadmium and zinc contamination in a natural environment. They were able to demonstrate measurable growth differences that were significantly correlated with cadmium levels.

Wilder and Stanley (1983) found that white muscle RNA-DNA ratios of brook trout were not reduced in a stream contaminated with low levels of carbaryl (Sevin). Barron and Adelman (1984) demonstrated that RNA, DNA, and protein content per larval fathead minnow and RNA-DNA ratio were sensitive to toxicant stress and followed a log-linear dose response. Larval RNA content appeared to be the 96-hour measurement most responsive to toxicant exposure. In a study of sublethal exposure of fathead minnow larvae to hydrogen cyanide (HCN), Barron and Adelman (1985) found RNA-DNA ratios to be responsive to both the growth rate of larvae and to HCN exposure. Cyanide toxicosis occurred within 24 hours of exposure as evidenced by significant reductions in protein content, RNA content, and RNA-DNA

ratio. After 96 hours of exposure, larvae had
recovered to the same growth rate and protein synthetic
rate (RNA/DNA) as control fish.

SUMMARY AND OUTLOOK
        RNA content, RNA concentration, and RNA-DNA ratio
of whole fish and of various tissues have been used as
indicators and predictors of rate of growth, rate of
protein synthesis, level of metabolic activity,
relative cell size, and relative condition of the fish
or the specific tissue. DNA content and DNA
concentration of whole fish and of various tissues have
been used as indicators of cell number. Protein-DNA
ratio has been used as an index to relative cell mass.
        These indicators have received wide application in
laboratory and field investigations involving the
biology of fish growth and the response of fish to
changes in their internal and external environments.
Complex interrelationships often exist, and caution
should be exercised in the interpretetion of these
indicators and their application in both intraspecific
and interspecific studies. It has been demonstrated,
for example, that the relationship between RNA-DNA
ratio and growth varies between species and with age
within a species. This relationship is also affected
by external factors such as temperature which can
affect the activity of RNA.
        In using RNA-DNA ratios as indicators of growth
rate in fish, the use of white muscle tissue is
advisable. RNA-DNA ratios of liver tissue, for
example, can rise during accelerated metabolic activity
that is unrelated to growth of the fish, and red muscle
appears to be metabolically similar to liver tissue in
this regard.
        Future investigations should be aimed at further
clarification of changes that occur in levels of DNA,
RNA, and RNA-DNA ratios during the life of a fish, how
these changes relate to the evolutionary adaptations of
fish, and how changes are influenced by various
environmental factors. There is also a need for
standardization of analytical procedures and perhaps a
simplification of procedures to encourage wider
application of these indicators in age and growth
studies.

REFERENCES
ASHWELL, G. 1957. Colorimetric analysis of sugars:
        cysteine reaction of DNA. Pages 102-103 in S.P.
        Colowick and N.O. Kaplan, editors. Methods in

enzymology. Volume 3. Academic Press, New York,
New York, USA.
BARRON, M. G. AND I. R. ADELMAN. 1984. Nucleic acid,
protein content, and growth of larval fish
sublethally exposed to various toxicants. Canadian
Journal of Fisheries and Aquatic Sciences 41:141-
150.
BARRON, M. G. AND I. R. ADELMAN. 1985. Temporal
characterization of growth of fathead minnow
(Pimephales promelas) larvae during sublethal
hydrogen cyanide exposure. Comparative
Biochemistry and Physiology, Part C Toxicology 81
C:341-344.
BOUCHE, G., Y. CREACH, AND A. SERFATY. 1977. Turnover
rate of hepatic RNA in carp (Cyprinus carpio L.)
after prolonged total starvation. Annales de
Biologie Animale, Biochimie, et. Biophysique
17:21-32.
BROWN, A. H. 1946. Determination of pentose in the
presence of large quantities of glucose. Archives
of Biochemistry and Biophysics 11:269-278.
BUCKLEY, L. J. 1979. Relationships between RNA-DNA
ratio, prey density, and growth rate in
Atlantic cod (Gadus morhua) larvae. Journal
of the Fisheries Research Board of Canada 36:1497-
1502.
BUCKLEY, L. J. 1980. Changes in ribonucleic acid,
deoxyribonucleic acid, and protein content during
ontogenesis in winter flounder, Pseudopleuronectes
americanus, and effect of starvation. Fishery
Bulletin 77:703-708.
BUCKLEY, L. J. 1981. Biochemical changes during
ontogenesis of cod (Gadus morhua L.) and winter
flounder (Pseudopleuronectus americanus) larvae.
Rapports et Proces-Verbaux des Reunions. Counseil
International pour l'Exploration de la Mer
178:547-552.
BUCKLEY, L. J. 1982. Effects of temperature on growth
and biochemical composition of larval winter
flounder Pseudopleuronectes americanus.
Marine Ecology Progress Series 8:181-186.
BUCKLEY, L. J. 1984. RNA-DNA ratio: an index of larval
fish growth in the sea. Marine Biology 80:291-298.
BUCKLEY, L. J. AND R. G. LOUGH. 1985. Recent growth,
chemical composition and prey field of haddock
(Melanogrammus aeglefinus) and cod (Gadus morhua)
larvae and post-larvae on Georges Bank, May 1983.
Unpublished Manuscript.
BUCKLEY, L. J., T. A. HALAVIK, G. C. LAURENCE, S. J.
HAMILTON, AND P. YEVICH. 1985. Comparative
swimming stamina, biochemical composition,

backbone mechanical properties, and histopathology
of juvenile striped bass from rivers and
hatcheries of the eastern United States.
Transactions of the American Fisheries Society
114:114-124.
BUCKLEY, L. J., S. I. TURNER, T. A. HALAVIK, A. S.
SMIGIELSKI, S. M. DREW, AND G. C. LAURENCE. 1984.
Effects of temperature and food availability on
growth, survival, and RNA-DNA ratio of larval sand
lance (Ammodytes americanus). Marine Ecology
Progress Series 15:91-97.
BULOW, F. J. 1970. RNA-DNA ratios as indicators of
recent growth rates of a fish. Journal of the
Fisheries Research Board of Canada 27:2343-2349.
BULOW, F. J. 1971. Selection of suitable tissues for
use in the RNA-DNA ratio technique of assessing
recent growth rate of a fish. Iowa State Journal
of Science 46:71-78.
BULOW, F. J. 1974. A review of the literature
concerning the relationship between nucleic acids
and the growth rates of fish. The Tennessee Tech
Journal 9:17-23.
BULOW, F. J., C. B. COBURN, JR., AND C. S. COBB. 1978.
Comparisons of two bluegill populations by means
of the RNA-DNA ratio and liver-somatic index.
Transactions of the American Fisheries Society
107:799-803.
BULOW, F. J., M. E. ZEMAN, J. R. WINNINGHAM, AND W.
F. HUDSON. 1981. Seasonal variations in RNA-DNA
ratios and in indicators of feeding, reproduction,
energy storage, and condition in a population of
bluegill, Lepomis macrochirus Rafinesque. Journal
of Fish Biology 18:237-244.
BURTON, K. 1956. A study of the conditions and
mechanism of the diphenylamine reaction for the
colorimetric estimation of deoxyribonucleic acid.
Biochemical Journal 62:315-323.
COWEY, C. B., D. A. BROWN, J. W. ADRON, AND A. M.
SHANKS. 1974. Studies on the nutrition of marine
flatfish. The effect of dietary protein content on
certain cell components and enzymes in the liver
of Pleuronectes platessa. Marine Biology
28:207-213.
DORTCH, Q., T. L. ROBERTS, J. R. CLAYTON, JR., AND S.
I. AHMED. 1983. RNA-DNA ratios and DNA
concentrations as indicators of growth rate and
biomass in planktonic marine organisms. Marine
Ecology-Progress Series 13:61-71.
EMMERSEN, B. K. AND J. EMMERSEN. 1976. Protein, RNA
and DNA metabolism in relation to ovarian
vitellogenic growth in the flounder Platichthys

flesu (L.). Comparative Biochemistry and
Physiology 55B:315-321.

GILES, K. W. AND A. MYERS. 1965. An improved
diphenylamine method for the estimation of
deoxyribonucleic acid. Nature 206:93.

GOOLISH, E. M., M. G. BARRON, AND I. R. ADELMAN. 1984.
Thermoacclimatory response of nucleic acid and
protein content of carp muscle tissue: influence
of growth rate and relationship to glycine uptake
by scales. Canadian Journal of Zoology 62:2164-
2170.

HAINES, T. A. 1973. An evaluation of RNA-DNA ratio as
a measure of long-term growth in fish populations.
Journal of the Fisheries Research Board of Canada
30:195-199.

HAINES, T. A. 1980. Seasonal patterns of muscle RNA-
DNA ratio and growth in black crappie, Pomoxis
nigromaculatus. Environmental Biology of Fishes
5:67-70.

HINEGARDNER, R. AND D. E. ROSEN. 1972. Cellular DNA
content and the evolution of teleostean fishes.
The American Naturalist 106:621-644.

HOLM-HANSEN, O., W. H. SUTCLIFFE, JR., AND J. SHARP.
1968. Measurement of deoxyribonucleic acid in the
ocean and its ecological significance. Limnology
and Oceanography 13:507-514.

KAYES, T. 1978. Effects of hypophysectomy and beef
growth hormone replacement therapy on morphometric
and biochemical indicators of growth in fed versus
starved black bullhead (Ictalurus melas). General
and Comparative Endocrinology 35:419-431.

KAYES, T. 1979. Effects of hypophysectomy and beef
growth hormone replacement therapy, over time and
at various dosages, on body weight and total RNA-
DNA levels in the black bullhead (Ictalurus
melas). General and Comparative Endocrinology
37:321-332.

KEARNS, P. K. AND G. J. ATCHISON. 1979. Effects of
trace metals on growth of yellow perch (Perca
flavescens) as measured by RNA-DNA ratios.
Environmental Biology of Fishes 4:383-387.

LIED, E., G. ROSENLUND, B. LUND, AND A. VON DER
DECKEN. 1983. Effect of starvation and refeeding
on in vitro protein synthesis in white trunk
muscle of Atlantic cod (Gadus morhua). Comparative
Biochemistry and Physiology 76B:777-781.

LIED, E. AND G. ROSENLUND. 1984. The influence of
the ratio of protein energy to total energy in the
feed on the activity of protein synthesis in
vitro, the level of ribosomal RNA and the RNA-DNA
ratio in the white trunk muscle of Atlantic cod

(Gadus morhua). Comparative Biochemistry and
    Physiology 77A:489-494.
LONE, K. P., AND B. W. INCE. 1983. Cellular growth
    responses of rainbow trout (Salmo gairdneri) fed
    different levels of dietary protein, and an
    anabolic steroid ethylestrenol. General and
    Comparative Endocrinology 49:32-49.
LONE, K. P., AND A. J. MATTY. 1980a. The effect of
    feeding methyltestosterone on the growth and body
    composition of common carp (Cyprinus carpio L.).
    General and Comparative Endocrinology 40:409-424.
LONE, K. P., AND A. J. MATTY. 1980b. The effect of oral
    administration of trenbolone acetate on the growth
    and tissue chemistry of common carp, Cyprinus
    carpio L. Pakistan Journal of Zoology 12:47-56.
LONE, K. P., AND A. J. MATTY. 1980c. The effect of
    feeding testosterone on the food conversion
    efficiency and body composition of common carp,
    Cyprinus carpio L. Proceedings, First Pakistan
    Congress of Zoology 1:455-467.
LONE, K. P., AND A. J. MATTY. 1982a. The effect of
    feeding 11-ketotestosterone on the food conversion
    efficiency and tissue protein and nucleic acid
    contents of juvenile carp, Cyprinus carpio L.
    Journal of Fish Biology 20:93-104.
LONE, K. P., AND A. J. MATTY. 1982b. Cellular effects
    of adrenosterone feeding to juvenile carp,
    Cyprinus carpio L., effect on liver, kidney,
    brain, and muscle protein and nucleic acids.
    Journal of Fish Biology 21:33-45.
LONE, K. P., AND A. J. MATTY. 1983. The effect of
    ethylestrenol on the growth, food conversion and
    tissue chemistry of the carp, Cyprinus carpio.
    Aquaculture 32:39-55.
LONE, K. P., AND A. J. MATTY. 1984. Oral administration
    of an anabolic-androgenic steroid dimethazine
    increases the growth and food conversion
    efficiency and brings changes in molecular growth
    responses of carp (Cyprinus carpio) tissues.
    Nutritional Reports International 29:621-638.
LOVE, R. M. 1970. The chemical biology of fishes.
    Academic Press, New York, New York, USA.
LOVE, M. R. 1980. The chemical biology of fishes.
    Volume 2. Academic Press, New York, New York, USA.
MARTIN, F. D., D. A. WRIGHT, J. C. MEANS, AND E. M.
    SETZLER-HAMILTON. 1985. Importance of food
    supply to nutritional state of larval striped bass
    in the Potamac River estuary. Transactions of the
    American Fisheries Society 114:137-145.
MATTY, A. J., M. A. CHAUDHRY, AND K. P. LONE. 1982.
    The effect of thyroid hormones and temperature

on protein and nucleic acid contents of liver and muscle of <u>Sarotherodon</u> <u>mossambica</u>. General and Comparative Endocrinology 47:497-507.

MUDGE, J. E., J. L. DIVELY, W. H. NEFF, AND A. ANTHONY. 1977. Interrenal histochemistry of acid-exposed brook trout, <u>Salvelinus</u> <u>fontinalis</u> (Mitchill). General and Comparative Endocrinology 31:208-215.

MUNRO, H. N., AND A. FLECK. 1966. The determination of nucleic acids. Pages 113-176 <u>in</u> D. Glick, editor. Methods of biochemical analysis. Volume 14. Interscience Publishers, New York, New York, USA.

MUNRO, H. N., AND A. FLECK. 1969. Analysis of tissues and body fluids for nitrogenous constituents. Pages 423-525 <u>in</u> H. N. Munro, editor. Mammalian protein metabolism. Volume 3. Academic Press, New York, New York, USA.

MUSTAFA, S. 1977a. Influence of maturation on the concentration of RNA and DNA in the flesh of the catfish <u>Clarias</u> <u>batrachus</u>. Transactions of the American Fisheries Society 106:449-451.

MUSTAFA, S. 1977b. Nucleic acid turnover in the dark and white muscles of some freshwater species of carps during growth in the pre-maturity phase. Copeia 1977:173-176.

MUSTAFA, S. 1978. Deoxyribose nucleic acid in the musculature of freshwater cat-fish <u>Heteropneustes</u> <u>fossilis</u> (Bloch). Broteria Ciencias Naturais 48:83-92.

MUSTAFA, S. 1979. RNA and synthesis of protein in relation to "biological condition" of freshwater teleost, <u>Channa</u> <u>punctatus</u>. Comparative Physiology and Ecology 4:118-120.

MUSTAFA, S., AND A. K. JAFRI. 1977. RNA and protein contents in the flesh of teleost <u>Channa</u> <u>punctatus</u> Bloch during growth. Annales de Biologie Animale, Biochimie, et. Biophysique 17:991-995.

MUSTAFA, S., AND A. MITTAL. 1982. Distribution of protein, RNA, and DNA in alimentary canal of freshwater catfish <u>Clarias</u> <u>batrachus</u> (Linn.). Journal of Animal Morphology and Physiology 29:157-161.

MUSTAFA, S., AND S. M. ZOFAIR. 1983. The relation of nucleic acids to condition factor in the catfish, <u>Heteropneustes</u> <u>fossilis</u>. Reproduction, Nutrition, Development 23:145-149.

NAG, A. C. 1972. Ultrastructure and adenosine triphosphatase activity of red and white muscle fibers of the caudal region of a fish, <u>Salmo</u> <u>gairdneri</u>. Journal of Cell Biology 55:42-57.

NASIRI, S. 1972. Growth of carpsuckers, <u>Carpiodes</u> spp. as indicated by RNA/DNA ratios. Doctoral

Dissertation, Iowa State University, Ames, Iowa, USA.

ROSENLUND, G., B. LUND, E. LIED, AND A. VON DER DECKEN. 1983. Properties of white trunk muscle from saithe Pollachius virens, rainbow trout, Salmo gairdneri and herring Clupea harengus: protein synthesis in vitro, electrophoretic study of proteins. Comparative Biochemistry and Physiology 74B:389-397.

SABLE, D. 1974. RNA-DNA ratios as indicators of short-term growth in channel catfish. Doctoral Dissertation. Iowa State University, Ames, Iowa, USA.

SATOMI, Y. 1966. Nucleic acid and phospholipid content of silver carp Hypophthalmichthys molitrix, crucian carp Carassius carassius and carp Cyprinus carpio, raised together in fertilized ponds. (In Japanese with German abstract). Bulletin of Freshwater Fisheries Research Laboratory (Tokyo) 16:113-132.

SATOMI, Y. 1969. Change of the chemical composition (nucleic acid, phospholipid, kjeldahl-nitrogen, total phosphorus, and water content) of larval carp under conditions of satiation, starvation, and refeeding. (In Japanese with German abstract). Bulletin of Freshwater Fisheries Research Laboratory (Tokyo) 19:47-72.

SATOMI, Y. 1972. Change in the nucleic acid and phospholipid content of carp tissue (white and red muscle, liver, intestine) during the growing phase. (In Japanese with German abstract). Bulletin of Freshwater Fisheries Research Laboratory (Tokyo) 22(2):127-135.

SATOMI, Y. 1978. RNA-DNA ratios in the tissues of three cyprinids reared in fertilized fish ponds. (In Japanese with English abstract). Bulletin of Freshwater Fisheries Research Laboratory (Tokyo) 28:15-20.

SATOMI, Y., AND R. ISHIDA. 1976. Seasonal changes of nucleic acid contents in liver, red and ordinary muscle of Ayu-fish (Plecoglossus altivelis T. and S.) reared in ponds. (In Japanese with English abstract). Bulletin of Freshwater Fisheries Research Laboratory (Tokyo) 26:35-44.

SATOMI, Y., AND H. TANAKA. 1973. Fluctuations in the content of nucleic acids and phospholipids in the liver of rainbow trout fingerlings during feeding with different protein diets. (In Japanese with German abstract) Bulletin of Freshwater Fisheries Research Laboratory (Tokyo) 23:75-86.

SATOMI, Y., AND H. TANAKA. 1978a. Nucleic acids, carbon, nitrogen and phosphorus contents of elvers raised on tubificids. (In Japanese with English abstract). Bulletin of Freshwater Fisheries Research Laboratory (Tokyo) 28:1-4.

SATOMI, Y., AND H. TANAKA. 1978b. Seasonal changes of nucleic acid contents in liver, red and ordinary muscle of cultured eel (Anguilla japonica). (In Japanese with English abstract). Bulletin of Freshwater Fisheries Research Laboratory (Tokyo) 28:5-14.

SCHMIDT, G., AND S. J. THANNHAUSER. 1945. A method for the determinatin of deoxyribonucleic acid, ribonucleic acid, and phosphoproteins in animal tissues. Journal of Biological Chemistry 161:83-89.

SCHNEIDER, W. C. 1957. Determination of nucleic acids in tissues by pentose analysis. Pages 680-684 in S. P. Colowick and N. O. Kaplan, editors. Methods in enzymology. Volume 3. Academic Press, New York, New York, USA.

SHIBKO, S., P. KOIVISTOINEN, C. A. TRATNYEK, A. R. NEWHALL, AND L. FRIEDMAN. 1967. A method for sequential quantitative separation and determination of protein, RNA, DNA, lipid, and glycogen from a single rat liver homogenate or from a subcellular fraction. Analytical Biochemistry 19:514-528.

SOWER, S. A., C. B. SCHRECK, AND M. EVENSON. 1983. Effects of steroids and steroid antagonists on growth, gonadal development, and RNA-DNA ratios in juvenile steelhead trout. Aquaculture 32:243-254.

SPIGARELLI, S. A., AND D. W. SMITH. 1976. Growth of salmonid fishes from heated and unheated areas of Lake Michigan -- measured by RNA-DNA ratios. Pages 100-105 in G. W. Esch and R. W. McFarlane, editors. Thermal Ecology II. ERDA Symposium Series Conference 750425. Augusta, Georgia, USA.

STROGANOV, N. S. 1977. Role of environment in the plastic metabolism of fish. Pages 35-47 in G. S. Karzinkin, editor. Metabolism and biochemistry of fishes. Indian National Scientific Documentation Center, New Delhi, India.

THORPE, J. E., C. TALBOT, AND C. VILLARREAL. 1982. Bimodality of growth and smolting in Atlantic salmon, Salmo salar L. Aquaculture 28:123-132.

UCHIYAMA, H., S. EHIRA, M. TAKEUCHI, AND H. KOBAYASHI. 1969. Biochemical studies on wintering of culture carp. VII. Seasonal changes in nucleic acid levels and activities of xanthine oxidase and

protease in hepatopancreas of carp. Bulletin
Tokai Regional Fisheries Research Laboratory
59:75-80.

WANNEMACHER, R. W., JR., W. L. BANKS, JR., AND W. H.
WUNNER. 1965. Use of a single tissue extract to
determine cellular protein and nucleic acid
concentrations and rate of amino acid
incorporation. Analytical Biochemistry 11:320-
326.

WEATHERLEY, A. H. 1972. Growth and ecology of fish
populations. Academic Press, New York, New York.
U.S.A.

WEBB, J. M., and H. B. Levy. 1955. A sensitive
method for the determination of deoxyribonucleic
acid in tissues and microorganisms. Journal of
Biological Chemistry 213:107-117.

WILDER, I. B., AND J. G. STANLEY. 1983. RNA-DNA
ratio as an index to growth in salmonid fishes in
the laboratory and in streams contaminated by
carbaryl. Journal of Fish Biology 22:165-172.

ZEITOUN, I. H., D. E. ULLREY, W. G. BERGEN, AND W. T.
MAGEE. 1977. DNA, RNA, protein, and free amino
acids during ontogenesis of rainbow trout (Salmo
gairdneri). Journal of the Fisheries Research
Board of Canada 34:83-88.

Uptake of radioactive amino acids as indices of current growth rate of fish: A review

IRA R. ADELMAN
DEPARTMENT OF FISHERIES AND WILDLIFE
UNIVERSITY OF MINNESOTA
ST. PAUL, MN 55108

ABSTRACT

In contrast to traditional determinations of growth rate of fish, methods using uptake or incorporation of radioactive amino acids provide an index of the current growth rate at the exact moment the fish is examined. Measurement of protein synthetic rate of fish muscle tissue by use of radioactive amino acids has been used as an index of fish growth rate. Also, uptake of radioactively labeled glycine by fish scales in vitro has been used as a relative estimate of protein synthetic rate and thus an index of growth rate. Under certain conditions, these two indices have been shown to be highly correlated with fish growth rate. Endogenous and environmental factors that affect whole body growth rate of fish generally affect these indices in a similar manner. Whereas measurements of protein synthetic rate in muscle tissue provide a theoretically supportable procedure, the complex methodology required makes it impractical for large sample sizes. Although less supportable theoretically and subject to several errors, glycine uptake by scales permits rapid processing of large sample sizes and is nonsacrificial. With further development, glycine uptake may provide a useful, practical index.

This paper summarizes studies on uptake of amino acids as they have been used to estimate growth rates of fishes and evaluates the efficacy of the methods for that purpose.

The traditional methods of determining fish growth rate by measurement of changes in length or weight must be done over intervals of time, and therefore do not provide an index of growth rate at the time the fish is examined. Back calculations from the body-scale relationship are retrospective. In contrast

The Age and Growth of Fish, edited by Robert C. Summerfelt and Gordon E. Hall © 1987 The Iowa State University Press, Ames, Iowa 50010.

to these traditional measurements, methods using uptake of radioactive amino acids or their incorporation into protein, do enable determination of instantaneous or current growth. It is advantageous to determine a current growth rate, particularly in natural populations where environmental variables affecting growth are uncontrolled and frequently change rapidly. Even in controlled laboratory settings, reductions in assessment time and the ability to measure rapid changes in growth rate are extremely useful.

Protein synthesis plays a major role in the growth of mammals (Munro 1970) and fish (Haschemeyer and Smith 1979; Haschemeyer et al. 1979); therefore, rates of protein synthesis should be related to fish growth rate. Radioactive amino acids have been used to estimate whole body growth rate of fish through actual, purported or relative determinations of protein synthetic rate of various tissues. Few of the numerous published papers on protein synthesis in fish deal directly with the relationship between protein synthesis and whole body growth rate. The bulk of the research effort has been devoted to absolute rates of synthesis in various tissues or factors affecting the rate of synthesis. A relative index of protein synthesis, amino acid uptake by fish scales, has been used almost exclusively for determination of growth rate.

PROTEIN SYNTHESIS

Methodology

Typical procedures for direct determination of protein synthesis involve administering a radioactive amino acid to an animal and subsequently analyzing the amount of radioactivity incorporated into protein. In addition to the actual protein synthetic rate, incorporation of labeled material will depend on the time course of isotope uptake by the tissue, dispersion throughout the tissue, intracellular specific activity, potential degradation of recently labeled proteins and subsequent reutilization of the isotope, and secretion of proteins in some tissues. Failure to consider and control these variables can lead to serious errors in the interpretation of results. Also, environmental conditions both in vivo and in vitro will affect results and interpretation (Haschemeyer 1976).

Administration of radioactive amino acid to fish in vivo has been by infusion or injection of a tracer dose (Haschemeyer and Smith 1979; Fauconneau et al. 1981), or by injection of a large dose of labeled amino acid (Haschemeyer and Smith 1979; Pocrnjic et al. 1983). In a novel approach, Fauconneau (1984) measured whole body protein synthesis in larval common carp Cyprinus carpio by dosing the aquarium water with the radioactive amino acid and measuring the amino acid absorbed from the water and incorporated into protein. The main differences among the various methods of administration involve the manner in which the amino acid is incorporated and the degree of stabilization of the specific activity of the amino acid in the precursor pool. Stabilization of activity is necessary for a valid estimate of the rate of incorporation into protein (Fauconneau 1984).

In vitro determinations of the rate of protein synthesis in fish have been accomplished by measurements of amino acid incorporation: into crude mitochondrial protein of isolated liver cells and slices (Kent and Prosser 1980); into protein of isolated hepatocytes (Saez et al. 1982); and by ribosomes isolated from muscle tissue and immersed in a liver cell sap (Lied et al. 1982). In addition to measuring the radioactivity incorporated into protein fractions of various tissues, Haschemeyer and co-workers (Haschemeyer 1969a; Nielsen et al. 1977) measured rates of elongation of polypeptide chains both in vitro and in vivo. Under certain physiological conditions the overall rate of protein synthesis may be controlled by the elongation stage of the protein synthetic process.

The most commonly used tissues for determination of protein synthesis seem to be liver and muscle but kidney, gill, spleen, heart, testis (Pocrnjic et al. 1983) and whole bodies (Fauconneau 1984) also have been analyzed. Haschemeyer and Smith (1979) and Smith (1981) suggest that protein synthesis in muscle tissue is probably the most closely correlated with whole body growth rate because there is a low turnover rate of muscle protein in comparison to other tissues.

Reviews by Haschemeyer (1973, 1976) and Rannels et al. (1982) have elucidated potential errors associated with several aspects and methodologies in the measurement of protein synthesis. The theoretical and methodological pitfalls in measuring protein synthesis are numerous and failure to follow appropriate procedures easily produces erroneous results. The methodology is complex and time consuming and is not easily applied to large sample sizes.

Relationship to Growth

Few of the numerous published papers on protein synthesis in fish deal directly with the relationship between protein synthesis and whole body growth rate. Much of the research has been devoted to measurement of absolute rates of synthesis in various fish tissues (Pocrnjic et al. 1983) and effects of temperature and temperature acclimation (Haschemeyer et al. 1979).

Jackim and La Roche (1973) found that amino acid incorporation into muscle protein of Fundulus heteroclitus increased with temperature up to 26-29 C and then declined at higher temperatures. Incorporation was reduced by starvation, low dissolved oxygen, and restricted water volume but was increased by insulin. These factors affected amino acid incorporation in the same manner as they affect whole body growth. Although procedures used were subject to the kinds of errors mentioned previously, results probably provide a relative estimate of protein synthetic activity but certainly not a true measure of protein synthetic rate.

Comparison of protein synthesis in liver, gill and muscle of rainbow trout Salmo gairdneri has shown that a greater proportion of muscle tissue protein synthesis contributes to growth than does synthesis by the other tissues (Smith 1981). Less than 6% of liver or gill protein synthesis contributed to growth with the rest presumably turnover necessary to sustain metabolism. Approximately 65% of muscle synthesis contributed to growth. Similar findings were obtained with two Antarctic fishes (Trematomus hansoni and T. bernacchii) in which approximately 76% of protein synthesis in muscle resulted in growth (Smith and Haschemeyer 1980).

Smith (1981) suggested that the difference in muscle protein synthetic rates between fed and starved fish might be used as a measurement of potential growth rate. This conclusion arose from his finding that the difference in protein synthetic rate between muscle of fed and starved rainbow trout (0.28% per day) was numerically similar to the whole body growth rate of the fed fish (0.25% per day).

Lied et al. (1982) examined the effect of starvation for 10 days on protein synthesis of ribosomes isolated from Atlantic cod Gadus morhua muscle tissue. Amino acid incorporating activity was reduced approximately 87%. Among cod which were fed diets containing several proportions of protein energy, there was a highly significant correlation with the quantity of protein in the feed and the three measured variables: protein synthesis of isolated ribosomes, ribosomal RNA content, and the rRNA/DNA ratio (Lied and Rosenlund 1984). Although growth rate of the fish was not measured in this instance, RNA/DNA ratio is an index of growth rate (Bulow 1986). In comparing protein synthetic activity of white muscle of pollock Pollachius virens, rainbow trout, and herring Clupea harengus, differences in growth rates between the species were inferred from different rates of protein synthesis (Rosenlund et al. 1983).

Protein synthetic rates of isolated muscle ribosomes were significantly correlated (r = 0.80) with growth rate of pollock (Rosenlund et al. 1984). Groups of fish were starved or fed different amounts of food to produce different growth rates.

Protein synthetic rate in fish has usually been measured on larger individuals (> 25 g) because of difficulties in administering the radioactive amino acid or recovery of labeled protein in tissues of small fish. Based on the findings of Pavillon and Tan Tue (1981) that dissolved amino acids would penetrate through body surfaces into intact fish, Fauconneau (1984) immersed juvenile carp (46 mg) in water containing a radioactively labeled amino acid. The protein synthetic rate of 300% (fraction of tissue protein replaced per day) seemed high in comparison to other species and methods of administration, for example, < 1-5% per day in rainbow trout muscle (Smith 1981; Fauconneau 1984) or 17% in liver (Smith 1981). However, protein synthetic rate most likely decreases during development as does metabolic rate and growth rate (Kaushik and Dabrowski 1983), so differences between the young carp and older or larger individuals of other species may be reasonable.

Haschemeyer and co-workers, in a long series of studies, (e.g., Haschemeyer 1969b; Haschemeyer et al. 1979; Haschemeyer and Mathews 1982) investigated effects of temperature and temperature acclimation on protein synthesis in fish. These studies provided insight into processes and mechanisms of temperature acclimation as well as methods for determining protein synthesis in fish, but in general did not deal directly with the relationship of protein synthesis to growth rate.

The numerous studies clearly establish that the protein synthetic rate of muscle tissue responds to growth-altering factors, such as ration size, dissolved oxygen, hormones, and stress, in the same manner as does whole body growth. Highly significant correlations between protein synthesis of muscle and whole body growth rate or RNA content are typical. Because the addition of new muscle protein is primarily responsible for increases in fish size, accurate measurement of protein synthetic rate, in vitro or in vivo, provides a theoretically supportable and technically feasible means of estimating current growth rates of fish. However, current procedures for accurate determination of protein synthesis are complex and are unlikely to be practical for routine biological studies, particularly field studies or where large sample sizes are necessary for population estimates. Where a current growth rate is of interest in laboratory experiments or where knowledge of the effect of a growth-altering factor on the protein synthetic process is desired, determinations of protein synthetic rate would be appropriate.

GLYCINE UPTAKE BY FISH SCALES

The rate of uptake of glycine by a fish scale in vitro has been used as a relative index of protein synthetic rate in the scale (Ottaway and Simkiss 1977a). The traditional practices of measuring distances between annular rings on fish scales to determine annual growth rates and the fundamental assumption that growth rate of the scale is proportional to growth rate of the fish suggests that protein synthetic rate of the scale might then provide an index of whole body fish growth rate.

Methodology

Methodology involves removing one or more scales from a fish, incubating the scales in a physiological saline containing ^{14}C-glycine, rinsing the scales in saline, and counting radioactivity taken up by the scale after digesting the scale in a tissue solubilizer (Busacker and Adelman 1986). This method differs from protein synthesis discussed in the previous section in that only uptake of the amino acid is measured and not its incorporation into protein. An assumption upon which the method is based is that living cells associated with scalar formation and overlying epidermal cells remain viable; thus the rate of uptake of the amino acid reflects the growth rate of the fish. Ottaway and Simkiss (1977a) provided the following evidence in support of that assumption:

(1) Histology and electron microscopy have shown that scale-forming cells remained attached to the fibrillary plate and ossified layers of the scales.

(2) Autoradiographic studies have shown accumulation of ^{14}C-glycine in the vicinity of the scale-forming cells and even greater accumulation in the region of epidermal cells.

(3) Highest glycine incorporating activity was among cells on the scale margins, presumably because the osteoblasts and fibroblasts located there incorporated glycine at a faster rate than did the epidermal and dermal cell types located on the posterior surface (Ottaway 1978). Osteoblasts and fibroblasts secrete the collagenous scalar matrix of which glycine constitutes approximately 30%.

(4) Although damage and disarray to cells on removal of the scale is inevitable, and tissue anoxia during incubation is possible, the consistency of glycine incorporation by different scales from the same fish suggests that variation in uptake due to these factors is minimal.

(5) Kinetic studies have shown that: glycine is incorporated at a constant rate for up to 24 hours; immersion in boiling water stops glycine uptake; glycine uptake is temperature dependent; and scales consume oxygen at a constant rate for many hours.

Goolish and Adelman (1983a) provided further evidence that scales incorporate glycine into protein by demonstrating that a protein synthesis inhibitor, cycloheximide, would reduce glycine uptake by up to 88%. Also, maximum uptake occurred when pH of the incubation medium was maintained at 7.7. This pH falls within the range 7.5 to 8.0 where maximum amino acid incorporating activity occurs in isolated muscle polyribosomes (Lied et al. 1982).

Unpublished data by Busacker and Adelman confirmed that protein synthesis actually occurred when scales were incubated in glycine. Groups of scales were incubated in glycine with or without cycloheximide and were removed from the incubation media at time intervals up to 12 hours. After rinsing, the scales were homogenized and protein was precipitated with trichloroacetic acid (TCA). Radioactivity present in the TCA precipitated protein pellet was a measure of glycine actually incorporated into protein and not just absorbed or adsorbed by the scale. Radioactivity in the protein pellet from cycloheximide exposed scales was reduced by 89% after 2 hours incubation and 91% after 12 hours. Whereas the amount of glycine incorporated into protein increased about seven fold between scales taken at 1 hour and 12 hours, and not exposed to cycloheximide, there was only a threefold difference in incorporation over that time period for cyclo-

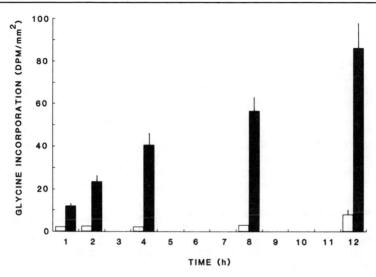

Figure 1. Mean glycine incorporation (+SE) in protein of common carp scales at various incubation times with (open) or without (solid) a protein synthesis inhibitor, cycloheximide.

heximide exposed scales (Fig. 1).

Other unpublished work of Busacker and Adelman determined that glycine was taken up by scales in a manner consistent with an enzymatic process. Uptake by carp scales became asymptotic at high concentrations of glycine and data were amenable to a Lineweaver-Burk plot with maximum velocity (V_{max}) at 55 pmol/mm^2/hr and a Michaelis-Menten constant (K_m) of 129 uM.

Finally, on-going studies of Busacker and Adelman have found that glycine was incorporated into mucus secreted by scales during incubation. To determine if glycine was being incorporated in mucopolysaccharides, scales were incubated in ^{14}C-glycine in saline for 2 hours. Dithiothreitol then was added to the incubation medium to separate mucus clumps, and the medium was eluted through a gel filtration column that separated molecules in the molecular weight range of 10,000-4,000,000 daltons. Some radioactively labeled molecules in that range were eluted from the column. These molecules absorbed ultraviolet (UV) light at the typical spectral peak for mucopolysaccharides (280 nm). To determine if the radioactivity associated with the mucopolysaccharides was simply due to adherence, control scales were incubated with unlabeled glycine, and radioactive glycine was added to the incubation medium only after the scales were removed. When this incubation medium was eluted through the gel column, UV absorption occurred in the same fractions as with the medium containing radioactive glycine during scale incubation, but no radioactivity was present in these fractions. The glycine incorporated into mucus during incubation can be a source of variability in measurement of radioactivity depending on the consistency with which scales are rinsed.

Although Ottaway (1978) reported a circadian rhythm in glycine uptake by scales of roach Rutilus rutilus, Goolish and Adelman (1983a) found no significant and repeatable rhythm in 12 laboratory and field trials on common carp or bluegills Lepomis macrochirus. Differences among sampling times were attributable to growth rate of the fish sampled. However, if a circadian rhythm in uptake occurs, then time of day becomes a variable in routine use of the methodology.

Originally, Ottaway and Simkiss (1977a) reported uptake of glycine in terms of scale weight, but Smagula and Adelman (1982) found that this procedure introduced a systematic error related to scale size. Uptake expressed in terms of scale area (pmol/mm^2) was better, although not entirely satisfactory, because glycine incorporating activity is not uniformly distributed on the scale surface. Glycine uptake between fish differing considerably in size may not be directly comparable. Also, glycine uptake methodology does not seem to work well with salmonids (Pereira 1982).

Glycine uptake by scales from fish taken from different temperatures is affected by both the fish's growth rate and the incubation temperature of the glycine solution. Ottaway and Simkiss (1977a) used a Q_{10} relationship to adjust glycine uptake of scales incubated at different temperatures to a constant 14 C. However, Adelman (1980) showed that the relationship between glycine uptake and incubation temperature was not constant for fish acclimated to different temperatures. Thus, comparisons between fish from different temperatures could not be standardized to a constant temperature. Smagula and Adelman (1982) tentatively proposed that when comparisons of fish acclimated to different temperatures are of interest, scales should be incubated at the temperature of maximum glycine uptake associated with the acclimation temperature of the fish. This criterion would not limit uptake by scales from fish acclimated to different temperatures, but the optimum temperature for uptake would need to be determined for a series of acclimation temperatures for each species of interest. In an experiment where fish were fed various-sized rations to produce fish growing at the same rate in different temperatures, Goolish and Adelman (1983b) showed that the correlation coefficient between glycine uptake and growth rate of individual fish was higher when scales were incubated at the fish's acclimation temperature than at the incubation temperature for maximum uptake. This finding eliminates the need to standardize incubation temperature. Scales should be incubated at the same temperature as the water from which the fish are taken (assuming that the fish are acclimated to that condition).

Relationship to Growth

Published information consistently supports a relationship between growth rate of groups of fish or individual fish and glycine uptake by scales.

Ottaway and Simkiss (1977a) originally used the method to compare growth rates of roach from three ponds, one with a slow growing population, one fast growing, and one intermediate. Previous growth rates of these fish were determined by back calculation. Glycine uptake by scales taken from fish from the three ponds on the same date showed a similar ranking as the back-calculated growth rates. Moreover, glycine uptake was highest for age 1 fish, in comparison to ages 2, 3 and 4 which were similar. Uptake varied seasonally with the highest rate in July.

Significant correlation coefficients between glycine uptake by scales and growth rate of individual fish have been found for sea bass <u>Dicentrarchus</u> <u>labrax</u> (Ottaway and Simkiss 1979); white sucker <u>Catastomus</u> <u>commersoni</u>, yellow perch <u>Perca</u> <u>flavescens</u>, bluegill and common carp (Adelman 1980; Goolish and Adelman 1983a); and for largemouth bass <u>Micropterus</u> <u>salmoides</u> (Smagula and Adelman 1982). The relationship between glycine uptake and growth rate of carp and the effect of fish acclimation temperature on that relationship are illustrated in Fig. 2. (Acclimation

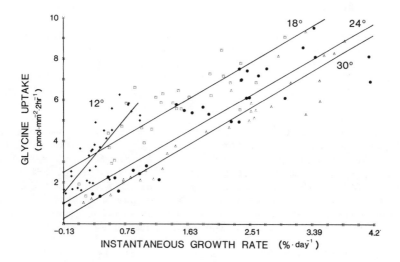

Figure 2. The relationship between the instantaneous growth rate of individual common carp over a 2-week period and the rate of glycine uptake by their scales for fish acclimated to 12 (♦), 18 (□), 24 (●), and 30 (△) C. Glycine uptake values are for scale incubation at the fish's acclimation temperature (Reproduced from Goolish and Adelman 1983b).

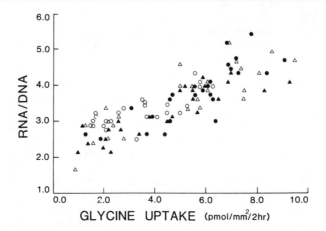

Figure 3. The relationship of the white muscle RNA/DNA ratio to
the rate of glycine uptake by the scales of common carp growing at rates from 0.0 to 4.2%/day. Fish were acclimated to
either 12 (o), 18 (•), 24 (△) or 30 C (▲) (Reproduced from
Goolish et al. 1984).

effects were explained earlier.) Glycine uptake by carp scales
and the RNA/DNA ratio in muscle tissue from the same fish also
were significantly correlated and effects of acclimation temperature on the RNA/DNA ratio versus growth rate relationship were
similar to effects on the glycine uptake and growth rate relationship (Goolish et al. 1984) (Fig. 3). These results suggest an
integration between thermoacclimatory changes in the RNA component of muscle protein synthesis and in the rate of amino acid
uptake by scales.

Glycine uptake is sensitive to environmental factors that
influence growth rate. After 12 hours, glycine uptake by scales
of roach exposed to a dissolved oxygen of 30% saturation was
significantly reduced in comparison to fish exposed to 85% saturation, but by 24 hours uptake by scales of fish exposed to the
lower oxygen level increased to become similar to fish at the
higher oxygen level   Handling and transportation stress significantly reduced glycine uptake by the roach scales sampled 24
hours after the stress (Ottaway and Simkiss 1977b). Continuous
confinement of small bluegills in a net resulted in a 25% reduction in glycine uptake within 1 hour and a 62% reduction by 4
hours (Goolish and Adelman 1983a). Increasing differences in
glycine uptake by scales of fed and starved roach occurred at 4,
14, and 32 days (Ottaway and Simkiss 1977b), and glycine uptake
by scales of bluegills starved for 2 weeks was reduced by about
80-85% over fish fed to satiation (Adelman 1980).

Ottaway and Simkiss (1979) and Smagula and Adelman (1983) applied the glycine uptake methodology to field populations of sea bass and largemouth bass, respectively, to ascertain its usefulness in assessing seasonal changes in growth rates. Both studies showed a bimodal seasonal pattern in glycine uptake by scales, with a peak in April (sea bass) or June (largemouth bass), a decline during the time of annulus formation (May for sea bass, July for largemouth bass) and a higher peak during the period of rapid growth (July-August for sea bass, August-September for largemouth bass). There was a significant correlation between glycine uptake and increase in length of sea bass throughout the sampling period; for largemouth bass a significant relationship between glycine uptake and scale growth rate occurred only during the period when the fish were actually growing. However, variation in glycine uptake of largemouth bass during May through July, when no positive growth occurred, caused a significant correlation over the entire study between glycine uptake and scope for growth (Brett 1979).

Correlation between glycine uptake and scope for growth suggests that glycine uptake may reflect an overall potential for growth as influenced by ration size or quality on a given day. The temporal proximity of feeding to analysis of glycine uptake seems to exert a strong influence on the relationship of uptake to growth rate, particularly when corroboratory measurements of growth rate can only be determined over intervals of time. During the period when the fish were apparently not growing, high glycine uptake by largemouth bass scales (Smagula and Adelman 1983) may have resulted from high growth rates on individual sampling dates. These high growth rates may have been of such short duration that measurement of increases in fish length or scale size over 2-week intervals did not reveal a change.

As with the more direct analyses of protein synthetic rate, glycine uptake by scales clearly responds in the expected manner to growth-influencing factors such as temperature, ration size, low dissolved oxygen, and handling stress. Correlation coefficients between glycine uptake and growth rate over 2 or 3 week periods before the glycine uptake determination have consistently been highly significant for many fish species. The methodology is nonsacrificial, relatively easy to perform, and large sample sizes can be handled quickly. Theoretically, the method is less supportable than some of the more recently developed methods for determining protein synthetic rate. There are a variety of cells and tissues on the scale, and uptake and distribution of the glycine and synthetic rates of these different cells and tissues are probably not uniform.

The standard method measures both glycine incorporated in protein and that remaining in the free amino acid pool, with the latter probably not responsive to fish growth rate. Measurement of the amount of glycine incorporated into protein only would eliminate that source of error but also would increase the complexity of the method. It may also be possible to reduce the error caused by uptake into the free amino acid pool by using the difference in uptake between cycloheximide-free scales and scales from the same fish incubated in the presence of the protein synthesis inhibitor. This procedure would double the required sample size but not increase complexity.

Differences in rates of glycine uptake between scales differing widely in size probably result in a systematic error, and loss of mucus from the scales causes a random error. In field situations, there is uncertainty as to whether glycine uptake is an index of actual growth or scope for growth. It would be useful to determine the glycine-uptake rate on a field population of fish on a daily basis for an extended period of time to determine the consistency of glycine uptake and its relationship to daily feeding rate.

Even with these caveats and uncertainties, measurement of glycine uptake by scales seems to be a useful, practical means of determining the influence of a variety of growth-affecting variables in laboratory situations. With further refinement and understanding its greatest utility may be realized in the determination of current growth rates in field populations.

ACKNOWLEDGMENT

I gratefully thank Greg P. Busacker for suggestions on content and editing and Jo Ann Schroeder for editing and typing the manuscript. This work was supported in part by the University of Minnesota Agricultural Experiment Station, project number 75, Scientific Journal Series Number 14,341.

REFERENCES

ADELMAN, I. R. 1980. Uptake of ^{14}C-glycine by scales as an index of fish growth: effect of fish acclimation temperature. Transactions of the American Fisheries Society 109:187-194.

BRETT, J. R. 1979. Environmental factors and growth. Pages 599-675 in W. S. Hoar, D. J. Randall and J. R. Brett, editors. Fish physiology. Volume VIII. Academic Press, New York, New York, USA.

BULOW, F. J. 1986. RNA-DNA ratios as indicators of growth in fish: a review. In R. C. Summerfelt and G. E. Hall, editors. Age and growth of fish. Iowa State University Press, Ames, Iowa, USA.

BUSACKER, G. P., AND I. R. ADELMAN. 1986. Uptake of ^{14}C-glycine by fish scales (in vitro) as an index of growth rate. In R. C. Summerfelt and G. E. Hall, editors. Age and growth of fish. Iowa State University Press, Ames, Iowa, USA.

FAUCONNEAU, B. 1984. The measurement of whole body protein synthesis in larval and juvenile carp (Cyprinus carpio). Comparative Biochemistry and Physiology 78B:845-850.

FAUCONNEAU, B., M. ARNAL, and P. LUQUET. 1981. Etude de la synthèse protéique in vivo dans le muscle de la truite arcen-ciel (<u>Salmo</u> <u>gairdneri</u> R.). Influence de la temperature. Reproduction, Nutrition, Développement 21:293-301.

GOOLISH, E. M., and I. R. ADELMAN. 1983a. 14C -glycine uptake by fish scales: refinement of a growth index and effects of a protein-synthesis inhibitor. Transactions of the American Fisheries Society 112:647-652.

GOOLISH, E. M., and I. R. ADELMAN. 1983b. Effects of fish growth rate, acclimation temperature and incubation temperature on in vitro glycine uptake by fish scales. Comparative Biochemistry and Physiology 76A:127-134.

GOOLISH, E. M., M. G. BARRON, and I. R. ADELMAN. 1984. Thermoacclimatory response of nucleic acid and protein content of carp muscle tissue: influence of growth rate and relationship to glycine uptake by scales. Canadian Journal of Zoology 62:2164-2170.

HASCHEMEYER, A. E. V. 1969a. Rates of polypeptide chain assembly in liver in vivo: relation to the mechanism of temperature acclimation in <u>Opsanus</u> <u>tau</u>. Proceedings of the National Academy of Science USA 62:128-135.

HASCHEMEYER, A. E. V. 1969b. Studies on the control of protein synthesis on low-temperature acclimation. Comparative Biochemistry and Physiology 28:535-552.

HASCHEMEYER, A. E. V. 1973. Control of protein synthesis in the acclimation of fish to environmental temperature changes. Pages 3-31 <u>in</u> W. Chavin, editor. Responses of fish to environmental changes. Charles C. Thomas, Springfield, Illinois, USA.

HASCHEMEYER, A. E. V. 1976. Kinetics of protein synthesis in higher organisms in vivo. Trends in Biochemical Science 1:133-136.

HASCHEMEYER, A. E. V., and R. W MATHEWS. 1982. Effects of temperature extremes on protein synthesis in liver of toadfish, <u>Opsanus</u> <u>tau</u>, in vivo. Biological Bulletin 162:18-27.

HASCHEMEYER, A. E. V., R. PERSELL, and M. A. K. SMITH. 1979. Effect of temperature on protein synthesis in fish of the Galapagos and Perlas Islands. Comparative Biochemistry and Physiology 65B:91-95.

HASCHEMEYER, A. E. V. and M. A. K. SMITH. 1979. Protein synthesis in liver, muscle and gill of mullet (<u>Mugil</u> <u>cephalus</u> L.) in vivo. Biological Bulletin 156:93-102.

JACKIM, E. and G. LA ROCHE. 1973. Protein synthesis in <u>Fundulus</u> <u>heteroclitus</u> muscle. Comparative Biochemistry and Physiology 44A:851-866.

KAUSHIK, S. J. and K. DABROWSKI. 1983. Post-prandial metabolic changes in larval and juvenile carp (<u>Cyprinus</u> <u>carpio</u>). Reproduction, Nutrition, Developpement 23:223-234.

KENT, J. and C. L. PROSSER. 1980. Effects of incubation and acclimation temperatures on incorporation of U- ^{14}C glycine into mitochondrial protein of liver cells and slices from green sunfish, <u>Lepomis</u> <u>cyanellus</u>. Physiological Zoology 53:293-304.

LIED, E., B. LUND, and A. VON DER DECKEN. 1982. Protein synthe-
    sis in vitro by epaxial muscle polyribosomes from cod, Gadus
    morhua. Comparative Biochemistry and Physiology 72B:187-193.
LIED, E., and G. ROSENLUND. 1984. The influence of the ratio of
    protein energy to total energy in the feed on the activity
    of protein synthesis in vitro, the level of ribosomal RNA
    and the RNA-DNA ratio in white trunk muscle of Atlantic cod
    (Gadus morhua). Comparative Biochemistry and Physiology
    77A:489-494.
MUNRO, H. N. 1970. A general survey of mechanisms regulating
    protein metabolism in mammals. Pages 3-130 in H.N. Munro,
    editor. Mammalian protein metabolism, Volume 4. Academic
    Press, New York, New York, USA.
NIELSEN, J. B. K., P. W. PLANT, and A. E. V. HASCHEMEYER. 1977.
    Control of protein synthesis in temperature acclimation. II.
    Correlation of elongation factor 1 activity with elongation
    rate in vivo. Physiological Zoology 50:22-30.
OTTAWAY, E. M. 1978. Rhythmic growth activity in fish scales.
    Journal of Fish Biology 12:615-623.
OTTAWAY, E. M., and K. SIMKISS. 1977a. "Instantaneous" growth
    rates of fish scales and their use in studies of fish popu-
    lations. Journal of Zoology, London 181:407-419.
OTTAWAY, E. M., and K. SIMKISS. 1977b. A method for assessing
    factors influencing "false check" formation in fish scales.
    Journal of Fish Biology 11:681-687.
OTTAWAY, E. M., and K. SIMKISS. 1979. A comparison of traditional
    and novel ways of estimating growth rates from scales of
    natural populations of young bass (Dicentrarchus labrax).
    Journal of the Marine Biological Association of the United
    Kingdom 59:49-59.
PAVILLON, J. F., and V. TAN TUE. 1981. Un aspect de l'absorption
    des substances organiques dissoutes chez les organismes
    marins et d'eaux saumâtres: données autoradiographiques aux
    premiers stades de développement chez Artémia spécies (ex
    salina) et Dicentrarchus labrax. Oceanis 7:705-718.
PEREIRA, D. L. 1982. Size, temperature and photoperiod effects on
    growth, smoltification and glycine uptake by scales of chi-
    nook salmon (Oncorhynchus tshawytscha). Master's thesis.
    University of Minnesota, St. Paul, Minnesota, USA.
POCRNJIC, Z., R. W. MATHEWS, S. RAPPAPORT, and A. E. V. HASCHE-
    MEYER. 1983. Quantitative protein synthetic rates in various
    tissues of a temperate fish in vivo by the method of phe-
    nylalanine swamping. Comparative Biochemistry and Physio-
    logy 74B:735-738.
RANNELS, D. E., S. A. WARTELL, and C. A. WATKINS. 1982. The
    measurement of protein synthesis in biological systems. Life
    Sciences 30:1679-1690.
ROSENLUND, G., B. LUND, E. LIED, and A. VON DER DECKEN. 1983.
    Properties of white trunk muscle from saithe Pollachius
    virens, rainbow trout, Salmo gairdneri, and herring, Clupea
    harengus: protein synthesis in vitro, electrophoretic study

of proteins. Comparative Biochemistry and Physiology 74B:389-397.

ROSENLUND, G., B. LUND, K. SANDNES, O. R. BRAEKKAH and A. VON DER DECKEN. 1984. Muscle protein synthesis in vitro of saithe (Pollachius virens) correlated to growth and daily energy intake. Comparative Biochemistry and Physiology 77B:7-13.

SAEZ, L., O. GOICOECHEA, R. AMTHAUER, and M. KRAUSKOPF. 1982. Behavior of RNA and protein synthesis during the acclimatization of the carp. Studies with isolated hepatocytes. Comparative Biochemistry and Physiology 72B:31-38.

SMAGULA, C. M., and I. R. ADELMAN. 1982. Temperature and scale size errors in the use of ^{14}C-glycine uptake by scales as a growth index. Canadian Journal of Fisheries and Aquatic Sciences 39:1366-1372.

SMAGULA, C. M., and I. R. ADELMAN. 1983. Growth in a natural population of largemouth bass, Micropterus salmoides Lacepede, as determined by physical measurements and ^{14}C-glycine uptake by scales. Journal of Fish Biology 22:695-703.

SMITH, M. A. K. 1981. Estimation of growth potential by measurement of tissue protein synthetic rates in feeding and fasting rainbow trout, Salmo gairdneri Richardson. Journal of Fish Biology 19:213-220.

SMITH, M. A. K., and A. E. V. HASCHEMEYER. 1980. Protein metabolism and cold adaptation in Antarctic fishes. Physiological Zoology 53:373-382.

Mathematical and biological expression of growth in fishes: Recent trends and further developments

JACQUES MOREAU
FISHERIES DEPARTMENT
ECOLE NATIONALE SUPÉRIEURE AGRONOMIQUE
145 AVENUE DE MURET
31076 TOULOUSE CÉDEX, FRANCE

ABSTRACT

This paper deals with the general problem of fitting growth curves and their biological interpretation, with special reference to the Von Bertalanffy growth function (VBGF). A possible generalization of this equation and its biological interpretation is proposed that avoids short-comings pointed out previously. Possibilities exist to describe the seasonal pattern of growth by introducing a SINE function in the VBGF. Some ways of using the VBGF for quantitative comparisons of growth are given with examples from African and marine fishes. Methods for rapid, approximate evaluation of growth parameters L_∞ and K are also proposed and suggestions for further research are given.

This paper aims to survey some approaches used to describe somatic growth of individual fishes. The main purpose is to review possibilities of fitting growth curves and recent trends in biological interpretation of growth parameter evaluations. We have to keep in mind that growth, as a phenomenon, will interest the cell biologist, bioenergeticist and physiologist, as well as mathematician and fishery manager.

The main criteria for choosing a growth curve are quality of fit and convenience, differing according to whether the need is for mathematical description of a detailed physiological growth process in young fishes or for fishery management. The suitability of a curve may also be different for length or weight growth fitting.

Historically, most of the curves in use have been proposed with some mathematical-physiological theory as to how growth might be regulated. Much ingenuity has been expended in trying to relate them to growth processes. Among these curves are the

The Age and Growth of Fish, edited by Robert C. Summerfelt and Gordon E. Hall © 1987 The Iowa State University Press, Ames, Iowa 50010.

logistic, the Gompertz, the Johnson, the Richards and the Von
Bertalanffy growth function (VBGF). Mainly the VBGF will be
considered in this review, but other possibilities of growth
fitting have to be systematically explored in the future.

MATHEMATICAL EXPRESSIONS OF GROWTH

General Approach
     Growth curves of fish in the wild are developed from size at
age using either length of weight. Length is usually easier to
use because only the part of the curve having decreasing curvature
needs to be described by a formula for yield predictions or evaluations. By contrast, the absolute rate of increase in weight
often increases for several years before decreasing. To fit the
whole weight curve, it is necessary to find an S-shaped curve
having the correct curvature on both sides of the inflection
point. Whether length or weight data are fitted depends on the
quality of the data.
     The curves most frequently fitted to size data are almost all
bipartite in their general differential form. Rate of increase in
size is proportional to the difference between a positive constant
times size already achieved "ay" and some function of that size,
"f(y)" (Ricker 1979). The differential form is:

$$\frac{dy}{dt} = |ay - f(y)|$$

It can be proved mathematically that, if y(t) is a continuous
solution to this differential equation, then the sign of ay - f(y)
is constant. This differential form has been generalized by Pauly
(1979) and Schnute (1981) who replaced "ay" by "ay^b" where b is
constant. Schnute (1981) showed that most common growth curves
were derived from the same second order differential equation:

$$\frac{d^2y}{dt^2} = \frac{dy}{dt} \left| -a + (1-b) \frac{d(\log_e y)}{dt} \right| \quad \text{where a and b are constants}$$

     This equation shows clearly the close relationship between
the various growth curves, and it appears that inflection points
and asymptotic growth are only particular situations in a general
context. Moreover, Schnute (1981) has showed that the Von Bertalanffy, logistic and Gompertz equations are different expressions of the same model.
     The general form of curves described by these equations have
an upper asymptote: where L and W are length and weight at any
age t, and L_∞ and W_∞ asymptotic sizes. Formulae are given in
terms of one or the other as seems more appropriate and all these
curves have given reasonable fits but not necessarily on the same
species over the complete range of ages.

Growth Models Other Than Von Bertalanffy
     Several growth models have been documented according to the

expected needs of biological and mathematical descriptions of
growth. These are:

Logistic growth curve. A logistic curve represents the auto-
catalytic law of physiology and chemistry. Its differential form
is:

$$\frac{dy}{dt} = ay - by^2$$

or with weight symbols and after rearrangement:

$$\frac{dW}{dt} = \frac{gW}{W_\infty} (W_\infty - W) \text{ where } g \text{ is a constant}$$

The instantaneous rate of increase in weight is proportional
to the difference between the asymptotic weight W_∞ and the actual
weight, W. The integral form of this equation may be written:

$$W = W_\infty / (1 + \exp(-g(t - t_0)))$$

This is the equation of a sigmoid curve with an inflection
point for which W is equal to $W_\infty/2$. The two halves of the curve
are antisymmetrical, where g is the instantaneous growth rate
when W = 0 and t_0 is the time at which the growth rate is maximum;
it is also the abscissa of the inflection point of the curve for
which $W = W_\infty/2$.

A logistic curve has rarely been used to describe individual
fish growth for the entire life span, but Miura et al. (1976)
described successfully the yearly seasonal growth pattern of four
age groups of redspot salmon Oncorhynchus rhodurus.

Gompertz curve. A Gompertz curve is S-shaped with both a
lower and a upper asymtote and an inflection point, but the two
halves of the curve are not antisymmetrical. It expresses the
decrease of growth rate throughout the adult phase of the life.
A number of investigators have interpreted the Gompertz curve as
reflecting the activity of two different and opposing types of
regulatory factors during growth (Ricker 1979). The differential
form of a Gompertz curve is written as follows:

$$\frac{dy}{dt} = ay - by(\log_e y)$$

Using weight symbols, we have:

$$\frac{dW}{dt} = aW - bW(\log_e W)$$

$$= gW(\log_e W_\infty - \log_e W) \quad \text{where } g = b$$

A possible integral form of this equation is

$$W = W_\infty \exp(-\exp(-g(t - t_0)))$$

where t_0 is the abscissa of the inflection point and g is
the instantaneous growth rate when $t = t_0$ (Fletcher 1973).

Methods for fitting this curve are given by Riffenburg
(1960), Ricklefs (1967), and Regner (1980). Zweifel and Lasker
(1976) used two successive Gompertz cycles to describe the growth
of Anchovy larvae Engraulis mordax. Gompertz curves are rarely
used in fishery problems because yield computations usually employ
the Von Bertalanffy models, but Silliman (1967) showed that they
might be suitable when using analog computers.

I find that a Gompertz curve describes weight-at-age quite
well, often including the early years of increasing increments
and length-at-age data when growth shows an inflection point.
This is the case for some anadromous fishes spending the first of
their life in rivers where growth is very slow compared to the
later years in the sea (Moore and Moore 1974 and Fig. 1).
Gompertz curves can also be used for short-lived species of
tropical flood-plains, where the food changes significantly from
juvenile to adult stages.

The use of Gompertz curves should be explored further because
yield calculation with any growth curve can now be generalized by
integral, numerical-value computation using micro computers
(Moreau et al. 1984).

Selection of the optimal fitting. Richards and Johnson
curves have rarely been used for individual fish growth studies.
According to Krüger (1973) quoted in Ricker (1979), Richards and
Johnson curves can fit length-at-age data well but no theoretical
basis is given for their applicability.

S-shaped curves should be tested routinely for fitting to
growth of diadromous fishes physiology of which changes with
habitat, or of fishes where food and feeding habits change marked-
ly in the course of their life. However, procedures for determining optimal curve fitting to a given data set need to be generalized. Some methods do exist, and Schnute (1981) showed how the
result of a simple variance analysis indicates which type of model
is most appropriate. Kappenman (1981) proposed predictive reuse
techniques previously developed by Gesser and Eddy (1979) to
select the most plausible model.

The Von Bertalanffy Growth Function (VBGF)

Because it has been widely used in length-growth studies and
fisheries management, the VBGF has been the most evaluated one.
The following points have been documented: sampling bias, limitations of validity, quality and confidence limits of the evaluated
parameter estimates, possible methods of computation and biological significance of the evaluations.

Initially, the VBGF looked easy to handle because of its
simplicity, lack of an inflection point, and its possible use in
yield computations (Beverton and Holt 1957). It can be also
fitted to tagging increment data and to size-at-age data (Kirkwood
1983).

The VBGF is written

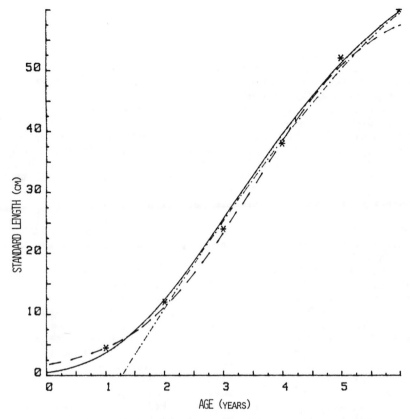

Figure 1. Fitting of length-at-age data for sea trout <u>Salmo</u> <u>trutta</u> of the Garonne (Labat, pers. comm.) using three types of growth curves:

Continuous line: Gompertz curve: L_∞ = 75 cm. g = 0.515

t_0 = 3.15 years

Dashed line: Logistic curve: L_∞ = 63 cm   g = 0.98

t_0 = 3.5 years

Mixed line: VBGF curve: L_∞ = 160 cm   K = 0.10

t_0 = 1.29 years

$$L = L_\infty \left| 1 - \exp(-K(t - t_0)) \right| \qquad \text{(with length symbols)}$$

$$W = W_\infty \left| 1 - \exp(-K(t - t_0)) \right|^b \qquad \text{(using weight symbols)}$$

where L_∞ and W_∞ are asymptotic sizes, K is a constant, t = t_0 when L or W = 0; b is the exponent of the length/weight relationship.

The following problems are usually mentioned when using the VBGF:

The limitations of validity. The VBGF is unable to describe the sigmoid length-growth resulting from an inflection point early in the life of fishes (Yamaguchi 1975). This limitation occurs in short-lived species, which show an inflection point and are exploited early in their life.

Pauly (1979) claimed that L_∞ was often overestimated. This occurs when slow-growing old fishes are not sampled such as in African freshwater fishes (Merona et al. 1985).

The concept of asymptotic size itself needs further clarification. Ricker (1975) defined it as the mean size the fishes in a given population would reach if they lived and grow indefinitely. However, various authors have contested the validity of this concept. For example, Paloheimo and Dickie (1965) wrote:

"In many cases the Von Bertalanffy growth curve is fitted to data consisting mainly of young fish well below the projected final size. Hence the value of L_∞ apparently reflects the early growth".

Knight (1968) contended that for fishes asymtotic size was a mathematical fiction. For similar reasons, Roff (1980) asked for the retirement of the VBGF. Others maintained that asymptotic growth was real but that estimates of the asymptotic values were subject to wide variation depending on the curve fitting. This is really a problem when considering the influence of these variations on yield computations using Beverton and Holt models (Gulland 1983). Asymptotic growth to an indefinite old age can be neither proved nor disproved, nor need all species of fish to be the same in this respect. However, asymptotic formulae are a convenient way of modelling many observed growth series, and we may expect them to be used into the indefinite future.

The necessity of setting a "correct" birth date. The computation of t_0 requires the setting of a birth date to coincide with the peak of the breeding season (Lopez Veiga 1979). When a single birth date is assigned, the quality of t_0 estimation is surely affected, particularly the variance around t_0. This problem occurs mainly with tropical fishes with a long breeding season: 3 to 6 months (Moreau 1979). However, even under these conditions, fishes which hatch at the beginning and end of the reproductive season have low larvae survival rates. Consequently, the efficient hatching period is considerably shorter than the breeding season (Philippart 1977); so, the maximum of sexual activity should be used for accurate setting of a "mean" birth date for each cohort. It is much easier to estimate t_0 in temperate waters, where the duration of the breeding activity is only a few weeks.

To improve the quality of t_0 evaluations, Lopez Veiga (1979) suggested the length of hatching L_h to be the length-at-age $t = 0$. Thus, if L_∞ and K have been previously computed:

$$t_0 = \frac{1}{K} \log_e \frac{L_\infty - L_h}{L_\infty}$$

then, for every set (L, t):

$$\log_e(L_\infty - L) = \log_e L_\infty - K(t - t_0)$$

New evaluations of the parameters L_∞ and K can be obtained by successive iterations.

Possibilities of reparameterization. Keeping in mind the possible biological inaccuracy of L_∞ and t_0 computations, Schnute and Fournier (1980) introduced into the growth equation the upper and lower limits, l_{max} and l_{min}, of the observed length-frequency distribution. If for l_{min}, the mean observed age is t = a_1, then

$$l_{min} = L_\infty \left| 1 - \exp(-K(a_1 - t_0)) \right|$$

If for l_{max} the mean observed age is t = a_M = a_1 + M, then

$$l_{max} = L_\infty \left| 1 - \exp(-K(a_1 + M - 1 - t_0)) \right|$$

Thus, for every age t, the length L is computed as follows:

$$L = l_{min} + (l_{max} - l_{min}) \frac{1 - \exp(-K(t - 1))}{1 - \exp(-K(M - 1))}$$

In this equation L_∞ and t_0 do not appear and the problem of their biological interpretation disappears. In fact, we have:

$$L_\infty = \frac{l_{max} - l_{min} \exp(-K(M - 1))}{1 - \exp(-K(M - 1))}$$

$$\text{and } t_0 = a_1 - \frac{1}{K} \log_e \left| \frac{L_{max} - L_{min}}{L_{max} - L_{min} \exp(-K(M - 1))} \right|$$

In a similar way, Schnute (1981) proposed a general versatile growth model without using the concept of asymptotic size. His expression of the VBGF is a generalization of the equation of Schnute and Fournier (1980).

Computation problems when fitting data to a VBGF. The usual method of estimating K, L_∞, and t_0 has involved plotting a Ford-Walford line without consideration of the variance properties of the estimates. Vaughan and Kanciruk (1983) used a Monte Carlo simulation to show that the Walford plot gives biased estimates, less precise than using non-linear procedures. The simulations included variations of standard deviation of the lengths, sample sizes, sampling time intervals, and Von Bertalanffy growth parameters.

When standard deviations and confidence intervals are computed from a Ford-Walford analysis, it is necessary to assume that the linear regression of the random variables L_t+1 on L_t has a normally distributed error structure with zero mean, constant variance, and uncorrelated errors. These conditions are generally satisfied. Other linear methods of fitting the VBGF are given by Fabens (1965), Allen (1966), Daget and Le Guen (1975), Bayley (1977), Ricker (1975), and developed by Gaschütz et al. (1980).

Gallucci and Quinn (1979) and Vaughan and Kanciruk (1983) recommended to replace linear methods by non-linear fitting techniques. Age would be used as a true, non-stochastic, independent variable, and the estimation would focus upon the data and the model, not on its linearized forms. This sometimes requires the user to specify both the function and its derivatives, and it provides estimates of parameters and the variance-covariance matrix of these estimates (Abramson 1971; Kimura 1980; Kirkwood 1983). Non-linear regression routines are also used which no longer require derivatives and provide independent evaluations of the parameters with simultaneous improvement of maximum likelihood procedures (Dixon 1977; Kimura 1980; Kirkwood 1983).

The effect of individual variability on parameter evaluations. The work of Sainsbury (1980) was the first known analysis of the effect of variations among individuals on the description of the growth of a group.

The variance of mean length-at-age increases with age to a maximum and declines for older age-groups. Individual variations of length-at-age data induce variations in the value of K, mainly when referring to observed periodical increment data. In some situations K might be overestimated and the predicted mean length at each age underestimated. This invites us to remember the necessity of the requirement of homogeneity of variances in the dependent variables in using the least-square methods for fitting growth curves to weight data, since the variability among individuals in their weight usually increases as their age or average weight increases. Some possibilities exist to deal with this problem (Sainsbury 1980).

Finally, fitting a model is not only a statistical procedure; it requires a decision about whether or not the data are suitable. This, however, complicates the interpretation of the results of the regression. A regression program always computes and minimizes a residual sum of squares (SSR) in the dependent variable to measure the quality of the fit to the data. In general, the lower the SSR, the better the statistical fit . However, a regression fit to inappropriate or invalid data may yield a very good statistical fit but unrealistic parameters. Many authors have encountered this difficulty ; it is the reason why Pauly (1979) proposed a generalized expression of the VBGF.

A GENERALIZATION OF THE VBGF (PAULY 1979)

The Mathematical Expression and Its Physiological Implications

In its usual form, the VBGF results from the integration of the relationship:

$$\frac{dW}{dt} = HW^d - KW^m \tag{1}$$

where dW/dt is the weight growth with HW^d and KW^m representing the synthesis (or anbolism) and the degradation (or catabolism) of body substance respectively, both being assumed proportional to a certain power (d and m respectively) of the fish weight (Von Bertalanffy 1957, 1964). Equation (1) was integrated by Beverton and Holt (1957) with values of d and m assumed equal to 2/3 and 1 respectively, which is a particular case.

If it seems appropriate to assume that catabolism in fishes should be proportional to weight (Von Bertalanffy 1964), then m = 1 in equation (1). Setting d at 2/3 implies, among other things, that the O_2 consumption of fish should be proportional to a 2/3 power of the weight (2/3 rule of metabolism); the link between anabolism and O_2 consumption is established by the fact that fishes are aerobic heterotrophs. It has been repeatedly demonstrated, however, that the power linking body weight and O_2 consumption (hence anabolism) generally ranges in fishes between 0.5 (in small fishes such as guppies) and 0.9 (Muir 1969). Also, the power linking gill surface area and body weight (which, since it determines the scope of O_2 consumption, also determines the scope of anabolism) has been shown to have values generally averaging 0.8 (De Jager and Dekkers 1975) and values up to 0.9 in large fishes such as tuna.

Pauly (1979) defined the parameter D = 3(m − d) and demonstrated that high values of d (or low values of D) occur mainly in fishes reaching a very larg size, while low values of d (and high values of D) occur in fishes in which the adult size remains very small (Fig.2). This provides a method for obtaining preliminary estimates of D from size data alone, and independently of growth data, by mean of the equation:

$$D = 3 \left[1 - (0.6472 + 0.3574 \log_{10} W_{max}) \right] \tag{2}$$

where W_{max} is the maximum weight recorded from the fishes of a given stock in grams (Fig. 2). The use of distinct value of D for each stock (rather than assumimg that D = 1 in all fishes) considerably extends the flexibility of the VBGF.

Pauly (1979) integrated equation (1) assuming that d $\neq$ 2/3 and m = 1. This produces a "generalized VBGF" of the following form in which

$$D = 3(m - d)$$

$$W^{D/b} = W_{\infty}^{D/b} \left| 1 - \exp\left(- \frac{3KD}{b} (t - t_0)\right) \right|$$

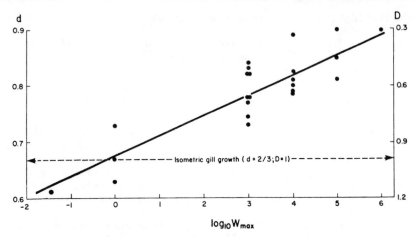

Figure 2. Oxygen consumption or gill surface area of fishes
increases as a power of maximum body weight. The low
values refer to very small fishes such as Cyprinodonts
and maximum values to Tunas. Intermediate values concern
"usual " fishes. Redrawn from Pauly (1982).

in which b is the exponent of the weight-length equation $W = aL^b$,
or in a form easier to handle:

$$W = W_\infty \left| 1 - \exp\left(- \frac{3KD}{b} (t - t_0)\right)\right|^{b/D}$$

Length-growth is expressed as follows:

$$L = L_\infty \left| 1 - \exp\left(- KD(t - t_0)\right)\right|^{1/D} \qquad (3)$$

The curve described by this equation has an inflection point
for which $t_i = t_0 + \log_e D/(KD)$: thus t_i moves toward t_0 when D
increases to 1 as shown on figure 3.

Definition and Discussion of the Parameters
        The generalized VBGF for length contains four explicit para-
meters: L_∞, K, D, and t_0. In addition, the VBGF implies a "hidden"
parameter l_x. This parameter refers to the smallest length from
which the VBGF begins to describe individual growth in a given
fish population. Definitions of the five parameters may help to
interpret the numerical values of these parameters obtained from
various fish stocks.
        Asymptotic size: W_∞ and L_∞. A reasonably good agreement
between values of the mean maximum observed length L_{max} and L_∞
has been demonstrated in fishes less than 50 cm L_{max} (Beverton
1963; Taylor 1962). Taylor proposed that:

$$L_{max} = 0.95 L_\infty \qquad (4)$$

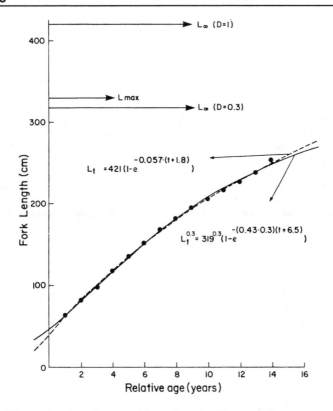

Figure 3. Length-at-age data for bluefin tuna <u>Thunnus thynnus</u> fitted by the usual (dashed line) and by the general VBGF as proposed by Pauly (1979). Redrawn from Pauly (1982).

which allows for the estimation of reasonable values of asymptotic length in small fishes. In large fishes, when the usual VBGF is used, the value of asymptotic size obtained from a set of size-at-age data differs for the same stock by an amount which increases with their maximal size. Pauly (1981) expresses that differently: "the most erroneous the assumption d = 2/3 is, the higher the difference between W_{max} and W_∞, or L_{max} and L_∞."

The generalized VBGF provides estimates of W_∞ and L_∞ very close to the values of W_{max} and L_{max} when the appropriate value of D is used . This allows the application of equation (4) to the stock of any fish species. It may be also suggested that:

$$\sqrt[3]{W_{max}} = 0.95 \sqrt[3]{W_\infty}$$

The surface factor: D. This parameter is defined as the difference between the power of length in proportion to weight increase and the power of length in proportion to gill surface
area increase.

The stress factor: K. This parameter is the most difficult
to visualize. K refers to the rate of degradation of body substance, especially proteins. However, this degradation in the body
of a living fish must be continuously compensated by synthesis
of new proteins. Thus, in addition to expressing protein degradation, K expresses abiotic and biotic factors which limit
oxygen availability for protein synthesis. Pauly (1979) argued
that K should be considered as a stress factor rather than as a
coefficient of catabolism. The word "stress" refers to the total
sum of adverse effects which raise the value of K, i.e., un-
favorable temperatures, salinities, population densities, food
supply, etc.

From the definition of K and stress, it may be derived that
fish never live stress-free, but that their growth performances
and their asymptotic size are highest when K (i.e., the associated stress) is lowest. This agrees with Rumohr (1975), who
insisted that there are no growth-enhancing exogenous factors but
only factors which depress growth.

The origin of the growth curve: t . The t_0 value is defined
as the hypothetical age the fish would have had at zero length,
had they always grown in the manner described by the VBGF
(Ricker 1975); t_0 is therefore not a biological parameter but
only a convenience parameter used to make the curve fit slightly
better. For many applications the exact value of t_0 is not
needed: yield per recruit computation, use of length-converted
catch curves for Z evaluations, and length-based cohort analysis.
Some methods (Gulland 1977, 1983; Munro 1982) estimate K and asymptotic size from tagging but they do not allow for an exact
age to be attributed to a certain length, although the growth
curve itself can be drawn.

The starting size: l_x and W_x. As t_0 cannot be interpreted
biologically, the length at age zero, or l_0, has no biological
significance, for example, as length at birth or length at hatching. At what lowest size can the VBGF describe the growth of
a fish? It often appears that the size at "metamorphosis" from
larva to juvenile corresponds to the starting size (l_x and W_x).
Figure 4 represents the beginning of growth curve; it shows (not
to scale) the age at fertilization (t_f) of an egg, the length of
embryo at hatching (L_h), and the growth of larvae up to the age
when fishes undergo their transition from a specific young growth
pattern to the one described by a VBGF with $D \neq 1$. It suggests
that positive values of t_0 are generally erroneous because they
imply negative values of L_h. This provides a criterion for
estimating the quality of some evaluations of K and t_0; when the
value of K is too high, due to biased data, this will result in
positive values of t_0 which are biologically impossible. Thus,
Pauly (1979) has suggested an empirical expression for a pre-

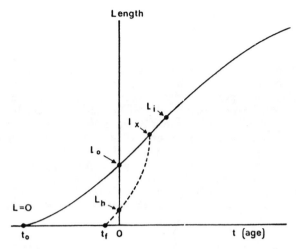

Figure 4. Biological and mathematical origin of the VBGF as
suggested by Pauly (1979); scales: arbitrary units. Terms
are defined in the text

liminary computation of t_0 from any estimate of K and L_∞. This
equation is:

$$\log_{10}(-t_0) = -0.3922 - 0.2752 \log_{10}L_\infty - 1.038 \log_{10}K$$

It is based on 153 triplets of values L_∞, K and negative values
of t_0 selected by the author, so as to cover a wide diversity of
fish taxa and sizes. The multiple correlation coefficient is
R = 0.685 for 150 degrees of freedom. This equation is used for
a preliminary computation of t_0 because the standard error of t_0,
given the predictors L_∞ and K, is:

$$SE = (1 - R^2)^{1/2} = 0.75$$

i.e., 75% of the standard error ignoring the predictors, which is
very high.
        Thus, the general VBGF can improve the biological definition
of the parameters but seasonal growth oscillations, which can
greatly affect the shape of a growth curve, have to be considered.

THE SEASONAL GROWTH PATTERN

The Ecological Problem and its Mathematical Expression
        Seasonality is a characteristic of growth in temperate
water fishes; e.g. growth tends to follow the cycle of the
seasons, usually faster in summer and slower in winter. In
tropical regions growth variation is usually related to seasonal
rainfall. However, in temperate waters, seasonal growth is

not always, perhaps never, entirely under temperature control.
In whitefish Coregonus clupeaformis, Hogman (1968) showed growth
was closely correlated with the seasonal cycle of day length.

The annual variations of fish growth in temperate and
boreal regions resemble an S-shaped cycle, as shown by Gerking
(1966) in sunfishes of Northern Indiana lakes, Fortin and Magnin
(1972) in Canadian perches Perca flavescens, Beall and Davaine
(1981) in trout Salmo trutta of the Kerguelen Islands, Abad (1982)
in Salmo trutta of some French rivers and Philippart (1977) for
cyprinid fishes in Belgian rivers (Fig. 5). A seasonal model is
needed in temperate waters to describe growth patterns shown
by frequent samples throughout the year (Philippart 1977).
Similarly, tropical biologists need also models to follow the
seasonal pattern in the progression of the modal length-frequency
data.

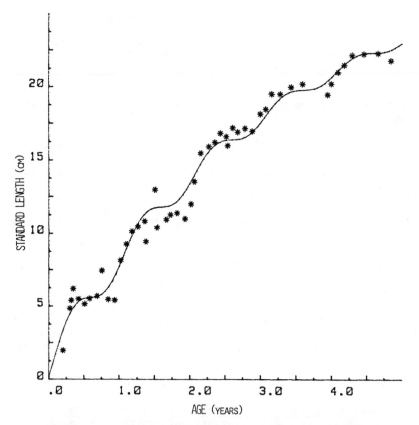

Figure 5. Length seasonal growth of dace Leuciscus leuciscus
recorded by monthly sampling for over 4 years (Philippart
1977). The line has been fitted by the equation of
Gaschütz et al (1980): L_∞ = 29.6 cm K = 0.298 t_0 = -0.1
C = 0.96 t_s = 0.1 year.

   Several versions of the VBGF have been published in recent
years to describe the seasonally oscillating growth pattern of
fishes (Ursin 1963; Pitcher and Macdonald 1973; Lockwood 1974;
Daget and Ecoutin 1976; Pauly and Gaschütz 1979).

   Pitcher and Macdonald (1973) suggested two modifications.
The first uses a cosine function to switch off growth in winter
and to turn it on again in spring. The second incorporated a
sine function that produced a smoother seasonal pattern; however,
it included some shrinkage in winter, which could be appropriate
for weight data but not for length. In either case, two parameters are added to the VBGF. On the same basis, Cloern and
Nichols (1978) rewrote the VBGF in the following form, in which
a_1 and Θ are constants and L_{min} the length-at-age t_0:

$$L = L_\infty - (L_\infty - L_{min}) \, A$$

$$A = \exp \left| -a_1(t - t_0) - \frac{180 \, a_1}{\pi}(\cos \frac{\pi(t_0 + \Theta)}{180} - \cos \frac{\pi(t + \Theta)}{180}) \right|$$

   As it is impossible to linearize this formula, the parameters
are very difficult to estimate except by means of computer programs and the added parameters cannot be biologically interpreted.
The one which determines the amplitude of the growth oscillations
cannot be visualized nor used comparatively, e.g., for assessing
the effects of winter conditions on the growth of different
stocks or species of fishes.

A Version of the Length VBGF Including a Seasonal Growth Pattern
   As shown by Pauly and Gaschütz (1979) a sine wave can be
incorporated into the VBGF such that:

$$L = L_\infty \left| 1 - \exp(-K(t - t_0) + A \sin 2\pi(t - t_s))) \right| \tag{5}$$

   Equation (5) is analogous to equation (6) of Pitcher and
MacDonald (1973) and to equation (6) of Cloern and Nichols (1978).
A is an empirical constant which yields no insight into the magnitude of growth oscillations which it modulates. This can be overcome as follows: length growth can be halted in winter such that:

$$\frac{dL}{dt} = L_\infty qr \qquad \text{where} \tag{6}$$

$$q = K + A2\pi \cos 2\pi(t - t_s) \qquad \text{and} \tag{7}$$

$$r = \exp(-K(t - t_0) + A \sin 2\pi(t - t_s))$$

   For equation (6) to have zero values, at least one of the
values of q or r must, at time, equal zero. However, L_∞ and r
must always remain $\neq 0$. Hence growth ceases if, and only if,
equation (7) equals zero. If we now define:

$$A = \frac{CK}{2\pi} \tag{8}$$

then equations (6) and (7) become zero once per year when
$t = t_s + 1/2$ or one of its multiples. There are no length growth
oscillations when $C = 0$; $C = 1$ when the oscillations produce a once
per year growth stop, i.e., one zero value of dL/dt.

Referring to published observations, Pauly (1982) described
the linear relationship between C and the difference between the
highest and the lowest monthly mean temperature in the relevant
water body (Fig. 6, from Pauly 1982). Small annual variations
of temperature, as in tropical areas, can generate growth
oscillations.

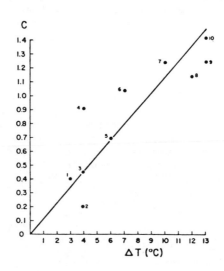

Figure 6. Relationship between the value of C and the difference between the highest and the lowest mean monthly
temperature to which the fishes were exposed. Redrawn
from Pauly (1982).

The parameter t_s expresses the time between birth (at $t = 0$)
and the onset of the first growth oscillation (which is modulated
by the sine wave curve of period one year).

After introducing the parameter C, equation (5) can be
rewritten in terms of the generalized VBGF:

$$L = L_\infty \left| 1 - \exp\left(-KD(t - t_0) + C\frac{KD}{2\pi} \sin 2\pi(t - t_s)\right)\right| \qquad (9)$$

and rearranged in e linear form:

$$\log_e \left(1 - \frac{L^D}{L_\infty^D}\right) = -K(t - t_0) + C\frac{KD}{2\pi} \sin 2\pi(t - t_s)$$

Since $\sin(\alpha - \beta) = (\sin \alpha)(\cos \beta) - (\sin \beta)(\cos \alpha)$, we have

$$\log_e (1 - \frac{L^D}{L_\infty^D}) - KD(t - t_0) + \frac{CKD}{2\pi}(\sin 2\pi t)(\cos 2\pi t_s)$$

$$- \frac{CKD}{2\pi}(\sin 2\pi t_s)(\cos 2\pi t)$$

which has the structure of a multiple linear regression of the form

$$y = a + b_1 x_1 + b_2 x_2 + b_3 x_3 \text{ in which:}$$

$a = KDt_0$ $y = \log_e(1 - \frac{L^D}{L_\infty^D})$

$b_1 = -KD$ $x_1 = t$

$b_2 = -\frac{CKD}{2\pi} \cos 2\pi t_s$ $x_2 = \sin 2\pi t$

$b_3 = \frac{CKD}{2\pi} \sin 2\pi t_s$ $x_3 = \cos 2\pi t$

where $t_s = \arctan(b_3/b_2)/(2\pi)$

The best value of L_∞ is the one which maximizes the multiple correlation between y and x_1, x_2, x_3 (Gaschütz et al. 1980).
The method of Pauly and Gaschütz has three advantages over its predecessors.
The parameters can be estimated easily, e.g., by means of a programmable pocket calculator because this modified VBGF can be linearized.
The parameter C expresses the amplitude of the growth oscillations and allows comparison of winter effects on species and stocks differing widely in their growth parameters.
The conditions which limit the applicability of the model are defined.

Seasonal Weight Growth
The authors cited above considered the seasonal growth pattern in length, but it is not misleading to deal with weight (Ursin 1979). Usually there is some yearly seasonal shrinkage in individual weight of fishes. Spawning often induces a significant cyclical loss of weight. Jones and Johnston (1977) also noted the possible influence of seasonal changes in food intake. The shrinkage may often reach 10-15% of total weight (Abad 1982; Moreau 1979).
A model similar to that of Pauly and Gaschütz (1979) has been developed by Shulman (1974) for the seasonally oscillating weight growth of Black Sea fishes. The Pauly and Gaschütz (1979) equation (9) can describe oscillating weight growth including a yearly shrinkage by using C values > 1. A possible way (Fig. 7; Moreau unpubl.) is to integrate values of C increasing with age; thus: $C = C_0 + a(W/W_\infty)$; a and C_0 are constants depending on the fish population.

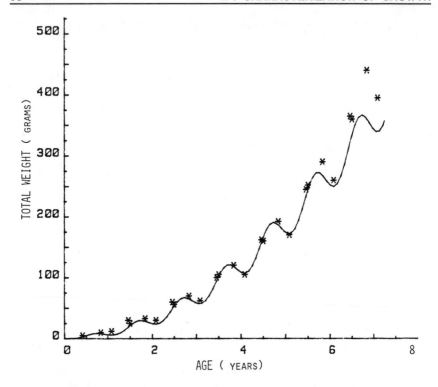

Figure 7. Seasonal weight growth of trout <u>Salmo</u> <u>trutta</u>
<u>fario</u> of the R. Viau, France (data from <u>Abad 1982)</u>. The
<u>VBGF</u> describes the yearly weight shrinkage by incorporat-
ing C values increasing with age (Moreau unpubl.).

After this review of the generalization of the VBGF two more
problems have to be mentioned which, in some ways, are related to
each other: the comparison of growth in fishes and the possibilities of rapid evaluation of growth parameters.

COMPARING GROWTH PERFORMANCES

Comparaison of the Growth of Different Fish Species
     Since the first systematic studies on fish growth, both
inter and intraspecific comparisons of growth have been made.
For instance Carlander (1969) and Merona (1983) compiled size
and age data of freshwater fishes of North America and Tropical
Africa, respectively. These compilations fail to explain why
certain fishes grow in a given environment, as they do.
     This problem persists for intraspecific comparisons between
different environments. Hofstede (1973), Szczwerbowsky (1976),
Hickley and Dexter (1979), Hickley and Sutton (1984) (Fig. 8)

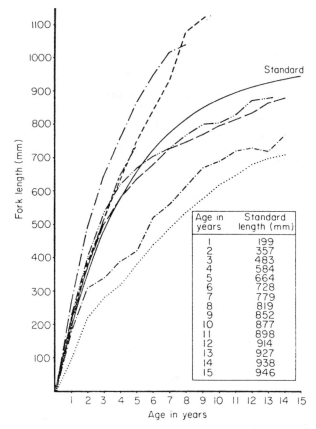

Figure 8. Standard length for age values and corresponding standard curve (continuous line) for Esox lucius as proposed by Hickley and Sutton (1984); the other lines show observed growth of Esox lucius, in British waters, to be compared to the standard curve.

developed standard curves for intraspecific growth comparison in limited geographical areas. Similary, Edwards (1984) showed that temperate demersal fishes grow faster than their tropical counterparts. A possible explanation is the higher metabolic rate in the tropics.

Comparison of the Growth Parameters

Various authors tried to compare the growth performances of fishes by referring to the values of L_∞ and K themselves (Allen 1976; Mirsa 1980). However, populations comparisons must use K and L_∞ values together as they are correlated. More generally, since growth is not linear, comparison of growth curves is inherently a multivariate problem. For instance, one stock can grow faster than another when younger and slower when older.

Thus, Bernard (1981) proposed the generalized use of Hotelling T^2 tests for growth curve comparisons. Other improvements have been suggested by Hoenig and Hammara (1983). These tests allow separation of effects of each parameter on observed differences in growth.

Kimura (1980) and Kirkwood (1983) suggested comparing growth curve parameters by likelihood ratio tests. Kimura (1980) assumed that t_0 was the same for two fish populations (males and females of Pacific hake <u>Merluccius productus</u>) and they showed that difference of growth could be pointed out graphically by referring to K and L_∞ estimates, simultaneously, and to their 95% confidence limits (Fig. 9).

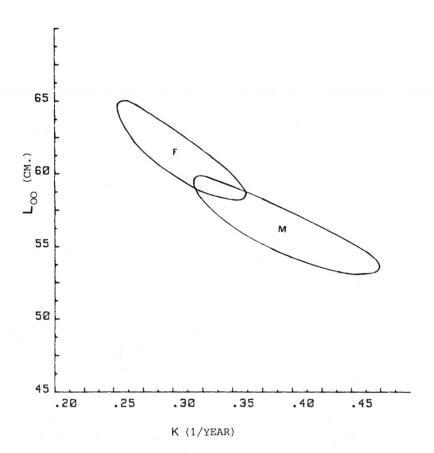

Figure 9. Cross sections of approximate 95% confidence
limits around least square estimates of L_∞ and K estimates
for males (M) and females (F) Pacific hake <u>Merluccius</u>
<u>productus</u>; letters M and F are centered on least square
estimates. Redrawn from Kimura (1980).

The concept of Index Growth Performance
     Even if it is clear that growth comparisons cannot be made
on a single parameter, some authors tried to identify an index of
overall growth (Gallucci and Quinn 1979; Pauly 1979).
     The parameter ω (Gallucci and Quinn 1979). Gallucci and
Quinn (1979) pointed out the necessity of independence between
the estimations of K and L_∞ to improve the flexibility of the
VBGF and, thus, the possibility of statistical comparisons of
growth. They suggested a new parameter ω = K L_∞, so that:

$$L = \frac{\omega}{K}(1 - \exp(-K(t - t_0)))$$

where ω , the growth rate near t_0, might be suitable for growth
comparisons. Gallucci and Quinn argued that ω is a stable parameter more normally distributed that K and L_∞ separately.
     Kingsley (1980) showed that two populations can have the
same ω value and different growth curves. He did not agree with
Gallucci and Quinn's claims of statistical robustness and suitability of ω. For Gulland (1983), ω is useful in distinguishing
differences in the early growth rate of different populations
and less suitable when these differences occur mainly in larger
fishes. Following Gulland, I think that ω might be helpful in
comparing growth when t_0 values are the same for the compared
fish-populations and when their longevity is similar. These two
conditions were valid in a comparison of growth of Esox lucius
in lake Windermere (Kipling 1983).
     The index of growth performance (Pauly 1979). According to
Pauly (1979) an index of growth performance should relate to the
weight growth and consist of a single value easy to compute. It
should be also applicable for any fish and biologically interpretable.
     The growth rate in weight dW/dt, or slope of the weight
growth curve , has in all fishes a maximum, whether plotted
against age or against size. Weight growth has only one maximum
value of growth rate. Therefore, Pauly (1979) proposed the
growth rate at the inflection point of a weight growth-curve be
used as a possible standard for comparison of growth performance
of different fishes. In a weight growth-curve, the slope at the
inflection point is given by:

$$(dW/dt)_{max} = \frac{4}{9} \ K \ W_\infty = \frac{4}{9} . 10^P \qquad \text{where:}$$

$$P = \log_{10} (K \ W_\infty) = \log_{10} K + \log_{10} W_\infty \qquad \text{(Pauly 1979)}$$

     The growth curves of different fishes cannot be directly
compared because the curves themselves are produced by growth
rates that are constantly changing with time and size. The value
of P, however, is directly related to (dW/dt) max, irrespective
of the values of W_∞ or K. The value of P can, therefore, be
used to compare the growth performance of fishes with different
values of asymptotic size.

The graphical use of P: the auximetric grid (Pauly 1979).
The character of the index P may be best demonstrated by
transposition into a graph called an auximetric grid. The axis
scales consist of value of $\log_{10} W_\infty$ and $\log_{10} K$ (1/year). The
range covered by both scales is such that average-sized commercial
fishes appear near the center of the grid. Also, lines connecting
some P values are drawn at regular intervals of P and a base line
is selected (at P = 0). On such a grid, the distance from a
point representing a pair of growth parameters (W_∞, K) to the
baseline represents P and is therefore a direct indication of
growth performance. A wide range of P values is evident in
commercial marine fishes (Fig. 10).

The auximetric grid also allows for the separation and
definition of taxa by means of their growth parameters. Within
one species it can help to quantify and compare the effects of
endogenous growth determinants (e.g., sex), as well as
environmental factors (e.g., salinities) on the growth of fishes.

Kingsley's (1980) comments on the parameter ω of Gallucci
and Quinn (1979) are surely applicable to the P index. The
auximetric grid shows clearly that P depends strongly on the
species.

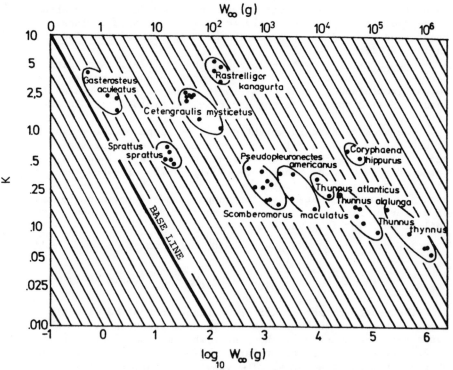

Figure 10. The auximetric grid: observed values of P are
quoted for 13 marine species. Redrawn from Pauly (1979).

Thus, P should be quoted as an "index of endogenous growth potentialities." These growth potentialities should be expressed by any combination of K and W_∞ that are compatible with the other biological characteristics of the species: longevity, ecological requirements. Current research on natural populations of Tilapia in Africa should enable a detailed discussion of this idea.

These different approaches to growth comparison have helped to develop some rapid and approximate means of evaluating growth parameters.

RAPID EVALUATION OF GROWTH PARAMETERS

Various authors have reported that in different stocks of a given species W_∞ and L_∞ tend to decrease as K increases. Several quantitative investigations of these interrelationships have been made for a rapid, approximative evaluation of growth parameters needed for multispecies fish stock management.

The Relationship Between the Longevity and the Parameters K and L_∞

In general, the larger the size a species of fish attains the greater its longevity: Taylor (1962) suggested that K could be considered as an index of theoretical longevity t_{max}. For these reasons, Merona (1983) and Moreau et al. (1985) considered the relationship between longevity, the maximum observed length L_{max} and the parameters K and L_∞.

Assuming that $t_0 = 0$ and that $L_{max} = 0.95\ L_\infty$, the usual VBGF can be written (Merona 1983):

$$0.95\ L_\infty = L_\infty(1 - \exp(-Kt_{max}))$$

$$0.05 = \exp(-Kt_{max}) \text{ and } t_{max} = 3/K$$

It means that for one population of a species a possible value of K may be computed only by knowing L_{max} with correct sampling of the oldest fishes. Thus, a relationship between K and L_∞ can be found for one species from K and L_∞ values of the literature. Such a relationship has been pointed out for all the fishes in a defined area, such as inland waters of Tropical Africa, for which the following relationship can be given (Merona 1983):

$$K = 153/L_\infty \qquad r = 0.466$$

$$L_\infty = 1.248\ L_{max} \qquad r = 0.917$$

with 111 sets of K, L_∞ and observed L_{max} data such that at each age t:

$$L = 1.248\ L_{max} \left| 1 - \exp(- 101\ t/L_{max}) \right|$$

This last equation results in a normogram allowing graphic determination of L_∞ and of the length at each age for any African freshwater fish (Fig. 11 and 12).

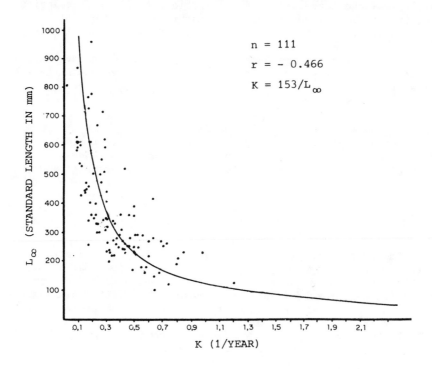

Figure 11.   Relationship between asymptotic length L_∞ and K
in the freshwater fishes of Africa as computed by Merona
(1983).

Reference to the Maximum Observed Weight (Munro and Pauly 1983)

For the similar purpose (i.e., to quantify the relationship between K and L_∞ and W_∞), a linear regression between $\log_{10}K$ and $\log_{10}L_\infty$ has been calculated for each of 126 species by Pauly (1979): a mean value of the slope b is -0.632 ± 0.386 (standard deviation). Similarly, a mean value of $b = -0.714$ is obtained when plotting $\log_{10}K$ against $\log_{10}W_\infty$ in various fish stocks. Therefore, Pauly (1979) proposed to take an unweighted mean of these two independently obtained mean values of b, such that:

$$b = \frac{0.632 - 0.714}{2} = -0.673$$

or, for simplicity, $b \simeq 2/3$. Thus, it can be generalized:

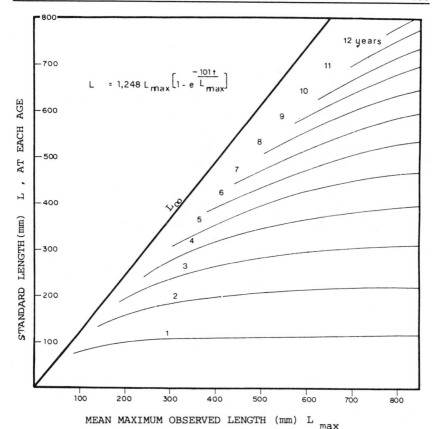

Figure 12. Abacus for estimations of L_∞, calculated length
at each age and longevity by reference to the mean
maximum observed length L_{max}. The continuous line "L_∞"
shows the relationship between L_{max} and L_∞. Redrawn
from Merona (1983).

$$\log_{10}K = a - (2/3)\log_{10}W_\infty$$

where the constant a depends on the fish stock. This equation can
be rearranged as:

$$a = \log_{10}K + (2/3)\log_{10}W_\infty \qquad (10)$$

Figure 13 (Munro and Pauly 1983) shows the frequency distribution of calculated values of "a" for different stocks of some
species of fishes for which W_∞ is expressed in grams. As can be
seen, the use of equation (10) to estimate "a" from published

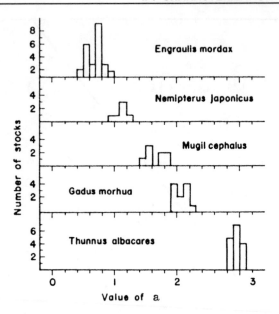

Figure 13. Frequency distributions of values of a = $\log_{10}K$ + 0.67 $\log_{10}W_\infty$ for species for which growth parameters are available. Redrawn from Munro and Pauly (1983).

growth parameters for various stocks of one species produces distributions of "a" values that are essentially normal, and rather sharply peaked, suggesting that equation (10) indeed describes the interrelationship between K and W_∞. This is also suitable for fish species of the same family.

Provided that a mean value of "a" is available from the literature-data on W_∞ and K for the relevant species or families, the growth parameters of a new population can be estimated by knowing only the mean maximum observed weight, keeping in mind that:

$$\sqrt[3]{W_{max}} = 0.95 \sqrt[3]{W_\infty} \qquad \text{or} \qquad W_{max} = 0.86 \ W_\infty$$

Use of the Auximetric Grid

Pauly (1979) has defined the index of growth performance:

$$P = \log_{10}K + \log_{10}W_\infty$$

which is relatively constant within species. The auximetric grid (Fig. 10) shows that there is only a limited range of values of P in a given species. This results in a limited range of possible

W_∞ and K combinations. Thus, given the growth parameters of some populations closely related to the one under investigation, a most likely value of K can be graphically selected from a reasonable estimate of asymptotic weight.

These methods of approaching the evaluation of the parameters are not perfect but particularly helpful in a tropical, multispecies stock context for a preliminary estimation of growth. They can also allow identifications of erroneous values of L_∞ and asymptotic size, through comparisons with other K/W_∞ or K/L_∞ pairs pertaining to the same species and available from the literature.

CONCLUSION

From biological point of view, describing fish growth by mathematical expressions is a complex problem and the contrast has been drawn many times between the so called indeterminate growth in fishes and the determinate one of other vertebrates. Fishes display a considerable intraspecific range of growth rates under different environmental conditions. For this reason, a definite, final adult size is not associated with a particular species. More complete understanding of factors affecting maximization of fish growth would contribute significantly to the understanding of fundamental biology of fishes and give improved insight into problems of modelling fish growth.

There are obviously upper limits to growth that are species-specific, i.e., genetic and other non-genetic factors influencing growth in fish. Figure 14 mentions the possible nature of the

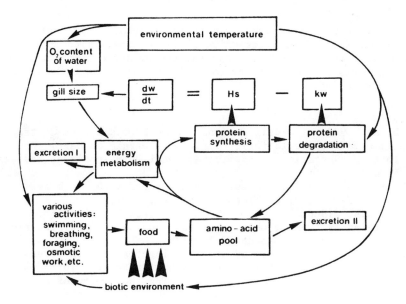

Figure 14. Biological approach of fish growth: the processes and factors involved. Redrawn from Pauly (1981).

relationships between trophic relations in the environment and
growth and depicts several major factors having a potential influence on growth rate. The complexity of the process suggests the
need for modelling using inputs from physiology, cellular biology,
ecology, ethology, as well as mathematics.

From mathematical point of view, several models and methods
exist for fitting a growth curve to data; the choice depends on
the quality of the data and on the aim of the study. However, as
shown by Schnute (1981), the proliferation of models and methods
disguise the basic simplicity of the problem of fitting a growth
curve. All that is really needed is a suitable general model to
encompass the usual special cases. Statistical tests such as F
tests can be used to decide among these cases. The best model
should minimize the residual sum of squares and, if the residuals
themselves show sign of periodicity, then a seasonal model should
be required. Even if new versions of the VBGF are rather flexible,
other models are surely more suitable for describing growth in
young stages, or in short-lived tropical fishes, and for management purposes. The availability of micro-computers now allow
yield calculations whatever the model of the growth curve. We
can expect these calculations to be generalized and, therefore,
I suggest that particular attention be paid to the statistical
properties of all models, bias in parameter evaluations, techniques for corrections and quality of the estimation of these
parameters. The word "quality" refers here, among other things,
to the possibilities of appropriate biological interpretation.

ACKNOWLEDGMENTS
The authors is extremely grateful to Professors Carlander
and Summerfelt (Iowa State University) for their kind invitation
to deliver this paper. The careful review and helpful criticisms
and comments of the draft by the referees also has to be gratefully acknowledged.

REFERENCES
ABAD, N. 1982. Ecologie et dynamique des populations de truites
    communes (S. trutta fario L.) dans le bassin du Tarn. Doc-
    toral Dissertation. Institut National Polytechnique, Toulouse,
    France.
ABRAMSON, N.J. 1971. Computer programs for fish stock assess-
    ment. FAO Fisheries Technical Papers. 101.
ALLEN, K.R. 1966. A method of fitting growth curves of the Von
    Bertalanffy type to observed data. Journal of the Fisheries
    Research Board of Canada 23:163-179.
ALLEN, R.L. 1976. A method of comparing fish growth curves. New
    Zealand Journal of Marine and Freshwater Resarch 10:687-692
BAYLEY, P.B. 1977. A method for finding the limits of applica-
    tion of the Von Bertalanffy growth model and statistical
    estimates of the parameters. Journal of the Fisheries Research
    Board of Canada 34:1079-1084.

BEALL, E., and P. DAVAINE. 1981. Acclimatation de la truite commune S. trutta en milieu subantarctique (I. Particularités scalimétriques). Colloque sur les Ecosystèmes Subantarctiques. Centre National de recherches sur les Terres australes 51:399-412.

BERNARD, D.R. 1981. Multivariate analysis as a mean of comparing growth in fishes. Canadian Journal of Fisheries and Aquatic Sciences 38:233-236.

BERTALANFFY, L. Von 1957. Quantitative laws on metabolism and growth. Quarterly Review of Biology 32:217-231.

BERTALANFFY, L. Von 1964. Basic concepts in quantitative biology of metabolism. Helgoländer Wissenchaft und Meeresuntersuch. 9:5-37.

BEVERTON, R.J.H. 1963. Maturation growth and mortality of Clupeid and Engraulid stocks in relation to fishing. Conseil Permanent de l'Exploration de la Mer. Rapport 154:44-67.

BEVERTON, R.J.H., and S.J. Holt. 1957. On the dynamics of exploited fish populations. U.K. Ministry of Agriculture and Fisheries. Fisheries Investigations (Ser. 2) 19.

CARLANDER, K.D. 1969. Handbook of freshwater fishery biology, volume 1. Iowa State University Press, Ames, U. S. A.

CLOERN, J.E., and F.H. NICHOLS. 1978. A Von Bertalanffy growth model with a seasonally varying coefficient. Journal of the Fisheries Research Board of Canada 35:1479-1482.

DAGET, J., and J.M. ECOUTIN. 1976. Modèles mathématiques de production applicables aux poissons tropicaux subissant un arrêt prolongé de croissance. Cahiers ORSTOM, série Hydrobiologie 10:59-69.

DAGET, J., and J.C. LE GUEN. 1975. Les critères d'âge chez les poissons. Pages 253-292. In LAMOTTE, M. and F. BOURLIERE, editors. La démographie des populations de Vertébrés. Masson, Paris, France.

DE JAEGER, S., and W.J. DEKKERS. 1975. Relations between gill structure and activity in fishes. Netherlands Journal of Zoology 25:276-308.

DIXON, W.J. editor. 1977. DMDP. Biomedical computer Programs. P series. University of California Press, Berkeley, California, U. S. A.

EDWARDS, R.R.C. 1984. Comparison of growth in weight of temperate and tropical marine fish countreparts. Canadian Journal of Fisheries and Aquatic Sciences 41:1381-1384.

FABENS, A.J. 1965. Properties and fitting of the Von Bertalanffy growth curve. Growth 29:265-289.

FLETCHER, R.I. 1973. A synthesis of deterministic growth laws. University Rhode Island, School of Oceanography. Kingston, Rhode Island.

FORTIN, R., and E. MAGNIN. 1972. Croissance en longueur et en poids de Perchaudes, Perca flavescens, de la Grande Anse de l'Ile Perrot, au lac St-Louis. Journal of the Fisheries Research Board of Canada 29:517-523.

GALLUCCI, V.F., and T.J. QUINN. 1979. Reparameterizing, fitting and testing a simple growth model. Transactions of the American Fisheries Society 108:14-25.

GASCHUTZ, G., D. PAULY, and N. DAVID. 1980. A versatile BASIC program for fitting weight and seasonally oscillating length growth data. International Council for the Exploration of the Sea. CM/1980/D:6 Statistics Comittee.

GEISSER, S., and W.E. EDDY. 1979. A predictive approach to model selection. Journal of the American Statistics Association. 74:152-160.

GERKING, S.A. 1966. Annual growth cycle, growth potential and growth compensation in the Blue-gill sunfish in northern Indiana lakes. Journal of the Fisheries Research Board of Canada 23:1923-1952.

GULLAND, J.A. editor. 1977. Fish population dynamics. J. Wiley & Sons, London, England.

GULLAND, J.A. 1983. Fish stock assessment, a manual of basic methods. FAO/Series on Agriculture 1. J. Wiley & Sons, London, England.

HICKLEY, P., and K.F. DEXTER. 1979. A comparative index for quantifying growth in length of fish. Fisheries Management 10:147-189.

HICKLEY, P., and A. SUTTON. 1984. A standard growth curve for Pike. Fisheries Management 15:29-30.

HOENIG, N.A., and R.C. HAMMARA. 1983. Statistical considerations in fitting seasonal growth models for fishes. International council for the Exploration of the Sea. Council meeting 1983. D:25

HOFSTEDE, A.E. 1973. Studies on growth, ageing and backcalculation of roach Rutilus rutilus L. and dace Leuciscus leuciscus L. Pages 137-147. In BAGENAL, T.B. editor. The ageing of fish. Union Brothers Limited, Surrey, England.

HOGMAN, W.J. 1968. Annulus formation on the scales of four species of coregonid reared under artificial conditions. Journal of the Fisheries Research Board of Canada 25:2111-2122.

JONES, R., and H.C. JOHNSTON. 1977. Growth, reproduction and mortality in Gadoid fish species. Pages 37-62. In J.H. STEELE. editor. Fisheries Mathematics, Academic Press, London, England.

KAPPENMAN, R.F. 1981. A method for growth curve comparisons. Fishery Bulletin 79:95-101.

KIMURA, D.K. 1980. Likelihood methods for the Von Bertalanffy growth curve. Fishery Bulletin 77:765-776.

KINGSLEY, M.C.S. 1980. Comments. Von Bertalanffy Growth Parameters. Transactions of the American Fisheries Society 109:252-253.

KIPLING, G. 1983. Changes in the growth of Pike Esox lucius in Windermere. Journal of Animal Ecology 52:647-657.

KIRKWOOD, G.P. 1983. Estimation of Von Bertalanffy growth curve parameters using both length increment and age length data. Canadian Journal of Fisheries and Aquatic Sciences 40:1045-1051.

KNIGHT, W. 1968. Asymptotic growth. An example of nonsense disguised as mathematics. Journal of the Fisheries Research Board of Canada 25:1303-1307.

LOCKWOOD, S.J. 1974. The use of the Von Bertalanffy equation to describe the seasonal growth of fish. Journal du Conseil International pour l'Exploration de la Mer 35:175-179.

LOPEZ VEIGA, E.C. 1979. Fitting Von Bertalanffy growth curves, a new approach. Investigacion Pesquera 43:179-186.

MERONA, B. de 1983. Modèle d'estimation rapide de la croissance des poissons : Applications aux poissons d'eau douce d'Afrique. Revue d'Hydrobiologie Tropicale 16:103-113.

MERONA, B. de, J. MOREAU, and T. HECHT 1985. La croissance. 32 pages. In LEVEQUE, C.M. BRUTON, and G.W. SSENTONGO. editors. Biology and Ecology of African freshwater fishes, recent trends. ORSTOM, Paris, France, in press.

MISRA, R.K. 1980. Statistical comparisons of several growth curves of the Von Bertalanffy type. Canadian Journal of Fisheries and Aquatic Sciences 37:920-926.

MIURA, T., N. SUZUDI, M. HAGOSHI, and K. YAMANURA. 1976. The rate of production of food consumption of the biwamasu Oncorhynchus rhodurus population in Lake Biwa. Research Population Ecology (Tokyo) 17:135-154.

MOORE, J.W., and I.A. MOORE. 1974. Food and growth of arctic char Salvelinus alpinus in the Cumberland sound area of Baffin Island. Journal of Fish Biology 6:79-92.

MOREAU, J. 1979. Biologie et évolution des peuplements de Cichlidés (Pisces) introduits dans les lacs malgaches d'altitude. Thèse de Doctorat d'Etat. Institut National Polytechnique, Toulouse, France.

MOREAU, J., C. BAYLE, and C. BRIERE. 1984. Emploi du modèle de Beverton et Holt : le cas d'une mortalité naturelle variable et d'une croissance décrite par une équation de Gompertz. Revue d'Hydrobiologie Tropicale 18:161-170.

MOREAU, J., A. BELAUD, F. DAUBA, and A. NELVA. 1985. A method for rapid growth evaluation in fishes:the case of french cyprinid fishes. Hydrobiologia 120:225-227.

MUIR, B.S. 1969. Gill size as a function of fish size. Journal of the Fisheries Research Board of Canada 26:165-170.

MUNRO, J.L. 1982. Estimation of the parameters of the Von Bertalanffy growth equation from recapture data at variable time intervals. Journal du Conseil International pour l'Exploration de la Mer 40:199-200.

MUNRO, J.L., and D. PAULY. 1983. A simple method for comparing the growth of fishes and invertebrates. Fishbyte 1:5-6.

PALOHEIMO, J.E., and L.M. DICKIE. 1965. Food and growth of fishes. I. A growth curve derived from experimental data. Journal of the Fisheries Research of Canada 22:521-542.

PAULY, D. 1979. Gill size and temperature as governing factors in fish growth:a generalization of Von Bertalanffy's Growth Formula. Berichte aus dem Institute für Meereskunde 63, Kiel University, Kiel, Western Germany.

PAULY, D. 1981. The relationships between gill surface area and growth performance in fish:a generalization of Von Bertalanffy's theory of growth. Meeresforschung 28:251-282.

PAULY, D. 1982. Studying species dynamics in tropical multispe-

cies context. Pages 33-70. In PAULY, D., and G.I. MURPHY
editors. Theory and Management of tropical fisheries. ICLARM
Conference Proceedings 9, Manila, The Philippines.

PAULY, D., and G. GASCHÜTZ. 1979. A simple method for fitting
oscillating length growth data with a programm for pocket
calculators. International Council for the exploration of the
Sea. Demersal Fish Comittee. CM 1979/G/24.

PHILIPPART, J.C. 1977. Ecologie, dynamique et production des
populations de poissons dans la zone à Barbeau supérieure de
l'Ourthe. Etude approfondie du Barbeau, Barbus barbus, du Che-
vaisne, Leuciscus leuciscus, du Hotu, Chondrostoma nasus et de
l'Ombre, Thymallus thymallus. Thèse de Doctorat ès Sciences
Zoologie, Faculté des Sciences, Liège.

PITCHER, T.J., and P.D.M. MACDONALD. 1973. Two models for
seasonal growth in fishes. Journal of Applied Ecology
10:599-606.

REGNER, S. 1980. On semigraphic estimation of parameters of
Gompertz function and its application on fish growth. Acta
Adriatica 21:227-236.

RICKER, W.E. 1975. Computation and interpretation of biological
statistics of fish populations. Bulletin of the Fisheries
Research Board of Canada 191.

RICKER, W.E. 1979. Growth rates and models. Pages 677-743. In
HOAR W.S., D.J. RANDALL, and J.R. BRETT. editors. Fish
Physiology, volume VIII. Bioenergetics and growth. Academic
Press, New-York, U.S.A.

RICKLEFS, R.E. 1967. A graphical method of fitting equations to
growth curves. Ecology 48:978-983.

RIFFENBURG, R. 1960. A new method of estimating parameters for
the Gompertz growth curve. Journal du Conseil International
pour l'Exploration de la Mer 23:285-293.

ROFF, D.A. 1980. A motion for the retirement of the Von Bertalanffy Function. Canadian Journal of Fisheries and Aquatic
Sciences 37:127-129.

RUMOHR, H. 1975. Der Einfluss von Temperatur und Salinitat auf
das Wachstum und die Geschlechtreife von nutzbaren Knochen-
fischen. Berichte aus dem Institute für Meereskunde, Kiel University, Kiel Western Germany.

SAINSBURY, K.J. 1980. Effect of individual variability on the
Von Bertalanffy growth Equation. Canadian Journal of Fisheries
and Aquatic Sciences 37:241-247.

SCHNUTE, J. 1981. A versatile growth model with statistically
stable parameters. Canadian Journal of Fisheries and Aquatic
Sciences 38:1128-1140.

SCHNUTE, J., and D. FOURNIER. 1980. A new approach to length-
frequency analysis:Growth structure. Canadian Journal of
Fisheries and Aquatic Sciences 37:1337-1351.

SCHUL'MAN, G.E. 1974. Life cycles of fishes. Physiology and
biochemistry Ist. Prog. Scientific Translations. Wiley &
Sons, New-York, U. S. A.

SILLIMAN, R.P. 1967. Analog computer models of fish populations. Fishery Bulletin 66:31-46.

SZCZWERBOWSKI, J.A. 1976. An attempt to establish the criteria of the assessment of fish growth. Revue des Travaux de l'Institut des Pêches Maritimes 40:750-751.

TAYLOR, C.C. 1962. Growth equation with metabolic parameters. Journal du Conseil International pour l'Exploration de la Mer 27:270-286.

URSIN, E. 1963. On the seasonal variation of growth rate and growth parameters in Norway Pout Gadus esmarkii in the Stagerrak. Meddelelser fra Danmarks Fiskeri-og Havundersogelser N. S. 4:17-29.

URSIN, E. 1979. Principles of growth in fishes. Symposium of the Zoological Society of London 44:63-87.

VAUGHAN, D.S., and P. KANCIRUK. 1982. An empirical comparison of estimation procedures for the Von Bertalanffy growth equation. Journal du Conseil International pour l'Exploration de la Mer 40:211-219.

YAMAGUCHI, M. 1975. Estimating growth parameters from growth rate data. Oecologia (Berlin) 20:321-332.

ZWEIFEL, J.R., and L. LASKER. 1976. Prehatch and posthatch growth of fishes. A general model. Fishery Bulletin 74:609-621.

Considerations for estimation and interpretation of annual growth rates

STEVE GUTREUTER
TEXAS PARKS AND WILDLIFE DEPARTMENT
4200 SMITH SCHOOL ROAD
AUSTIN, TX 78744

ABSTRACT
     The logic and derivation of growth estimators is treated in a unified manner. Various estimators of length-at-age based on different body-scale relations are shown to be members of a general class differing in detail, but not in principle. Growth estimators for linear body-scale relations that incorporate information on individual scale radii and lengths-at-capture seem to be more efficient than simple regression-based estimators. Restriction of estimation of lengths-at-age to the most recent annuli minimizes uncertainty introduced by Lee's phenomenon and size-biased sampling, and yields growth estimates that are more comparable among successive years than those incorporating information from all annuli. Body size explained approximately eight percent more of the variation in annual length increments than did age for largemouth bass <u>Micropterus</u> <u>salmoides</u> from Texas Reservoirs.

     Age and growth studies occupy a prominent position in the fishery literature. Often these studies are descriptive in character, and at least some have been conducted simply to characterize fish populations. At other times, reliable estimates of growth are required for stock assessments. Growth in length is often estimated as the difference between back-calculated lengths at successive ages. Two general approaches to back calculation are in common use, the regression method and the proportional method. Both methods begin with estimation of parameters of an equation that expresses length-at-capture as some function of the radius of a structure that shows periodic marks such as an otolith, scale, vertebra, fin ray, or opercular bone; such structures and this equation will be referred to as scales and the

The Age and Growth of Fish, edited by Robert C. Summerfelt and Gordon E. Hall © 1987 The Iowa State University Press, Ames, Iowa 50010.

body-scale equation, respectively. The regression method uses the body-scale equation to estimate lengths-at-age directly from periodic marks on scales without concern for variation in scale size among individual fish. The proportional method incorporates information on individual lengths and scale radii and therefore accounts for some variation among individuals.

The regression method is suspect for two reasons. First, as Carlander (1981) noted, some regression-based estimates of back-calculated lengths will inevitably be longer than actual lengths-at-capture. In contrast, the proportional method assures that all estimates of prior lengths will be less than length-at-capture. Second, data from a single fish suggests that back-calculated lengths obtained from the regression method may be more variable than those obtained from the proportional method (Whitney and Carlander 1956). Marques et al. (1982) attempted to justify the regression method by claiming that estimates of individual length increments are not important because fishery managers are usually concerned with means.

Often, published presentations of back calculations for various linear and nonlinear body-scale relations list formulas for special cases at the expense of a more general account of underlying principles or explanations of their derivation (Lagler 1956; Jones 1958; Hile 1970; Bagenal and Tesch 1978; Everhart and Youngs 1981). Except for Jones (1958), those authors that mention nonlinear body-scale relations imply use of the regression method as if proportional counterparts do not exist. Further, accounts that do attempt to explain the rationale of proportional back calculation for linear body scale relations are largely verbal (Whitney and Carlander 1956; Carlander 1981), and, if translated exactly to mathematical terms, will yield the desired equation only if the body-scale relation has an intercept of zero. Apparently we are left without a general derivation of proportional methods. Admittedly, this is more of a pedigogical problem than a practical one; back-calculation formulas can be used without understanding underlying principles. However, some comprehension of principles is necessary to assure effective and efficient application of methods.

This paper addresses four objectives. First, the derivation of proportional back-calculation equations will be treated in a general, unified manner in order to demonstrate the rationale behind the approach. The second objective is to assess relative efficiencies of estimates of annual length increments resulting from the proportional and regression approaches to back calculation using a large data set. Third, I propose an alternative to the traditional practice of back-calculating lengths at all annuli; this procedure minimizes uncertainty resulting from Lee's phenomenon. Last, I compare size and age-related expressions of growth.

DERIVATION OF PROPORTIONAL BACK-CALCULATION EQUATIONS

The logic underlying proportional back-calculation equations can be explained using simple algebraic and geometric arguments. Consider an individual fish from a random sample for which estimated length-at-capture, $\hat{1}_c$, and scale radius, s_c, satisfy the polynomial body-scale equation

$$\hat{1}_c = b_0 + b_1 s_c + \ldots + b_k s_c^{\ k} \tag{1}$$

where b_0, ..., b_k are parameter estimates obtained from some rule such as least-squares, geometric mean regression (Ricker 1973), or mixture methods (Schnute 1984). A polynomial body-scale equation is used for generality. Note that 1_c and s_c could also be redefined as any monotonic transformations of lengths-at-capture and scale radii such as log 1 and log s; therefore equation (1) would satisfy any body-scale equation that is intrinsically linear in the parameters, i.e., the power function or "Monastyrsky" method (Bagenal and Tesch 1978). All back-calculation methods assume that b_0, ..., b_k hold for all prior lengths and scale radii so that $\hat{1}_t$ and s_t, the regression estimate of length at age t and the radius of the t-th age mark, respectively, can be substituted into equation (1).

Proportional methods incorporate functions of lengths based on equation (1) that are linear with respect to radial measurements on individual scales. Define L_c as $1_c - b_2 s_c^2 - \ldots - b_k s_c^{\ k}$ and define $\hat{L}'_t$ as $\hat{1}'_t - b_2 s_t^2 - \ldots - b_k s_t^{\ k}$, where $\hat{1}'_t$ is the proportional estimate of length at age $\underline{t}$. Note that $L_c - b_0$ and $\hat{L}'_t - b_0$ are directly proportional to s_c and s_t, respectively, with constant of proportionality equal to the slope of the line segment containing points $\{s,L\}$ equal to $\{0,b_0\}$ and $\{s_c,L_c\}$ (Fig. 1). That is, triangles ABC and AB'C' are similar, or in algebraic terms $(\hat{L}'_t - b_0)/(L_c - b_0) = s_t/s_c$. Solving for $\hat{1}'_t$ yields

$$\hat{1}'_t = 1_c(s_t/s_c) + \sum_{j=0}^{k} b_j s_t (s_t^{\ j-1} - s_c^{\ j-1}) \tag{2}$$

the general proportional formula for the polynomial series. This formula can also be derived by noting that line segment AC, unique for each fish, has intercept b_0 and slope $(L_c - b_0)/s_c$, then solving the equation of this line for $\hat{1}'_t$.

Proportional back-calculation formulas for specific cases in the polynomial series of body-scale relations are obtained by

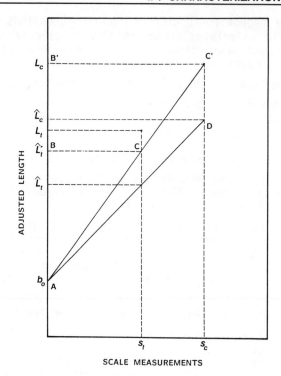

Figure 1. Geometric interpretation of the origin of proportional back calculations for polynomial body-scale
relations. Line segment AD represents the regression of
lengths-at-capture l_c, on a polynomial of scale radii s_c,
for a sample of fish. Information from only one fish is
displayed for clarity. Line segment AC' contains the
points $(0,b_0)$, $(s_t,\hat{L}'_t)$ and (s_c,L_c) where L'$_t$ contains the
proportional estimate of length-at-capture, $\hat{l}'_t$.

setting appropriate values of b_j from equation (2) to zero and
estimating the remainder. For example, for a linear body-scale
relationship $b_2 = \ldots = b_k = 0$, so that equation (2) becomes

$$\hat{l}_t = l_c(s_t/s_c) + b_0 (1 - s_t/s_c)$$

an algebraic variant of the well-known "Lee formula" (the
intercept of the body-scale equation is traditionally denoted by $\underline{a}$
for this linear case; for consistency in equation (2) I use b_0).

Proportional estimators for quadratic and cubic body-scale
relations are obtained by setting parameters higher than b_2 and b_3

respectively, equal to zero. Although final forms of various proportional back-calculation equations for linear and nonlinear body-scale equations may appear distinct, they are all simply special cases of a single, more general approach.

Jones (1958) noted that differences in back-calculated lengths that would be obtained assuming various body scale equations can be obtained from differences between proportional formulas. For example, the difference between back-calculated lengths from linear body-scale relations that have zero and positive intercepts are given by $b_0(1 - s_t/s_c)$. For values of s_t near s_c, i.e., the most recent annuli, this difference is small, but approaches b_0 as back calculations are taken to the first annulus.

RELATIVE EFFICIENCY OF GROWTH ESTIMATORS

One of the primary uses of back-calculated lengths is the estimation of growth rates. Annual length increments for individual fish are the differences between successive lengths-at-age and can be estimated using either the proportional or regression approach. For present purposes, most-recent increments of the form $\Delta \hat{l}_{t-1} = \hat{l}_t - \hat{l}_{t-1}$ will be considered, where t is age-at-capture for an individual fish. If the linear proportional estimator of back-calculated lengths is used, annual increments at age $t - 1$ are given by

$$\Delta \hat{l}'_{t-1} = (s_t - s_{t-1})(l_c - b_0)/s_c$$

and if the linear regression approach is used they are

$$\Delta \hat{l}_{t-1} = b_1(s_t - s_{t-1})$$

Variances of these length-increment estimators provide a basis for comparison of regression and proportional methods of back calculation. Because proportional estimators are constrained by l_c and s_c, they should account for some variation among individual fish, and therefore might generally have lower variances than regression estimators. Proofs of conditions under which $Var(\hat{l}'_t) < Var(\hat{l}_t)$ or $Var(\Delta \hat{l}'_{t-1}) < Var(\Delta \hat{l}_{t-1})$ are not apparent, nor are exact expressions or asymptotic approximations of variances of the proportional estimators. Instead, sample mean-squared errors $(\widehat{MSE} = \sum \{x - \bar{x}\}^2/n)$ of $\Delta \hat{l}_{t-1}$ and $\Delta \hat{l}'_{t-1}$, based on empirical distributions generated by samples of fish, are compared. Sample MSE is always a viable measure of the spread of an empirical distribution, but estimates variance of the true sampling distribution of a random variable only in well-behaved cases.

The data used to evaluate relative variability of proportional and regression estimators consist of measurements from the Texas Parks and Wildlife Department (TPWD) age and growth data base representing 2,196 largemouth bass <u>Micropterus</u> <u>salmoides</u> for which observations from at least five fish appeared in any unique combination of reservoir, year, and age. These data provided 172 samples from which $\widehat{MSE}$s of proportional and regression estimators of length increments were computed. Ad hoc relative efficiency of proportional length-increment estimates relative to the regression estimates is defined as the ratio of the $\widehat{MSE}$ of the latter to the $\widehat{MSE}$ of the former. Efficiencies greater than 1.0 indicate that proportional estimates are less variable within a sample than those from the regression approach. More efficient estimators require fewer data to achieve a given level of precision and can result in reduced sampling costs. Efficiencies ranged between 0.14 and 7.84 (Fig. 2) with a mean of 1.20 and a median of 1.10. Efficiency was at least 1.0 in 66% of the cases. A t-test on log transformed efficiencies indicated that median efficiency (the retransformed mean of a symmetric transformed variable is an unbiased estimator of the median of the original distribution) was significantly greater than one (P < 0.005). Although the differences in variability

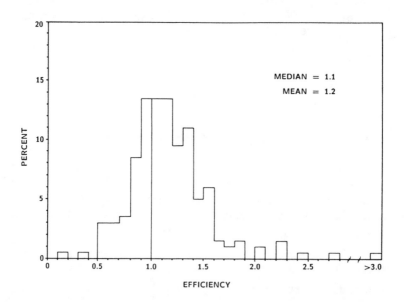

Figure 2. Distribution of 172 estimates of ad hoc relative efficiency of annual length increments based on linear proportional estimators of length-at annuli compared to regression-based estimators for largemouth bass from Texas reservoirs.

between these proportional and regression estimates are not large, they do indicate some advantage for the proportional method, and certainly do not support general use of the regression method.

WHAT DATA SHOULD BE USED TO ESTIMATE GROWTH?

It is valid and useful to ask what data should be used to estimate growth rates. The traditional approach to the estimation of annual growth incorporates measurements from all annuli and requires that scale readers identify and measure each annulus on each scale. If scales have been collected at, or after, the end of a growing season, measurements of scale radii and most-recent annuli are the only scale measurements needed for estimation of annual growth rates. If scales are collected during the growth season, radii of penultimate annuli are also required. If estimation of annual growth is the sole objective, all other scale measurements are unnecessary.

There are logical reasons for restricting measurements to the most-recent growing season. Biologists sometimes wish to evaluate changes in growth rates over several years or may want to evaluate effects on growth of a particular management policy or ecological change. If growth estimates are based on differences between back-calculated lengths for every pair of annual marks, comparisons among years may not be valid because, inevitably, growth information from prior years will be incorporated into estimates for latter years, if the biologist fails to refer growth rates to specific calendar years as well as ages. For example, a biologist may wish to evaluate potential changes in growth associated with a change in harvest regulations implemented three years prior. The biologist has collected samples during each of three years before and after the change. Age-six fish collected during the third year after the change contain growth information from three years before and three years after the change. By back calculating to all annuli, then averaging associated age-specific increments over all years, information from three years prior to the change will be incorporated into estimates of mean growth increments for ages 1+, 2+, and 3+ for the third year following the change. As a result, growth estimates from years following the regulation change will be contaminated with information from years before. Clearly, this approach is not rational. Instead, the biologist must be careful to assign historic growth rates to corresponding calendar years. The most conservative policy is to estimate only the most-recent annual growth increment for each fish.

A second reason for restriction of growth estimation to most-recent increments is uncertainty resulting from size-selective mortality, or more generally, Lee's phenomenon. Jones (1958) explained Lee's phenomenon by showing a result that can be generalized as follows; under reasonable assumptions, mean lengths of a cohort change by an amount equal to $b\sigma^2\Delta t$, where b is a measure of size-selective mortality operating over the time

interval Δt, and σ^2 is the variance of length. An important rami-
fication for back-calculating lengths, or estimating prior length
increments, is obtained from a posterior view of this result; bias
of back-calculated lengths associated with Lee's phenomenon is
proportional to Δt. That is, bias increases as back calculations
proceed farther back in time. Ricker (1969) also considered Lee's
phenomenon and concluded that if growth rates are estimated using
scales, then the two most-recent annuli provide the best
estimates. Because intensities of size-selective sampling bias
and mortality are usually unknown and the direction of the latter
may change over length, we should generally expect back-calculated
lengths and their associated increments to become more uncertain
as back calculations are carried back in time.

SIZE-RELATED EXPRESSIONS OF GROWTH

Because growth estimation usually depends on age data to
measure time, age has become an intuitive basis for expression of
these rates. Although expressions of mean lengths-at-age or age-
specific growth rates are required for some tasks, there is no
reason to preclude other options for more general purposes.
Larkin et al. (1956) showed that body size can provide a more
interpretable basis than age for expression of growth rates.
Further, metabolic requirements of fish of similar lengths but
different ages are more likely to be similar than those of fish of
similar ages but different lengths. Size-related expressions of
growth are ecologically interpretable and have application in
fishery management and population analysis (Parker and Larkin
1959; Reynolds and Babb 1978; Gutreuter and Anderson 1985).

Because growth is correlated with both age and body size, the
magnitudes of these correlations may reveal whether age or size is
a better covariate for general application. To compare the amount
of variation in growth explained by age and size, all samples for
which there were at least ten observations in each unique
combination of reservoir and year were selected from the TPWD
largemouth bass data base. This yielded 80 samples from which
most-recent annual length increments and lengths at the start of
the growth year were calculated from each fish. Length increments
were log transformed to linearize their relations with ages and
initial lengths. Within each sample, correlation coefficients
between log length increments and ages, and log length increments
and initial lengths were calculated. These 80 pairs of
correlation coefficients were $\tanh^{-1}$ transformed to achieve
approximate normality (Sokol and Rohlf 1981). A one-tailed paired
t-test indicated that the length correlations were significantly
greater than the age correlations ($P < 0.001$). Initial length
explained an average of 40% of the variation in log length
increments while age explained an average of only 32%. Although
the difference in variance explained by length and age is not
large, it does indicate that if just one of these covariates is to
be used, initial length should perform better.

Growth increments expressed as functions of length exhibit

distinctive patterns. These patterns are demonstrated using the TPWD largemouth bass growth data, from which length increments from the year preceding capture and lengths at the start of that year were calculated for each fish. The median, interquartile bounds, and extremes of annual length increments from successive 2.5-cm groups of initial lengths decreased with increasing median initial length (Fig. 3A). Smaller fish exhibited a broader scope for growth in length than larger fish. Similar statistics for weight increments, computed using the length-weight relation from each sample, increased with increasing median initial length (Fig. 3B). Because anglers value body weight as a measure of fish quality, as indicated by reporting procedures for record fish, size-related patterns in weight increments indicate that large fish accrue large increments of quality.

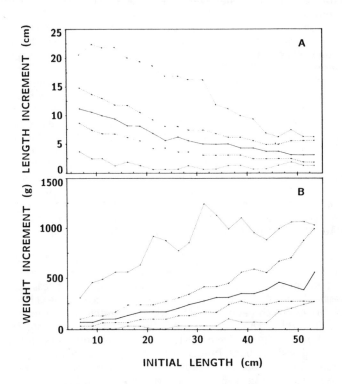

Figure 3. Length-specific expressions of growth; A) annual length increments (cm), B) annual weight increments (g). Dotted lines indicate maxima and minima, broken lines indicate 25th and 75th percentiles, and solid lines indicate medians -- see text for interpretation.

CONCLUSIONS

Arguments presented by Carlander (1981) and the distribution of efficiencies from this study indicate that the regression estimator of back-calculated lengths is generally inferior to the proportional estimator for linear body-scale relations. Proportional estimators use more information contained in samples and are easily derived for novel body-scale relations, although the linear case should suffice in most instances. Because computers are commonly used, computational convenience should never justify sole use of regression estimators. Proportional estimators of annual length increments should be used whenever they are more efficient than regression-based estimators. Because proportional estimates of back-calculated lengths are always less than lengths-at-capture for individual fish, they should be preferred over regression estimates whenever individual length increments from the current growing season are desired.

The logic of growth estimation can be improved by careful selection of annuli. For routine analyses of growth associated with annual stock assessments, the most-recent annuli provide the most generally useful information. Back calculation to every annulus risks unnecessary confusion of Lee's phenomenon with growth, and risks temporal contamination of data when comparisons are to be made over time. This does not imply that there is no reason to back calculate lengths to all annuli. Back calculation to all annuli provides a valuable means to assess the strength of size-selective mortality (Ricker 1969). It could also be argued that it is necessary to back calculate lengths to every annulus if one wishes to estimate historic growth during years a population was not sampled. Although useful, that practice yields unnecessarily biased measures of growth when mortality or sampling are size-selective. A safer, though more costly, alternative is to sample populations for each year that growth estimates are required.

Finally, although age-related expressions are traditional, size-related expressions merit reconsideration. Initial length explained more variation in growth than did age from many samples of largemouth bass. A partial explanation might be that body size is associated with ecological and bioenergetic mechanisms that influence growth, and age is merely a non-decreasing function of length. More likely than this, age also entails physiological effects on growth, but length effects may dominate. Mechanisms for size-related modification of growth have been proposed (Paloheimo and Dickie 1965, 1966; McComish 1971).

Perhaps it is time to question our age-specific tradition in fishery analysis. Size-specific expressions may have greater interpretability than their age-specific counterparts. The challenge remains to develop and refine size-related models and stock assessment procedures. For example, comparison of size-specific length increments with standard or desirable increments for fish of the same length is a viable qualitative and quantitative component of a stock assessment. This approach has been developed and used to advantage in Missouri (D. Neuswanger and G. Novinger, Missouri Department of Conservation, personal communication). Standard length-specific increments provide a common basis of comparison regardless of age because fish of similar initial lengths but of different ages are more likely to be ecological equivalents than fish of a given age.

ACKNOWLEDGMENTS
   I thank R. O. Anderson, C. W. Caillouet, Jr., K. D. Carlander, P. P. Durocher, G. P. Garrett, W. C. Guest, R. W. Luebke, D. J. Neuswanger, K. E. Palachek, J. A. Prentice, J. B. Reynolds and C. L. Smith for discussion, suggestions and/or contributions.

REFERENCES
BAGENAL, T. B., and F. W. TESCH. 1978. Age and growth. Pages 101-136 in T. B. Bagenal, editor. Methods for assessment of fish production in freshwater, 3rd edition. Blackwell Scientific Publications, Oxford, UK.
CARLANDER, K. D. 1981. Caution on the use of the regression method of back-calculating lengths from scale measurements. Fisheries 6(1):2-4.
EVERHART, W. H., and W. D. Youngs. 1981. Principles of fishery science, 2nd edition. Cornell University Press, Ithaca, New York, USA.
GUTREUTER, S. J., and R. O. ANDERSON. 1985. Importance of body size to the recruitment process in largemouth bass populations. Transactions of the American Fisheries Society 114:317-327.
HILE, R. 1970. Body-scale relation and calculation of growth in fishes. Transactions of the American Fisheries Society 86:468-474.
JONES, R. 1958. Lee's phenomenon of "apparent change in growth-rate" with particular reference to haddock and plaice. Pages 229-242 in International Commission for the Northwest Atlantic Fisheries, Special Publication 1. Halifax, Canada.
LAGLER, K. F. 1956. Freshwater fishery biology, 2nd edition. W. C. Brown, Dubuque, Iowa, USA.
LARKIN, P. A., J. G. TERPENNING, and R. R. PARKER. 1956. Size as a determinant of growth rate in rainbow trout Salmo gairdneri. Transactions of the American Fisheries Society 86:84-96.
MARQUES, K., G. PARDUE, R. NEVES, L. FORTUNATO, and A. TIPTON. 1982. Computer program for the computation of age and growth statistics of fish populations. Management Series Number 1, Department of Fisheries and Wildlife Sciences, Virginia Polytechnic Institute and State University, Blacksburg, Virginia, USA.
MCCOMISH, T. S. 1971. Laboratory experiments on growth and food

conversion by the bluegill. Ph.D. Dissertation, University
of Missouri, Columbia, Missouri, USA.
PALOHEIMO, J. E., and L. M. DICKIE. 1965. Food and growth of
fishes. I. A growth curve derived from experimental data.
Journal of the Fisheries Research Board of Canada 22:521-
542.
PALOHEIMO, J. E., and L. M. DICKIE. 1966. Food and growth of
fishes. II. Effects of food and temperature on the relation
between metabolism and body weight. Journal of the Fisheries
Research Board of Canada 23:869-908.
PARKER, R. R., and P. A. LARKIN. 1959. A concept of growth in
fishes. Journal of the Fisheries Research Board of Canada
16:721-745.
REYNOLDS, J. B., and L. R. BABB. 1978. Structure and dynamics of
largemouth bass populations. Pages 50-61 in G. Novinger and
J. Dillard, editors. New approaches to the management of
small impoundments. Special Publication 5, North Central
Division American Fisheries Society, Washington, DC, USA.
RICKER, W. E. 1969. Effects of size-selective mortality and
sampling bias on estimates of growth, mortality, production,
and yield. Journal of the Fisheries Research Board of Canada
26:479-541.
RICKER, W. E. 1973. Linear regressions in fishery research.
Journal of the Fisheries Research Board of Canada 30:409-434.
SCHNUTE, J. 1984. Linear mixtures: a new approach to bivariate
trend lines. Journal of the American Statistical Association
79:1-8.
SOKOL, R. R., and F. J. ROHLF. 1981. Biometry, 2nd edition.
W. H. Freeman, San Francisco, USA.
WHITNEY, R. R., and K. D. CARLANDER. 1956. Interpretation of
body-scale regression for computing body length of fish.
Journal of Wildlife Management 20:21-27.

Linear models for the growth of fish

SANFORD WEISBERG
DEPARTMENT OF APPLIED STATISTICS
RICHARD V. FRIE[1]
DEPARTMENT OF FISHERIES AND WILDLIFE
UNIVERSITY OF MINNESOTA
ST. PAUL, MN 55108

ABSTRACT
 A class of flexible, and extendable, linear models are proposed for modeling the growth of fish using data on scale increments and lengths. The analysis partitions into parts due to the age of a fish in a given year and to variation in the environment. These can in turn be used to understand the effects of management procedures, changes in the environment, or to describe the growth that would take place if the environment could be held constant. The method is compared to the more usual Fraser-Lee method of back-calculation.

A record of the growth history of an individual fish is contained in the increments between annuli on a scale. The purpose of this paper is to present new methodology for extracting the information in scale measurements concerning growth. The usual method (Hile 1970; Begenal and Tesch 1978) is to obtain a body-scale relationship between the length at capture, which we call L_c for an age c fish, and the distance S_c from focus to scale edge. This relationship is generally assumed to be linear. The length L_i of a fish at age i, $i \leq c$, is then estimated from the equation

$$\hat{L}_i = a + b \times S_i$$

[1] Present address: Missouri Department of Conservation, Fish and Wildlife Research Center, 1110 College Avenue, Columbia MO 65201.

where S_i is the lateral distance from focus to annulus i and a and b are constants. In the Fraser-Lee method, a separate value of b is computed for each fish from the formula $b = (L_c - a)/S_c$, which is the slope of the line joining the two points $(0, a)$ and (S_c, L_c). The intercept a can be computed in any of several ways; Carlander (1982, 1985) recommends using tabled values of a to facilitate comparison of studies. Clearly, the choice of a is very influential in determining the $\hat{L}_i$, especially for slower growing stocks. In regression methods, both a and b are estimated from the linear regression of the L_c on the S_c.

Both the regression and Fraser-Lee methods suffer from several fundamental difficulties. First, both assume that models fit to total lengths and scales are appropriate for earlier lengths at age. In particular, none of the information in the observed S_i is used in estimating a and b. Second, this method does not permit modeling, in the usual statistical sense, in which growth is partitioned into portions due to environmental changes, age-group effects, year-class effects, and so on. Also, reliable measures of precision of the back-calculated lengths are not easily obtained in either the Fraser-Lee or the regression methods.

As an alternative to this approach we propose reversing the order of the analysis by first fitting models to the scale increments and then translating the results to statements concerning length when necessary. This reversal of the order has several desirable properties. First, scale measurements, not lengths, are the observed data. Translating to lengths must introduce additional variation in the fitted models. Second, we will be able to fit useful models that will describe growth as a function of the age of a fish and of yearly variation in the environment. The fitted coefficients will provide a basis for comparing populations, environments and management policies. Third, usual statistical theory can be used to estimate standard errors and obtain tests. Fourth, the focus of the analysis can shift from the individual fish that were actually included in the sample to the whole age-group, which is a more natural unit for summarization of results. Finally, all underlying assumptions can be made explicit, so the user can decide if assumptions make sense for any particular population, and in general diagnostic checks for their validity can be performed.

MODELS FOR GROWTH FROM SCALE MEASUREMENTS

A sample of n fish from several age-groups is taken at random from a population. On each fish of age c, we measure total length

L_c, and from a key scale we measure the radii S_i to each annulus. In this section, we study a growth model that uses only the observed S_i and ages. The information from lengths is incorporated later in this paper.

The unit of analysis will be <u>yearly growth increments</u>, which we will call A_i (the A should have several more subscripts to indicate age-group, year of growth, and perhaps a specific individual in that age-group; however, for readability we state only the subscript for increment number). Thus, A_i is the increment in scale, in mm, for the ith year of life, and is the only observed indicator of the growth that took place in that year. Study of growth increments allows us to concentrate on modeling growth in each year for each age-group. The models we propose will result in estimated age and year or environment effects. A_1 is the radius to the first annulus (and is therefore equal to S_1), A_2 is the increment between the first and second annulus (= $S_2 - S_1$); A_3 = $S_3 - S_2$, and so on. As an example, Table 1 gives the length at capture L_c, and the S_i and A_i for fish of ages c = 1, 3, and 5. On a five year old fish we will observe five growth increments, while on a three year old fish we observe three increments.

Table 1. Scale radii and growth increments for sample fish of ages c = 1, 3, 5

c = 1				
S_1 = 1.9 mm				
A_1 = 1.9 mm				
c = 3				
S_1 = 1.6 mm	S_2 = 2.9 mm	S_3 = 4.5 mm		
A_1 = 1.6 mm	A_2 = 1.3 mm	A_3 = 1.6 mm		
c = 5				
S_1 = .9 mm	S_2 = 2.1 mm	S_3 = 3.5 mm	S_4 = 4.5 mm	S_5 = 5.3 mm
A_1 = .9 mm	A_2 = 1.2 mm	A_3 = 1.4 mm	A_4 = 1.0 mm	A_5 = .8 mm

To illustrate the methodology, we use data from several pooled samples of walleye <u>Stizostedion vitreum vitreum</u> collected in various parts of Lake of the Woods, Minnesota, with gill nets in July 1984. For each fish, length, age and the S_i (hence the A_i) were determined, using the methodology given in Frie(1982).

Table 2. Sample size, increment averages and increment
          standard deviations for Lake of the Woods, MN,
          walleye data.

2a. Number of fish in each age-group

c	1	2	3	4	5	6	7	8	9
n_c	97	53	197	42	96	98	52	66	14

2b. $\bar{A}_i$

Age				Year	of	growth			
i	1975	1976	1977	1978	1979	1980	1981	1982	1983
1	1.098	1.364	1.283	1.230	1.178	1.395	1.481	1.075	1.742
2		1.196	1.107	1.216	1.205	1.161	1.299	.965	1.697
3			1.141	.996	1.039	.986	1.237	.835	1.289
4				.931	.810	.817	1.036	.722	.963
5					.664	.657	.723	.582	.817
6						.561	.647	.530	.656
7							.508	.450	.521
8								.435	.438
9									.386

2c. SD_i

Age				Year	of	growth			
i	1975	1976	1977	1978	1979	1980	1981	1982	1983
1	.280	.278	.259	.200	.182	.244	.259	.241	.327
2		.248	.216	.213	.217	.227	.285	.240	.302
3			.305	.188	.202	.229	.250	.218	.264
4				.134	.154	.172	.174	.156	.259
5					.116	.157	.164	.123	.194
6						.151	.176	.146	.155
7							.157	.126	.149
8								.109	.130
9									.104

Reported in Table 2 is a summary of the A_i. Table 2a gives the
number of fish n_c in each age-group. We assume that analysis is
to be done using a single sample of fish, so distinction between
age-groups and year classes is unnecessary. While extension of
the method to several years of data is possible, the notation
required is more cumbersome, and is not used here. Table 2b gives
the average $\bar{A}_i$ of the A_i. In this table, the rows correspond to
age of the fish, indexed by the subscript $\underline{i}$. The first row gives
the $\bar{A}_1$. The value of $\bar{A}_1$ for the 14 fish from the 9 year old age-
group was 1.098, for the 66 fish from the 8 year old age-group was
1.364, and so on. The columns in the table correspond to year in
which growth took place. Thus, in 1975, only the 14 fish of age
nine at capture were observed when these fish were in their first

year of growth. For the year 1980, scale increment observations are available for all age-groups then alive. The $\bar{A}_i$ for a particular year-class are given along diagonals in the table. For example, the $\bar{A}_i$ for the 1979 year-class:

$$\bar{A}_1 = 1.178 \quad \bar{A}_2 = 1.161 \quad \bar{A}_3 = 1.237 \quad \bar{A}_4 = .722 \quad \bar{A}_5 = .817$$

Table 2c has the same layout as Table 2b, except it gives the standard deviation SD_i of the A_i rather than their average. These tables are produced by the most recent version of DISBCAL (Frie 1982).

All the averages in a row in Table 2b would be the same, except for sampling error, if growth at age $\underline{i}$ depended only on age of the fish in that year but not on prior growth or on changes in the environment. If the averages increased or decreased, we could suspect either a systematic sampling bias or some other time trend that could be attributed to an environmental or management change. If sampling were selective for smaller fish, then the averages in each row should tend to increase because fish captured from older age-groups would tend to be smaller representatives of their group, and therefore the A_i would be smaller for early years than for later years. If there were a steady degradation in the environment, then the averages may show a steady decrease in each row. Similarly, increasing exploitation may cause an opposite trend in each row. If the averages differed, but in a haphazard manner, as seems to be the case in Table 2b, we would attribute the differences to otherwise unquantified year-to-year variation in environment, such as changes in water temperature or available food supply. Finally, management changes made during the period covered by the data might result in the averages from the "before" period differing from the averages in the "after" period.

The models we propose estimate year or environment effects by essentially averaging down columns to get typical values for each year, and then the comparisons described above are made on these average values. Because the table is not rectangular, this averaging is not a straightforward adding up and dividing by sample size, but usual linear model calculations will produce the correct values.

The values within a column in Table 2b give the amount of growth for fish of all ages in a given year. These values would be the same only if growth did not depend on age or size of the fish, but only on environmental variation. This is generally an unreasonable assumption. The patterns in each column would be the same if growth were similar in each year. If the patterns are different, we would have an age × environment interaction in which

environmental influences effect age-groups unequally. Our models
produce typical values by averaging in each column in Table 2b to
get overall environmental effects. These average values can then
be used to describe year-to-year variation that cannot be
attributed to the growth of fish in a constant environment.

This discussion suggests fitting a model that expresses
growth, as given by the averages in Table 2b, as a function of age
(or row effects) and year/environment (or column effects) in much
the same way as is done with a two-way analysis of variance. Because the data form a triangular, not square table, and because
the number of fish sampled from each age-group is different, there
are several minor complications in the statistical analysis.
First the simple textbook formulas for fitting a two-way analysis
of variance will not apply, and a more general analysis of variance or linear regression program, such as the GLM procedure in
SAS (SAS 1982) is required. Second, weighted least squares must
be used for the analysis of variance, with the weight for an A_i
from age-group $\underline{c}$ equal to n_c. Third, because of the lack of balance, we must be concerned with the order of fitting of effects;
we choose to fit age effects first, and then fit environmental
effects adjusted for age effects; in the language of SAS-GLM, this
is Type I fitting. Fourth, it is convenient to fit models without
an overall intercept, using the NOINT option in SAS-GLM. This
will allow direct interpretation of the parameters that SAS
estimates. Finally, we can compute a "pure error" estimate of
residual variance $\hat{\sigma}^2_p$ by pooling the standard deviations in Table
2c using the following formula:

$$\hat{\sigma}^2_p = \frac{\sum [\sum (n_c - 1)SD^2_i]}{\sum c(n_c - 1)} \tag{1}$$

where the outer sum in the numerator is over all age-groups, the
inner sum is over all growth increments $\underline{i}$, and the sum in the
denominator is over all age-groups. This is just the usual formula for the pooled within class variance. The pure error estimate
of variance should be used as the denominator in all tests, and to
compute all standard errors. For the example, $\hat{\sigma}^2_p = 0.045$ with
2733 degrees of freedom (the number in the denominator of (1)).

For the walleye data, the fit of the additive model,

growth increment = age effect + environment effect

is summarized in Table 3. Table 3a gives the analysis of
variance, which has been modified from SAS by adding a line for
the pure error computed using (1). The line marked lack-of-fit
would be labelled the residual in SAS-GLM (see Weisberg 1985,

Table 3. Linear model for scale increments. Effects
are in mm of annular scale increments.

3a. Analysis of variance.

Source	df	SS	MS	F
Age	8	241.039	30.130	663.59
Year/envr.	8	46.027	5.753	126.71
Lack of fit	28	20.296	.725	15.96
Pure Error	2733	124.090	.045	
Residual	2761	144.386	.052	
Total	2777	431.452		

3b. Estimated coefficients for scale annular increments.

	Age effects				Year/environment effects		
	Estimate	SE	t		Estimate	SE	t
1	1.5670	.0128	122.53	1975	-.4693	.0584	-8.04
2	1.4436	.0135	107.29	1976	-.2109	.0266	-7.92
3	1.2854	.0129	99.97	1977	-.2957	.0216	-13.71
4	1.0847	.0145	74.62	1978	-.2874	.0175	-16.46
5	.8765	.0140	62.44	1979	-.2903	.0156	-18.63
6	.7394	.0154	47.96	1980	-.2571	.0147	-17.46
7	.6692	.0199	33.58	1981	-.0863	.0135	-6.39
8	.4992	.0239	20.86	1982	-.3526	.0130	-27.03
9	.3864	.0569	6.79	1983	0		0

Section 4.3 for discussion of lack-of-fit tests). Table 3b gives
the estimates of effects as produced by SAS-GLM and their standard
errors; to use the pure error estimate $\hat{\sigma}_p^2$ in these, the standard
errors produced by the program must be multiplied by $(\hat{\sigma}_p$/mean
square for lack-of-fit$)^{1/2}$; this has been done in Table 3b.

The overall F statistics from Table 3a exceed any conventional critical values, providing clear evidence that growth
differs both by age of fish and by year. Because the intercept
was not included in the model, and the age variables were fit
before environmental variables, the parameter estimates for age
effects correspond to estimated yearly scale increments in mm per
year for this population in a constant environment. Typically,
annular increments decrease with age, with a roughly exponential
drop. If of interest, one could use these fitted growth values to
estimate a growth curve such as the (derivative of the) von
Bertalanffy; the result would be quite good.

Interpretation of environmental effects requires a different
approach because of the parameterization. The coefficient for the
most recent year of complete growth, 1983, is automatically set to
zero by the program, and the parameters for all other years are
differences from 1983. As it happens, 1983 was a better year for
growth than was any of the previous years, so all the coefficients
are negative (Table 3b). The t-values for these coefficients
compare each of the other years to 1983. These year or
environmental effects do not seem to follow any pattern, suggesting that variation in the environment does have an important
effect on growth, but without a clear pattern or trend.

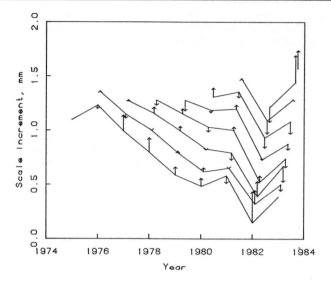

Figure 1.   Fitted scale increments for each age class, joined
   by a solid line.   The arrows point from the fitted length
   increment to the observed $\overline{A}_i$.

These results can be illustrated by any of several graphical
summaries.   In Figure 1, for each age-group we have computed the
fitted growth increments (age-effect + year-effect) and joined
them by a solid line.   The x-axis in the plot is the year of
growth (the arrows in the plot are discussed later).   If these
lines were parallel, growth would be entirely determined by age.
If they were coincident, there would be no age differences, only
environmental differences.   Neither of these is observed here.

Goodness of fit

We can judge the goodness of fit of the model using the
analysis of variance (Table 3a).   The sum of squares for lack-of-
fit or equivalently for an age × environment interaction, will be
large if the additive age plus environment model is not adequate
for the data.   This would indicate that the single description of
age effects applicable to all age-groups would be inadequate.   An
F test can be used to check for adequacy.

In all the F-tests, we use (1) for the denominator of the
test.   While in the example the F-test for interaction indicates
the likely existence of interactions, and hence of age-group
differences, these differences are likely to be comparatively
small because the sum of squares for interaction is small compared
to the main effects of age and year.   Although the additive model
is not entirely adequate for the example, it is probably good
enough for many purposes of the fishery manager.

Figure 1 can be used to explore the interactions further. The vertical arrows connect the observed values of the $\bar{A}_i$ at the arrowhead to the corresponding fitted values at the tail of the arrow. Where the arrow is short, the fit of the additive model is very good. The model seems to fit least well for the 9 year old age-group and for the year 1982. Indeed, in 1982, the additive model predicts too high a value for the three youngest age-groups, and predicts too low a value for the oldest age-groups. Apparently 1982 was a poor year only for the recent age-groups.

Technical Details

The use of standard analysis of variance and linear models requires several fundamental assumptions concerning growth that may have questionable validity in any given population of fish. The primary assumption is that the scale increments A_i are independent, both between fish and within a single fish. The assumption of independence will generally be more plausible for increments than for total attained growth. We also assume that the A_i are normally distributed with mean that depends on age and environment (not, for example on the relative size of a fish in its age-group), and variance that is the same for any age i and all age-groups. From examination of the standard deviations in Table 2c, it is apparent that variability of the A_i decreases with i, but the effects of nonconstant variance are likely to be relatively minor (Box 1953). Similarly, other assumptions can be checked using more or less standard regression diagnostics (surveyed by Cook and Weisberg 1982 and by Weisberg 1985, Chapters 5 and 6), and the modeling can be modified to account for failure of assumptions. For brevity and clarity, however, these diagnostics have been generally omitted.

USING INFORMATION ON LENGTH

Most of the information on growth of a single fish is contained in the scale increments. However, some information is available in the lengths at capture, and it may be desirable to use it. Also, since it is usual to translate fitted growth to lengths, we provide a methodology to do so.

Suppose we let B_i be the length increment of a fish in the ith year of life, so $B_1 = L_1$, $B_2 = L_2 - L_1$, $B_3 = L_3 - L_2$, and so on. Of course, the B_i are unobserved, but we shall state assumptions we need in terms of the joint distribution of the observed scale increment A_i, and the unobserved length increment B_i. This will then allow us to add the A_i to get the scale radius to the edge S_c, and add the B_i to get the length at capture L_c; we will then have a model for the body-scale relationship that is somewhat different from the standard model. For the moment, we assume that the data were collected at the start of the growing season, so neither L_c or S_c contains a part due to current season growth.

The fundamental assumptions are as follows. The growth of a fish in the ith year of its life (A_i, B_i) has a joint normal distribution, with means that depend on age and environment effects. The variance $Var(A_i)$ of the scale increments is constant for all years of growth i and all age-groups. The variance $Var(B_i)$ of length increments is constant for all i and all age-groups. The correlation between A_i and B_i is a constant ρ for all i and all age-groups. Finally, the growth increments are independent both between fish and within fish.

With these assumptions, standard linear model calculations (Weisberg 1986) can be used to add the yearly increments to summarize the relationship between total length L_c and total scale radius S_c. We get a linear regression model given by

$$L_c = \beta_{0c} + \beta_1 S_c + u \tag{2}$$

where $\beta_1 = \rho \times Var(B_i)/Var(A_i)$ is the slope of the body-scale relationship, and β_{0c} is the intercept for the cth age-group, so each age-group will have a different intercept. In addition u is a normally distributed random error with mean 0 and variance given by $c \times Var(B_i) \times (1 - \rho^2)$. This differs from the usual calculations in the Fraser-Lee methodology, or the methodology given in Bartlett et al. (1984), in that (1) each age-group has its own regression intercept, but common slope, and (2) the residual variance increases with the age of the fish, so weighted least squares, with weights given by the inverse of age c at capture, is required.

If the sample of fish is taken in summer or early autumn, current partial-season growth must be included in the model, because length of the fish when the last annulus was formed is not observable. Then the L_c and the S_c can be defined as before by adding the A_i and B_i, and also adding the current season increments. Model (2) remains appropriate, but the residual variance will increase to $(c+f)Var(B_i)(1-\rho^2)$, where f is the ratio of the variance of a scale increment for the partial growing season to $Var(A_i)$; usually $0 < \underline{f} \leq 1$. If SD_0 are the standard deviations of the current season scale increments for each age-group, we estimate $\underline{f}$ by the ratio

$$f = minimum\{1, \frac{[\sum(n_c - 1)SD_0^2]/\sum(n_c - 1)}{\hat{\sigma}_p^2}\}$$

and then proceed as if the estimated value of $\underline{f}$ were the true value. Since $\underline{f}$ will typically be estimated with a large number of observations, this substitution should not effect the analysis appreciably (Carroll 1983). For the example, $\underline{f} = 0.294$.

Goodness of fit

If the constant correlation assumption does not hold, then in (2) β_1 must be replaced by a separate slope for each age-group (variances will also change). A test for equal slopes in the length-scale relationships for each age-group is a check for the constant correlation assumption. Methodology for this test is given by Weisberg (1985, Section 7.4). For the walleye data, the test for a common slope has value $F = 5.154$, which, with 8, 607 df, has a very small corresponding p-value. However, if the one year old age-group is excluded and the test is repeated, we find $F = 1.40$, with 7, 512 df, and a p-value = .20. Thus, the constant correlation assumption seems plausible, except for the one year old age-group, and the body-scale increment model seems different in the first year than in other years. This can be explained by the allometric relationship between body and scale increments in the first year of growth. Table 4 gives the estimated slope $\hat{\beta}_1$ and intercepts $\hat{\beta}_{0c}$, along with their standard errors, as fit by weighted least squares with weights $(c + f)^{-1}$, assuming (2) for all age-groups.

Table 4. Regression of length on total scale radius.

	Estimate	SE
Intercept (1)	112.30	3.56
Intercept (2)	135.65	5.46
Intercept (3)	155.80	6.89
Intercept (4)	174.51	8.40
Intercept (5)	195.67	9.01
Intercept (6)	222.37	9.92
Intercept (7)	224.55	10.86
Intercept (8)	249.97	11.29
Intercept (9)	255.62	13.87
Slope	38.25	1.64
SE (Length)	9.60	

We now turn to the question of back-calculation of length. As with scale increments, it is desirable to model length increments with just a few parameters. We are therefore led to fit the same additive model for length as we did for scale increments, namely,

$$\text{length increment} = (\text{age effect}) + (\text{year effect}) \qquad (3)$$

Unfortunately, this model has too many parameters to estimate from the length measurements alone from a single sample of fish. We propose the following process that gives maximum likelihood estimates: (a) assume additive linear models for both scale and length increments; (b) estimate the environment effects for length and scale to be proportional; that is, set

(environment effect for length) =

$$\hat{\beta}_1 \times (\text{environment effect for scale})$$

(c) estimate age effects for length using the differences in intercepts in the fitted model (2) for lengths, adjusted by age effects for scale. In the simplest case in which all age-groups have at least one fish sampled, the age effects are estimated by the following equations. For $c = 1$,

$$(\text{age effect for length})_1 = \hat{\beta}_{01} + \hat{\beta}_1 (\text{age effect for scale})_1$$

and for $c = 2, 3, \ldots$, maximum age in the sample,

$$(\text{age effect for length})_c =$$

$$(\hat{\beta}_{0c} - \hat{\beta}_{0,c-1}) + \hat{\beta}_1 (\text{age effect for scale})_c$$

Assuming environmental effects are small relative to age effects, we can approximate the variances of these estimates by

Var(environmental effect for length) $\approx$

$$\hat{\beta}_1^2 \text{ Var(environmental effect for scale)}$$

$$+ \text{Var}(\hat{\beta}_1)(\text{estimated environmental effect for scale})^2$$

Var(age effect for length)$_1$ $\approx$

$$\hat{\beta}_1^2 \text{ Var(age effect for scale)}_1$$

$$+ \text{Var}(\hat{\beta}_1)(\text{estimated age effect for scale})_1^2$$

$$+ \text{Var}(\hat{\beta}_{01}) + (\text{estimated age effect for scale})\text{Cov}(\hat{\beta}_{01}, \hat{\beta}_1)$$

$$\text{Var(age effect for length)}_c \approx$$

$$\hat{\beta}_1 \text{Var(age effect for scale)}_c$$

$$+ \text{Var}(\hat{\beta}_1)(\text{estimated age effect for scale})_c^2$$

$$+ \text{Var}(\hat{\beta}_{0c}) + \text{Var}(\hat{\beta}_{0,c-1}) - 2\text{Cov}(\hat{\beta}_{0c}, \hat{\beta}_{0,c-1})$$

With the exception of the covariance term in the above expressions which may be expected to be negligible, all the terms are easily obtained from standard regression output.

Table 5 gives the estimated age and environmental effects using the above method. These coefficients are in the units of mm of length increase per year. As with the scale increments, the age effects generally decrease exponentially with age. These coefficients are very precisely determined, as one might hope from such a large sample. Of course these standard errors are only appropriate if the assumptions made are correct; otherwise they may underestimate the variability in the estimates. The year/environment effects are again comparisons to the last full year of growth in the sample 1983. These are just a multiple of the corresponding values for scale increments, and can be interpreted as for the scale increments. Back-calculated total lengths can be obtained by (1) estimating length increments by adding age and environment effects and (2) adding length increments to get total length. These have been computed, and are graphed in Figure 2.

Table 5. Linear model for length assuming that environmental effects are proportional to those for scale increments.

	Age effects			Year/environment effects			
	Estimate	SE	t		Estimate	SE	t
1	172.2435	1.3936	123.59	1975	-17.9535	2.5124	-7.15
2	78.5722	2.2726	34.57	1976	-8.0689	1.1450	-7.05
3	69.3302	2.5757	26.92	1977	-11.3113	1.0084	-11.22
4	60.2018	3.4796	17.30	1978	-10.9934	.8569	-12.83
5	54.6870	3.7696	14.51	1979	-11.1042	.7968	-13.94
6	54.9927	3.3090	16.62	1980	-9.8346	.7364	-13.35
7	27.7706	4.3494	6.38	1981	-3.3000	.5710	-5.78
8	44.5170	4.9632	8.97	1982	-13.4872	.7882	-17.11
9	20.4406	8.7095	2.35	1983	0	0	

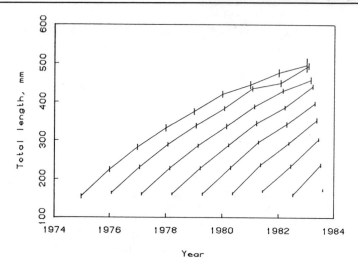

Figure 2. Back—calculated lengths, using the model given in
Table 4. The vertical lines correspond to ± 2 standard
errors.

DISCUSSION

Reliable measurement of scales is central to this method.
Either "key" scales must be measured or else the growth of various
scales on the same fish must be more similar than is the growth of
scales on different fish. Newman and Weisberg (1986) report that
the standard deviation between increments in different scales on a
single brown trout Salmo trutta is small enough for early increments (i = 1, 2, or possibly 3) for measurement from one scale per
fish to be adequate. For later increments, however, it would be
desirable to use the average of two or more scales per fish. In
addition, the accuracy of scale measurements may be improved by
the use of computer-aided devices (Frie 1982).

Fit of the model will also be affected by inaccurate aging.
Scale ages are often underestimated from older walleye (Campbell
and Babaluk 1979, Erickson 1979, 1983) as may be evident for the
age 9 age-group in the example (Figure 1), where observed
increments may have been too large because of overlooked annuli.

Modeling increments of growth directly from scales allows
standardized comparison between samples from different environments and time periods, and before and after management manipulations. Some workers have tried to compare average back-
calculated lengths between age-groups, stocks, and year classes,
but the questions of statistical and biological significance
remained unanswered. Differences found by comparison of average
back-calculated lengths generated from the traditional Fraser-Lee
algorithm may be due to the body-scale intercept (Carlander 1982,
1985), sampling, differing environment effects, differing genetic

growth capacity, variation inherent in the estimation methods or
any combination of these. Our model allows direct comparison of
growth after the effects of good and bad growth years are removed.
Comparisons of age effects via F and t tests can be used to compare growth rates in different populations.

As another example, we have applied our method to scale
samples from an experiment in which stunted and nonstunted
bluegill sunfish Lepomis macrochirus were placed into a controlled
pond environment (Murnyak et al. 1984). The experimenters
concluded that the stunted population grew differently even under
the same controlled environmental conditions; we came to the same
conclusion from the analysis using scale samples from their fish.
Since our method allows for separation of age and environment
effects, we hypothesize that the conclusion could have been
reached without artificially manipulating the environment.

The Fraser-Lee method of modeling fish growth requires
fitting separate models to each fish and then averaging over those
models. The approach here uses the age-group as the unit of
analysis, not the individual. We can therefore obtain meaningful
parameters that can describe the growth of an age-group in a
useful way.

In obtaining lengths in previous years, our model suggests
that the residual variance in length will be higher for older fish
than for younger ones, since growth is a sum of yearly increments.
Our model proposes a separate regression of length on scale for
each age-group, but with a common slope. The common slope can be
traced to the assumption that the correlation between scale increment and length increment is constant for all years. The assumption is biologically sound because the proportionate lateral increase in scale relative to length should be the same no matter if
body growth is fast or slow. In contrast, the Fraser-Lee method
requires a separate slope for each fish, but with a common intercept, a. The validity of this approach has not been demonstrated
(Carlander 1981, 1982); yet, a universal value of a for each
species is suggested. In our method, the increasing intercept
with age-group is again a result of adding yearly growth increments. This is consistent with the findings of Lea(1933) and
Weese(1951), cited in Carlander(1982); they fit simple regression
lines to each age-group and reported that the intercept did
increase with age.

If the assumptions of the Fraser-Lee method are accepted,
confusion concerning the body-scale relationship within an age-
group is inevitable. Carlander (1985, personal communication)
states, "The body-scale regressions for an age-group represents
the change in size of scale with length of the slower growing fish
of the age-group to the faster growing fish of the age-group and
do not necessarily represent the change with increase in length of
the individual or average fish." We have largely avoided this
problem by fitting models to the observed scale increments, which
contain information on yearly changes, and not relying on the
information on totals from any given fish.

REFERENCES
Bagenal, T. B. and F. W. Tesch. 1978. Age and growth. In T. B.
    Bagenal, editor, Methods for Assessment of Fish Production in
    Fresh Waters. Blackwell Scientific Publications, Oxford,
    101-136.
Bartlett, J. R., P. F. Randerson, R. Williams, and D. M. Ellis.
    1984. The use of analysis of covariance in the
    backcalculation of growth in fish. Journal of Fish Biology
    24:201-213.
Box, G. E. P. 1953. Non-normality and tests on variances.
    Biometrika 40:318-335.
Campbell, J. S. and J. A. Babaluk. 1979. Age determination of
    walleye Stizostedion vitreum vitreum (Mitchill), based on the
    examination of eight different structures. Canada Fisheries
    and Marine Service Technical Report 849, Winnipeg, Manitoba.
    23 pp.
Carlander, K. D. 1981. Caution on the use of the regression
    method of back-calculating lengths from scale measurements.
    Fisheries (Bethesda) 6(1):2-4 (corrections 1983 8(5):25).
Carlander, K. D. 1982. Standard intercepts for calculating
    length from scale measurements for some centrarchid and
    percid fishes. Transactions of the American Fisheries
    Society 111:332-336.
Carlander, K. D. 1985. Example of need for a standard 'a' value.
    In R. C. Summerfelt and G. E. Hall, editors. Age and Growth
    of fish. Iowa Statue University Press, Ames, IA, USA.
Carroll, R. J. 1982. Adapting for heteroscedasticity in linear
    models. Annals of Statistics 4:1224-1233.
Cook, R. D. and Weisberg, S. 1982. Residuals and Influence in
    Regression. Chapman-Hall, London, England.
Erickson, C. M. 1983. Age determination of Manitoban walleyes
    using otoliths, dorsal spines, and scales. North American
    Journal of Fisheries Management 3:176-181.
Erickson, C. M. 1979. Age differences among three hard tissue
    structures observed in fish populations experiencing various
    levels of exploitation. Manitoba Department of Natural
    Resources Report 79-77, Winnipeg, Canada.
Frie, R. V. 1982. Measurement of fish scales and backcalculation
    of body lengths using a digitizing pad and microcomputer.
    Fisheries (Bethesda) 7(6):5-8.
Hile, R. 1970. Body-scale relation and calculation of growth in
    fishes. Transactions of the American Fisheries Society
    99:468-474.
Lea, E. 1933. Connected frequency distributions, a preliminary
    account. Report of the Norwegian Fisheries and Marine
    Investigations 4:1-12.
Murnyak, D. F., M. O. Murnyak, and L. J. Wolgast. 1984. Growth
    of stunted and nonstunted bluegill sunfish in ponds.
    Progressive Fish Culturist 46:133-138.
Newman, R. M. and S. Weisberg. 1986. Among fish and within fish
    variation of scale growth increments. In R. C. Summerfelt

and G. E. Hall, editors. Age and Growth of Fish. Iowa State University Press, Ames, IA, USA.

SAS. 1982. SAS User's Guide: Statistics. SAS Institute, Cary, NC, USA.

Weese, A. O. 1951. Age and growth of Lepibema chrysops in Lake Texoma. Proceedings of the Oklahoma Academy of Science 30:45-48.

Weisberg, S. 1985. Applied Linear Regression, 2nd Edition. Wiley, New York, USA.

Weisberg. S. 1986. A linear model approach to backcalculation of fish length. Journal of the American Statistical Association (in press).

III

Variability, error, and bias

Variability in growth parameter estimates from scales of Pacific cod based on scale and area measurements

GEORGE HIRSCHHORN AND
GREGG J. SMALL[1]
NATIONAL MARINE FISHERIES SERVICE
NORTHWEST AND ALASKA FISHERIES CENTER
SEATTLE, WA 98115-0070

ABSTRACT

The time required to generate and analyze fish scale measurements has been substantially reduced with the advent of computer-related technology. Using a digitizer, we examined the correlations of fish length and the size of key scales in a sample of 104 Pacific cod Gadus macrocephalus from the eastern Bering Sea. Scale size was expressed as the square root of scale surface area or as selected scale radii. Highest r-values of the log-transformed variables were obtained with square root of scale area representing scale size. The variability of von Bertalanffy parameters from scale annuli was examined directly be generating scale annulus measurements in each of 47 scales removed from a single female specimen of this species. The material was divided into 3 body regions of scale removal; on each scale, the annuli were measured along each of 6 selected radii, or expressed as square root of scale area. One of the regions was defined as near a vertical between the midpoint of the insertion of the second dorsal fin and the lateral line. For this region, the standard deviation of the residuals from fit based on area measurements was less variable (CV = 14.4%) than the SD's for any of the 27 other combinations of body region and scale measure examined. This body region is easily identified in field sampling. Because parameter estimates for single fish are "unweighted," routine collection of this type of information may provide an alternative to present limitations to growth estimates based on mean-length data.

---

[1] Present address: Natural Resources Consultants, 4055 21st Avenue West, Seattle, Washington 98199.

The Age and Growth of Fish, edited by Robert C. Summerfelt and Gordon E. Hall © 1987 The Iowa State University Press, Ames, Iowa 50010.

INTRODUCTION

Scales have been widely used to estimate fish age by identifying rings considered to form annually. Age readers are trained to recognize patterns formed by annuli, but validation of such agings is difficult in many species (Carlander 1974), and frequently lacking in fisheries literature (Beamish and McFarlane 1983). In a recent evaluation of methods of aging Pacific cod Gadus macrocephalus from scales, Westrheim and Shaw (1982) concluded that annulus criteria developed by Kennedy (1970) for samples from Hecate Strait, B.C., were unsatisfactory for samples from Georgia Strait, less than 300 miles to the south, which raised problems in estimating both age composition and growth.

For management purposes, fish growth is described most often by the parameters of the von Bertalanffy equation (labeled VB in this paper), viz.

$$1_t = L_\infty\{[1 - \exp(-k(t - t_0)]\} \tag{1}$$

where 1_t is linear size at age t, L_∞ is asymptotic size in linear units, and k the "growth completion rate" of Beverton and Holt (1959). The latter is the rate at which a fish approaches asymptotic length; t_0 is the age when fish size is estimated to have been zero. The data fitted by (1) consist typically of a set of mean fish lengths, one per age, obtained from age-length keys. The resulting parameters are then said to represent the catch, stock or population of a species, depending on circumstances. VB estimates obtained in this manner affect estimates of other important factors such as natural mortality, age at recruitment, yield per recruit, or biomass. Use of such estimates is still a recommended procedure; however, a number of caveats have appeared in fisheries literature for either statistical reasons (Chapman 1961; Sainsbury 1980), or biological reasons (Taylor 1958). The effect on estimated numbers at age due to certain combinations of age and length data was discussed by Westrheim and Ricker (1978). It is of historical interest that von Bertalanffy (1951) had individual, rather than group growth descriptions in mind when presenting his "organismic" theory; the application of the equation to means of grouped lengths which subsequently became so useful in fisheries management, was considered by him as "metaphorical." However, VB parameters hae also found use as a criterion by which to judge whether rings counted by age readers were likely annular in deposition. One such procedure was developed by Frost and Kipling (1959) for reducing error in aging of pike Esox lucius from Lake Windermere, England; they considered the method useful, but too slow for general use. More recently, computer-assisted methods employing digitizing tablets have been described (Frie 1982). Also, Small and Hirschhorn (1986) developed a software package to generate and display tracings of selected scale radii and areas, together with the associated sets of VB parameters. The time needed to acquire corroborative information is sufficiently short so the method could be used routinely if necessary.

Because of the traditional importance of certain allometric relations in fisheries research we first utilized the method to compare correlations of fish length with scale size, the latter expressed as any of 8 selected radii, or as the square root of the surface area of scales of Pacific cod. Secondly, we applied the system to 47 scales from a single specimen of Pacific cod to examine the differences in the VB parameter estimates, and their variability with respect to 4 scale-location regions of body surface. This was done to define a "preferred" body region for routine scale sample collecting in the field.

MATERIALS AND METHODS

For examining relations of fish length and scale length, a sample of 104 Pacific cod was collected in 1979 in the eastern Bering Sea. Information from each specimen included fork length to nearest cm, weight, sex, and body position of scales (series A, positions 9-1 in Figure 1a). Fork lengths ranged from 30 to 95 cm. Surface areas and 8 selected radii were measured by digitizing photographic enlargements of plastic scale impressions (Small and Hirschhorn 1986).

Within-fish comparisons of scale parameters were based on a total of 47 unregenerated scales removed from a single 96 cm female Pacific cod aged 6 years; body locations and the direction of scale radii measured are shown in Figures 1a and b. Both scale area and radii were measured by digitizing from photographic scale enlargements as described in Small and Hirschhorn (1986). Absolute measurements of linear size at age were standardized by setting the radius measurement of the highest numbered annulus equal to 100 percent. In fitting the data to equation 1 the method of Fabens (1965) was chosen: L_∞ and $\exp(-k)$ are estimated first by an iterative procedure such that consecutive estimates of them differ by no more than 0.001 percent. Following this the x-intercept t_0 is estimated from

$$\exp(-kt_0) = \Sigma_t (L_\infty - 1_t) * \exp(-kt)/\Sigma_t \exp(-2kt)$$

This estimation method differs from the iterative Asymptotic Linear Regression method of Stevens (1951) which yields a least squares solution for all 3 parameters simultaneously. It has been applied frequently to mean length data developed from age-length keys. Ensuing estimates of t_0 from such applications are often strongly negative and supply no useful management information; t_0 is sometimes considered a nuisance parameter and ignored in synoptic treatments of relations between growth rates and natural mortality rates (Pauli 1980). Alternatively, heuristic adjustments may be found (Archibald et al. 1981). On

1a

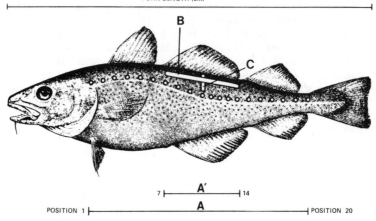

FORK LENGTH (cm)

B

C

7 ⊢ A' ⊣ 14

POSITION 1 ⊢ A ⊣ POSITION 20

1b

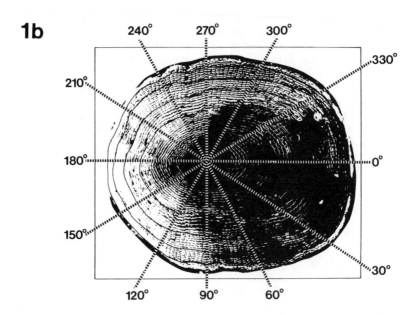

240° 270° 300°

330°

210°

180° 0°

150°

30°

120° 90° 60°

Figure 1a. Scale series collected from Pacific cod: A and A',
along lateral line; B. Horizontal Series; C. vertical series.

1b. Directions of scale radii for which measurements and
parameter estimates are generated routinely for comparative
purposes. Estimates are generated simultaneously with those
of scale area when each annulus is traced.

the other hand, if the growth data from a single animal result in a plausible fit of the parameters (including t_0), then the validity of the model is supported. In Pacific cod from Bering Sea, scale formation is thought to commence shortly after completion of the larval stage with body coverage completed within a month. Therefore, the fraction of the year when scale deposition occurs is likely to be small or smaller than subsequent periods of annulus formation.

RESULTS

Association between fish length and selected scale radii (Figure 1b), or the square root of scale surface area (SSA), was measured by simple correlation (r) between the log transforms. In Table 1 the correlation coefficients are seen to be higher for the SSA correlate than for any radius. This is believed due to the averaging of a large number of radius measurements generated by digitizing about 250 measurements per scale, on which the SSA estimate is based. Corresponding r values for untransformed variables were found slightly lower than those shown for the transforms in Table 1.

Table 1. Correlation coefficients of natural logarithms of fork length and linear scale size (x58), by sex, body position, and radius direction as in Figure 1b.

| | n | SSA | Scale Radius | | | | | | | |
			0°	30°	330°	90°	270°	150°	210°	180°
Males										
Position 9	33	0.85	0.80	0.72	0.76	0.81	0.80	0.72	0.76	0.70
Position 10	55	0.85	0.76	0.75	0.78	0.83	0.88	0.62	0.73	0.50
Position 11	49	0.91	0.81	0.89	0.86	0.87	0.89	0.69	0.68	0.52
Females										
Position 9	45	0.90	0.84	0.82	0.85	0.77	0.85	0.69	0.81	0.67
Position 10	51	0.79	0.73	0.74	0.69	0.79	0.79	0.63	0.62	0.52
Position 11	42	0.87	0.67	0.69	0.78	0.82	0.81	0.75	0.74	0.66

The SSA measurements of scale size in left side scales are shown in Figure 2, both by scale series (A, B, C in Figure 1a) and position within series. The total size of each scale is approached in descending annual increments, forming similar patterns with differences in absolute size becoming more pronounced with age. The basic measurements are presented in Table 2 for each combination of scale series and scale measure. The parameter means are seen to be similar within methods while the associated CV's are not. The latter are always largest for series A or A'; they are lowest in series C scales for the majority of scale radii examined, and for the SSA method. Parameter estimates by the SSA method lie always within the range of corresponding radius-based estimates and the standard deviations of residuals from the VB fits were least variable in series C data generated by the SSA method.

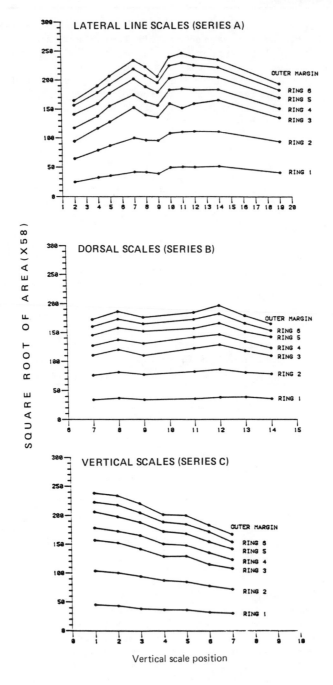

Figure 2. Absolute size of scale annuli and outer margin by body position. Left side scales only.

Table 2. Coefficients of variation of scale parameter estimates and related quantities, by scale series and method (specimen 82-1). The number of scales in scale series A, A', B and C was 23, 14, 10 and 14, respectively. Means of estimates shown in parentheses.

		Scale Radius						
		30°	90°	180°	270°	330°	0°	SSA
L_∞	A	5.0	10.7	17.2	6.8	5.4	5.2	6.4
		(116.3)	(133.0)	(123.9)	(135.8)	(117.8)	(114.9)	(121.7)
	A'	5.3	11.5	5.0	6.1	4.4	2.9	2.9
		(117.0)	(130.2)	(113.0)	(133.1)	(116.0)	(114.0)	(118.7)
	B	3.8	4.1	8.0	7.7	2.6	3.2	2.7
		(117.0)	(124.7)	(122.1)	(132.8)	(112.1)	(114.2)	(119.5)
	C	2.4	2.7	10.8	5.2	2.6	2.2	1.8
		(116.4)	(129.9)	(117.7)	(132.7)	(115.2)	(114.3)	(119.3)
e^k	A	5.0	5.5	11.2	3.8	5.6	6.6	4.7
		(0.70)	(0.77)	(0.71)	(0.79)	(0.70)	(0.69)	(0.73)
	A'	5.1	5.8	7.8	4.7	4.9	3.6	2.8
		(0.70)	(0.76)	(0.67)	(0.78)	(0.70)	(0.69)	(0.72)
	B	5.0	3.5	7.1	4.7	4.5	4.2	2.5
		(0.71)	(0.75)	(0.73)	(0.78)	(0.67)	(-0.69)	(0.72)
	C	2.4	2.7	8.7	3.0	3.0	2.8	1.8
		(0.70)	(0.77)	(0.70)	(0.77)	(0.69)	(0.69)	(0.72)
t_0	A	15.5	21.8	23.1	18.3	11.9	15.2	9.1
		(0.45)	(0.38)	(0.42)	(0.39)·	(0.50)	(0.43)	(0.43)
	A'	18.5	22.4	21.1	11.5	12.4	16.2	11.3
		(0.43)	(0.41)	(0.41)	(0.41)	(0.48)	(0.40)	(0.42)
	B	14.5	22.7	16.7	19.0	10.0	15.4	9.8
		(0.43)	(0.36)	(0.40)	(0.36)	(0.46)	(0.38)	(0.39)
	C	9.4	15.7	16.2	16.0	10.6	9.1	6.7
		(0.46)	(0.40)	(0.44)	(0.41)	(0.48)	(0.40)	(0.43)
S.D.	A	40.0	43.7	36.6	49.1	38.4	41.1	35.0
		(1.9)	(2.3)	(2.2)	(2.3)	(1.9)	(1.7)	(1.7)
	A'	44.7	46.6	39.4	51.2	40.7	48.8	35.9
		(1.6)	(2.3)	(2.3)	(2.6)	(1.9)	(1.6)	(1.7)
	B	31.2	39.5	22.9	30.7	28.7	33.7	21.5
		(1.9)	(2.1)	(1.7)	(1.8)	(2.8)	(2.2)	(1.7)
	C	21.1	25.5	23.2	31.2	37.8	26.5	14.4
		(2.1)	(2.1)	(2.3)	(1.8)	(1.8)	(1.8)	(1.8)
$A_{.95}$	A	14.0	22.2	40.3	14.4	14.9	16.1	15.2
		(9.0)	(12.5)	(10.3)	(13.1)	(9.2)	(8.6)	(10.1)
	A'	14.7	24.9	17.4	13.4	13.0	11.4	7.8
		(9.1)	(12.0)	(8.2)	(12.6)	(8.9)	(8.5)	(9.6)
	B	11.6	10.6	21.0	17.1	10.6	10.4	7.4
		(9.1)	(10.8)	(10.2)	(12.5)	(8.1)	(8.6)	(9.7)
	C	6.7	10.4	27.9	11.2	8.0	6.9	4.9
		(8.9)	(11.9)	(9.2)	(12.3)	(8.6)	(8.5)	(9.6)
kL_∞	A	9.5	11.9	21.1	9.0	10.9	13.2	9.3
		(41.4)	(33.8)	(40.5)	(32.5)	(41.2)	(42.9)	(38.1)
	A'	9.8	12.2	15.2	8.7	9.5	6.8	5.5
		(40.9)	(34.8)	(44.4)	(43.1)	(41.9)	(42.5)	(39.0)
	B	10.2	8.0	14.9	10.7	8.3	8.5	5.1
		(40.8)	(36.1)	(38.4)	(33.4)	(44.5)	(42.0)	(38.5)
	C	4.4	6.0	16.8	6.6	5.7	5.4	3.7
		(41.3)	(34.1)	(42.1)	(33.6)	(42.6)	(42.5)	(38.9)

Associated coefficients of skewness of the SD distributions (not shown in Table 2) from SSA data were considered low, ranging from −1.39 for series A data to −0.2 for series C. The coefficient of variation of SD's for the SSA method−C series combination was 14.4 percent, lower than for any other combination.

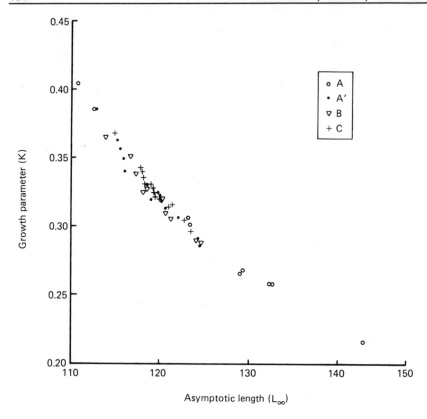

Figure 3. Parameter pairs (k, L∞) from specimen 82-1, by scale
and series (n = 47).

The distributions around parameter means of k from scales
of the midbody regions show a central tendency, which is
strongest in series C (Figure 3); k-estimates of 0.32-34 were
found in 37, 40, and 64% of the material from series A', B and
C, respectively. Converting to curvature units of exp(-k), this
range becomes 0.73-0.71, or 0.02. This appears to be small in
relation to the total range this parameter can assume under the
VB relation for growth in fishes, viz. $0 < exp(-k) < 1$; thus, it
suggests that the method may be of use in describing growth
differences between individual fish.

DISCUSSION
    Applying the equipment and procedure described in Small and
Hirschhorn (1986), the results here indicate that the
variability of growth parameter estimates from scale annulus
measurements depends both on scale location and scale measure.

Scales from region C, i.e., near a vertical between the center
of the insertion of the second dorsal fin and the lateral line,
can be readily identified and collected (by scraping) in the
field. Parameter means from this region were found less
variable than from other regions. The variability of group
means from any body region was lower from scale area
measurements than from scale radii. Therefore, parameter
estimates based on the square root of annulus areas in scales
from region C are likely to be of most use to age readers
because aging errors due to the inclusion of false annuli, or
exclusion of true ones, can be reduced. Other sources of error
remain, e.g., annulus recognition at the scale margin during
certain times of year, or inability to distinguish between
consecutive annuli at high ages; these, of course, affect
conventional age reading results as well. A limitation of the
present method is that the shape parameter $\exp(-k)$ cannot be
estimated from fewer than three annuli (and only without
estimate of variance in age three fish). However, aging
difficulties are frequently encountered in fish aged 4 years or
older, and in these instances the limitation would not apply.

SSA measurements of scale size were also found more highly
correlated with fish length than were radius measurements. This
suggests that the usual assumptions about relations between fish
growth and scale growth continue to be justified. However,
these results cannot be considered a confirmation of the VB
hypothesis as such: scales are not known to represent catabolic
surfaces, nor were their surface areas found to increase (with
respect to fish weight) at a rate of 2/3, as was implied for
catabolically effective surfaces in fishes by von Bertalanffy
(1951). The reasons why area-based measures produced less
variable parameter estimates are not clear: one possibility is
that measurements of single radii are more variable with respect
to size at age than large numbers of radius measurements such as
are generated by the digitizing method and utilized to
approximate the scale area included by each annulus. In any
event, the ability to generate an individual growth estimate of
apparently low variability from the scales of a fish, coupled
with the speed of this method suggests applicability to most,
possibly all members of catch samples; one such study is now in
progress. With respect to entire catch samples, such growth
estimates could be considered "unweighted" in a statistical
sense and suitable for computing an average growth rate. As
shown by Ricker (1979) the difference between "true average
growth rate" and "population growth rate" based on VB fits to
mean lengths at age, can be quite large. If growth estimates of
the latter kind are subsequently used to estimate other vital
rates, e.g., natural mortality, these may be affected as well.

REFERENCES

ARCHIBALD, C. P., W. SHAW, AND B. H. LEAMAN. 1981. Growth and mortality estimates of rockfishes (Scorpaenidae) from B.C. coastal waters, 1977-1979. Canadian Technical Report on Fisheries and Aquatic Sciences Number 1048, Resource Services Branch, Pacific Biological Station, Nanaimo, B.C., V9R5K6

BEAMISH, R. J., AND G. A. McFARLANE. 1983. The forgotten requirement for age validation in fisheries investigations. Transactions of the American Fisheries Society 112(6):735-743.

BEVERTON, R. J. H., AND S. J. HOLT. 1957. On the hynamics of exploited fish populations. Fishery Investigations, Ministry of Agriculture and Fisheries, Series II, Volume 19, 533 p.

BEVERTON, R. J. H., AND S. J. HOLT. 1959. A review of the life span and mortality rates of fish in nature and their relation to growth and other physiological characteristics. Pages 142-177 in G. E. Wostenholme and M. D. Connor, editors, CIBA Foundation Colloquia on Aging, Volume 5. Lifespan of Animals, Little, Brown and Company, Boston.

CARLANDER, K. H. 1974. Difficulties in aging fish in relation to inland fisheries management. Pages 200-205 in T. B. Bagenal, editor, Aging of fish, Unwin Brothers Ltd., Surrey, England.

CHAPMAN, D. G. 1961. Statistical problems in dynamics of exploited fish populations. Fourth Berkeley Symposium of Probability and Statistics 4:153-168.

FABENS, A. J. 1965. Properties of fitting the von Bertalanffy growth curve. Growth 29:276-289.

FRIE, C. V. 1982. Measurement of fish scales and back-calculation of body lengths using a digitizing pad and microcomputer. Fisheries 7(6):5-8.

FROST, W. E., AND C. KIPLING. 1959. The determination of age and growth of pike (Esox lucius) from scales and opercular bones. Journal of the International Council for Exploration of the Sea 24(3):321-341.

KENNEDY, W. A. 1970. Reading scales to age Pacific cod (Gadus macrocephalus) from Hecate Strait. Journal of the Canadian Fisheries Research Board of Canada 27:915-922.

PAULY, D. 1980. On the relationship between natural mortality growth and environmental temperature in 175 fish stocks. Journal of the International Council for Exploration of the Sea 39:175-192.

RICKER, W. E. 1979. Growth rates and models. Pages 678-743 in Fish Physiology, Volume 8. Academic Press, Inc.

SAINSBURY, K. J. 1980. Effect of individual variation of the von Bertalanffy growth equation. Canadian Journal of Fisheries and Aquatic Sciences 37:241-247.

SMALL, G. J. AND G. HIRSCHHORN. 1986. Computer assisted age and growth pattern recognition of fish scales using a digitizing tablet. In R. C. Summerfelt and G. E. Hall, editors. Age and growth of fish. Iowa State University Press, Ames, Iowa, USA.

STEVENS, W. L. 1951. Asymptotic regression. Bimetrics 7:247-267.

TAYLOR, C. C. 1958. A note on Lee's phenomenon in Georges Bank haddock. International Commission for North Atlantic Fisheries, Special Publication Number 1:243-251.

VON BERTALANFFY, L. 1951. Theoretische Biologie, Volume 2. A. G. Franke, Berne, Switzerland.

WESTRHEIM, S. J., AND W. E. RICKER. 1978. Bias in using an age-length key to estimate age-frequency distributions. Canadian Journal of Fisheries and Aquatic Sciences 35(2):184-189.

WESTRHEIM, S. J., AND W. SHAW. 1982. Progress report on validating age determination methods for Pacific cod (Gadus macrocephalus). Canadian Manuscript Report of Fisheries and Aquatic Sciences Number 1670.

Among- and within-fish variation of scale growth increments in brown trout

RAYMOND M. NEWMAN
DEPARTMENT OF FISHERIES AND WILDLIFE
SANFORD WEISBERG
DEPARTMENT OF APPLIED STATISTICS
UNIVERSITY OF MINNESOTA
ST. PAUL, MN 55108

ABSTRACT
 Scale increments were measured on several scales from each of
over 1200 brown trout Salmo trutta, collected from April 1980 to
June 1983. Among-fish variance of scale increments was generally
much greater than the variance between scales on the same fish.
Both mean length of increments and variances for given annuli
within a year-class were generally constant over the sampling
dates and indicated a lack of Lee's phenomenon. The analyses did
detect significantly larger first annular increments for age-1
fish sampled in the spring than on later sampling dates. A table
is provided to aid in determining the number of scales to be
measured per fish, given the variation among and within fish.

 Measurements of bony structures have been used extensively
for age determination and estimation of fish growth (Bagenal and
Tesch 1978). When scales are used, investigators typically
examine a sample from each fish, since many are available and some
scales will often be unreadable or regenerated (Bagenal and Tesch
1978). Numerous authors (Whitney and Carlander 1956; Hile 1970;
Bagenal 1974) have found a strong linear correlation between fish
length and scale radius, and the general conclusion is that scale
length (or increment length) is a reliable indicator of growth,
justifying the use of scales in growth analysis.
 Consideration of within-sampling unit variation is a common
concern in the design of experiments and sampling procedures
(Snedecor and Cochran 1967). Because several readable scales are
often obtained from the same fish, it is desirable to know the
between-scale (within-fish) as well as the among-fish variability
in these measurements; nonetheless, this issue has rarely been
addressed in scale-growth studies. Various investigators (Bagenal
1974) have measured several scales from the same fish to estimate

The Age and Growth of Fish, edited by Robert C. Summerfelt and Gordon E. Hall © 1987 The Iowa
State University Press, Ames, Iowa 50010.

a mean scale radius; however, we know of only one published study
that considered between-scale variance (Kipling 1962). Some
authors (Scarnecchia 1979; Hudson and Bulow 1984) have examined
differences due to sampling location on a fish.

Estimates of scale variation are necessary to determine the
sample size required for a given level of precision. If between-
scale variability is high, less confidence can be placed in
estimates based on one or few scales. Consideration of between-
scale variation is most important when numbers of fish sampled are
low. An analysis of variance technique for scales provides an
estimate of the among- and within-fish variance and, therefore,
the precision of growth estimates based on one or several scales.
Direct analysis of scale growth increments permits interpretation
that is independent of the additional assumptions of back-
calculation techniques.

The purpose of this paper is to: 1) describe and apply an
appropriate methodology for the determination of the among- and
within-fish variation in scale growth increments; 2) determine the
constancy of increment estimates over time; and 3) determine
appropriate sample sizes of scales for growth studies.

METHODS

This paper is based on six samples of brown trout Salmo
trutta, 10 cm and larger, from South Branch Creek, Minnesota.
About 200 fish were sampled for scales in each April and August of
1980 and 1981 and in August 1982 and June 1983. Fish were
captured by electrofishing; generally, two fish from each 0.2 cm
size-group were sampled for scales. Ten to 50 scales were taken
from a selected area (above the lateral line, ventral and
posterior to the dorsal fin, but anterior to the adipose fin) on
each fish. The fish were aged according to the number of annuli,
and the annulus increments (distance between successive annuli)
were measured using the computer program DISBCAL (Frie 1982). For
each fish sampled, scale measurements were made on n = three
scales, even though typically 5-10 readable scales were examined.
Additional sampling details are given in Newman (1985).

Each annulus increment was analyzed separately for each
year-class, so no assumption concerning equality of variances for
the different annuli was required. Edge increments were not
analyzed because these depend on sampling date. Random effects
analysis of variance models were used to obtain estimates of the
variance components for between scales on the same fish (σ_s^2) and
among fish (σ_f^2). For each sampling date and year-class, a random
effects analysis of variance (Snedecor and Cochran 1967, p 279)
was computed for each annulus increment. Mean scale increments
for each fish comprised the main effects for fish or among fish,
and among scales on a fish was the error term (scale effect). The
among-fish variance can be estimated as $\hat{\sigma}_f^2 = (MS_{fish} - MS_{scale})/n$;
scale variance (σ_s^2) is estimated by MS_{scale} (Snedecor and Cochran
1967). This analysis was used for single sample date estimates of

scale increment parameters.

A separate analysis was performed for each annulus of each year-class over sampling dates, to permit an examination of constancy of annulus increments. This analysis was essentially a random effects model with subsampling (Steel and Torrie 1980, p 161), with sampling dates as treatments, fish as units, and scales as subsamples. From this analysis, variance components for fish (σ_f^2) and for scales (σ_s^2) were estimated and examined. In addition, standard F-tests for differences between sampling dates, and a test for a linear trend due to sampling dates, were computed.

For all analyses, the intraclass correlation was estimated as $\hat{\rho} = \hat{\sigma}_f^2 / (\hat{\sigma}_f^2 + \hat{\sigma}_s^2)$. Values of the intraclass correlation near one imply that the information on any one scale is a good estimator of the average scale increment for that fish; conversely, if the intraclass correlation is near zero, any one scale would provide a poor estimate of the average increment. Thus, large values of the intraclass correlation are required for reliable analyses based on a small number of scales per fish.

RESULTS

An example of the results for analysis of a given combination of year-class (1979) and first annulus is given in Table 1. Members of the 1979 year-class were included in all 6 samples. The F-ratio shown tests $\sigma_f^2 = 0$; this test gave very small P-values for all but the age-4 fish in June 1983 (Table 1). The data for all other year-classes also gave very small P-values for nearly every year-class, sampling date and annulus combination (63 combinations with $P < 0.05$, 58 with $P < 0.01$, 70 total tests). All of the non-significant values occurred for fish age-4 or older when the number of fish sampled was ≤ 6. A complete presentation of these results is given by Newman (1985 Appendix II).

The estimates of the variance components ($\hat{\sigma}_f^2$ and $\hat{\sigma}_s^2$) were remarkably similar between sampling dates, and the estimates of intraclass correlation were relatively high (Table 1). For a

Table 1. Mean scale increment lengths ($\bar{X}_i$) and summary statistics for the 1st annulus of the 1979 year-class of brown trout. Data are for 3 scales per fish ($\underline{N}$ = number of fish). Significance ($P<0.01$) indicated by *.

Sample date		age	$\bar{X}_i$ (mm)	F ratio	$\hat{\sigma}_s$	$\hat{\sigma}_f$	$\hat{\rho}$	$\underline{N}$
Apr	1980	1	0.603	14.51*	0.052	0.110	0.818	38
Aug	1980	1	0.491	36.29*	0.036	0.124	0.922	82
Apr	1981	2	0.478	18.85*	0.042	0.102	0.856	60
Aug	1981	2	0.479	23.40*	0.040	0.109	0.882	71
Aug	1982	3	0.460	12.28*	0.051	0.098	0.790	42
Jun	1983	4	0.387	6.89	0.031	0.044	0.662	5
Pooled			0.494	21.00*	0.043	0.110	0.869	298

given year-class and annulus, none of the between-scale variances were significantly different from each other ($P > 0.05$; 18 year-class and annuli combinations); only two of 18 sets of among-fish variances had significant differences (Newman 1985 Appendix II). Therefore, the analysis of variance with subsamples was computed, including sampling date as an additional factor. These pooled results for the first annulus of the 1979 year-class are shown in Table 1. The high pooled intraclass correlation (0.869) suggests that variation between scales on a fish was relatively unimportant for estimating scale increments for these data.

Sampling-date effects can be tested by comparing the mean increments for the various sampling dates (overall F-test). At the suggestion of a referee, a test for a linear trend was also used to maximize power against time-dependent bias. For the results of Table 1, the linear trend $F = 9.3$ with 1,292 df ($P < 0.01$). The April 1980 sample of the 1979 year-class had increments significantly longer than other sampling dates. The same results were observed for first samples of other year-classes and were probably due to sampling bias; the mean length of age-1 fish sampled in the spring was generally higher than the mean length in the population (Newman 1985).

Table 2 gives the summary statistics for each year-class and

Table 2. Mean scale increment lengths ($\bar{X}_i$) and summary statistics for annulus number and year class. Data are for 3 scales per fish (N = number of fish). Estimates that include the biased 1st spring samples indicated by @. F (fish) is the F ratio for among fish; F (date) is the ratio for the sampling date effect. The slope is for a linear trend over sampling dates. Slopes were not computed for 2 date estimates. Significance (P<0.05) is indicated by *.

Annulus number	Yr. class	$\bar{X}_i$ (mm)	$\hat{\rho}$	$\hat{\sigma}_s$	$\hat{\sigma}_f$	F (fish)	slope	F (date)	N
1	76	0.456	0.758	0.045	0.080	10.4*	-0.029	1.1	26
1	77	0.468	0.898	0.036	0.108	27.4*	0.004	0.2	36
1	78	0.432	0.883	0.036	0.099	23.5*	0.005	1.6	240
@ 1	79	0.494	0.869	0.043	0.110	21.0*	-0.026*	9.3*	298
1	79	0.478	0.878	0.041	0.110	22.5*	-0.011	1.4	260
@ 1	80	0.526	0.818	0.045	0.095	14.5*	-0.042*	19.0*	231
1	80	0.502	0.835	0.044	0.099	16.1*	-0.023	3.6*	181
1	81	0.410	0.897	0.042	0.123	27.2*	-	4.6*	188
2	76	0.469	0.592	0.058	0.070	5.4*	0.005	0.5	26
2	77	0.486	0.829	0.043	0.094	15.6*	-0.033	1.6	36
2	78	0.556	0.838	0.048	0.109	16.5*	0.007	2.2	240
2	79	0.473	0.839	0.047	0.106	16.6*	0.008	0.5	178
2	80	0.443	0.853	0.045	0.108	18.4*	-	0.3	107
3	76	0.272	0.760	0.036	0.065	10.5*	-0.008	0.5	26
3	77	0.263	0.822	0.040	0.086	14.9*	0.014	0.7	36
3	78	0.237	0.733	0.041	0.068	9.3*	0.003	0.8	80
3	79	0.237	0.746	0.049	0.084	9.8*	-	3.2	47
4	76	0.174	0.612	0.032	0.040	5.7*	-0.253	4.0*	26
4	77	0.178	0.728	0.032	0.052	9.0*	-	0.4	7
4	78	0.166	0.076	0.039	0.036	0.8	-	-	7

annulus combination. Significant differences in mean increments
over sampling dates were observed only 5 times. Only two
combinations had significant linear trends (the first increments
for the 1979 and 1980 year-classes). It will be shown later that
these trends can be explained by the spring sampling biases;
elimination of the biased samples greatly reduced sampling date
effects and eliminated the significant linear trends. The
intraclass correlations were generally high, but they appeared to
decline with increasing annulus number.

Because of the consistency in results for a given annulus,
all samples on a particular annulus were further combined by
assuming constant σ_f^2 and σ_s^2, using usual analysis of variance
formulas (Table 3). Annulus 1 and annulus 2 both showed high
intraclass correlations, and for these annuli a constant variance
assumption seems reasonable. Measurment of only one or two scales
should be adequate for these annuli. The intraclass correlations
for annuli 3 and 4 were smaller (Table 3), and it may be worthwhile to measure several scales per fish for these annuli. The
estimates of σ_s were very similar for the four annuli, but $\hat{\sigma}_f$ was
much lower for annuli 3 and 4; even the coefficient of variation
was not constant. Thus modeling of scale increments (e.g.
Weisberg and Frie 1986) may require modeling of variances as well.

Table 3. Pooled estimates of mean scale increment length and summary
statistics for annuli 1-4. The mean increments are weighted by
sample size.

	Annulus			
	1	2	3	4
Number of fish ($\underline{N}$)	1019	587	189	40
Mean Increment ($\overline{mm}$)	0.469	0.502	0.247	0.173
Intraclass correlation	0.869	0.831	0.760	0.609
Pooled $\hat{\sigma}_s$	0.0414	0.0472	0.0423	0.0330
Pooled $\hat{\sigma}_f$	0.1060	0.1060	0.0754	0.0413
coefficient of variation per observation	0.243	0.231	0.350	0.306

DISCUSSION

Between-scale variance was generally not a major source of
variation in mean scale measurements. For older fish however, the
proportion of total variance due to scale variation was larger.
Although these results are specific to the South Branch Creek
brown trout population, they do provide a framework for analyses
using scales from other fish populations. Similar analyses will
be essential to the design of appropriate sampling schemes.

Annular scale increments were used rather than the radius of
each annulus, since annular radii are not statistically
independent of preceeding ones. For example, a true significant
difference in the first annular radius may result in similar

conclusions for succeeding annular radii, even when the succeeding
increments are not significantly different. Measurement errors of
the growth increments are also not independent and are included in
the scale variance, but not the among-fish variance.

The general constancy of mean increments of a year-class over
several sample dates and the lack of significant linear trends
indicated that essentially the same population was being sampled.
Increment length of a given annulus did not generally change with
fish age or sample date, which suggests that the scale increments
themselves are good indicators of growth. The obviously high
incremental values for spring-sampled age-1 fish indicated that
the analysis was sensitive to detection of sampling bias. Fish <
12 cm were generally not included in the spring scale sampling, so
the mean length of age-1 fish sampled in the spring was notably
higher than the mean length in the population (typically 2 cm
longer; Newman 1985).

The consistency of the mean annular increments of a year-
class over several years, also indicated that Lee's phenomenon
(Van Oosten 1929; Hile 1970; Bagenal and Tesch 1978) was not
evident for scale increments of this trout population. Lee's
phenomenon technically refers to back-calculated lengths (Hile
1970); however, the phenomenon may be due to biased sampling,
size-selective mortality or improper back-calculation methods
(Bagenal and Tesch 1978). We suggest examining scale increments
before basing conclusions on back-calculated values, since the
sampling and selective mortality questions can be addressed
independently of failure of the back-calculation technique.
Significant differences between increments can be more readily
attributed to sampling bias (which can be tested if the population
size structure can be estimated) or selective mortality. If
significant Lee's phenomena only appeared after back-calculation,
it would likely be due to a procedural problem in back-
calculation, such as an inappropriate body-scale intercept or
relationship. Experience with the present data set indicated that
the appearance of Lee's phenomenon (and a reverse effect) was
dependent on the body-scale relationship used (Newman 1985).

Estimates of among- and within-fish variances can be used to
estimate appropriate sample sizes needed to achieve desired
precision. The appropriate number of fish to sample can be
estimated by the usual methods (Snedecor and Cochran 1967, p 516).
For example, using the data in Table 3 for annulus 1, 82 fish
should be sampled to detect a 5% difference in the first increment
at the 95% confidence level, provided an adequate number of scales
were measured (94 fish if few scales were used). The intraclass
correlation can be used to determine whether more scales per fish
should be taken. The number of scales per fish needed to give the
same level of precision as a corresponding increase in the number
of fish sampled (with one scale per fish), was computed for a
range of intraclass correlations (Table 4). This table can be
used to determine the number of scales that should be measured,
given an estimate of the intraclass correlation (ρ). For example,

Table 4. The increase in number of fish sampled (one
scale per fish) that is equivalent to an increase
in the number of scales sampled($\underline{n}$) per fish for a
range of values of the intraclass correlation.

$\underline{n}$	Intraclass correlation				
	0.5	0.7	0.8	0.9	0.95
1	1	1	1	1	1
2	1.33	1.18	1.11	1.05	1.03
3	1.50	1.25	1.15	1.07	1.03
4	1.60	1.29	1.18	1.08	1.04
5	1.67	1.32	1.19	1.09	1.04
10	1.82	1.37	1.22	1.10	1.05
∞	2.00	1.43	1.25	1.11	1.05

with $\rho = 0.9$, taking five scales per fish would be equivalent to a
9% increase in the number of fish sampled with one scale per fish.
In general, one scale per fish is probably adequate with $\rho \geq 0.9$,
two scales with ρ about 0.8, and many scales with $\rho \leq 0.5$.

The homogeneity of variance over sample dates for both $\hat{\sigma}_f^2$ and
$\hat{\sigma}_s^2$ also supports the conclusions regarding constancy. These
results also suggest that only a few sampling dates should be
adequate to estimate σ_f^2 and σ_s^2, as long as sampling is not
size-biased; however, studies with other populations and species
will be needed before this generalization can be extended.

Values for the edge increments are not presented; however,
F-ratios for edge effects were generally lower than for the
annular increments. Coefficients of variation ($\hat{\sigma}_f/\overline{X}_i$ and $\hat{\sigma}_s/\overline{X}_i$)
were typically much higher for the edge increments than for
annular increments. There may be more variability in the timing
of edge growth, and edge increments may be more affected by
removal of scales from the fish and scale handling.

Many researchers have commented about among-scale variability
and among-fish variability (Van Oosten 1929; Whitney and Carlander
1956; Hile 1970). These discussions have concerned back-
calculated lengths, which are several steps removed from the
actual scale increments, the primary units of measurement. The
general conclusion was that among-scale variability was probably
not very large. Kipling (1962) examined among-fish and within-
fish variance of total scale radius for a limited sample of 13
fish, and she concluded that among-scale variance was lower. Her
estimates of among- and within-fish variance are comparable to the
lowest variance estimates found in this study.

Results in this study are independent of any further
assumptions associated with various back-calculation techniques,
since they are dependent only on scale measurements. Other
researchers should use similar methods to determine the confidence
that can be placed in mean increment values. If the analysis
indicates that among-scale variability is substantial, then larger
numbers of scales can be measured for each fish and precision
increased. Further, comparison of scale measurement parameters
over several sampling dates may indicate sampling biases or
important population shifts which might otherwise go undetected.

ACKNOWLEGMENTS

   T. Polomis recorded the measurements. R.V. Frie, E.M.
Goolish and T.F. Waters made helpful comments on the manuscript.
Support was provided by T.F. Waters and a Graduate School
Dissertation Fellowship; computer time was provided by the
University Computer Center. Paper number 14,275 of the Scientific
Journal Series, Minnesota Agricultural Experiment Station.

REFERENCES

BAGENAL, T., editor. 1974. The aging of fish. Unwin Brothers,
   Old Woking, England.
BAGENAL, T.B. and F.W. TESCH. 1978. Age and growth. Pages
   101-136 in T. Bagenal, editor. Methods for assessment of
   fish production in fresh waters, 3rd edition. IBP Handbook
   No. 3, Blackwell Scientific Publications, Oxford, England.
FRIE, R.V. 1982. Measurement of fish scales and back-calculation
   of body lengths, using a digitizing pad and microcomputer.
   Fisheries 7:5-8.
HILE, R. 1970. Body-scale relation and calculation of growth in
   fishes. Transactions of the American Fisheries Society
   99:468-474.
HUDSON, W.F. and F.J. BULOW. 1984. Relationship between squamation
   chronology of the bluegill, Lepomis macrochirus Rafinesque,
   and age-growth methods. Journal of Fish Biology 24:459-469.
KIPLING, C. 1962. The use of scales of the brown trout (Salmo
   trutta L.) for the back-calculation of growth. Journal du
   Conseil, Conseil International pour l'Exploration de la Mer
   27:304-315.
NEWMAN, R.M. 1985. Production dynamics of brown trout in South
   Branch Creek, Minnesota. Doctoral dissertation. University
   of Minnesota, St. Paul, MN, USA.
SCARNECCHIA, D.L. 1979. Variation of scale characteristics of
   coho salmon with sampling location on the body. Progressive
   Fish-Culturist 41:132-135.
SNEDECOR, G.W. and W.G. COCHRAN. 1967. Statistical methods, 6th
   edition. Iowa State University Press, Ames, Iowa, USA.
STEEL, R.G.D. and J.H. TORRIE. 1980. Principles and procedures
   of statistics, 2nd edition. McGraw-Hill Company, New York,
   New York, USA.
VAN OOSTEN, J. 1929. Life history of the lake herring
   (Leucichthys artedi Le Sueur) of Lake Huron as revealed by
   its scales, with a critique of the scale method. Bulletin of
   the United States Bureau of Fisheries 44:265-428.
WEISBERG, S. and R.V. FRIE. 1986. Linear models for the growth
   of fish. In R.C. Summerfelt and G.E. Hall, editors. Age and
   growth of fish. Iowa State University Press, Ames, Iowa,
   USA.
WHITNEY, R.R. and K.D. CARLANDER. 1956. Interpretation of body-
   scale regression for computing body length of fish. Journal
   of Wildlife Management. 20:21-27.

Reliability of age and growth-rate estimates derived from otolith analysis

JAMES A. RICE[1]
CENTER FOR LIMNOLOGY
UNIVERSITY OF WISCONSIN
MADISON, WI 53706

ABSTRACT
    Otolith rings are usually assumed to be daily when the slope
of ring count regressed on known age is not significantly
different from one. This "acceptance" of the null model greatly
increases the risk of type II error, i.e., mistakenly failing to
reject a false null hypothesis. By calculating statistical power
(probability of not committing type II error), one can critically
evaluate the conclusion that rings are daily. My estimates show
that power varies widely among published ring count–age
relationships; power can be increased by manipulating several
aspects of experimental design.
    Growth rate over all or part of the life history of an
individual fish can be estimated using ages derived from ring
counts and sizes backcalculated from the fish size–otolith size
relationship. Reliability of these growth–rate estimates depends
on confidence limits of the age estimates and back-calculated
lengths used in the growth rate calculation. Combined effects of
these uncertainties on growth–rate estimates can be substantial,
particularly when duration of the growth interval is short or the
initial growth–rate point estimate is low.

    Analysis of "daily rings" on otoliths is becoming
increasingly common in ecological and fisheries studies where data
such as age, growth rate or hatching date of individual fish are
needed. Inferences drawn from such studies are directly
contingent on the reliability of information obtained by otolith
analysis.

---

[1] Present address: North Carolina State University, Department
of Zoology, Box 7617, Raleigh, NC 27695-7617.

The Age and Growth of Fish, edited by Robert C. Summerfelt and Gordon E. Hall © 1987 The Iowa
State University Press, Ames, Iowa 50010.

All ring count-age relationships exhibit variability, and under some conditions ring counts clearly do not correspond to age in days (see Geffen 1982; Nielson and Geen 1982; Rice et al. 1985), either because deposition does not occur daily, or because it does not produce detectable rings. Hence, validation that the rings we observe are, in fact, daily is a necessary prerequisite for using otolith analysis in age and growth-rate determination. Only one study (Wilson and Larkin 1982) has investigated the potential for error in fish size backcalculated from otolith size, and the implications of uncertainty in age and growth-rate estimates derived from ring counts have not been addressed. This paper explores statistical power analysis as a way to critically evaluate the assumption that rings are daily and examines the confidence of age and growth-rate estimates derived from otolith analysis.

AGE DETERMINATION

Otolith rings are usually assumed to be daily when the slope of ring count regressed on known age is not significantly different from one. This approach to validation, i.e., taking non-rejection of the null hypothesis as positive evidence that the null model is correct, ignores the potential for Type II error--concluding rings are daily when in fact they are not. The regression slope of ring count on age will not be significantly different from 1.0 when the relationship is actually daily (Fig. 1A). However the same result may also occur when the true slope is very different from 1.0 but the data are too variable to distinguish it from a slope of 1.0 (Fig. 1B).

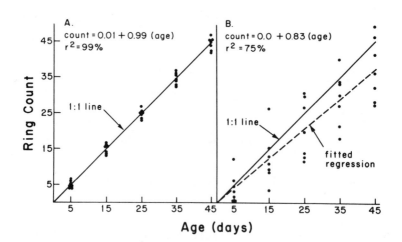

Figure 1. Hypothetical data sets with low (A) and high (B) variability. The regression slopes are not significantly different from each other (P > 0.05) or from the 1:1 line (P > 0.05).

Statistical power measures the probability of avoiding Type II error. By calculating power for the regression of ring count on age, one can determine the likelihood that a significant difference in slope would be detected if one really existed, and thus critically evaluate the conclusion that observed rings represent daily increments. Power depends on three factors (Fig. 2): the critical level of α used, the magnitude of deviation from a slope of 1.0 that the experimenter wants to be able to detect (γ), and σ_b, the standard deviation of the regression slope $\underline{b}$.

The statistical basis for these relationships and example power calculations are documented in Table 1. The influence of these

Table 1. Power calculation for regressions.

let $\underline{b}$ = sample regression slope
let $\bar{\beta}$ = population regression slope
H_0: $\beta = 1$

For two-sided tests:

Power = P(reject H_0|H_1 true)

$$= P\left\{t = \frac{b-1}{s_{\underline{b}}} > t_\mu | H_1 \text{ true}\right\} + P\left\{t = \frac{b-1}{s_{\underline{b}}} < -t_\mu | H_1 \text{ true}\right\}$$

When $\beta \neq 1$, $\dfrac{b-\beta}{s_{\underline{b}}}$ has a central t distribution, and

Substituting $\dfrac{b-\beta}{s_{\underline{b}}} + \dfrac{\beta-1}{s_{\underline{b}}}$ for $\dfrac{b-1}{s_{\underline{b}}}$

$$\text{Power} = P\left\{\frac{b-\beta}{s_{\underline{b}}} + \frac{\beta-1}{s_{\underline{b}}} > t_\mu | H_1 \text{ true}\right\} + P\left\{\frac{b-\beta}{s_{\underline{b}}} + \frac{\beta-1}{s_{\underline{b}}} < -t_\mu | H_1 \text{ true}\right\}$$

let $\gamma = \beta - 1$ = deviation of true population regression slope from H_0, and

let $\theta = \dfrac{\gamma}{s_{\underline{b}}}$

Subtracting θ from each side of the inequalities,

Power = P$\{t > t_\mu - \theta | H_1 \text{ true}\}$ + P$\{t < -t_\mu - \theta | H_1 \text{ true}\}$

When n is large (>25–30), t can be approximated by z, the normal distribution statistic. For example, for $\alpha = 0.05$, power = P($z > 1.96 - \theta$) + P($z < -1.96 - \theta$)

For one-sided tests:

$$\text{Power} = P\left\{t = \frac{b-1}{s_{\underline{b}}} \begin{array}{c} > t_\mu \\ \text{or} \\ < -t_\mu \end{array} \middle| H_1 \text{ true}\right\}$$

Following the same substitutions as for a two-sided test,

$$\text{Power} = P\left\{\begin{array}{c} t > t_\mu - \theta \\ \text{or} \\ t < -t_\mu - \theta \end{array} \middle| H_1 \text{ true}\right\}$$

Again, when n is large t can be approximated by z. For example, with $\alpha = 0.05$,
lower tail: power = P($z < -1.645 - \theta$)
and
upper tail: Power = P($z > 1.645 - \theta$)

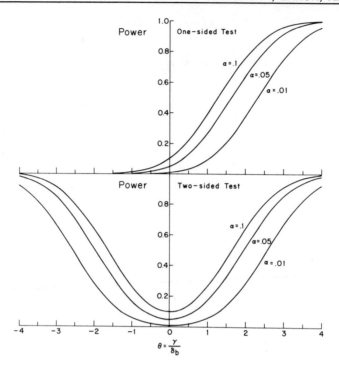

Figure 2. Power curves from one—sided and two—sided tests at
    critical α levels of 0.01, 0.05 and 0.1. Explained in
    text.

variables on power is illustrated in Figure 2; as α decreases, the
probability of rejecting H_0 when it is true (Type I error)
decreases, but the probability of accepting H_0 when it is false
(Type II error) increases, hence power decreases. Large
deviations from a slope of one are easier to detect than small
ones so, as γ increases, θ, the ratio of γ to σ_b (abscissa in
Fig. 2), increases and therefore power increases. The more
reliable the sample slope estimate is, the less likely both Type I
and Type II error are; decreasing σ_b increases θ and therefore
power (Fig. 2).
    The experimenter can manipulate all three of these factors to
control power in an otolith analysis regression. The choice of α
level and the magnitude of deviation from a slope of 1.0 to be
detected both directly affect power. The standard deviation of b
can be influenced by the number of data points used in the
regression, and their x values (x = age in days). Note that

$$\sigma_b = \frac{\sigma}{\{\Sigma(x_i - \bar{x})^2\}^{1/2}}$$

The standard deviation of $\underline{b}$ will decrease, and power will therefore increase, as data points with x values at the extemes of the range of x are added to the regression. Given an estimate of σ (the square root of the variance about the regression) from the literature or other experimental results, and a chosen α level and value of γ, the investigator can determine the number of otoliths of different ages required to achieve a desired power level. Because a low estimate of σ may result in a gross underestimate of the necessary sample size, resulting in low power, it is better to err on the high side when estimating σ. A more efficient approach to this problem, if sampling logistics allow, is to use a sequential test (see Wetherill 1966).

   Once a regression equation has been calculated its power for testing whether the slope differs from 1.0 can be estimated by assuming that σ_b equals the value of s_b estimated from the regression; because σ_b is not known exactly this provides only an approximation of power. For example the regressions shown in Fig. 1A and B have s_b = 0.0164 and 0.0836 respectively. Their power to detect a deviation of 0.1 from a slope of 1.0 (i.e., probability of detecting such a deviation if it existed) at a critical level of α = 0.05 (two-sided test) would be about 0.99 and 0.33, respectively.

   A survey of the literature identified four papers with ring count-age regressions for which power could be estimated from published statistics, or by analysis of data digitized from figures. For each of these regressions I calculated power to detect a fairly large deviation of 0.1 from a slope of 1.0 at the α = 0.05 level (two-sided test). Barkman's (1978) relationship for Atlantic silversides Menidia menidia had an estimated power of 0.99. Wilson and Larkin (1980) reported two regressions for juvenile sockeye salmon Oncorhynchus nerka; one was constrained to go through an intercept of zero rings at zero days post-emergence and had power of about 0.17, and the other was unconstrained (i.e., intercept was estimated from the data) and had power of about 0.84. Campana and Neilson (1982) calculated ring count-age regressions for juvenile starry flounder Platichthys stellatus raised under various photoperiod and temperature regimes; for those experiments power ranged from 0.2 to 0.4. In laboratory experiments with bloater Coregonus hoyi larvae at different ration levels (Rice et al. 1985) power ranged from 0.92 to 0.99 except for the lowest ration experiment where power was estimated at 0.55. Clearly the power of ring count-age regressions can vary substantially.

CONFIDENCE OF AGE ESTIMATES
   One of the primary advantages of otolith analysis is that it yields information on age and growth rate of individuals rather than the average fish. While making age estimates for

individuals, confidence intervals for the ring count on age
regression should be calculated for an individual estimate rather
than for the mean estimate. Once the regression of ring count on
age has been validated, age (the independent variable) can be
estimated from ring count (the dependent variable) using inverse
regression (Draper and Smith 1966), i.e., rearranging the
regression equation, ring count = $\underline{a}$ + $\underline{b}$(age), to solve for age,
given ring count: age = (ring count - $\underline{a}$)/$\underline{b}$. The calculation of
confidence limits for age, given ring count, is straightforward
(Draper and Smith 1966, p. 47). The age values at which an
observed ring count intersects the fitted regression and
individual confidence intervals of ring count regressed on age
indicate the age estimate corresponding to that ring count, and
the confidence limits associated with it (Fig. 3). Even very
strong ring count-age relationships typically exhibit 95%
confidence intervals for individual age estimates of about ±3 days
(Barkman 1978; Rice et al. 1985), and in many cases the intervals
are much wider.

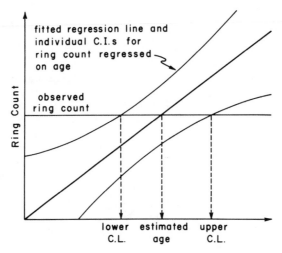

Figure 3. Confidence limits for an age estimate determined
using inverse regression.

CONFIDENCE OF GROWTH-RATE ESTIMATES
     Growth-rate estimates have two components; the time interval
(Δt), and the growth that occurred during that interval (ΔL).
Because these components are non-independent random variables,
Var($\Delta L/\Delta t$) actually is complicated beyond the empirical estimates.
But for the purposes of this paper empirical comparisons provide
an adequate means of assessing confidence of growth-rate
estimates. Variability in estimates of either or both components
will cause variability in the growth-rate estimate. In addition,

variability in the growth-rate estimate decreases as the magnitude of the time interval increases or the initial point estimate of growth rate decreases. The interactions of age variability, size variability, time interval and initial growth-rate estimate are illustrated in Figure 4 for growth rates expressed in mm/d; the principles are the same whether size is expressed in length or weight. Variability in age estimates causes uncertainty in the time interval length; for a growth-rate estimate of 0.3 mm/d and an interval of 10 d, a 95% confidence interval (C.I.) for age of ±1 d yields confidence limits for the growth-rate estimate ranging from only 0.27 to 0.33 mm/d, but a 95% C.I. for age of ±3 d (comparable to regression in Fig. 1A) causes the limits to range from 0.23 to 0.43 mm/d, and a ±5 d 95% C.I. for age results in

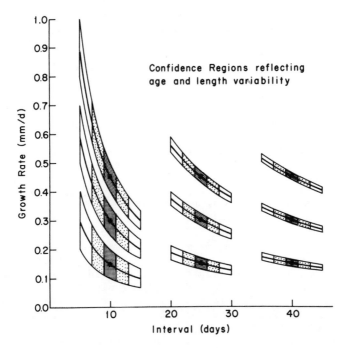

Figure 4. The influence of time interval and initial growth-rate estimate on approximate 95% Confidence Regions for growth rate is illustrated using three time intervals (10, 25, 40 d) and three growth rates (0.15, 0.30, 0.45 mm/d). Circular symbols indicate point estimates assuming no variability in age or length. Solid lines through these symbols indicate variation due to uncertainty in age estimate only. Vertical error indicates the effect of a ±0.5 mm 95% C.I. for length estimates. Shaded regions indicate the combined effect of a ±0.5 mm 95% C.I. for length estimates and uncertainty in age estimates: dark shaded area ±1 d 95% C.I., medium shaded area ±3 d 95% C.I., light area ±5 d 95% C.I.

confidence limits for growth rate ranging from 0.2 to 0.6 mm/d.

The effect of variability in age estimates decreases as the time interval increases; a ±1d 95% C.I. for age has the same effect on growth rate calculated over a 5 d interval as a ±5 d 95% C.I. for age has over a 25 d interval. Length 95% C.I.s of ±0.5 and 1.0 mm will cause confidence limits for a growth-rate point estimate of 0.3 mm/d at a time interval of 5 d to range from 0.2 to 0.4 and 0.1 to 0.5 mm/d, respectively. As the time interval increases, the effect of length variability also decreases; at a time interval of 10 d, the same confidence limits for length result in only half the uncertainty in growth rate. When the effects of variability in age and length estimates are combined, confidence limits for growth-rate estimates vary even more dramatically (Fig. 4).

Length estimates used in growth-rate calculations are frequently backcalculated from a fish size–otolith size regression. In such cases, variability in the length estimate should be determined using confidence intervals for an individual estimate from the regression. It is important that the data fit the linear model or have been transformed appropriately. Also, the fish size–otolith size relationship is not necessarily constant within a species. Significant differences in this relationship were observed for fed and starved sockeye salmon (Marshall and Parker 1982), bluegill Lepomis macrochirus populations from two different lakes (Taubert and Coble 1977) and groups of bloater fed different ration levels in the laboratory (Rice et al. 1985). Therefore, it is important to confirm that this relationship is constant, or to use the observed relationship for the population whose growth rates are being calculated.

Variability in length estimates backcalculated from otolith size can be substantial relative to fish size. Marshall and Parker (1982) reported 95% confidence intervals of ±3.4 to 4.7 mm for sockeye salmon ranging from 25 to 55 mm fork length, held under different feeding and temperature regimes. Analysis of data digitized from published figures of other fish size–otolith size relationships showed 95% confidence intervals ranging from ±2.5 to 3.8 mm for fish between 5 and 55 mm in length (Struhsaker and Uchiyamma 1976; Taubert and Coble 1977; Radtke and Dean 1982), 5–10x larger than those used in Figure 4.

Only the study by Wilson and Larkin (1982) has compared actual measurements of fish size with estimates backcalculated from otolith size. They recorded initial weights of sockeye salmon (range about 0.4 to 2.7 g) and then backcalculated estimates of those weights 2–4 weeks later using their observed fish weight–otolith radius relationship and assuming daily ring deposition. They stated that errors arising from this technique were on the order of 15%.

The examples in Figure 4 illustrate growth-rate variability in the simple case, such as average growth rate of an individual, where only one age estimate is involved and variability of only one length estimate is substantial. However, one of the more

appealing aspects of otolith analysis is the potential for
determining growth rate over a segment of an individual's life
history, for example from day 20 to day 30 for a fish 50-d old at
capture. In this situation, variability in two age and length
estimates is involved: ages and back-calculated lengths at the
beginning and end of the time interval. Growth-rate estimates in
such cases may be even more uncertain than those illustrated in
Figure 4.

CONCLUSION
        Information provided in this paper allows one to design
otolith validation experiments to achieve a specified power level
and to easily estimate the power of a regression after the
experiment. By calculating and reporting power for otolith-
analysis validation regressions the investigator can critically
evaluate the conclusion that observed rings represent daily
increments, and allow the reader to do the same.
        Traditionally, the emphasis in statistical analysis has been
on avoiding Type I error by choosing a relatively small α value.
However, in otolith analysis, where positive assertions are drawn
from negative statistical results, Type II error assumes much
greater importance. In this case, a larger α value of 0.1 or 0.15
may be justified to increase power and reduce the probability of
Type II error. Toft and Shea (1983) illustrate how the relative
costs of Type I and Type II error can be weighed in selecting an
appropriate   value.
        Clearly, variability in age and length estimates can have a
large impact on reliability of growth-rate estimates, especially
when calculated over short time intervals. A difference of 0.3
mm/d between two growth-rate estimates may be substantial when
reliability of growth-rate estimates is high, or inconsequential
when reliability is low. By considering the effects of
variability in age and length estimates on growth-rate
calculations, one can match the strength of conclusions to
reliability of the data.

ACKNOWLEDGMENTS
        Power calculatons presented in this paper would not have been
possible without the generous assistance of Dr. Norman Draper of
the Statistics Department, University of Wisconsin-Madison. Steve
Gutreuter, Richard Methot, Kenneth Carlander, James Kitchell and
Larry Crowder provided helpful comments on the manuscript.
Cheryle Hughes prepared the figures and Ann Hill typed the
manuscript. This work was funded by the University of Wisconsin
Sea Grant College Program under grants from the Office of Sea
Grant, National Oceanic and Atmospheric Administration, U.S.
Department of Commerce, and from the State of Wisconsin. Federal
Grant NA800AA-D, project R/LR-22.

REFERENCES

BARKMAN, R. C. 1978. The use of otolith growth rings to age young Atlantic silversides, Menidia menidia. Transactions of the American Fisheries Society 107:790-792.

CAMPANA, S. E. AND J. D. NEILSON. 1982. Daily growth increments in otoliths of starry flounder (Platichthys stellatus) and the influence of some environmental variables in their production. Canadian Journal of Fisheries and Aquatic Sciences 39:937-942.

DRAPER, N. R. AND H. SMITH. 1966. Applied regression analysis. John Wiley and Sons, New York, NY, USA.

GEFFEN, A. J. 1982. Otolith ring deposition in relation to growth rate in herring (Clupea harengus) and turbot (Scophthalmus maximus) larvae. Marine Biology 66:317-326.

MARSHALL, S. L. AND S. S. PARKER. 1982. Pattern identification in the microstructure of sockeye salmon (Oncorhynchus nerka) otoliths. Canadian Journal of Fisheries and Aquatic Sciences 39:542-547.

NEILSON, J. D. AND G. H. GEEN. 1982. Otoliths of chinook salmon (Oncorhynchus tshawytscha): daily growth increments and factors influencing their production. Canadian Journal of Fisheries and Aquatic Sciences 39:1340-1347.

RADTKE, R. L. AND J. M. DEAN. 1982. Increment formation in otoliths of embryos, larvae and juveniles of the mumichog, Fundulus heteroclitus. U. S. National Marine Fisheries Service Fishery Bulletin 80:201-215.

RICE, J. A., L. B. CROWDER AND F. P. BINKOWSKI. 1985. Evaluating otolith analysis for bloater Coregonus hoyi: do otoliths ring true? Transactions of the American Fisheries Society 114:532-539.

STRUHSAKER, P. AND J. H. UCHIYAMA. 1976. Age and growth of the nehu, Stolephorus purpureus (Pisces: Engraulidae), from the Hawaiian islands as indicated by daily growth increments of sagittae. U. S. National Marine Fisheries Service Fishery Bulletin 74:9-17.

TAUBERT, B. D. AND D. W. COBLE. 1977. Daily rings in otoliths of three species of Lepomis and Tilapia mossambica. Journal of the Fisheries Research Board of Canada 34:332-340.

TOFT, C. A. AND P. J. SHEA. 1983. Detecting community-wide patterns: estimating power strengthens statistical inference. The American Naturalist 122:618-625.

WETHERILL, G. B. 1966. Sequential methods in statistics. Chapman & Hall, New York, NY.

WILSON, K. H. AND P. A. LARKIN. 1980. Daily growth rings in the otoliths of juvenile sockeye salmon (Oncorhynchus nerka). Canadian Journal of Fisheries and Aquatic Sciences 37:1495-1498.

WILSON, K. H. AND P. A. LARKIN. 1982. Relationship between thickness of daily growth increments in sagittae and change in body weight of sockeye salmon (Oncorhynchus nerka) fry. Canadian Journal of Fisheries and Aquatic Sciences 39:1335-1339.

Lack of first-year annuli on scales: Frequency of occurrence and predictability in trout of the western United States

LEO D. LENTSCH
U.S. FISH AND WILDLIFE SERVICE
YELLOWSTONE TECHNICAL ASSISTANCE OFFICE
P.O. BOX 184
YELLOWSTONE NATIONAL PARK, WY 82190

JACK S. GRIFFITH
DEPARTMENT OF BIOLOGICAL SCIENCES
IDAHO STATE UNIVERSITY
POCATELLO, ID 83209

ABSTRACT
        Lack of first-year annuli on scales is a common occurrence for salmonids in the Intermountain West. We examined 24 salmonid populations divided among two forms (cutthroat trout Salmo clarki and rainbow-cutthroat hybrid trout Salmo gairdneri x S. clarki); of these, 21 contained individual fish that failed to form a first-year annulus. These fish were identified by determining the maximum number of circuli formed by young fish during the first growing season and counting the number of circuli to the first annual mark on scales of older fish. Number of circuli formed and percentage of fish not forming an annulus at the conclusion of the first growing season varied among populations from 0 to 15 and 0 to 100, respectively. This variability was largely explained by the number of degree-days to which fish were exposed: when this number was less than 720, no fish formed a first-year annulus; when it exceeded 1,500, all fish formed the annulus.

        Under certain naturally occurring environmental conditions, salmonids do not form annuli on their scales until the conclusion of their second growing season. This observation was reported more than 50 years ago by Curtis (1934) who noted that the scales of up to 50% of the golden trout Salmo aquabonita in Cottonwood Lakes, California, lacked a first-year annulus. Although this phenomenon was later reported for five other salmonids--cutthroat

trout <u>Salmo</u> <u>clarki</u>, rainbow trout <u>Salmo</u> <u>gairdneri</u>, brook trout
<u>Salvelinus</u> <u>fontinalis</u>, brown trout <u>Salmo</u> <u>trutta</u>, and Atlantic
salmon <u>Salmo</u> <u>salar</u> (Table 1)--its frequency of occurrence and
factors influencing its occurrence have not been documented.

In past studies, the absence of first-year annuli on scales
was determined by the examination of scales from young-of-the-year
(YOY) collected in late fall or yearlings collected in early
spring (Curtis 1934; Brown and Bailey 1952; Laakso and Cope 1956).
Formation of a first-year annulus requires that enough circuli be
laid down on the scale to allow the occurrence of dissimilarity in
the growth pattern. The number of circuli to the first annulus
has, therefore, been used as the criterion for differentiating
between scales that have and those that lack first-year annuli.
Once the criterion was established for a population, it was
commonly applied to all year-classes. Generally, populations that
showed a wide range in numbers of circuli to the first annulus
contained some individuals that failed to form a first-year
annulus. Scales with a first-year annulus had fewer circuli to
the first annual mark than did those that lacked this annulus
(Table 1).

In many fishery management situations, capture of YOY to
develop a criterion for distinguishing fish that lack a first-year
annulus is difficult and even impossible. Laakso (1955) used the
"number of circuli" criterion developed for one cutthroat trout
population (Yellowstone Lake, Wyoming), to evaluate 11 disjunct
populations of this species in Montana, Wyoming, and Utah for lack
of a first-year annulus. This practice generally continued for
age and growth determinations of cutthroat trout in Yellowstone
National Park through 1981. Because of the variability in growth
between populations (Carlander 1969), however, the validity of
this approach may be questionable.

Identification of factors that prevent young fish from
growing large enough to form a first-year annulus might provide a
means of determining if this phenomenon occurs in populations,
without sampling YOY. Laakso (1955) found that growth was faster
and the percentage of fish forming first-year annuli was higher in
eutrophic than in oligotrophic waters. However, he did not
identify specific environmental factors that caused first-year
annulus formation to be more frequent in some populations than in
others. Jensen and Johnsen (1982) documented the influence of a
single environmental variable--temperature--on the absence of
first-year annuli on scales. They showed that the percentage of
riverine YOY brown trout and Atlantic salmon forming scales was
correlated ($r = 0.75$ and $r = 0.81$, respectively) with the number
of degree-days from May through October.

We examined a number of lake- and stream-dwelling populations
of salmonids in the Intermountain West to (1) assess the frequency
of failure of first-year annulus formation, (2) evaluate the use
of counts of circuli to identify fish scales that lack first-year
annuli, and (3) evaluate the use of degree-day accumulation to
predict whether a salmonid population contains individuals with
scales that lack first-year annuli.

Table 1. Summary of information on lack of a first-year annulus in salmonids, 1934-1981.

Species	Location	Source	Without a first-year annulus			With a first-year annulus	
			N	Number of circuli[a]	% of fish	Number of circuli	% of fish
Cutthroat trout							
	Colorado						
	Forest Canyon Creek	Cope (1959)	--	> 0	--	--	0
	Trappers Lake	Drummond (1966)	--	--	--	--	--
	Idaho						
	Gold Creek	Averett (1962)	--	> 6	100	3-6	0
	Simmons Creek	Averett (1962)	--	> 6	100	3-6	0
	St. Joe River	Rankel (1971)	347	> 6	60	< 7	40
	Montana						
	Arrow Lake	Laakso (1955)	17	> 7	81	< 8	19
	Bear Creek	Laakso (1955)	29	> 7	69	< 8	31
	Fish Creek	Laakso (1955)	11	> 7	64	< 8	36
	Flathead Lake	Graham et al. (1980)	--	> 8	69	< 9	31
	Flat Top Lake	Laakso (1955)	7	> 7	63	< 8	37
	Howe Lake	Laakso (1955)	29	> 7	59	< 8	41
	Lower Two Medicine Lake	Laakso (1955)	7	> 7	57	< 8	43
	Red Eagle Lake	Laakso (1955)	7	> 7	29	< 8	71
	Utah						
	Logan River	Fleener (1952)	306	7-9	--	--	--
	Provo River	Laakso (1955)	20	> 7	50	< 8	50
	Scofield Reservoir	Laakso (1955)	11	> 7	42	< 8	58
	Strawberry Reservoir	Laakso (1955)	32	> 7	34	< 8	66
	Weber River	Laakso (1955)	9	> 7	44	< 8	56
	Wyoming						
	No Name Lakes	Robertson (1947)	139	6-11	67	2-4	33
	Yellowstone Lake	Brown and Bailey (1952)					
	Yellowstone Lake	Laakso (1955)	59	> 7	56	< 8	44
	Yellowstone Lake	Laakso and Cope (1956)					
Rainbow trout							
	West Virginia						
	Big Spring Creek	Surber (1937)	--	--	20-30	--	70-80
Golden trout							
	California						
	Cottonwood Lakes	Curtis (1934)	270	--	50-77	--	23-50
Brook trout							
	Montana						
	Round Lake	Domrose (1963)	41	8-12	40	4-7	60
	Timberline Lake	Domrose (1963)	82	8-12	16	4-7	84
Brown trout							
	Norway[b]	Jensen and Johnsen (1981)	--	--	2-50	--	50-98
Atlantic salmon							
	Norway[b]	Jensen and Johnsen (1981)	--	--	15-79	--	21-85

a) Number of circuli to first annulus on scales collected from the area just ventral and posterior to the dorsal fin origin.
b) Both studies in Norway were in Beiarelva and Saltdalselva Rivers and scales for the Atlantic salmon were collected from the caudal peduncle.

METHODS

Scales were collected from salmonids in 24 populations in Colorado, Idaho, and Yellowstone National Park (Table 2). The body location from which scales were removed was based on standardized procedures for age and growth determinations in trout (Bagenal and Tesch 1978; Jerald 1983; Lagler 1956), not from the area where the first scales appear (Brown and Bailey 1952). All scales, therefore, were removed from a position just ventral and posterior to the dorsal fin origin.

Total length (TL) of fish was measured to the nearest 1 mm. Scale examinations included determinations of the number of annuli and of the number of circuli from the focus to the first annulus. Annuli were identified by the presence of one or more discontinuous circuli between two continuous ones, circuli crossing other circuli, or two or more rows of closely spaced circuli followed by widely spaced circuli (Lagler 1952).

Water temperature data were obtained for four of the populations examined. Water temperature was measured with a constant recording thermograph in two rearing areas for trout--a cold-water inlet and a warmer outlet--in Emerald Lakes, Colorado, and in Sedge Creek, Yellowstone National Park. For Clear Creek, a tributary to Yellowstone Lake, Yellowstone National Park, maximum and minimum daily water temperatures from the start of spawning through the first emergence of the young were compiled for the 5-year period 1979-1983. In West Fork of Bannock Creek, Idaho, water temperature was measured periodically, about five times during the growing season.

Estimates of the total number of degree-days accumulated for the first growing season were determined for the period when eggs were spawned until the end of the growing season when average daily water temperature reached 0 C. Degree-days were calculated from the average temperature in centigrade for a 24-h period. One degree-day equaled the average daily temperature of 1 C.

RESULTS AND DISCUSSION

First-year Growth and Circuli Formation

Determination of the maximum number of circuli formed during the first growing season is probably the most reliable method of identifying fish that lack a first-year annulus. In 9 of the 24 populations sampled, YOY were captured and identification criteria developed (Table 2). The ranges in number of circuli to the first annulus for 1,059 Emerald Lake hybrid trout Salmo gairdneri x S. clarki were 2-23 for all adult fish, 3-23 for adult fish captured in the inlet, and 2-20 for adult fish captured in the outlet. A significant correlation existed between number of circuli to the scale margin and length of YOY ($r = 0.89$, $N = 37$). The relationship was

$$C1 = -4.57 + 0.188 \ TL1 \hspace{3cm} [1]$$

where C1 is the number of circuli formed on a scale during the first growing season and TL1 is fish length in millimeters at the end of the first growing season. The largest YOY fish captured was taken on 14 October. It was 68 mm long, and its scales had 8 circuli. Because we believe that a somewhat larger size could be attained in Emerald Lakes in the first growing season, we arbitrarily selected 80 mm as a maximum attainable TL1. Equation [1] predicts that the number of circuli to annulus I on scales of a fish that reached 80 mm TL in its first growing season should not exceed 10.

Mean numbers of circuli to the first and second annuli for 733 Emerald Lake fish with 10 or fewer circuli to the first annulus were 7 (SD = 1.9) and 19 (SD = 3.8), respectively. Thus, about 12 circuli were laid down on scales of these fish during their second growing season. Mean C1 for fish with more than 10 circuli to their first apparent annulus was 13.4 (SD = 2.6). Circuli to the first apparent annulus on fish with a C1 value above 10 were probably formed during the second growing season.

In eight stream-residing populations of cutthroat trout, the identification of fish lacking a first-year annulus was based on the maximum number of circuli formed on scales of YOY captured in late fall or early spring (Table 2). This number varied between populations. In three of the populations (Bear Creek, Sedge Creek, and Unnamed stream No. 5016-02), yearling fish captured in July had no annulus. We concluded that all members of these populations failed to form a first-year annulus and the identification criterion was set at C1 > 0. The criterion for identifying cutthroat trout in Idaho that lacked a first-year annulus ranged from > 4 to > 10. In West Fork of Bannock Creek, the mean TL of 22 YOY captured on 20 February 1984 was 118 mm. Scales on none of these fish had an annual mark, although the fish were large enough for one to form. The number of circuli to the scale margin averaged 12.1 and ranged from 10 to 15.

Lack of First-year Annuli

Failure to form a first-year annulus appears to be a common occurrence in salmonid populations in the Intermountain West. All individual fish formed a first-year annulus in only one of the populations where YOY were captured. By applying Laakso and Cope's (1956) Yellowstone Lake population criterion to the remaining 15 populations, we estimated that fish in 21 of the 24 populations sampled contained individuals that failed to form a first-year annulus. Furthermore, the percentage of fish not forming a first-year annulus varied within and between trout populations (Table 2).

The percentage of individual fish not forming a first-year annulus ranged from 0 to 100% for the 9 populations where YOY were sampled (Table 2) and from 0 to 50% for the 15 populations where Laakso and Cope's (1956) criterion was applied. Overall, an average of 46% of the individuals of all populations sampled lacked a first-year annulus.

Table 2. First-year annulus formation in 24 trout populations of the Intermountain West.

	All fish			Fish without a first-year annulus		
	Number of circuli[a]					
Species and location	Mean	Range	Sample size	Number of circuli[a]	Percentage of all fish	Determination method[b]
Cutthroat trout						
Idaho						
Gibson Jack	8.7	7-9	21	8-9	86	YOY
Hoodoo Creek	7.1	6-10	40	7-10	85	YOY
Mink Creek	11.1	10-13	24	10-13	100	YOY
Pack Creek	6.1	5-7	19	5-7	100	YOY
West Fork Bannock	12.1	10-15	22	--	0	YOY
Yellowstone National Park						
Alum Creek	4.3	2-6	4	--	0	YOY
Bacon Rind Creek	6.1	3-8	19	8	26	LC
Bear Creek	4.9	2-8	21	2-8	100	LC
Cascade Lake	6.0	3-13	45	8-13	13	LC
Fan Creek	6.4	5-8	11	>7	27	LC
Heart River	5.8	4-9	35	>7	17	LC
Lamar River	7.8	4-13	46	>7	50	LC
McBride Lake	6.9	2-12	47	>7	36	LC
Riddle Lake	7.6	5-14	23	>7	39	LC
Rose Creek	5.3	3-10	18	>7	11	LC
Sedge Creek						
(1982)	4.7	2-7	36	2-7	100	YOY
(1984)	5.9	4-8	7	4-8	100	YOY
Snake River	5.9	3-10	57	8-10	11	LC
Soda Butte Creek (lower)	7.1	4-12	45	8-12	29	LC
Soda Butte Creek (upper)	5.5	4-8	19	8	11	LC
Sylvan Lake	7.0	4-13	79	8-13	27	LC
Teepee Creek	7.2	4-13	6	8-13	33	LC
Thistle Creek	4.6	4-7	10	--	0	YOY
Unnamed stream No. 5016-02	5.0	3-10	17	3-10	100	LC
Rainbow-cutthroat hybrid trout						
Colorado						
Emerald Lakes	8.8	2-23	1,059	>10	37	YOY

a) Number of circuli to first annulus.
b) YOY = sampling of young-of-the-year; LC = criterion of Laakso and Cope (1956).

In Yellowstone Lake, the average percentage of fish lacking a first-year annulus (Table 3) has decreased from 51% in 1970 and 1974 to 22% in 1976-1984. This decrease was reflected in the percentage of fish within each year-class that had scales without a first-year annulus. Mean number of fish lacking a first-year annulus was 53% for the 1964-1970 year-classes and 23% for the 1971-1979 year-classes.

Table 3. Percentage of cutthroat trout collected in Yellowstone Lake lacking a first-year annulus arranged by sampling year and year class, 1970-1984.

| Year Class | a) Year of sampling | | | | | | | | | | | Mean |
	1984	1983	1982	1981	1980	1979	1978	1977	1976	1974	1970	
1983	0	--	--	--	--	--	--	--	--	--	--	0
1982	17	0	--	--	--	--	--	--	--	--	--	9
1981	13	5	0	--	--	--	--	--	--	--	--	6
1980	12	17	24	0	--	--	--	--	--	--	--	13
1979	11	14	22	18	0	--	--	--	--	--	--	13
1978	22	16	20	21	49	0	--	--	--	--	--	21
1979	0	30	12	19	25	28	0	--	--	--	--	16
1976	--	80	7	26	38	25	28	0	--	--	--	29
1975	--	100	33	40	52	34	18	30	0	--	--	39
1974	--	--	--	0	33	17	17	32	30	--	--	22
1973	--	--	--	--	0	57	16	24	19	11	--	21
1972	--	--	--	--	--	--	0	21	14	67	--	26
1971	--	--	--	--	--	--	--	0	13	36	--	16
1970	--	--	--	--	--	--	--	--	--	62	--	62
1969	--	--	--	--	--	--	--	--	--	86	0	43
1968	--	--	--	--	--	--	--	--	--	--	48	48
1967	--	--	--	--	--	--	--	--	--	--	33	33
1966	--	--	--	--	--	--	--	--	--	--	59	59
1965	--	--	--	--	--	--	--	--	--	--	65	65
1964	--	--	--	--	--	--	--	--	--	--	60	60
Mean	11	18	19	22	40	27	20	23	18	56	46	

a) No samples collected in 1971-1973 or 1975.

This decrease might be partly explained by a change in fishery management practices--switching from a minimum-size limit of 356 mm to a maximum-size limit of 330 mm. Laakso and Cope (1956) indicated that year-classes with a high percentage of fish that lack a first-year annulus developed in years when late spawners produced a large number of YOY. Data from Clear Creek indicated that larger fish spawn earlier than small ones (Jones et al. 1985). Establishment of a maximum-size limit of 330 mm in 1975 increased the proportion of larger fish in the population. The peak migration period was 2 weeks earlier in 1984 than during the 1950's. These early spawners were probably producing more progeny in the 1980's than they did in the 1960's. This increase in production of early progeny apparently resulted in the formation of a first-year annulus by a higher percentage of fish.

Water Temperature and First-year Annulus Formation
Duration of the embryonic period has been shown to be negatively correlated with number of degree-days (Bagenal and Braum 1978). Furthermore, Dwyer et al. (1981, 1983, 1984) demonstrated that growth rate of YOY in a hatchery was correlated with the rate of degree-day accumulation. Fewer degree-days were

required to produce growth in length of 1 cm within than outside
an optimal range of temperatures. For rainbow trout, the optimum
range was 10 to 16 C or 10 to 16 degree-days per day. Maximum
growth rate occurred at 13 C or 13 degree-days per day.

In Emerald Lakes, degree-day exposure rate in the outlet
(10.9) was within the optimal range for growth efficiency of
rainbow trout (Figure 1). Total number of degree-days in this
area was about 1,500 and all fish formed a first-year annulus. In
the inlet, where total number of degree-days was 555 (5.1
degree-days per day), no fish formed a first-year annulus (Figure
1). In the outlet, the 330 degree-days required for hatching
(Bagenal and Braum 1978, Shapovalov 1937, Snyder and Tanner 1960)
were accumulated between 13 June and 20 July. By 4 August, enough
degree-days accumulated (555) to allow YOY to emerge from spawning
gravels (Shapovalov 1937). This level of degree-day exposure was
not reached in the inlet until after 15 October.

Influences of water temperature on development and growth
rate explain some of the variability in first-year annulus
formation within and between cutthroat trout populations in
Yellowstone National Park. When water temperature was measured in
Sedge Creek in 1984, about 220 degree-days accumulated between ice
breakup and 15 July, and about 720 degree-days between peak
spawning (15 July) and the end of the growing season (15 October).
On average, 6.1 degree-days per day accumulated during the growing
season. In this population, fish exposed to 720 or fewer
degree-days failed to form a first-year annulus.

In Clear Creek, a tributary to Yellowstone Lake, spawning has

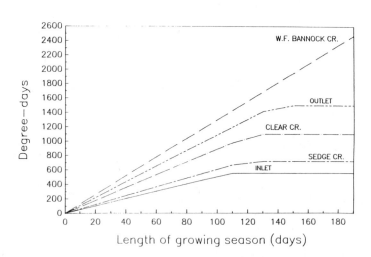

Figure 1. Comparison of the number of degree-days accumulated
    in the inlet and outlet of Emerald Lakes (1980), Colorado;
    Sedge Creek (1984), Yellowstone National Park; Clear Creek
    (1978-1983), Yellowstone National Park; and West Fork of
    Bannock Creek (1984), Idaho.

generally peaked during the last week in June and first week in
July (Jones et al. 1985). From 1978 to 1983, 610 degree-days
accumulated annually from the time the first spawners entered this
creek until the first YOY migrated out of it. On average, 8.9
degree-days per day accumulated in this stream. From the peak of
spawning to the end of the growing season, 1,100 degree-days
accumulated. This level of degree-day accumulation represents
about 74% of the 1,500 degree-days that was required for the
formation of a first-year annulus by all fish in Emerald Lakes.
This intermediate value of degree-day accumulation may explain, in
part, the apparent shift in the percentage of Yellowstone Lake
cutthroat trout forming a first-year annulus in response to a
change in a fishing regulation.

The influence of water temperature on formation of a
first-year annulus was also evident in West Fork of Bannock Creek.
This stream is a spring creek with a nearly constant water
temperature of 13 C. Total degree-day accumulation for the YOY
growing season in this stream probably exceeds 2,300 (13
degree-days per day). All of the scales collected from YOY in
this stream had formed enough circuli for growth pattern
dissimilation. The average number of circuli to the scale margin
was 12.1. About 148 degree-days accumulated for each circulus
formed on the scales of these fish while 135 degree-days
accumulated for each circulus formed on the scales of fish in the
outlet of Emerald Lakes.

Among salmonid populations in the Intermountain West, the
percentage of fish lacking first-year annuli on their scales
appears to be largely influenced by the number of degree-days
(Figure 2). When this number was 720 or fewer all fish lacked the
annulus, and all fish formed the annulus when the number was 1,500

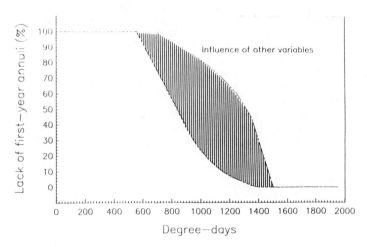

Figure 2. Theoretical relation between number of degree-days
and percentage of fish lacking a first-year annulus for
salmonid populations in the Intermountain West.

or greater. The percentage of fish not forming a first-year annulus in areas where 720 to 1,500 degree-days accumulate may not be a direct relationship (Figure 2), and between these two extremes, the percentage is probably influenced by other environmental variables. These factors may cause the frequency of failure of first-year annulus formation to fluctuate at any given level of degree-day accumulation. Jensen and Johnsen (1982) developed regression equations that described the relationship between degree-day accumulation and scale development in brown trout and Atlantic salmon. Their equation predicted that 1,300 degree-days were required for all YOY fish to form scales, and none of the YOY formed scales when 800 or fewer degree-days accumulated. Further, they reported that errors in age determination due to lack of a first-year annulus were minor when YOY were exposed to more than 1,100 degree-days. This close similarity in results suggests that degree-day accumulation can be a useful indicator in predicting lack of first-year annuli in salmonid populations.

We recommend that biologists conducting age and growth studies on salmonids in the Intermountain West make a determination as to whether the scales of fish they are studying form a first-year annulus. The best method to make this determination is to examine scales from YOY collected in late fall or from yearling fish collected in early spring. If YOY or yearling fish cannot be collected, then the number of degree-days accumulated during the growing season can be a useful indicator for the occurrence of this phenomenon.

Future studies should address an important question: does the location where a scale is collected on a trout influence whether or not it has a first-year annulus? We collected scales from the area ventral and posterior to the dorsal fin origin because this location is commonly used for age and growth studies of trout. However, this is not the location where scales first form on trout. Brown and Bailey (1952) showed that scales first form on cutthroat trout along the lateral line and caudal peduncle. This area developed scales when fish were 40 mm TL, while the area ventral to the dorsal fin was not scaled until fish were 60 mm TL. Scales collected from the caudal peduncle area, therefore, should better represent the first-seasons growth and might provide a more accurate method of determining age and growth characteristics of salmonids in the Intermountain West.

ACKNOWLEDGMENTS
        Research at Emerald Lakes, Colorado, was conducted through Colorado State University and funded by the Colorado Division of Wildlife. Support for analysis of populations in Yellowstone National Park and historical data on Yellowstone Lake were provided by the U.S. Fish and Wildlife Service and the National Park Service.

REFERENCES

AVERETT, R. C. 1962. Studies of two races of cutthroat trout in
    Northern Idaho. Idaho Department of Fish and Game.
    Dingell-Johnson completion report, F-47-R-1, Boise, Idaho,
    USA.
BAGENAL, T. B., and E. BRAUM. 1978. Eggs and early life history.
    Pages 165-201 in T. B. Bagenal ed. Methods for assessment of
    fish production in fresh waters. Blackwell Scientific
    Publications Ltd., Oxford, England.
BAGENAL, T. B., and F. W. TESCH. 1978. Age and growth. Pages
    84-100 in T. B. Bagenal ed. Methods for assessment of fish
    production in freshwaters. Blackwell Scientific Publications
    Ltd., Oxford, England.
BROWN, C. J. D., and J. BAILEY. 1952. Time and pattern of scale
    formation in Yellowstone cutthroat trout Salmo clarki.
    Transactions of American Microscopial Society 81:120-124.
CARLANDER, K. D. 1969. Handbook of freshwater fishery biology,
    volume 1. Iowa State University Press, Ames, Iowa, USA.
COPE, O. B. 1959. Rocky Mountain Sport Fishery Investigations.
    Pages 20-28 in United States Fish and Wildlife Service
    Circular 57.
CURTIS, B. 1934. The golden trout of Cottonwood Lakes (Salmo
    aquabonita Jordan). Transactions of the American Fisheries
    Society 64:259-265.
DOMROSE, R. J. 1963. Age and growth of brook trout, Salvelinus
    fontinalis, in Montana. Proceedings of the Montana Academy
    of Sciences 23:47-62.
DRUMMOND, R. A. 1966. Reproduction and harvest of cutthroat
    trout of Trapper's Lake, Colorado. Colorado Fisheries
    Research Division, Special Report 10. Denver, Colorado, USA.
DWYER, W. P., C. E. SMITH, and R. G. PIPER. 1981. Rainbow trout
    growth efficiency as affected by temperature. United States
    Fish and Wildlife Service, Bozeman Fish Cultural Center,
    Information Leaflet 18, Bozeman, Montana, USA.
DWYER, W. P., and C. E. SMITH. 1983. Brook trout growth
    efficiency as affected by temperature. Progressive
    Fish-Culturist 45:161-163.
DWYER, W. P., and R. G. PIPER. 1984. Atlantic salmon growth
    efficiency as affected by temperature. United States Fish
    and Wildlife Service, Bozeman Fish Cultural Center,
    Information Leaflet 30, Bozeman, Montana, USA.
FLEENER, G. C. 1952. Life history of cutthroat trout in the
    Logan River, Utah. Transactions of the American Fisheries
    Society 81:235-248.
GRAHAM, P. J., D. READ, S. LEATHE, J. MILLER, and K. PRATT. 1980.
    Flathead River Basin fishery study. Montana Department of
    Fish, Wildlife, and Parks. Kalispell, Montana, USA.
JENSEN, A. J., and B. O. JOHNSEN. 1982. Difficulties in aging
    Atlantic salmon (Salmo salar) and brown trout (Salmo trutta)
    from cold rivers due to lack of scales as yearlings.

Canadian Journal of Fisheries and Aquatic Sciences
39:321-325.

JERALD, A. Jr. 1983. Age determination. Pages 301-324 in L. A.
Nielsen and D. L. Johnson ed. Fisheries techniques.
Southern Printing Company Inc., Blacksburg, Virginia, USA.

JONES, R. D., D. CARTY, R. E. GRESSWELL, and L. D. LENTSCH. 1985.
Annual project report, Fishery and Aquatic Management
Program, calendar year 1984. United States Fish and Wildlife
Service Mimeo. Yellowstone National Park, Wyoming, USA.

LAAKSO, M. 1955. Variability in scales of cutthroat trout in
mountain lakes. Proceedings of the Utah Academy of Science,
Arts, and Letters 32:81-87.

LAAKSO, M., and O. B. COPE. 1956. Age determination in the
Yellowstone cutthroat trout by the scale method. Journal of
Wildlife Management 20:138-153.

LAGLER, K. F. 1956. Freshwater fishery biology, 2nd edition.
Wm. C. Brown Co., Dubuque, Iowa, USA.

RANKEL, G. 1971. Life history of St. Joe cutthroat trout. Idaho
Department of Fish and Game. Dingell-Johnson completion
report F-60-R, Boise, Idaho, USA.

ROBERTSON, O. H. 1947. An ecological study of two high mountain
lakes in the Wind River Range, Wyoming. Ecology 28:87-112.

SHAPOVALOV, L. 1937. Experiments in hatching steelhead trout
eggs in gravel. California Fish and Game 23:208-214.

SNYDER, G. R., and H. A. TANNER. 1960. Cutthroat trout
reproduction in the inlet to Trappers Lake. Colorado
Department of Game and Fish, Fish Technical Bulletin 7,
Denver, Colorado, USA.

Sources of back-calculation error in estimating growth of lake whitefish[1]

MARTIN A. SMALE[2] AND
WILLIAM W. TAYLOR
DEPARTMENT OF FISHERIES AND WILDLIFE
MICHIGAN STATE UNIVERSITY
EAST LANSING, MI 48824

ABSTRACT
    Scale radius measurements from populations of Northern Lake Michigan whitefish Coregonus clupeaformis did not vary between populations, between sexes, or with weight or lateral-line scale count. Variance of radius measurements between individuals of the same length was 3 to 4 times larger than variance between replicate scales from the same individuals. Scale radius variance increased proportionally with increasing length. Distributions of scale radii at discrete lengths and of lengths at discrete radii were most likely bivariate normal.
    A plot of mean radii calculated for consecutive length increments was linear over the full-sampled range. Regression line-fitting techniques consistently deviated from the line through these means with the degree of deviation influenced by the range, span and balance of data included in the regressions. Sample truncation and an undefinable dependent variable were considered as sources of regression error.
    Statistical properties of scale radius and length observations were predictable from a hypothesis that the rate of scale growth with respect to growth in length is constant for individual scales, but rate constants vary within populations of scales and fish. A method for distinguishing Lee's phenomenon due to back-calculation error from true Lee's phenomenon was suggested as a test for back-calculation error.

[1]Agricultural Experiment Station Journal Article Number 11567.

[2]Present address: School of Forestry, Fisheries and Wildlife, University of Missouri-Columbia, 112 Stephens Hall, Columbia, Missouri 65211.

The Age and Growth of Fish, edited by Robert C. Summerfelt and Gordon E. Hall © 1987 The Iowa State University Press, Ames, Iowa 50010.

Measurements of the spacing between annual marks on body structures has been widely used to backcalculate the growth history of fishes.   Information about growth differences between individuals, age-groups, years and year-classes can often be obtained simultaneously from a single sample of fish.   In addition, back calculation has been used in studies of growing season and production (Gerking 1962) and of size-selective mortality and sampling bias (Ricker 1969).   Although back calculation is a useful and powerful technique, confidence in results can be inhibited by uncertainty about the procedures used and by difficulty in testing results.

In initial attempts to describe growth in stocks of Northern Lake Michigan whitefish Coregonus clupeaformis by back calculation we encountered a problem.   Enough variety exists among details of recommended back calculation procedures (Nikolsky 1963; Hile 1970; Bagenal and Tesch 1978; Carlander 1981) that selecting an ideal procedure involves uncertainty.   In particular, we often found large inconsistencies among different procedures for estimating parameters of slope and intercept for the relationship between scale size and length.   Given different methods and different results from initial trials, we found it impossible to justify one method against any other with confidence.

Two general steps are common to all back-calculation methods. In the first step, statistical line-fitting or curve-fitting procedures are used to derive an algebraic description of the relationship between scale size and length measurements.   The second step converts annular scale-radius measurements into lengths using parameters estimated in the first step.   Direct proportion corrected for intercept (Lee 1920) is the most commonly employed conversion procedure, and Carlander has reviewed this method in comparison to alternate procedures.   An accurate description of the scale-to-length relationship is a prerequisite for reliable back calculation with any conversion procedure, and it is this first step which is the primary focus of this investigation.

METHODS

Most whitefish were sampled from the Northern Lake Michigan trap-net fishery (432 mm minimum size limit); additional samples were obtained from trawls operated by the Michigan Department of Natural Resources and the United States Fish and Wildlife Service.   Total length and weight were measured and a patch of scales taken from midway between the center of the dorsal fin and the lateral line.   Three symmetrical scales were selected for reading and cleaned.   Anterior scale radius and annular radii were measured at 22X magnification on a microfiche reader.

Scale radius and length measurements were treated as interval data by tabulating lengths at each 5 mm scale radius increment and scale radii at each 10 mm length increment.   For each variable and each increment, estimates of the mean, median, variance and coefficient of variation were made.   Frequency histograms were plotted for several increments of both variables and distributions tested for normality with the Kolmogorov-Smirnov one-sample test (Siegel 1956).

Comparisons of mean radii at length and variance were made among the study populations and between sexes. For fish of similar length, scale radius was plotted against weight and against lateral line scale count. Variance in scale radius at several lengths was partitioned into between-scale and between-individual fish factors by treating the three scales measured from each of 15 fish as replicate measurements.

Parameters estimated by several common line-fitting procedures were compared using identical data for each procedure. Comparisons were made between a regression of mean radii at consecutive lengths; least squares linear regressions (LSR) using both length and scale radius as the dependent variable; geometric mean functional regressions (GMR); least squares regressions following natural log transformation of the dependent variable; and lines fitted by eye. Calculations were repeated using the full range of lengths sampled (130 mm - 750 mm), the lower half of the range only, and the upper half only.

RESULTS

No differences in mean scale radius at length or variance were found either between sexes or between whitefish from the three principal study populations. Therefore, calculations were made with pooled data to increase overall sample size.

No evidence was found to suggest anterior scale radius changed with weight for fish of the same length. A possible trend of decreasing scale radius with increasing numbers of lateral line scales was suggested by plots of these variables. But the trend was inconsistent between individuals and not significant with a small sample size (N = 47).

Both between-scale and between-fish variance in scale radius were significant factors (F test, $P < 0.01$) at all lengths tested. Mean square error for between-fish variance was 3 to 4 times larger than for between-scale variance. Differences in scale size between relatively coarse-scaled and fine-scaled individuals of similar length were often apparent to the eye, while differences between neighboring scales on the same fish were not. Averaging of measurements from the three scales was used to reduce between-scale variance, but the larger nuisance factor, between-fish variance, remained.

A plot of variance in scale radius calculated for each consecutive length increment (Figure 1) showed variance increased geometrically with increasing length. Standard deviations increased arithmetically with length but coefficients of variation did not change over the range of sampled lengths.

Frequency distributions of scale radii at length were mildly, but consistently, skewed towards smaller scales at all lengths examined. Mean radii were typically 2 to 3 mm larger

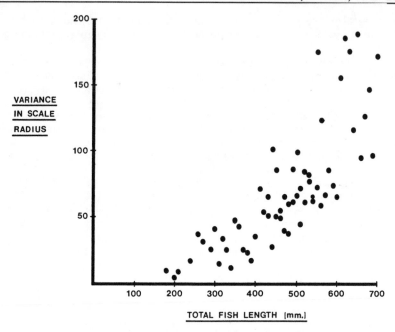

Figure 1.   Estimated variance of scale radii at consecutive
            10 mm increments of length. Sample sizes vary from 11 to
            47 fish per increment.

than the median.   Significant departures from normality
(P < 0.05) were not detectable with the relatively conservative
Kolmogorov-Smirnov test, unless 200 or more observations per
increment were available.   Standardizing and pooling neighboring
length increments were done to increase the sample size used to
test normality of the distributions.   On the other hand,
frequency distributions of lengths at a particular scale radius
were not consistent.   For radius increments near the middle of
the range, length distributions were mildly skewed towards
smaller fish.   For the larger scale radii, skewing towards small
fish was much stronger, while for smaller scales distributions
were skewed in the opposite direction.   Frequency distributions
of lengths at a particular scale radius were dependent on the
overall distributions of lengths within either the sample or the
population.   Any increase or decrease in the numbers of small
fish or large fish included in the total sample was reflected in
the distributions of length, particularly with scale sizes near
either end of the range.
        Plots of mean scale radii at consecutive lengths and of mean
lengths at consecutive radii were distinctly different when calculated from the same data (Figure 2).   Both series of mean values

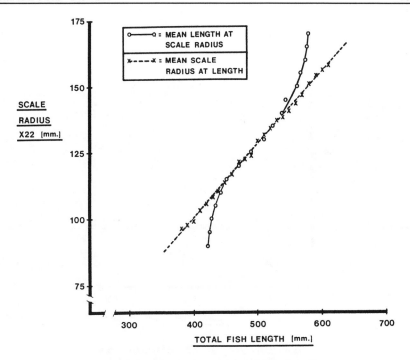

Figure 2. Estimated means of scale radii length and of
lengths at scale radius using 10 mm length increments and
5 mm scale radius increments.

were similar near the middle of the range. But mean radii at
length followed a consistent linear trend at all lengths, and mean
lengths at radius deviated sharply away from the linear trend at
both ends of the range. Further manipulation of the total sample
that extended the span of included lengths in either direction
simply shifted this curvature further along the line, but did not
eliminate it.

The function of an LSR line is to estimate the expected mean
value of the dependent variable at a given value of the independent variable. We bypassed the regression procedure and estimated
mean scale radii at length for consecutive increments over almost
the full range of lengths sampled (Figure 3). Stein's two-stage
sampling formula (Steel and Torrie 1980) was used to stratify the
sample, so that sample size in each increment increased with length
in proportion to increasing variance. Thus, each mean was estimated with similar precision, and problems with unbalanced data
were reduced. A regression through these means was calculated to
further smooth remaining variability. For reasons considered in
the discussion, the slope and intercept parameters from this
regression of means were judged to be the best possible estimates

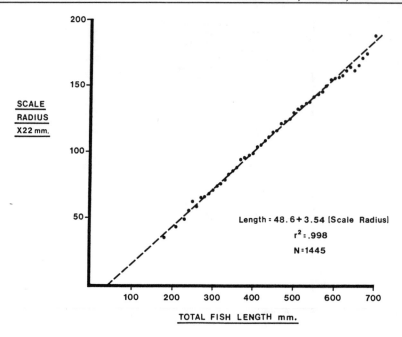

Figure 3.   Mean scale radii at 10 mm length increments for
the range of lengths from 130 to 620 mm.

from our data.   Regression of means was the only procedure tested
which was robust with respect to incompleteness of the sampled
range.   Regressions of shorter segments of the range extrapolated
cleanly through the remainder of the range as long as spans of
over 150 mm in length were used.

In comparison, none of the regression procedures which used
unaveraged data yielded similar results.   LSR following natural
log transformation of the dependent variable consistently produced the greatest error in the length-axis intercept estimates.
For non-transformed regressions, LSR which treated length as the
dependent variable consistently resulted in overestimation of the
intercept.   Treatment of length as the dependent variable has
been the most common and recommended procedure for estimating
scale-to-length parameters (Whitney and Carlander 1956), but we
found it to be one of the least accurate procedures.   LSR which
treated scale radius as the dependent variable underestimated the
intercept and GMR overestimated the intercept, but error with
these two procedures was less than with length designated as the
dependent variable.

Manipulation of the range and span of data included in the
regressions influenced the accuracy of regression parameters.
For either LSR and for the GMR, error in estimating intercept was
smallest when only the smaller half of the range was included.

Parameter error increased when the full range was included in the regressions and increased further when only the upper half of the range was used. Adding larger fish to the regression sampling, with their larger associated variance in radius, typically increased the divergence between the paired LSR lines. However, the paired lines did not diverge symmetrically away from the line fitted through the mean radii. Regressions which treated scale radius as the dependent variable were always closer to the line of means, and GMR lines were approximately halfway between the paired LSR lines.

In three trials in which lines were fitted by eye to a typical scatter plot of scale and length observations, results were closer to the line of means than any regression estimate made with similar data. We did not test whether lines fitted by eye varied with skill and mood of the participant, or with dimensions of the scatter plot. We suspect the human eye can be a fine discriminator of relationships but we hesitate to recommend by-eye fitting as a standard practice. We do, however, suggest comparing scatter plots of the data with calculated lines and curves. If the calculated line does not look right, we are not convinced the calculated relationship would be superior to a line fitted by eye.

DISCUSSION

The dominant source of error in estimating parameters of the scale-to-length relationship was an effect described as sample truncation (Cohen 1950). Typically, the range of lengths sampled from a population of fish is restricted at either end. At the lower end, limits to the gear efficiency restrict the lengths sampled; at the upper end, the cumulative effects of mortality place a limit on the range. But sampling of scale radii is not truncated independently of length. At any given length, all scale sizes found on fish of that length are available to the gear, and we see no reason to suspect that coarse-scaled and fine-scaled individuals die at different rates. Although sample truncation is usually not as abrupt as shown in Figure 4, it is the cause of the curvature found in the plot of mean lengths at scale radius. Truncation excludes both large fish with relatively fine scales and small fish with relatively coarse scales from the sample. The effect of truncation is to shorten and skew the distributions of lengths-at-scale sizes near either extreme, and bias estimates of the mean.

Sample truncation also influences parameters of the regression. If sampling of the dependent variable is truncated and sampling of the independent variable is not, the regression slope becomes steeper than would be expected in the absence of truncation. Ideally, the truncated variable (length in this case) is the proper choice for the independent variable for LSR calculations. We found that treating length as the independent rather than as the dependent variable reduced the degree of parameteri-

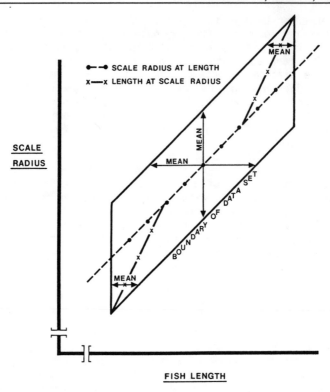

Figure 4.   The effect of sample truncation with respect to
length on estimates of the bivariate means.

zation error but did not eliminate it.   We then considered other
sources of error.

    We found variance of scale radii was distinctly non-homogen-
ous over the sampled range of lengths, and distributions of each
variable at fixed values of the other were not strictly normal.   A
likely distribution of the data is the bivariate normal distribution (Figure 5), in which observations are distributed symmetrically with respect to the underlying central trend, but are not
symmetrical at fixed values of either variable.   Direct testing
for bivariate normalcy was thwarted by the effects of sample
truncation on the distributions.   But the asymmetry found in the
histograms of both variables near the middle of the range is
consistent with bivariate normalcy.

    Although variance which increases geometrically with the size
of the fish and non-normality of distributions are a nuisance as
far as LSR is concerned, we also found that these features provided information about the underlying nature of the relationship
between scale size and length.   For instance, the observation that

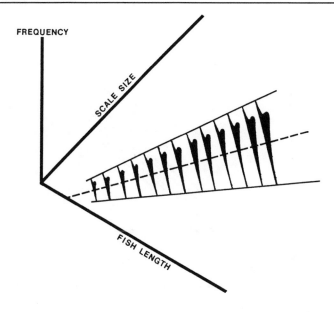

Figure 5. An example of the bivariate normal distribution,
which is symmetrical to the central trend line.

coefficients of variation were constant at all lengths indicates
that as fish grow the degree of deviation between fine- and
coarse-scaled fish increases in proportion to the increase in
length. This type of trend in variance, along with bivariate
normalcy, can be symptomatic of a type of relationship in which
variables co-vary (Kendall and Stuart 1963). If so, the relationship between variables is best described as interdependence rather
than dependence and independence, and this is a type of relationship for which LSR was not intended.

We were able to predict statistical features of the scale-
to-length relationship from a simple conceptual model about that
relationship. We began by hypothesizing what the relationship
would look like for a single scale on the same individual measured at several intervals over its growth history. A scale
which forms at some initial fish length and then grows at a
constant rate of growth with respect to the rate of growth in
length of the fish will follow a linear relationship. Scale
growth and growth in length will be synchronous, and the scale
will cover the same fraction of the fish's length throughout its
life. Given a population of fish, we would expect them to share
a similar initial length, but the rate constants which describe
the rate of scale growth with respect to growth in length would
vary amongst individuals and be normally distributed. Under such
conditions, lifetime measurements on scale size vs. length for

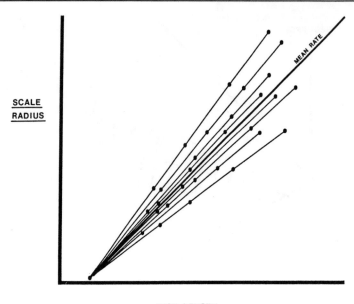

SCALE
RADIUS

FISH LENGTH

Figure 6.   A hypothesized scale-to-length relationship for a
sample of individuals whose scale growth relative to
length is constant, but with varying rates between individuals.

individuals in the populations would diverge outwards from the
common initial value (Figure 6), but each individual relationship
would be linear. Most relationships would cluster around a mean
rate. Essentially, the hypothesis suggests that neither scale
size nor length are dependent on the other. Instead, the hypothesis describes a relationship between variables in which each
is mutually interdependent on the underlying physiological processes which govern growth, and proportionality of growth, of the
organism as a whole.

   In most studies, scale and length data consists only of once
per lifetime observations of each fish. Without consecutive measurements from a sample of individuals, we were unable to directly
test the hypothesis. However, a simple indirect test was feasible
by simulating the relationship. We calculated values of X and Y
for a simulated population in which the largest value of X at
which Y equalled zero was the same for all individuals, the rate
of change in Y with X was constant for each individual, and rate
constants varied but were normally distributed amongst individuals. For this simulation, we found that variance of Y increased
geometrically with X and the joint distribution of observations
was bivariate normal. The ability to predict statistical features

found in the observed scale-to-length relationship from a simulation derived from a hypothesis about the relationship suggests there is some truth to the hypothesis.

In all likelihood, the rate-constant hypothesis is also an oversimplification of some details of the scale-to-length relationship. For instance, scale growth and growth in length may not be entirely synchronous throughout the growing season. Individual relationships would then be rippled rather than strictly linear, but the average relationship through time could still be described by a line. It is also unlikely that scale growth during the early growth history is typical of that in the remainder of life. Scales do not overlap when first formed (Van Oosten 1929), and a period of accelerated scale growth with respect to length is expected. Hogman (1970) suggested the transition from the rapid, early scale growth stanza to the normal stanza corresponded to the formation of the first circulus for two species of Coregonus. It is entirely possible that extrapolation of the scale-to-length relationship back to the length axis, such that scale size equals zero at some positive length, represents a fiction of convenience. To be strictly correct, back-calculation may require subtracting both a scale size correction and a length correction before estimating length in proportion to annular radii.

In any case, the objective for our scale-to-length hypothesis is to provide a biological understanding of the relationship so that parameterization is not an abstract, line-fitting exercise. Whether biological issues such as seasonal non-linearity or early stanzas in scale growth relative to length are unnecessarily fine details, or are detectable sources of error in back calculation, will require further study. We found two statistical problems influenced the accuracy of estimated parameters of the relationship: sample truncation and an undefinable dependent variable. Ricker (1984) has recommended the use of the GMR in part because it does not require designation of a dependent variable. For regression analysis, we cannot support designating either variable as dependent on either statistical or biological grounds. But parameters of the GMR are calculated from parameters of the LSR. Since sample truncation influences the slope and intercept of the LSR, we would expect parameters of the GMR to be influenced as well. Unless averaging of scale sizes at length is undertaken to minimize variance, we would expect any regression procedure to contain some parameterization error. If variance in scale size relative to length is less than encountered in this study, regression error may be negligible.

If intercept-corrected, proportional back calculation is used to calculate length from annual radii, it is possible to test for intercept error using Lee's phenomenon (LP). If false LP is defined as an apparent decline in length-at-age with increasing age of capture, which is due entirely to back calculation error, it can be distinguished from true LP which is due to either size-selective sampling or mortality. If back calculation is performed with an intercept which is either too large or too small (Figure

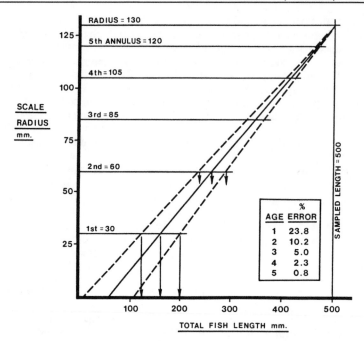

Figure 7.   The percentage of back-calculation error at each
age resulting from an overestimated and underestimated
intercept by 50 mm.

7), the percent error in the length estimate will increase with
the age-span of the back calculation. This is one possible source
of LP (or the reverse effect) in the estimates (Duncan 1980). If
sampling selects the larger members of a cohort, or if the younger
members die at an earlier age, true LP will be present in estimates of mean length at age. But larger than average members of a
cohort will also possess larger than average scales for their age.
Therefore sampling or mortality which selects for fish size will
also select for scale size. If true LP is present, mean annular
scale radius calculated separately for each age and age at capture
will decline with increasing age at capture. False LP, on the
other hand, is an artifact of back-calculation error and will not
be present in the scale size measurements. If the intercept is
estimated correctly, the degree of change in estimated mean length
at age with age at capture should correspond to the degree of
change in mean annular radius across the same ages. For example,
if mean annular radius of the third annulus declines by 3 percent
between age at capture 3 and age 4, but mean back-calculated
length increases by 2 percent for the same fish, an overestimated
intercept is present. Linearity of the scale-to-length relationship is necessary for this test of back-calculation error.

ACKNOWLEDGMENTS
    This research was sponsored by the Michigan Sea Grant College
Program, Grant number NA-80AA-D-00072 and NA-85AA-D-SG045, Project
number R/GLF-16 from the National Sea Grant College Program,
National Oceanic and Atmospheric Administration, U.S. Department
of Commerce and funds from the State of Michigan.   The U.S.
Government is authorized to produce and distribute reprints for
governmental purposes notwithstanding any copyright notation
appearing hereon.
    Special appreciation is due to members of the Michigan
Fish Producer's Association, the Michigan Department of Natural
Resources and the U.S. Fish and Wildlife Service for sampling
assistance.

REFERENCES
BAGENAL, T. B. AND F. W. TESCH.   1978.   Age and growth.   Pages
    101-136 in T. B. Bagenal, editor, Methods for assessment of
    fish production in fresh waters, third edition.   Blackwell
    Scientific Publications, Oxford, England.
CARLANDER, K. D.   1981.   Caution on the use of the regression
    method of back calculating lengths from scale measurements.
    Fisheries 6(1):2-4.
COHEN, A. C.   1950.   Estimating the mean and variance of normal
    populations from singly truncated and doubly truncated samples.
    Annals of Mathematical Statistics 21:557-569.
DUNCAN, K. W.   1980.   On the back calculation of fish lengths;
    modifications and extensions to the Fraser-Lee equation.
    Journal of Fisheries Biology 16:725-730.
GERKING, S. D.   1962.   Production and food utilization in a
    population of bluegill sunfish.   Ecological Monographs 32.7.
HILE, R.   1970.   Body-scale relation and calculation of growth in
    fishes.   Transactions of the American Fisheries Society 99:468-
    474.
HOGMAN, W. J.   1970.   Early scale development on the Great Lakes
    coregonids Coregonus artedii and C. kiyi.   Pages 429-436 in C.
    C. Lindsey and C. S. Woods, editors, Biology of coregonid
    fishes.   University of Manitoba Press, Canada.
KENDALL, M. G. AND A. STUART.   1963.   The advanced theory of
    statistics, second edition, Volume I.   Hafner, New York.
LEE, R. M.   1920.   A review of the methods of age and growth
    determination by means of scales.   Fishery Investigations
    Series II, Marine Fisheries of Great Britain, Ministry of
    Agriculture, Fisheries and Food 4, 2.
NIKOLSKY, G. V.   1963.   The ecology of fishes.   Academic Press,
    New York.
RICKER, W. E.   1969.   Effects of size-selective mortality and
    sampling bias on estimates of growth, mortality, production and
    yield.   Journal of the Fisheries Research Board of Canada
    26:479-541.

RICKER, W. E.   1984.   Computation and uses of central trend lines.
    Canadian Journal of Zoology 62:1897-1905.
SIEGEL, S.   1956.   Non-parametric statistics for the behavioral
    sciences.   McGraw-Hill, New York.
STEEL, R. G. D. AND J. H. TORRIE.   1980.   Principles and
    procedures of statistics, second edition.   McGraw-Hill, New
    York.
VAN OOSTEN, J.   1929.   Life history of the lake herring
    (Leucichthys artedii Le Sueur) of Lake Huron as revealed by its
    scales with a critique of the scale method.   Bulletin of the
    United States Bureau of Fisheries 44:265-448.
WHITNEY, R. R. AND K. D. CARLANDER.   1956.   Interpretation of the
    body-scale regression for computing body length of fish.
    Journal of Wildlife Management 20:21-27.

Discrepancies between ages determined from scales and otoliths for alewives from the Great Lakes[1]

ROBERT O'GORMAN
U.S. FISH AND WILDLIFE SERVICE
OSWEGO BIOLOGICAL STATION
17 LAKE STREET
OSWEGO, NY 13126

D. HUGH BARWICK[2] AND
CHARLES A. BOWEN
U.S. FISH AND WILDLIFE SERVICE
GREAT LAKES FISHERY LABORATORY
1451 GREEN ROAD
ANN ARBOR, MI 48105

ABSTRACT

Discrepancies between ages determined from otoliths and those determined from scales were common and, sometimes, quite large in alewives Alosa pseudoharengus collected in fall 1983 from Lakes Ontario, Huron, and Michigan. Among fish with "otolith ages" of 4 or more, the percentages having identical "scale ages" were 1% in Lake Ontario, 35% in Lake Huron, and 56% in Lake Michigan. Among alewives with different otolith and scale ages, the percentages with discrepancies of 3 years or more were 51% in Lake Ontario, 23% in Lake Huron, and 6% in Lake Michigan. Among the three populations, variation in the magnitude of age disagreements were perhaps due to the different mortality rates in each lake, whereas variation in the frequency of disagreements appeared to be due to different ratios of food to fish in each lake. Previously reported age compositions and rates of mortality and production for Great Lakes alewives derived from scales have probably been inaccurate to various degrees.

Ages of alewives Alosa pseudohargenus in the Great Lakes have historically been determined from scales (Pritchard 1929; Brown

---

[1]Contribution 656 of the Great Lakes Fishery Laboratory, United States Fish and Wildlife Service, Ann Arbor, Michigan 48105.

[2]Present address: Production Environmental Services, Duke Power Company, Route 4, Box 531, Huntersville, North Carolina 28078.

1972; Argyle 1982). However, O'Gorman and Schneider (1986) recently discovered that large, and possibly intermediate-sized, alewives in Lake Ontario did not form an annulus on their scales each year, and thus many were older than indicated by year-marks on their scales. Consistent underaging causes overestimates of growth, recruitment, mortality, and production, and could result in management procedures that unknowingly lead to overexploitation (Beamish and McFarlane 1983).

An alternative to the traditional scale method of aging was clearly needed for alewives in Lake Ontario and perhaps for the other Great Lakes as well. Otoliths seemed to offer potential as an alternative because accuracy of aging by this structure was validated for at least the younger age-groups of another clupeid, the Atlantic herring Clupea harengus harengus (Watson 1964), and also because annuli are often formed on otoliths several years longer than on scales from the same fish (Erikson 1979; Sikstrom 1963). Therefore, we compared the ages of alewives from Lakes Ontario, Huron, and Michigan, determined from otoliths versus scales.

METHODS

Alewives were collected with bottom trawls between mid-October and mid-November 1983 in southern Lake Ontario, from Olcott east to Mexico Bay; in western Lake Huron from Detour south to Harbor Beach; and in eastern Lake Michigan from Ludington south to Benton Harbor. Scales and sagittal otoliths were usually removed from fresh specimens, but some were obtained from alewives that had been frozen for several months. The 296 alewives used in this experiment were measured (total length, in millimeters), and 285 were weighed to the nearest 1.0 g (Table 1).

We removed scales from an area above the lateral line but below the base of the dorsal fin, and either mounted them between glass slides or made impressions of them on soft plastic. Annuli

Table 1. Percentages of alewives in different length intervals in which scale ages differed from otolith ages, Lakes Ontario, Huron, and Michigan, fall 1983.

Length interval (mm)	Lake Ontario			Lake Huron			Lake Michigan		
	Fish aged (N)	Mean weight (g)	Discrepancies (%)	Fish aged (N)	Mean weight (g)	Discrepancies (%)	Fish aged (N)	Mean weight (g)	Discrepancies (%)
50–120	11	–	0	1	13	0	5	10	0
120–129	3	14	0	5	14	20	3	14	0
130–139	5	19	0	5	18	0	1	24	0
140–149	6	23	0	5	25	20	5	27	0
150–159	10	27	20	10	30	20	10	32	10
160–169	15	29	87	15	34	20	15	38	13
170–179	17	34	94	15	41	47	15	48	20
180–189	16	37	100	15	47	47	15	55	20
190–199	16	43	100	15	53	73	4	60	75
200–209	5	45	100	12	63	75	5	71	40
210–229	5	64	100	5	69	40	6	79	33
Total	109		67	103		42	84		19

on scales, identified by features described by Rothschild (1963), were counted with the aid of a commercial scale projector at a magnification of 41X. We dried the otoliths and mounted them in depressions in black plastic trays, distal (concave) side up, in a histological medium; this technique was described by Libby (1982), who modified it from Watson (1965).

Annuli on otoliths were identified and counted using reflected light under a binocular microscope at magnifications ranging from 35X to 210X. The winter hyaline (transparent) zones were used as annular marks as described for herring otoliths by Watson (1964) and Hourston (1968). Two of us counted the annuli on each scale and otolith independently; when the counts differed for an individual scale or otolith, the third person independently counted the annuli on the structure in question and the value agreed upon by two of the three readers was accepted. The terms "scale ages" and "otolith ages" as used here refer to the number of annuli on the respective structure.

RESULTS

Ages of individual alewives determined from scales often differed from those determined from otoliths (Tables 1 and 2). The

Table 2. Ages of alewives from Lakes Ontario, Huron, and Michigan in fall 1983, as determined from otoliths and scales. Underlined are the numbers of alewives whose otolith age and scale age were identical; to the right of them are those whose otolith age was greater than their scale age; and to the left those whose otolith age was less than their scale age.

Scale age	Otolith age											
	0	1	2	3	4	5	6	7	8	9	10	11
Lake Ontario												
0	12											
1		12										
2			1		4	5	6					
3				10	10	14	13	1				
4					1	2	4		2	1	1	1
5							2		1			1
6										1	4	
Lake Huron												
0	2	1										
1		13										
2		3	16	2								
3			4	11	5	2	1					
4					12	4	3	1	1			
5					1	4	2	5	4			
6							2	1		1	2	
Lake Michigan												
0	6											
1		15										
2			1									
3				26	4	1						
4					13	3	3	1				
5						6	4					
6												
7							1					

discrepancies were almost entirely among fish longer than 149 mm, their frequency increased with fish length, and they were much more common and greater in fish from Lake Ontario than in those from Lakes Huron and Michigan (Fig. 1). Repeatability of age assignments was high for both structures. The initial two readings disagreed for only 3 or 4 scales or otoliths from any one lake.

For alewives from Lake Ontario, scale and otolith ages were identical for fish shorter than 150 mm; but otolith ages were greater than scale ages for 20% of the fish 150-159 mm long, for 91% of those 160-179 mm long, and for all of those longer than 179 mm. When ages differed, it was usually by 2 or 3 years but occasionally by 5 years or more. The otolith age was greater than the scale age for 86% of the fish with two or more annuli on their scales. Only 8% of the scales, but 54% of the otoliths, had more than four annuli. For fish with no annuli (young-of-the-year) and one annulus, there was perfect agreement between scale age and otolith age (Table 2).

Scale and otolith ages of alewives from Lake Huron commonly differed, with the discrepancies occurring over a wider range of lengths than in Lake Ontario (Fig. 1). Also, Lake Huron fish occasionally had one less annulus on their otoliths than on their

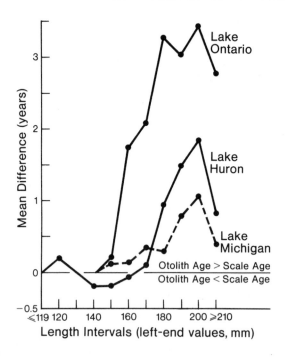

Figure 1. Mean differences between the otolith age and the scale age of alewives in various length intervals collected from three Great Lakes, fall 1983.

scales--a condition not observed in Lake Ontario fish (Table 2).
Among 56 alewives shorter than 180 mm, otolith ages were identical
to scale ages for 42 fish, younger than scale ages for 8 fish in
four length intervals (140-149 mm, 1 fish; 150-159 mm and 160-169
mm, 2 each; and 170-179 mm, 3), and older than scale ages for 6
fish in three length intervals (120-129 mm and 160-169 mm, 1 each;
and 170-179 mm, 4). Among alewives longer than 179 mm, otolith
ages were always equal to or older than scale ages; the percentages disagreeing increased from 47% for fish 180-189 mm long to
69% for those longer than 189 mm. The otolith age never exceeded
the scale age by more than 4 years and the most frequent difference was 1 year. Scales from 21% of the alewives and otoliths
from 32% had more than four annuli.

     For Lake Michigan alewives, age discrepancies between scales
and otoliths occurred throughout the same range of lengths as in
Lake Ontario; however the discrepancies were neither as frequent
nor as large (the two age determinations differed by more than 2
years only once). As in Lake Ontario, but not Lake Huron, the
otolith ages of Lake Michigan fish were never younger than the
scale ages, and both structures yielded identical ages for young-
of-the-year and yearlings (Table 2). Only 13% of the scale ages
exceeded 4, compared with 23% of the otolith ages.

DISCUSSION

     Even though the frequency and magnitude of discrepancies be-
tween ages from scales and those from otoliths varied among alewives from the three lakes, our results agree well with most published studies in which ages of fish determined from these two
structures were compared. Erickson (1979), who studied three
species of fish in eight Manitoba lakes, and Sikstrom (1983), who
worked with Arctic grayling Thymallus arcticus in Alaska, both
reported that ages determined from otoliths were consistently
greater than those determined from scales. Barnes and Power
(1984) found that otolith ages of lake whitefish Coregonus clupeaformis from western Labrador were commonly greater than scale
ages, but they were occasionally less. On the other hand, Norden
(1967) found that ages of Lake Michigan alewives determined from
scales and otoliths were similar. However, his fish were preserved in formalin, which is now known to degrade otoliths (Jearld
1984). Thus, although it is possible that discrepancies were not
present among Lake Michigan alewives in the mid-1960's, it is more
likely that the formalin obliterated fine, closely spaced annuli
near the edge of the otoliths of older fish.

     We believe that the variation in magnitude and frequency of
discrepancies between scale ages and otolith ages in fish from the
three lakes in this study was caused by interlake differences in
mortality rates and in ratios of food to fish. Clearly the magnitude of age discrepancies depends on past mortality rates; differences as great as those in Lake Ontario could occur only when mortality had been sufficiently low to allow some alewives to reach

an advanced age. Erickson (1979) found that differences between otolith ages and scale ages were slight among fish from heavily exploited populations and large among fish from unexploited ones. Age discrepancies should be more frequent in lakes with low food to fish ratios, if it is assumed that the reason for the fewer annuli on alewife scales was the cessation or marked slowing of growth, and that this growth change affected more fish at a low than at a high level of food availability.

In Lake Ontario during 1978-1982 alewives failing to form scale annuli became more common as the stock increased and condition of the individuals declined after a massive die-off (O'Gorman and Schneider 1986). In the present study, mean weights of fish in nearly all length groups were lowest in Lake Ontario, intermediate in Lake Huron, and highest in Lake Michigan (Table 1)--indicating that the food to fish ratio was highest, and thus the potential for growth greatest, in Lake Michigan, followed by Lake Huron and Lake Ontario. Alewives in Lake Michigan should then have exhibited detectable growth on their scales for a longer time than fish from either Lakes Huron or Ontario, and this difference may explain why discrepancies between scales and otoliths were fewer in Lake Michigan than in the other lakes. The magnitude and frequency of age discrepancies in each lake seems to be related to mortality rates and food availability.

Validation of scale and otolith ages was not attempted, but a natural marker indicated that ages determined from otoliths were more accurate than those from scales. Rapid growth of Lake Ontario alewives in 1977, after a massive die-off the previous winter, left a wide band of growth on the scales (O'Gorman and Schneider 1986). If the fish surviving the die-off had formed an annulus on both their scales and otoliths each year, the number of annuli present in fall 1983 would have been seven plus the number formed before 1977. Of the 12 fish examined during this study that had the natural marker, all were age 8 or older as determined from their otoliths, but considerably less as determined from scales (Table 2).

In the Great Lakes the modal scale-age group of alewives in fall trawl catches was invariably 3 or 4 (Brown 1972; Argyle 1982; O'Gorman and Schneider 1986). We found that the percentages of fish older than the scale-age 3 or 4 group to which they were assigned were 82%, 39%, and 24% in Lakes Ontario, Huron, and Michigan, respectively (Table 2). Non-random error rates of this magnitude among age-groups that compose the bulk of the catch would severely bias age statistics and ultimately result in mortality estimates that are much too high.

Previously published scale-age frequencies of fall-caught alewives indicated annual mortality rates of 75% after the fourth growing season in Lake Ontario (O'Gorman and Schneider 1986) and 79% and 71% after the fifth growing season in Lakes Michigan (Brown 1972) and Huron (Argyle 1982), respectively. But when mortality was determined independent of scale age (i.e. from changes in abundance of the group of naturally marked alewives in Lake Ontario during 1978-1982), the resulting estimate of 8% was, indeed, much lower (O'Gorman and Schneider 1986).

We hypothesize that in previous studies the predominant groups of middle-aged fish were made up not only of three and four-year old fish but also of fish from many older cohorts as well. These alewives had all ceased growing after four or five summers and, having similar numbers of annuli on scales, were all assigned to one of the two common scale-age groups. In previous years, just as in 1983, abundance of middle-aged fish was probably less, and that of older fish greater, than indicated by scales. Consequently, the steep right-hand limb of the scale-age catch curve did not represent the true mortality rate.

Because of the seeming predominance of middle-aged alewives in bottom-trawl catches, the younger fish were thought not fully recruited into the population near bottom (Brown 1972). Therefore, when determining production, the biomass of alewives not fully recruited was approximated by back-calculating from the biomass of those that were (Eck and Brown 1985). If the true modal age in bottom-trawl catches was not always 3 or 4, recruitment to the bottom population was not delayed until middle age and there was no basis for assuming that the younger cohorts were more numerous (and thus, the population more productive) then indicated by assessment catches.

The occurrence of discrepancies between scale ages and otolith ages of alewives in Lake Ontario, Huron, and Michigan indicate that scale ages should be used with caution (unless validated) in calculations of growth, survival, and production. Furthermore, because this discrepancy is probably not a recent phenomenon, previously published age-frequencies based on scales for Great Lakes populations of alewives are suspect--as are some of the age-related aspects of the accepted population biology of those stocks. Alewives in the Great Lakes may not be short-lived and have an intrinsically high mortality rate, but rather may be somewhat longer-lived and have an inherently lower mortality rate --which implies lower productivity than previously projected.

REFERENCES
ARGYLE, R. L. 1982. Alewives and rainbow smelt in Lake Huron: midwater and bottom aggregations and estimates of standing stocks. Transactions of the American Fisheries Society 111: 267-285.
BARNES, M. A., AND G. A. McFARLANE. 1984. A comparison of otolith and scale ages for western Labrador lake whitefish, Coregonus clupeaformis. Environmental Biology of Fishes 10:297-299.
BEAMISH, R. J., AND G. A. McFARLANE. 1983. The forgotten requirement for age validation in fisheries biology. Transactions of the American Fisheries Society 112:735-743.
BROWN, E. H. JR. 1972. Population biology of alewives, Alosa psuedoharengus, in Lake Michigan, 1949-70. Journal of the Fisheries Research Board of Canada 29:447-500.

ERICKSON, C. M. 1979. Age differences among three hard tissue structures observed in fish populations experiencing various levels of exploitation. Manitoba Department of Natural Resources, Report 79-77, Winnipeg, Canada.

ECK, G. W. AND E. H. BROWN, JR. 1985. Lake Michigan's capacity to support lake trout Salvelinus namaycush and other salmonines: an estimate based on status of prey populations in the 1970s. Canadian Journal of Fisheries and Aquatic Sciences 42:449-454.

HOURSTON, A. S. 1968. Age determination of herring at the biological station, St. John's, Newfoundland. Fisheries Research Board of Canada, Technical Report 49.

JEARLD, A., JR. 1984. Age determination. Pages 301-326 in L. A. Nielsen and D. L. Johnson, editors. Fisheries techniques. American Fisheries Society, Bethesda, Maryland, USA.

LIBBY, D. A. 1982. Decrease in length at predominant ages during a spawning migration of the alewife, Alosa pseudoharengus. United States National Marine Fisheries Service Fishery Bulletin 80:902-905.

NORDEN, C. R. 1967. Age, growth, and fecundity of the alewife, Alosa pseudoharengus (Wilson), in Lake Michigan. Transactions of the American Fisheries Society 96:387-393.

O'GORMAN, R. AND C. P. SCHNEIDER. 1986. Dynamics of alewives in Lake Ontario following a mass mortality. Transactions of the American Fisheries Society 115:1-14.

PRITCHARD, A. L. 1929. The alewife (Pomolobus pseudoharengus) in Lake Ontario. University of Toronto Studies Biological Series Publication 33. Publication of the Ontario Fisheries Research Laboratory 38:37-54, Toronto, Canada.

ROTHSCHILD, B. J. 1963. A critique of the scale method for determining the age of the alewife, Alosa pseudoharengus (Wilson). Transactions of the American Fisheries Society 92:409-413.

SIKSTROM, C. B. 1983. Otolith, pectoral fin ray, and scale age determinations for Arctic grayling. Progressive Fish-Culturist 45:220-223.

WATSON, J. E. 1964. Determining the age of young herring from their otoliths. Transactions of the American Fisheries Society 93:11-20.

WATSON, J. E. 1965. A technique for mounting and storing herring otoliths. Transactions of the American Fisheries Society 94:267-286.

Bias in growth estimates derived from fish collected by anglers[1]

L. ESTEBAN MIRANDA, WILLIAM M. WINGO,[2] ROBERT J. MUNCY, AND TERRY D. BATES[3]

MISSISSIPPI COOPERATIVE FISH AND WILDLIFE RESEARCH UNIT
MISSISSIPPI STATE, MS 39762

ABSTRACT
    Empirical data and angling simulations were used to explore
bias in growth determinations derived from largemouth bass
Micropterus salmoides collected by anglers. Back-calculated
lengths of largemouth bass were greater in fish caught by anglers
than in those collected by electrofishing and cove rotenone
sampling; however, backcalculations converged as age of fish
increased. Apparently, anglers took the largest members of the
youngest age-groups but a more representative sample of older fish.
Angler catch was simulated in largemouth bass populations having
three different rates of growth, equal average length at
corresponding annuli but two different length spreads, and three
different minimum-length limits. In the simulated population,
angling bias was inversely proportional to rate of growth, directly
proportional to the length range of a year-class, and directly
proportional to increments in minimum-length limit. Both empirical
data and angling simulations suggested that fish obtained from
anglers tend to overestimate a population's average rate of growth,
particularly at younger ages. Fish obtained from anglers could
provide unbiased estimates of growth, if year-classes not fully
recruited into the fishery are not excluded from the calculations.

    An important problem in determination of age and growth is that
of obtaining an unbiased sample of the population under study.

---

[1]Mississippi Agricultural and Forestry Experiment Station
Publication No. 6054.
[2]Present address: Mississippi Department of Wildlife
Conservation, Route 3, Box 99, Canton, Mississippi 39046.
[3]Present address: Farm Fish Company, Route 1, Box 108-A, Louise,
Mississippi 39097.

Most fishing gears are size selective (Backiel and Welcomme 1980).
Additionally, fish do not distribute randomly; many species show
widely differing habitat choice depending on age and size.
Consequently, samples of fish rarely include all year-classes, or
sizes within a year-class, in proportion to their true abundance in
the population. The use of diverse gears increases the probability
of obtaining a representative sample from each year-class, and may
give unbiased estimates of growth. However, it does not ensure
proportional representation of fish of all year-classes, precluding
reliable estimates of a population's age structure.

Age and growth of popular game fishes are often determined
using specimens captured by recreational anglers (e.g., Beckman
1946; Webb and Reeves 1975; Lee et al. 1983). Angling, like most
sampling methods, is size selective; selectivity depends on
behavioral differences over a species' size range, length limits
set by law, size standards self-imposed by individual anglers, and
fish population characteristics. Angling selectivity is directed
primarily toward the faster-growing portion of a given year-class
because faster-growing fish are the first to become available for
harvest. Thus, fish caught by anglers may tend to overestimate
average length of individual age-groups (Ricker 1969). The extent
of overestimation depends on angling selectivity and population
characteristics, and has not been investigated. In this study, we
compared growth histories of largemouth bass Micropterus salmoides
caught by recreational anglers with those collected by
electrofishing and cove rotenone sampling. We also simulated
growth determinations on the basis of angler harvest by varying
population parameters and harvest restrictions.

METHODS

Fish Collections And Growth Determinations
       This study was conducted in Columbus Lake, Mississippi, a
3,600-hectare impoundment of the U.S. Army Corps of Engineers on
the Tennessee-Tombigbee Waterway. Largemouth bass scales were
collected from fish taken by electrofishing surveys, cove rotenone
sampling, and angler's catches during 1982 and 1983.
Electrofishing was conducted during the fall at randomly selected
locations. Electric current was supplied by a boat-mounted 5,000
watt, 240-volt AC generator; rotenone samples were obtained during
late summer and early fall at each of three fixed coves in both
years, using methods described by Chance (1958) and Hall (1974).
Data from fish caught by anglers were obtained during a roving
creel survey, similar in design to that described by Malvestuto et
al. (1978), conducted from June 1982 through December 1983. All
measurements represent total lengths recorded to the nearest
millimeter.
       Annual growth estimates were calculated using the Lee method
of back-calculating length of fish from scales (Everhart et al.
1975). The standard a value proposed by Carlander (1982) was used

in the backcalculations. Scales were read at least twice. If
agreement was not reached after several readings, the fish in
question was excluded from the growth calculations. Most such
rejections represented older fish (28 total) and were evenly
distributed among the sampling methods. This practice undoubtedly
biased age composition, but probably did not significantly affect
estimates of growth. Age validation was attempted through analysis
of length-frequency modes.

Catch Simulations

We simulated angler catch from largemouth bass populations
having (1) three different growth rates, (2) equal average length
at corresponding annuli, but unequal length spreads, and (3) three
different minimum-length limits.

Populations representing three rates of growth--slow,
moderate, and fast--were simulated (Fig. 1). For convenience,

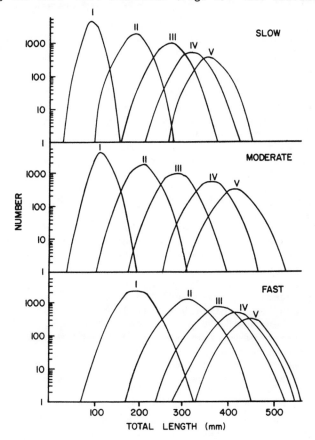

Figure 1. Length-frequency distribution of largemouth bass
    populations representing slow, moderate, and fast
    growth.

simulations were made under the assumption that lengths were
normally distributed within the age-groups. Lengths were grouped
into 20-mm intervals. Using mean length (μ) and standard deviation
(σ) for each age-group, we computed expected frequencies for every
length interval and year-class as follows: (1) we converted lower
limit of each length interval (X_{li}) to corresponding standard
normal value (Z_{li}), using the equation $Z_{li} = (X_{li} - \mu)/\sigma$; (2) we
obtained expected relative frequency of each length interval from a
standard normal distribution table using values of Z_{li} as
boundaries of length intervals; and (3) we multiplied these
expected relative frequencies by total number of fish in the
age-group to obtain expected frequencies in each length interval.
We assumed steady-state populations with stable recruitment of
10,000 yearlings and a constant annual mortality of 40% among all
sizes within an age-group. Slow growth is illustrated by the
average growth of largemouth bass in Wisconsin (Bennett 1937);
moderate growth by the rate observed in Columbus Lake (present
study); and fast growth by average growth rate in Texas reservoirs
(Prentice and Durocher 1978). Standard deviations of mean lengths
of age-groups observed in Columbus Lake were applied to all three
models. This procedure allowed us to examine the effect of three
diverse rates of growth on estimates derived from angling data
while eliminating variation due to size range.

The effect of growth rate on growth determinations based on
fish collected by anglers was assessed by assuming that anglers
harvested only largemouth bass $\geq$ 300 mm long. Thus, the average
length of the fish $\geq$ 300 mm long in each age-group simulated in
Fig. 1 represents the average length of angler-caught fish at their
last annuli. The proportion of fish $\geq$ 300 mm long in each
age-group was used to backcalculate average length of that portion
of the age-group in previous years. For example, in a population
where 5% of the larger fish of age 2 were $\geq$ 300 mm long, average
length of those fish at the end of their first year of life was
backcalculated as the average length of the larger 5% of the fish
of age 1. This was possible because we assumed growth to be
constant through the years and mortality uniform over all sizes
within age-groups. Bias of growth estimates (B) derived from
angler-harvested fish was calculated as: $B = 100(A/P - 1)$
where A = mean length of fish available for harvest at a given age,
and P = mean length of the population at that age. B represents
the percentage by which A exceeds P, and was computed for all
annuli in each year-class.

The time required for fish of a given year-class to appear in
the creel is also influenced by the amplitude of its length
distribution. We simulated two populations whose corresponding
year-classes had equal average lengths but length distributions
with standard deviations differing by 50%. Small length spread was
illustrated by the moderate growth model (Fig. 1), and large length
spread by magnifying (x 1.5) the standard deviation of each
age-group in the model while preserving the same mean lengths.
Again, it was assumed that anglers harvested only fish $\geq$ 300 mm

long. Mean length of the part of each age-group available to
anglers under each spread, growth history of that part of the
age-group, and expected bias at different annuli were calculated as
previously described.

The effects of minimum-size regulations on growth estimates
derived from angler-collected fish were similarly examined. Three
length limits frequently imposed on largemouth bass fisheries were
considered: 250, 300, and 350 mm. These size regulations were
applied independently to the moderate growth model (Fig. 1). Mean
length of the portion of each age-group available for harvest under
the three size regulations, growth histories, and expected bias at
different annuli were calculated as described above.

RESULTS AND DISCUSSION

Growth Histories In Columbus Lake

A total of 416 largemouth bass were used to determine growth
of the species in Columbus Lake. Length of fish collected by
electrofishing and cove rotenone sampling were 60 to 480 mm and
those of fish harvested by anglers were 200 to 520 mm. As
expected, electrofishing and rotenone provided scale samples from a
wider range of sizes than did angler catches. Although there was
no length limit regulation on angler harvest, 93% of the largemouth
bass harvested were longer than 250 mm. Absence of largemouth bass
smaller than 200 mm in creels might have resulted from selectivity
of both angling and anglers toward the larger fish.

Validity and accuracy of the scale method in aging largemouth
bass from Columbus Lake were questioned by inspecting length
frequencies of fish collected by electrofishing and rotenone
sampling in fall 1983. Average lengths at ages I and II, as
estimated from length-frequency distributions, were slightly less
than those obtained by backcalculations. Considering that the fish
were collected before the growing season was over, agreement can be
inferred. Analysis of length-frequency modes did not enable us to
determine accuracy of the technique as applied to older age-groups.
However, the validity of the scale method in age determination of
largemouth bass of ages I-IV was demonstrated in a warmer latitude
by Prentice and Whiteside (1975).

Results showed that back-calculated lengths of young fish
caught by anglers exceeded those of fish collected by electrofishing and cove rotenone sampling; however, backcalculations
converged as age increased (Table 1). This pattern developed from
anglers taking the largest members of the younger ages but a more
representative sample of the older fish. Similar conclusions could
have been reached if electrofishing and cove rotenone data had not
been available for comparison; the angling data exhibited Lee's
phenomenon in that, at a given annulus, back-calculated lengths
tended to be larger in younger fish. This suggests that anglers
took more of the larger representatives of the younger ages.
However, because Lee's phenomenon is also induced by size-selective
mortality, angling selectivity would not have been obvious.

Table 1. Back-calculated lengths (mm) of largemouth bass
collected by anglers (A) and by electrofishing and
rotenone sampling (E) in Columbus Lake, Mississippi,
1982-1983. Standard errors are given below each length.

Year-class	N		Annulus								
			I		II		III		IV		V
	A	E	A	E	A	E	A	E	A	E	A	E
					1982 Collection						
1982	0	9										
1981	2	14	126	107								
			6	5								
1980	48	19	138	121[a]	235	220						
			3	5	3	5						
1979	31	12	125	121	220	215	291	287				
			4	5	5	11	5	10				
1978	7	3	136	134	223	238	299	309	365	364		
			8	5	4	1	7	3	7	4		
1977	2	1	108	136	219	214	286	279	357	346	433	425
			5		6		6		15		4	
All	90	58	132	118[a]	228	220	292	290	363	358	433	425
			2	3	3	5	4	7	9	6	4	
					1983 Collection						
1983	0	43										
1982	1	57	138	118								
				2								
1981	18	56	133	111[a]	228	206[a]						
			6	2	6	4						
1980	70	21	126	119	225	208	296	281				
			2	5	3	6	3	7				
1979	31	6	121	118	216	200	288	288	359	360		
			3	9	4	8	4	13	4	16		
1978	13	4	118	116	200	199	283	292	363	360	421	416
			4	14	6	9	9	9	8	3	9	9
All	133	187	125	115[a]	221	206[a]	293	284	360	360	421	416
			2	2	2	3	2	5	4	9	9	9

[a]A t test indicated that length of fish caught by anglers
was significantly greater (P < 0.05) than length of
fish caught by electrofishing and rotenone.

Catch Simulations

Simulations of angling harvest indicated that growth histories
derived from creel samples would often be biased. Direction of the
bias was always positive, overestimating average growth; its
magnitude, however, depended on age of fish, rate of growth, length
spread at age, and length-limit regulation.

Bias was greatest among the younger age-groups and decreased
with increases in age. As fish in given year-classes advanced in
age, they became increasingly vulnerable to harvest, allowing
anglers to take a more representative sample of fish in the
year-class. Within each year-class, bias was greatest at the
earlier annuli.

Rate of growth affected angling bias inversely, bias being
greater in slower-growing populations (Table 2). A greater portion
of the younger age-groups was recruited to the fishery each year in
faster-growing than in slower-growing populations. Therefore, mean

Table 2. Percentage bias (B, as defined in Methods) at back-calculated annuli during the growth histories of largemouth bass harvested by anglers, under a 300-mm minimum length-limit regulation, from populations representing the three rates of growth illustrated in Figure 1. N indicates the number of harvestable size fish at a given age, a dash indicate no harvestable size fish present, and 0 indicate no bias.

Growth rate	Age at capture	N	Annulus				
			I	II	III	IV	V
	1	–	–				
	2	–	–	–			
Slow	3	487	30	22	19		
	4	1541	9	6	5	4	
	5	1237	2	1	1	1	1
	1	–	–				
	2	–	–	–			
Moderate	3	1068	18	15	13		
	4	2093	1	1	1	1	
	5	1291	0	0	0	0	0
	1	–	–				
	2	3726	6	5			
Fast	3	3582	0	0	0		
	4	2158	0	0	0	0	
	5	1297	0	0	0	0	0
Slow	All	3265	10	7	6	3	1
Moderate	All	4452	5	4	4	1	0
Fast	All	10763	2	2	0	0	0

length of fish of the younger age-groups could be more accurately estimated from angling data derived from the faster-growing populations.

Angling bias was directly proportional to the range of sizes within individual age-groups, the bias being greater in populations with broader ranges (Table 3). Broadening of length distribution produced an increase in average length of harvestable size fish within each age-group. Factors influencing broadening of length distribution include extended spawning periods, advantage of larger fish in obtaining food, and differential growth between localities within a body of water. Skewed and bimodal length distributions often develop within cohorts, significantly spreading sizes among fish in an age-group and further complicating growth determinations based on angler catches. In a study by Shelton et al. (1979) in West Point Reservoir, Alabama and Georgia, total length of fish of the first year-class of largemouth bass ranged from 50 to 400 mm by the end of the first growing season, and length distribution was sharply bimodal with modes at 100 and 275 mm. Minimum-length limit for largemouth bass in West Point Reservoir was 300 mm. Average growth of fish of this year-class would have been greatly overestimated, if it had been determined from angling data alone.

Angling bias increased with increments in length limit (Table 4). Higher length limits magnified angling bias by making harvest of the younger year-classes more selective toward the

Table 3. Percentage bias (B, as defined in Methods) at back-calculated annuli during the growth histories of largemouth bass harvested by anglers, under a 300-mm minimum length limit regulation, from two populations exhibiting equal rates of growth but unequal length spreads at corresponding age groups. N, dash, and 0 are as defined in Table 2.

Length spread	Age at capture	N	Annulus				
			I	II	III	IV	V
	1	–	–				
	2	–	–	–			
Small	3	905	22	17	14		
	4	2052	2	1	1	1	
	5	1261	0	0	0	0	0
	1	–	–				
	2	36	70	53			
Large	3	1166	27	21	18		
	4	1880	6	5	4	3	
	5	1279	1	1	1	0	0
Small	All	4218	5	4	3	1	0
Large	All	4361	10	8	7	3	0

Table 4. Percentage bias (B, as defined in Methods) at back-calculated annuli during the growth histories of largemouth bass harvested under three length-limit regulations, from the moderate growth population model illustrated in Figure 1. N, dash, and 0 are as defined in Table 2.

Length limit (mm)	Age at capture	N	Annulus				
			I	II	III	IV	V
	1	–	–				
	2	251	37	28			
250	3	2986	5	4	3		
	4	2158	0	0	0	0	
	5	1261	0	0	0	0	0
	1	–	–				
	2	–	–	–			
300	3	405	22	17	14		
	4	2052	2	1	1	1	
	5	1261	0	0	0	0	0
	1	–	–				
	2	–	–	–			
350	3	36	39	35	28		
	4	1180	12	10	8	7	
	5	1214	2	1	1	1	1
250	All	10874	4	3	1	0	0
300	All	4218	5	4	3	1	0
350	All	2430	8	6	5	4	1

faster-growing individuals. Distribution of the bias among year-classes also varied with length limit. As length limit increased, progressively older age-groups showed bias in backcalculations. Length-limit regulations are becoming a popular tool for manipulating fish populations (Anderson 1980). Their

effect on growth determinations must be considered when age and
growth studies are designed to include or to consist entirely of
angling data.

MANAGEMENT IMPLICATIONS
     Empirical data and angling-catch simulations indicated that
fish collected by anglers misrepresented a population's rate of
growth, especially at the younger ages. When vulnerability
increases with size, fish in year-classes not fully recruited into
the fishery are not representatively sampled, and average lengths
at successive annuli, as determined from backcalculation from
scales, are overestimated.
     Significance of an overestimate, however, depends on the
effectiveness with which a manager is able to manipulate a fish
population. In Columbus Lake, overestimates of growth based on
fish collected by angling--although some were statistically
significant (Table 1)--were of little practical importance. Fine
adjustments to regulate fish populations in large bodies of water
are seldom feasible. In this particular lake and under the
available management options, recommendations based on growth of
largemouth bass captured by biologists would not have disagreed
with those based on anglers' catches.
     Bias in growth estimates derived from fish collected by
anglers could be eliminated, if year-classes not fully recruited to
the fishery are excluded from the analyses. The number of
year-classes that should be excluded depends on rate of growth,
length distribution within age-groups, and size regulations imposed
on the catch. Because rate of growth and length distribution
within age-groups are often unknown beforehand, in practice, bias
could be reduced by impartially excluding some of the newer
year-classes from the backcalculations.

ACKNOWLEDGMENTS
     The Mississippi Cooperative Fish and Wildlife Research Unit is
sponsored jointly by Mississippi Agricultural and Forestry
Experiment Station, Mississippi Department of Wildlife
Conservation, U.S. Fish and Wildlife Service, and the Wildlife
Management Institute.
     We gratefully acknowledge editorial assistance provided by
Kenneth Carlander, Paul Eschemeyer, John Heinen, Larry Mitzner, and
Randall Robinette.

REFERENCES
ANDERSON, R. O. 1980. The role of length limits in ecological
     management. Pages 41-45 in S. Gloss and B. Shupp, editors.
     Practical fisheries management: more with less in the 1980's.
     New York Chapter American Fisheries Society, New York, USA.
BACKIEL, T., and R. L. WELCOMME, editors. 1980. Guidelines for
     sampling fish in inland waters. Food and Agriculture
     Organization of the United Nations, Rome, Italy.

BECKMAN, W. C. 1946. The rate of growth and sex ratio for seven
    Michigan fishes. Transactions of the American Fisheries Society
    76:63-81.
BENNETT, G. W. 1937. The growth of the largemouth black bass,
    Huro salmoides (Lacepede) in the waters of Wisconsin. Copeia
    1937:104-118.
CARLANDER, K. D. 1982. Standard intercepts for calculating
    lengths from scale measurements for some Centrarchid and Percid
    fishes. Transactions of the American Fisheries Society
    111:332-336.
CHANCE, C. J. 1958. How should population surveys be made?
    Proceedings of the Annual Conference Southeastern Association of
    Game and Fish Commissioners 11:84-89.
EVERHART, W. H., A. W. EIPPER, and W. D. YOUNGS. 1975. Principles
    of fishery science. Cornell University Press, Ithaca, New York,
    USA.
HALL, G. E. 1974. Sampling reservoir fish populations with
    rotenone. Pages 249-259 in R. L. Welcomme, editor. Symposium
    on methodology for the survey, monitoring, and appraisal of
    fishery resources in lakes and rivers. European Inland
    Fisheries Advisory Commission, Technical Paper Number 23.
LEE, D. W., E. D. PRINCE, and M. E. CROW. 1983. Interpretation of
    growth bands on vertebrae and otoliths of Atlantic bluefin tuna,
    Thunnus thynnus. Proceedings of the International Workshop on
    Age Determination of Oceanic Pelagic Fishes: Tunas, Billfishes,
    and Sharks. U.S. Department of Commerce, National Oceanic and
    Atmospheric Administration Technical Report, National Marine
    Fisheries Service 8:49-59.
MALVESTUTO, S. P., W. D. DAVIES, and W. L. SHELTON. 1978. An
    evaluation of the roving creel survey with nonuniform
    probability sampling. Transactions of the American Fisheries
    Society 107:255-262.
PRENTICE, J. A., and B. G. WHITESIDE. 1975. Validation of aging
    techniques for largemouth bass and channel catfish in central
    Texas farm ponds. Proceedings of the Annual Conference
    Southeastern Association of Game and Fish Commissioners
    28:414-428.
PRENTICE, J. A., and P. P. DUROCHER. 1978. Average growth rates
    for largemouth bass in Texas. Proceedings of the annual meeting
    of the Texas Chapter, American Fisheries Society 1:49-57.
RICKER, W. E. 1969. Effects of size-selective mortality and
    sampling bias on estimates of growth, mortality, production, and
    yield. Journal Fisheries Research Board of Canada 26:479-541.
SHELTON, W. L., W. D. DAVIES, T. A. KING, and T. J. TIMMONS. 1979.
    Variation in the growth of the initial year class of largemouth
    bass in West Point Reservoir, Alabama and Georgia. Transactions
    of the American Fisheries Society 108:142-149.
WEBB, J. F., and W. C. REEVES. 1975. Age and growth of Alabama
    spotted bass and northern largemouth bass. Pages 204-215 in R.
    H. Stroud and H. Clepper, editors. Black bass biology and
    management. Sport Fishing Institute, Washington, D.C., USA.

IV
Validation

Methods of validating daily increment deposition in otoliths of larval fish

A. J. GEFFEN
ENVIRONMENTAL SCIENCES CENTER
UNITY COLLEGE
UNITY, ME 04988

ABSTRACT

Early literature on larval fish otoliths has documented the presence of otolith increments, but only a few attempts have successfully validated a daily rhythm of ring deposition. Utilization of otolith data is limited by the need to validate rates of increment formation. Full-scale validation studies are time consuming, labor intensive, and often technically difficult. Some basic assumptions for daily increment validations are outlined, and several validation techniques are summarized in a framework of alternative methods for rapid identification of daily or non-daily deposition rhythms. New approaches to validation may provide faster and easier, but less reliable, alternatives. The ultimate choice of a validation method will always be a compromise between desired reliability and available resources.

The development of techniques to age fish larvae using otolith microstructure is one of the most significant of recent advances in larval fish studies. The ability to obtain accurate ages of individual, field-caught, larvae can lead to improved estimates of population growth rates, age-specific growth and mortality functions, individual variation in growth and survival, and the effect of specific meteorological events on larval growth. Measures of individual growth rates may provide information on daily changes in growth and the effects of developmental events and varying food availability on growth rate.

The otolith microstructural units used in aging fish larvae are layers of calcium carbonate crystals which are

The Age and Growth of Fish, edited by Robert C. Summerfelt and Gordon E. Hall © 1987 The Iowa State University Press, Ames, Iowa 50010.

deposited periodically on an organic matrix. The layers
of crystal deposition are termed incremental zones and
breaks in crystal formation are called discontinuous
zones (Campana and Neilson 1985). Together, two consecutive zones form each otolith increment, or ring, visible
under transmitted light as adjacent bands of light
(incremental) and dark (discontinuous) material. The
periodicity of increment formation may be daily, meaning
that one complete two-part increment is formed during
each 24 h period, or other than daily. Increments may be
referred to as primary increments, especially in the
European literature, to distinguish them from otolith
features of longer (tidal, lunar monthly) (Pannella 1971)
or shorter (sub-daily) (Brothers 1978) periodicities.

Numerous publications report on the aging of field-
caught larvae of many species by counts of primary
increments. In most of these instances, population
growth rates were calculated based on length-at-age data,
where otolith ring count was used as an indicator of
larval age (e.g., Brothers et al. 1976; Brothers and
McFarland 1981; Gjøsaeter and Tilseth 1982; Laroche et
al. 1982). The incorporation of otolith aging techniques
into ecological research is less common (Townsend and
Graham 1980; Rosenberg and Laroche 1982; Victor 1982;
Methot 1983; Rice 1985) because, for the most part, they
rely upon validation of the rate of increment formation,
determination of the confidence limits for aging wild
larvae, and information on environmental variables that
are likely to affect ring deposition.

The validation of ring-deposition rates in the
otoliths of fish larvae is thus a critical step toward
further developments of this technique. Although complete understanding of the deposition process is the ideal
precursor to applications of aging techniques in field-
caught larvae, there are realistic constraints. Full-
scale validation studies require the rearing of larvae,
the manipulation of growth conditions, and the processing
of large numbers of otoliths; years may be spent on
validation before actual application of the technique to
field-caught specimens can be implemented. For many pro-
jects, lengthy preliminary studies are an impediment to
this application. In addition, the successful rearing of
many species is difficult, placing methodological limits
on validation studies for these larvae.

Fortunately, much of the existing otolith literature
has successfully laid the foundation for assessing the
rate of ring deposition in many species. These studies
include full-scale validation, which entails comprehensive testing of increment formation, and experiments or
observations leading to inference of daily ring deposition. Alternative validation methods are needed to reach

a realistic compromise between available resources and technology, and a level of reliable testing of the assumptions used for larval fish. Incorporation of these abbreviated validation techniques may, at least, provide early warning of situations where otolith ring counts may not provide reliable aging of fish larvae.

The literature cited here represents examples of validations applied to fish larvae. The selection of examples and topics, as well as their arrangement and treatment, is designed to provide a framework for the planning of validation studies, not as a critique of the existing literature. For this reason, numerous works on the microstructure of juvenile and adult fish otoliths have not been included. Traditional reviews of the extensive otolith microstructure literature pertaining to larvae, juveniles and adults have been provided by Campana and Neilson (1985) and Brothers (1979, 1982). A recent reminder of the importance of validating general aging techniques was presented by Beamish and McFarlane (1983).

VALIDATING DAILY INCREMENT FORMATION

Short History of Larval Fish Otolith Studies
        The aging of fish larvae on the basis of primary otolith increments relies on four assumptions, or criteria for validation: 1) ring-deposition rate is constant with respect to time throughout the larval period, such that ring number can be related to larval age; 2) initiation of ring deposition is constant between individuals; 3) ring-deposition rate is independent of larval growth rate, age, and season; and 4) deviations which occur in otolith ring deposition can be predicted and modeled, based on some understanding of the deposition process.

        Larval otolith studies may be readily divided into validation and application components. The majority of application studies rely upon previous validation of ring-deposition rate, specifically daily ring deposition. Many studies, however, continue to be published without validation, mostly because of shortages of time, money, and rearing technology. Validation studies in the early literature focused on determining the timing of initial ring deposition (age at which all larvae had at least one ring, mean age of larvae with one ring, age at which 50% of the larvae deposited the first ring) and the average ring-deposition rate for populations of reared larvae (Brothers et al. 1976; Taubert and Coble 1977; Barkman 1978; Methot and Kramer 1979; Laroche et al. 1982). Scattered throughout these studies are investigations

into environmental factors which control and influence
ring deposition. The latter have become more prominent
recently, as investigators begin to consider the biochemical and physiological mechanisms of ring deposition
in fish larvae and juveniles (e.g., Tanaka et al. 1981;
Campana and Neilson 1982; Geffen 1982a; Neilson and Geen
1982; Radtke and Dean 1982; Wilson and Larkin 1982;
Campana 1983; Radtke 1984; Volk et al. 1984).

Framework for Selecting Alternative Validation Methods
        A summary of techniques useful for validating daily
ring deposition in fish larvae is compiled in Table 1.
These methods are described and the reliability evaluated
in the following paragraphs. Validation by the three
methods listed first is probably more reliable than by
the last four methods, from which daily ring deposition
can only be inferred. Quantitative comparison of the
reliability of each method is probably feasible, but not
yet available for any species. The inference of daily
ring deposition should replace proper validation, but
even less reliable methods might provide a preliminary
go-ahead signal for field applications in conjunction
with full-scale validation. More importantly, violations
of the assumptions that form the basis of larval otolith
work might be identified very rapidly for new species or
new situations by using the faster, but less reliable,
alternative validation methods, alerting investigators to
the need for more intensive background research.

Rearing and sequential sacrifice. The most desirable
method of validating ring-deposition rate is to rear
larvae through the entire larval stage, from eggs to
metamorphosis, sampling reasonable numbers of individuals
at frequent intervals throughout this period (Fig. 1).
The appropriate sample size can be determined by the
requirement of placing 95% confidence limits on the
estimation of individual values of ring counts (Y) from
the regression of ring counts on age (X). Extreme
variability in ring counts reduces the confidence for
estimates of Y and limits the applications of otolith
aging for individuals, but statistically acceptable
results can still be produced for populations (Rice
1985). The otoliths of embryos should be checked for the
presence of rings (Wilson and Larkin 1980; Geffen 1983;
McGurk 1984), and sampling should be intensive during the
period from hatching to yolk sac absorption to identify
the timing and variability of initial ring deposition.
The individuals sampled should be obtained from groups of
larvae hatched over a discrete time unit, usually one 24
h period.
        Many early investigators dismissed observed non-

Table 1. A framework for evaluating the suitability of different methods of validating daily otolith increment formation in fish larvae.

Validation method	Level of resource commitment	Level of reliability (evidence for daily rings)	Criteria examined	Selected examples
Rearing and sequential sacrifice	high—larval rearing otolith processing	high	ring deposition rate time of initial ring formation individual variability	Geffen 1982a Radtke & Dean 1982
Marking otoliths	high—preliminary studies otolith processing detecting marks	high	ring depositon rate individual variability	Campana & Neilson 1982 Hettler 1984
Estimate of hatching dates	moderate—sample collecting otolith processing	medium/high	inference of daily deposition if corroborated by known hatching dates	Brothers et al. 1983 Rice 1985
Short term changes in growth rate	moderate to high—some larval rearing otolith processing	medium/high	short term deposition rate insight into mechanisms of increment formation	Neilson & Geen 1982 Wilson & Larkin 1982
Natural otolith marks	moderate to high—some larval rearing otolith processing	medium	short term deposition rate cross-check assumed daily rate	Brothers 1979 Geffen in press
Comparison with other growth rate estimates	moderate—sample collecting otolith processing	medium	inference of daily deposition for population if corroborated by known growth rates	Townsend & Graham 1980 Laroche et al. 1982
Synchronization of increment formation	moderate—sample collecting otolith processing	low/medium	inference of daily deposition	Tanaka et al. 1981 Geffen 1982b

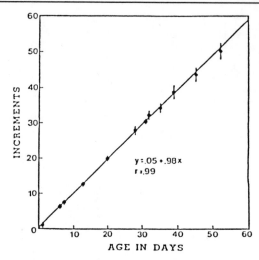

Figure 1. Example of validation by rearing and
sequential sacrifice. Comparison of age (x) and
number of increments (y) formed in sagittae of
striped mullet Mugil cephalus larvae. Each point
is the mean of 10 individuals + SD. From Radtke
(1984) reproduced with permission of author and
publisher.

daily deposition rates as the result of poor growth under
laboratory conditions (e.g., Laroche et al. 1982; Lough
et al. 1982), and assumed that daily ring deposition
would be observed in larvae sampled in the field because
selective mortality will remove slower-growing individuals from the population. The scarcity of field-caught
larvae that can be identified as starving or slow-growing
(Hunter 1984) may support this argument for some species,
although not for others (Brothers et al. 1983). Optimal
growth conditions providing for maximum growth rates may
be useful in validation studies to indicate best-case
scenarios for measuring ring-deposition rates. Rearing
conditions conducive to high growth rates are often
difficult to provide, but they have been improved for
several species by using large outdoor enclosures (Gjøsaeter and Øiestad 1981; Geffen 1982a; Bergstad 1984).
But because maximum and natural growth rates may differ,
validation studies that are preliminary to field applications should also include larvae reared under manipulated
growth conditions. Many characteristics of the rearing
environment, such as feeding level, feeding periodicity
and temperature, affect the growth rate of larvae and may
also influence ring deposition to some extent (Geffen
1982a; Neilson and Geen 1982; Radtke and Dean 1982;

McGurk 1984). The amount of variation in ring-deposition rate that can be explained by growth conditions may limit the accuracy of field applications.

Manipulation of environmental parameters, even outside those normally experienced by the larvae, such as 24L or 6L6D photoperiods, is valuable in elucidating the controlling factors in ring deposition (Tanaka et al. 1981; Campana 1983; Geffen 1983; Campana and Neilson 1985). Such studies also indicate the extent to which metabolic and physiological constraints may influence the process of ring deposition and control its rhythm.

Marking otoliths. Although the capture, marking, and subsequent release or holding of larvae has had only limited success, this technique is suitable for species with hardy larvae and for juveniles (Struhsaker and Uchiyama 1976; Miller and Storck 1982). Fish can be captured from the wild, marked using compounds which are incorporated into the otolith, and maintained in controlled conditions or in enclosures. After marking, periodic samples can be taken and the number of rings deposited between the marking and the sacrifice date recorded (Fig. 2). Otoliths of even small marine larvae can be

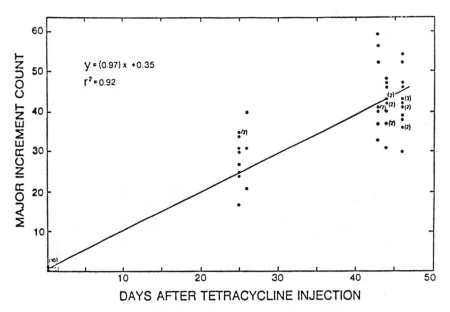

Figure 2. Example of validation by marking otoliths. Counts of major otolith growth increments distal to the origin of the flourescent tetracycline band. Data are pooled from all laboratory environments. From Campana and Neilson (1982).

marked by submerging the larvae in tetracycline (Hettler 1984). The otoliths of juvenile fish have been marked using tetracycline, as well as by variations in temperature, feeding level, or stresses such as cold shock or starvation (Victor 1982; Campana and Neilson 1985; Brothers pers. comm). The efficiency of the marking method, as well as the number of rings and the interval of time represented by the mark, should be determined by prior investigation.

The advantage of the marking method when used on fish captured from the wild is that the larvae marked represent a sample of surviving larvae, thereby avoiding one of the biases resulting from the use of reared populations. Studies involving species such as cod Gadus morhua or rainbow smelt Osmerus mordax, which are more difficult to rear in early stages but may be maintained when captured from the wild, may benefit by this method. Multiple marking might provide a measurement of ring-deposition rate based on known individual histories, rather than on sequential samples, since the date of marking is known for the entire group. Also, marking, release, and recapture from the wild assures that growth conditions during the experimental period are natural. When combined with manipulated growth conditions and sequential sacrifice studies, this may provide the most accurate information on individual histories of ring deposition. Unfortunately, errors may be introduced by stresses induced by the marking or transfer of larvae.

Estimate of hatching dates. Otolith ring counts representing larval age may be used to calculate individual hatching dates (Brothers et al. 1983; Methot 1983; Rice 1985). Comparison of the calculated hatching date with known periods of hatching or spawning activity could help evaluate the strength of the assumption of daily ring production (Fig. 3). However, selective mortality of hatching cohorts may bias the results of these comparisons. This could become an increasing problem with more protracted spawning periods and as larvae grow older. In the absence of any other information on ring-deposition rate, major discrepancies between calculated hatching dates and expected hatching dates might indicate periods of cessation of ring deposition, or deviations from daily ring deposition (Geffen and Nash 1986).

Short term manipulations, ring width vs. ring number. Two conceptual models may be developed for the relationship between ring deposition and larval growth. If otolith size and larval length show a significant relationship, changes in the growth rate of an individual should be accompanied by changes in the growth rate of

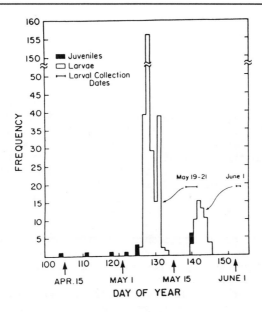

Figure 3.   Example of hatching date calculations.
Back-calculated hatching dates for larvae (open
bars) and juveniles (solid bars).   Cross bars
indicate the dates of larval collections.   Otolith
ages (daily growth unit counts) were corrected by
adding 4 d before determining hatching times.
Adapted from Brothers et al. (1983).

the otolith (Methot 1981; Wilson and Larkin 1982;
Brothers et al. 1983; Volk et al. 1984).   Alternative
models suggest that changes in otolith growth might
be accomplished by altering the width of increments
or by changes in the number of increments deposited.
Larval growth rate may be manipulated in the laboratory
using feeding levels, temperature, or tank size.   In
short-term experiments, it may be possible to determine
the response of otoliths to changes in larval growth
rate, and to test the assumption of constancy and independence of ring-deposition deposition rate (Neilson and
Geen 1982; Wilson and Larkin 1982; McGurk 1984).

Developmental and natural markers in otolith microstructure.   In some species, specific rings or microstructure
patterns seem to be associated with developmental events.
Examples of these developmental marks include the yolk
sac absorption ring in herring Clupea harengus larvae
(Geffen 1982a; McGurk 1984), rings associated with embryo
development or swim-up in Atlantic salmon Salmo salar
(Geffen 1983) and sockeye salmon Oncorhynchus nerka

(Marshall and Parker 1982), or groups of rings associated
with metamorphosis and changes in habitat (Brothers and
McFarland 1981). The age at which these events occur has
already been reported in the literature for a range of
temperatures and growth rates; in these instances, it is
possible to cross-check the age and the event represented. In addition, the assumption of daily ring deposition may also be tested by assuming that ring deposition
is daily, and by backcalculation, obtaining lengths and
individual growth rates. The estimated individual growth
rate can then be used to calculate the length of time
required for that individual to grow from the length at
hatching to the length at a specific ring associated with
a specific developmental event. That time interval is
compared with values reported in the literature. If the
assumption of daily ring deposition is not violated, the
estimated growth rate should result in close agreement
between the estimated age represented by the otolith mark
and the previously established age for the event with
which that mark is associated (Geffen in press).

Patterns in the otolith microstructure which are
known to reflect tidal or lunar monthly periodicities
might be used to check deposition rates of primary increments. While not commonly reported in larval otoliths,
such marks have been observed on the otoliths of older
fish (Pannella 1971).

Comparison of growth rates with other field estimates.
Growth rates calculated from otolith age data can be compared with other field estimates of growth rate (Townsend
and Graham 1980; Laroche et al. 1982; Brothers and
McFarland 1981; Brothers et al. 1983; Geffen in press).
For larvae, these field estimates are usually made by
analyzing the progressions of length frequency modes for
repeated samples of the same population. Because otolith
aging techniques were developed for larvae in order to
avoid some of the biases and interpretational difficulties inherent in modal progression analysis, this method
may be less reliable for first investigations of new
species. In addition, estimates using otoliths are often
compared with modal estimates from the same larvae so
that selectivity of the sampling gear may introduce undetected sampling biases into both methods (Brothers et al.
1983). However, in situations where larval growth rates
have been estimated for many years, and researchers have
a good idea of the expected growth rates, the reliability
is greatly enhanced. The additional information provided
by otolith aging may help to define more clearly the
variations in growth and mortality which affect the modal
progressions. Daily ring deposition inferred only from
comparison with other field estimates of larval growth

may limit applications of otolith data to population estimates. Further validation may be required to make use
of the major advantage of otolith analysis, that is, the
production of individual statistics.

Synchronization of ring deposition. The assumption that
ring-deposition rate is daily and independent of larval
growth implies that deposition is controlled by cyclic
external environmental cues, or is an endogenous rhythm
of diel periodicity. Larvae experiencing the same environmental conditions, such as diel changes in light
levels, temperature, or feeding activity should show synchronization of ring deposition. Individuals would be
expected to initiate and cease increment deposition at
approximately the same time of day, or in response to the
same portion of some environmental cycle. The outermost,
or ultimate, otolith increment of larvae sampled throughout a 24 h cycle may be examined, and categorized as to
the percentage completion when compared to the previous,
adjacent increment (Tanaka et al. 1981)(Fig. 4). The
categories of completion may be designated more subjectively as newly-formed, half-formed, or completed (Geffen

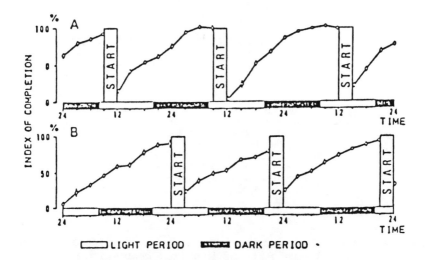

Figure 4. Example of examination of synchronization
of ring deposition. Daily growth of Tilapia
nilotica otoliths as represented by changes in the
index of completion for current increment every 3 h
for 72 h. The fish were maintained under two
contrasting photoperiods: A) 12L-12D (light phase,
0800-2000h): B) 12L-12D (light phase, 2000-0800h).
Each circle represents mean + SE for six fish.
START = start of new incremental zone. From Tanaka
et al. (1981).

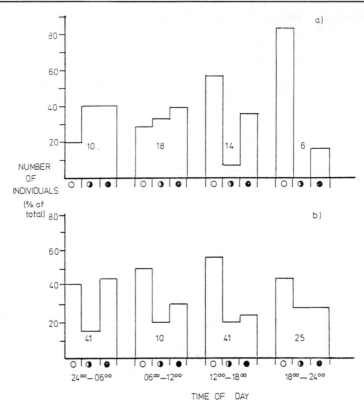

Figure 5. Example of examination of synchronization
of ring deposition. Distribution of the state of
completion of the most recent increment in herring
Clupea harengus larvae sampled over 24 h periods.
a) laboratory-reared larvae, b) field-caught
larvae. Numbers within each block indicate sample
size. Symbols represent O newly-formed, ◑
half-formed, or ● completed increments. From
Geffen (1982b).

1982b). If ring deposition is synchronous throughout the
population, the state of increment completion should progress with time through the cycle (Fig. 5). From this it
may be possible to infer that ring-deposition rates are
daily, and controlled by or correlated with cyclic environmental cues such as photoperiod, temperature, or in
some situations, with food availability or vertical migration behavior. The reliability of this method may be
reduced in cases where sub-daily increments are common,
since these structures could cause errors in determining
the state of completion of the outermost increment.

Otolith Preparation
    Although not strictly within the topic of this
paper, the techniques of sample preparation and viewing
are equally important in validation work. The otoliths
of many larvae may be viewed as whole mounts, but others
should be ground and polished, or sectioned for proper
analysis. An assessment of the suitability of routine
use of otolith ring count data may be made using standard
light microscopy, but resulting ring counts represent
only the number visible by transmitted light (usually increments > 1 μm in width) (Campana and Nielson 1985), and
possibly not all the rings actually present. The determination of true ring number (all rings, including those
< 1 μm) might only be assured through the use of high re-
solution techniques, such as scanning electron microscopy.

CONCLUSIONS
    Many methods exist for validating daily otolith
increment formation in fish larvae. Full-scale validation involves comprehensive testing of the major assumptions basic to larval otolith work. Validation studies
should be designed to test the criteria that ring deposition is constant or predictable in time, that ring number
reflects larval age, that ring deposition is independent
of larval growth rate and feeding, and that it is controlled by the same factors throughout the population.
Realistic constraints may limit the feasibility of full-
scale validation, but in some circumstances alternative
methods may be used. These methods could lead to inference of daily ring deposition, but are more probably
limited to indicating instances of non-daily deposition
rates. Future studies may quantify the differences between established methods of full-scale and alternative
validations, encouraging an objective rationale for
selection of validation methods. Investigators may be
able then to evaluate the consequences of delaying full-
scale validation in favor of preliminary work using
methods that may be faster, easier, less expensive, but
also less reliable. Although objective measures of reliability are not yet quantified, full-scale validation,
and marking combined with sequential sacrifice, could
subjectively be considered methods of good reliability.
Intermediate reliability may characterize the short-term
experiments on the effects of growth-rate changes on
increment formation and those methods that use developmental and natural otolith marks to cross check ring-
deposition rates. Reduced levels of reliability would
probably accompany validations based on methods which
compare estimates of population growth rates from otolith
data to growth estimates made by other means, or based on

diurnal synchronization of increment formation. The reliability of using alternative validation methods varies, but reduced reliability may be a necessary compromise for preliminary investigations, especially for new species or in new conditions.

ACKNOWLEDGMENTS
    I wish to thank E. Brothers and J. Neilson for their constructive comments, and the many investigators cited for the use of published and unpublished results referred to in this paper.

REFERENCES
BARKMAN, R.C. 1978. The use of growth rings to age young
    Atlantic silversides, Menidia menidia.
    Transactions of the American Fisheries Society
    107:790-792
BEAMISH, R.J. and G.A. McFARLANE. 1983. The forgotten
    requirement for age validation in fisheries
    biology. Transactions of the American Fisheries
    Society 112:735-743.
BERGSTAD, O.A. 1984. A relationship between the number
    of growth increments on the otoliths and age of
    larval and juvenile cod, Gadus morhua L. Flødevigen
    rapportser 1:251-272.
BROTHERS, E.B. 1978. Exogenous factors and the formation
    of daily and subdaily increments in fish otoliths.
    American Zoologist 18:631.
BROTHERS, E.B. 1979. Age and growth studies on tropical
    fishes. Pages 119-136 in S.B.Saila and P.M. Roedel
    (editors). Stock assessment for tropical small
    scale fisheries. Proceedings of an international
    workshop, September 1979, International Center for
    Marine Resources Development, University of Rhode
    Island, Rhode Island, USA.
BROTHERS, E.B. 1982. Aging reef fishes. Pages 2-23 in
    G.R. Huntsman, W.R. Nicholson and W.W. Fox Jr.
    (editors). The biological basis for reef fishery
    management. Proceedings of a workshop, October
    7-10, 1980, St. Thomas, United States Virgin
    Islands.   (National Oceanographic and Atmospheric
    Administration Technical Memorandum, National
    Marine Fisheries Service-Southeast Fisheries
    Center-80).
BROTHERS, E.B. and W.N. McFARLAND. 1981. Correlations
    between otolith microstructure, growth, and the
    life history transitions in newly recruited French
    grunts (Haemulon flavolineatum (Desmarest)
    Haemulidae). Rapport et Proces verbeaux de la

Reunion de Conseil International d'Exploration de la Mer 178:369-374.

BROTHERS, E.B., C.P. MATTHEWS, and R. LASKER. 1976. Daily growth increments on otoliths from larval and adult fishes. United States National Marine Fisheries Service Fishery Bulletin 74:1-8.

BROTHERS, E.B., E.D. PRINCE, and D.W. LEE, 1983. Age and growth of young-of-the-year Bluefin Tuna, Thunnus thynnus, from otolith microstructure. Pages 49-59 in E.D. Prince and L.M. Pulos (editors), Proceedings of the international workshop on age determination of oceanic pelagic fishes: Tunas, billfishes, and sharks. National Oceanographic and Atmospheric Administration United States National Marine Fisheries Service Technical Report 8.

CAMPANA, S.E. 1983. Feeding periodicity and the production of daily growth increments in otoliths of steelhead trout (Salmo gairdneri) and starry flounder (Platichthys stellatus). Canadian Journal of Zoology 61:1591-1597.

CAMPANA, S.E. and J.D. NEILSON. 1982. Daily growth increments in otoliths of starry flounder (Platichthys stellatus) and the influence of some environmental variables in their production. Canadian Journal of Fisheries and Aquatic Sciences 39:937-942.

CAMPANA, S.E. and J.D. NEILSON. 1985. Microstructure of fish otoliths. Canadian Journal Fisheries and Aquatic Sciences 42:1014-1032.

GEFFEN, A.J. 1982a. Otolith ring deposition in relation to growth rate in herring (Clupea harengus) and turbot (Scophthalmus maximus) larvae. Marine Biology 71:317-326.

GEFFEN, A.J. 1982b. Growth and otolith ring deposition in Teleost larvae. Doctoral dissertation. University of Stirling, Stirling, Scotland.

GEFFEN, A.J. 1983. The deposition of otolith rings in Atlantic salmon, Salmo salar L., embryos. Journal of Fish Biology 23:467-474.

GEFFEN, A.J. in press. The growth of herring larvae (Clupea harengus L.) in the Clyde, an assessment of the suitability of otolith ageing methods. Journal of Fish Biology

GEFFEN, A.J. and R.D.M. NASH. 1985. Prolonged recruitment of young-of-year rainbow smelt to bottom trawls: migration or over winter cessation of growth. International Symposium on Age and Growth of Fish, Des Moines, Iowa, June 9-12, 1985, Abstract 9:3.

GJØSAETER, H. and V. ØIESTAD. 1981. Growth patterns in otoliths as an indication of daily growth

variations of larval herring Clupea harengus from
an experimental ecosystem. International Council
for the Exploration of the Seas. Council Meeting
Papers H/31:1-9.

GJØSAETER, H. and S. TILSETH. 1982. Primary growth
increments in otoliths of cod larvae (Gadus morhua
L.) of the Arcto-Norwegian cod stock.
Fiskeridirektoratets Skrifter. Serie
Havundersøkelser 17:287-295.

HETTLER, W.F. 1984. Marking otoliths by immersion of
marine fish larvae in tetracycline. Transactions of
the American Fisheries Society 113:370-373.

HUNTER, J.R. 1984. Inferences regarding predation on the
early life stages of cod and other fishes.
Flødevigen rapportser 1:533-562.

LAROCHE, J.L., S.L. RICHARDSON, and A.A. ROSENBERG.
1982. Age and growth of a pleuronectid, Parophrys
vetulus, during the pelagic larval period in Oregon
coastal waters. United States National Marine
Fisheries Service Fishery Bulletin 80:93-104.

LOUGH, R.G., M. PENNINGTON, G.R. BOLZ, and A. ROSENBERG.
1982. Age and growth of larval Atlantic herring,
Clupea harengus, in the Gulf of Maine-Georges Bank
region based on otolith growth increments. United
States National Marine Fisheries Service Fishery
Bulletin 80:187-200.

McGURK, M.D. 1984. Ring deposition in the otoliths of
larval Pacific herring, Clupea harengus pallasi.
United States National Marine Service Fishery
Bulletin 82:113-120.

MARSHALL, L.S. and S.S. PARKER. 1982. Pattern
identification in the microstructure of Sockeye
salmon (Oncorhynchus nerka) otoliths. Canadian
Journal of Fisheries and Aquatic Sciences
39:542-547.

METHOT, R.D., Jr. 1981. Growth rates and age
distribution of larval and juvenile northern
anchovy, Engraulis mordax, with inferences on
larval survival. Doctoral dissertation. University
of California, San Diego, California, USA.

METHOT, R.D., Jr. 1983. Seasonal variation in survival
of larval northern anchovy Engraulis mordax
estimated from the age distribution of juveniles.
United States National Marine Fisheries Service
Fishery Bulletin 81:741-750.

METHOT, R.D. and D. KRAMER. 1979. Growth of northern
anchovy Engraulis mordax larvae in the sea. United
States National Marine Fisheries Service Fishery
Bulletin 77:413-423.

MILLER, S.J. and T. STORCK. 1982. Daily growth rings in
otoliths of young-of-the-year largemouth bass.

Transactions of the American Fisheries Society
111:527-530.

NEILSON, J.D. and G.H. GEEN. 1982. Otoliths of chinook
salmon (Oncorhynchus tshawytscha): daily growth
increments and factors influencing their
production. Canadian Journal of Fisheries and
Aquatic Sciences 39:1340-1347.

PANNELLA, G. 1971. Fish otoliths: Daily growth layers
and periodical patterns. Science 173:1124-1127.

RADTKE, R.L. 1984. Formation and structural composition
of larval striped mullet otoliths. Transactions of
the American Fisheries Society 113:186-191.

RADTKE, R.L. and J.M. DEAN. 1982. Increment formation in
the otoliths of embryos, larvae, and juveniles of
the mummichog, Fundulus heteroclitus. United States
National Marine Fisheries Service Fishery Bulletin
80:41-55.

RICE, J.A. 1985. Mechanisms regulating survival of
larval bloater Coregonus hoyi in Lake Michigan.
Doctoral dissertation, University of Wisconsin,
Madison, Wisconsin, USA.

ROSENBERG A.A. and J.L. LAROCHE. 1982. Growth during
metamorphosis of English sole, Parophrys vetulus.
United States National Marine Fisheries Service
Fishery Bulletin 80:150-153.

STRUHSAKER, P. and J.H. UCHIYAMA. 1976. Age and growth
of the nehu, Stolephorus purpureus (Pisces,
Engraulidae), from the Hawaiian islands as
indicated by daily growth increments of sagittae.
United States National Marine Fisheries Service
Fishery Bulletin 74:9-16.

TANAKA, K., Y. MUGIYA, and J. YAMADA. 1981. Effects of
photoperiod and feeding on daily growth patterns in
otoliths of juvenile Tilapia nilotica. United
States National Marine Fisheries Service Fishery
Bulletin 79:459-466.

TAUBERT, B.D. and D.W. COBLE. 1977. Daily growth rings
in otoliths of three species of Lepomis and Tilapia
mossambica. Journal of the Fisheries Research
Board of Canada 34:332-340.

TOWNSEND, D.W. and J.J. GRAHAM. 1980. Growth and age
structure of larval Atlantic herring, Clupea
harengus harengus, in Sheepscot River estuary, as
determined by daily growth increments in otoliths.
United States National Marine Fisheries Service
Fishery Bulletin 79:123-130.

VICTOR, B.C. 1982. Daily otolith increments and
recruitment in two coral-reef wrasses, Thalassoma
bifasciatum and Halichoeres bivittatus. Marine
Biology 71:203-208.

VOLK, E.C., R.C. WISSMAR, C.A. SIMENSTAD and D.M.

EGGERS. 1984. Relationship between otolith microstructure and the growth of juvenile chum salmon (Oncorhynchus keta) under different prey rations. Canadian Journal of Fisheries and Aquatic Sciences 41:126-133.

WILSON, K.W. and P.A. LARKIN. 1980. Daily growth rings in the otoliths of juvenile sockeye salmon. Canadian Journal of Fisheries and Aquatic Sciences 37:1495-1498.

WILSON, K.W. and P.A. LARKIN. 1982. Relationship between thickness of daily growth increments in sagittae and changes in body weight of sockeye salmon (Oncorhynchus nerka) fry. Canadian Journal of Fisheries and Aquatic Sciences 39:1335-1339.

Validity of the otolith for determining age and growth of walleye, striped bass, and smallmouth bass in power plant cooling ponds

ROY C. HEIDINGER AND
KENNETH CLODFELTER
FISHERIES RESEARCH LABORATORY
AND DEPARTMENT OF ZOOLOGY
SOUTHERN ILLINOIS UNIVERSITY
CARBONDALE, IL 62901

ABSTRACT
   The age determined from the otolith agreed 100% of the
time with the known-age of 0+, 1+, 2+, 3+, and 4+ smallmouth
bass Micropterus dolomieui, striped bass Morone saxatilis,
and walleye Stizostedion vitreum from two power cooling ponds
in northern Illinois. Ages determined from the scales agreed
with known-ages 71%, 80%, and 75% of the time for smallmouth
bass, striped bass, and walleye, respectively.
   Walleye, smallmouth bass, and striped bass were collected
twice a year for 5 years and empirical mean total lengths
were compared with the back-calculated mean total lengths
obtained from the otoliths and scales. For all three species,
back-calculated mean total lengths at each age from both the
otoliths and the scales compared favorably with the corresponding
empirical mean total lengths.

   The ability to age fish accurately from scales is a very
useful tool in fisheries management. Unfortunately, the validity
of the technique is often unknown. In 1941, Van Oosten stated
that fisheries investigators had "taken too much for granted,"
by assuming that the validity of this method did not need
to be checked on a case by case basis. Beamish and McFarlane
(1983) also concluded that the "requirement to validate age
determinations continues to be neglected."
   Scale annuli from known-age fish have been inconsistent
indicators of age, even in relatively young fish. Prather
(1967) checked known-age largemouth bass Micropterus salmoides

The Age and Growth of Fish, edited by Robert C. Summerfelt and Gordon E. Hall © 1987 The Iowa
State University Press, Ames, Iowa 50010.

and bluegill <u>Lepomis</u> <u>macrochirus</u> from Alabama against their
scale-derived ages. Of 272 largemouth bass that were known
to be ages 0, 1, or 2, 80% were aged correctly, and of 264
age 1, 2, or 3 known-age bluegill only 76% were aged correctly.
Incorrect ages from scales are particularly prevalent in southern
latitudes and in power cooling lakes (Taubert and Tranquilli
1982).

        This paper compares the scale- and otolith-derived age
of walleye <u>Stizostedion</u> <u>vitreum</u>, striped bass <u>Morone</u> <u>saxatilis</u>,
and smallmouth bass <u>Micropterus</u> <u>dolomieui</u> with their known
age. In addition, back-calculated mean total lengths derived
from the scales and otoliths are compared with empirical mean
total lengths.

METHODS AND MATERIALS
        This study was conducted in two bermed cooling ponds
(Dresden Pond and Collins Pond) in northern Illinois. The
805 ha Collins Pond serves five, 500 MWe oil-fired peaking
units. The 516 ha Dresden Pond serves two, 800 MWe nuclear-fueled
base-loaded units. Thermal regimes in the upper, middle,
and lower end of each cooling pond were continuously monitored
with Ryan temperature recorders placed 1 m below the surface.
Thermal profiles indicated that lateral and vertical temperatures
did not differ by more than 3 C from those recorded by the
continuous temperature monitors.

        Weekly mean temperature in the cool end and in the middle
of Collins Pond typically reaches 0 C for several months during
the winter; in the vicinity of the warmwater discharge,
temperatures remain above 5 C. Summertime weekly mean maximum
temperatures in the cool end of the pond ranged between 25
and 30 C, whereas in the warm end they were frequently above
30 C. Dresden Pond tended to be warmer than Collins Pond;
weekly mean temperature in the cool end of Dresden Pond typically
reached 0 C for 6 weeks in the winter; in the middle of the
pond it exceeded 5 C, and in the warm end 10 C. Weekly mean
summer maximum temperature in the cool end of Dresden Pond
exceeded 27 C; the middle and the warm end 35 C. On occasion,
in both cooling ponds, considerable (10 C) daily fluctuations
in water temperature occurred because of the start-up or shut-down
of electric power generating units.

        Marked fingerling walleye and striped bass were stocked
in Collins Pond; marked smallmouth bass were stocked in Dresden
Pond. Finclipped fingerlings were stocked in late summer
each year from 1980 through 1984. Marked fish were recovered
in April and October from 1980 through 1984 by electrofishing,
gillnetting, and rotenone sampling. The total length of each
fish was recorded, and a scale sample and the sagittal otoliths
were removed for age analysis. Several scales were removed
from the fish below the lateral line near the point of the
pectoral fin when the fin is pressed against the body.

        Otoliths were prepared for aging by breaking them in

half perpendicular to the longest axis. The unbroken end
of the otolith was imbedded in a block of clay, and the freshly
broken surface coated with immersion oil and examined under
a binocular dissecting scope. Walleye and smallmouth bass
otoliths were examined with 50 power magnification and those
of striped bass with 25 power magnification. Contrast between
opaque (summer) and hyaline (winter) zones was enhanced by
side illumination. The hyaline zones were treated as annuli
and counted. In walleye and striped bass from Collins Pond,
and smallmouth bass from Dresden Pond, the last annulus on
the otolith was not distinguishable before April, and, in
some cases, not until late May. Therefore, since the samples
were collected in April, we measured the distance, using an
ocular micrometer, from the center of the kernel to the center
of each hyaline zone, rather than the distance from the center
of the kernel to the beginning or end of each hyaline zone.

Scale impressions were made on heated 0.08-cm-thick cellulose
acetate slides in a roller press, and magnified images of
the impression were examined for annuli with a microprojector.
Scales and otoliths were examined on two different days and
age determinations were made without reference to fish length
or known age.

Walleye and smallmouth bass length at age was back-calculated
from scale measurements using a linear regression model with
the intercept values recommended by Carlander (1982). A standard
intercept value has not been proposed for striped bass, thus
a value of 30 mm was assumed. Only fish whose number of annuli
on the scales agreed with the known age were used in the backcalculations. Because standard intercept values are not available
for back-calculating lengths from the otolith, intercept values
were derived from the linear regression model and also an
assumed zero intercept. Visual examination of the plots of
the otolith radius versus fish length for walleye (Fig. 1),
striped bass (Fig. 2), and smallmouth bass (Fig. 3) suggests
that the relationship may be curvilinear. Therefore, an intercept
based on the log of the radius and the log of the fish length
(log-log) was also calculated by regression analysis. The
back-calculated mean total lengths at each age derived from
the scales and the otoliths were statistically compared to
each other and to the empirical mean total length by the analysis
of variance technique with Scheffe post hoc procedure. Significant
differences were set at the 0.05 level.

The Scheffe post hoc procedure was also used to test
for significant differences between mean total lengths of
each year-class at each age for all three species. The only
significant difference ($P = 0.0230$) was between the age 1,
1982, walleye year-class and the age 1, 1983, year-class.
Inasmuch as the various year-classes were growing at essentially
the same rate, the data were combined for the back-calculations
and for calculating mean empirical total lengths at each age.
For example, the mean total length of age 1 fish was calculated

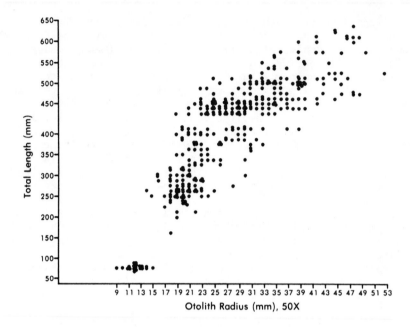

Figure 1. The relationship between walleye total lengths and otolith radius.

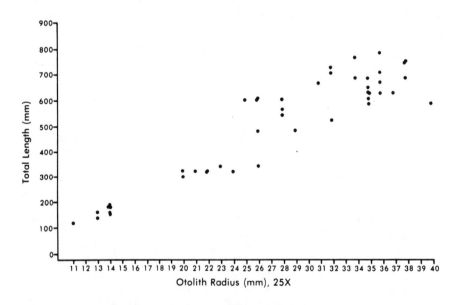

Figure 2. The relationship between striped bass total length and otolith radius.

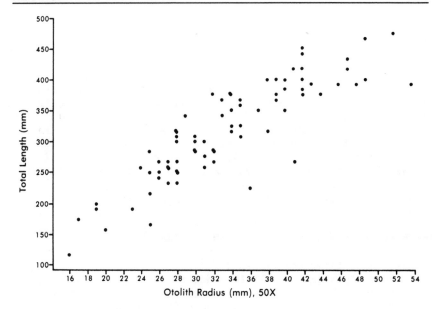

Figure 3. The relationship between smallmouth bass total
length and otolith radius.

from four year-classes: 1980 fish collected in 1981; 1981
fish collected in 1982; 1982 fish collected in 1983; and 1983
fish collected in 1984. The mean total length of age 4 fish
was calculated from fish stocked in 1980 and collected in
1984.

RESULTS
       The calculated age of all striped bass, walleye, and
smallmouth bass (ages 0+ through 4+) matched the corresponding
age derived from the otolith. Sample size, known-age, and
mean total length of these fish at the time of recapture are
given in Table 1. The number of annuli on the scales agreed
with the known-age of the fish 80%, 75%, and 71% of the time
for striped bass, walleye, and smallmouth bass. Of those

Table 1. Mean total length (mm) at the time of recapture
of finclipped striped bass and walleye from Collins
Pond and smallmouth bass from Dresden Pond.

Known age	Striped bass		Walleye		Smallmouth bass	
	Number	Length	Number	Length	Number	Length
0	5	132	14	161	16	156
1	6	218	10	304	18	268
2	1	485	5	475	12	368
3	12	535	3	488	7	391
4	7	656	12	506	5	383

striped bass that were mis-aged, 66% were under-aged by 1
year and 33% were over-aged by 1 year. Twenty-seven percent
of the walleye were over-aged by 1 year and 73% were under-aged
by 1 or 2 years. All of the smallmouth bass were under-aged
by 1 year.

The back-calculated mean total lengths at each age from
both the otoliths and the scales of all three species compared
favorably with the corresponding April mean empirical total
lengths. No significant differences between back-calculated
mean total lengths and empirical mean total lengths were found
for the striped bass (Table 2). There was a significant difference
at age 1 between the back-calculated mean total length obtained
using the otolith and the log-log derived intercept of 4 mm,
and the back-calculated mean total length obtained using the
scales with an assumed intercept of 30 mm. There was also
a significant difference at age 1 between the back-calculated
mean total length derived from the scales with an assumed
intercept of 30 mm and that obtained from the otolith using
an assumed 0 intercept (Table 2).

The only significant difference in mean total length
of smallmouth bass was between the empirical age 1 fish and
the back-calculated mean total length from the otolith with
a linearly derived intercept of 39 mm (Table 3).

Table 2. Comparison of striped bass total length (mm) in Collins Pond as
determined by the scale method and otolith method to the observed
total length after 1 to 4 years of age.

	\multicolumn{7}{c}{Age}						
	\multicolumn{2}{c}{1+}	\multicolumn{2}{c}{2+}	\multicolumn{2}{c}{3+}	4+
	April	October	April	October	April	October	April
	\multicolumn{7}{c}{Empirical length}						
Sample size	3	9	4	10	7	9	7
Mean	231	322	421	548	621	678	717
Standard deviation	15	14	43	44	29	34	58
	\multicolumn{7}{c}{Back-calculated length from otolith with log-log intercept of 4 mm}						
Sample size	37		30		19		7
Mean	288[a]		482		623		717
Standard deviation	56		86		42		58
	\multicolumn{7}{c}{Back-calculated length from otolith with assumed intercept of 0 mm}						
Sample size	37		30		19		7
Mean	286[b]		481		623		717
Standard deviation	56		56		42		58
	\multicolumn{7}{c}{Back-calculated length from otolith with a linearly derived intercept of -121 mm}						
Sample size	37		30		19		7
Mean	228		455		614		717
Standard deviation	53		55		44		58
	\multicolumn{7}{c}{Back-calculated length from scale with assumed intercept of 30 mm}						
Sample size	49		37		23		7
Mean	203[a,b]		466		623		716
Standard deviation	52		62		38		60

[a,b] Means with the same superscript are significantly different at the
0.05 level.

Table 3. Comparison of smallmouth bass total length (mm) in Dresden Pond as determined by the scale method and otolith method to the observed total length after 1 to 4 years of age.

	Age							
	1+		2+		3+		4+	
	April	October	April	October	April	October	April	October
				Empirical length				
Sample size	18	60	20	22	11	9	4	4
Mean	176[a]	264	300	360	358	422	418	407
Standard deviation	38	34	37	35	42	24	25	50
			Back-calculated length from otolith with log-log intercept of 11.5 mm					
Sample size	72		40		17		8	
Mean	194		292		350		390	
Standard deviation	33		33		35		42	
			Back-calculated length from otolith with assumed intercept of 0 mm					
Sample size	72		40		17		8	
Mean	189		289		348		390	
Standard deviation	34		34		36		42	
			Back-calculated length from otolith with a linearly derived intercept of 39 mm					
Sample size	72		40		17		8	
Mean	205[a]		298		354		392	
Standard deviation	32		32		44		41	
			Back-calculated length from scale with standard intercept of 35 mm					
Sample size	148		70		28		8	
Mean	187		294		364		395	
Standard deviation	43		41		40		37	

[a]Means with the same superscript are significantly different at the 0.05 level.

A number of significant differences were found among the mean total lengths of walleye (Table 4). At age 1 there was a significant difference between the back-calculated mean total length derived from the otolith using the log-log intercept of 8.1 mm and the mean total length derived from the otolith using the linear intercept of 71 mm. There was also a significant difference between the latter and the back-calculated mean total length derived from the otolith using the assumed 0 intercept. At age 2 the back-calculated mean total length from the scales and the empirical mean total length were not significantly different. Regardless of the intercept used, there were no significant differences among the back-calculated mean total lengths from the otolith at age 2, but all other comparisons at age 2 between otolith-derived mean total lengths and scale-derived mean total lengths were significantly different (Table 4).

Compared with other populations throughout the United States, the walleye and striped bass in Collins Pond and the smallmouth bass in Dresden Pond were growing rapidly (Table 5).

Table 4. Comparison of walleye total length (mm) in Collins Pond as determined by the scale method and otolith method to the observed total length after 1 to 4 years of age.

	Age							
	1+		2+		3+		4+	
	April	October	April	October	April	October	April	October
			Empirical length					
Sample size	43	43	70	58	26	29	18	11
Mean	278	362	424[a,b,c,d]	462	489	520	522	586
Standard deviation	25	41	22	29	32	45	49	48
			Back-calculated length from otolith with log-log intercept of 8.1 mm					
Sample size	298		212		84		29	
Mean	269[a]		399[d,g]		474		532	
Standard deviation	37		36		41		49	
			Back-calculated length from otolith with assumed intercept of 0 mm					
Sample size	298		212		84		29	
Mean	266[b]		398[b,f]		473		532	
Standard deviation	38		37		41		49	
			Back-calculated length from otolith with a linearly derived intercept of 71 mm					
Sample size	298		212		84		29	
Mean	290[a,b]		408[a,c,e]		479		534	
Standard deviation	34		32		40		49	
			Back-calculated length from scale with standard intercept of 55 mm					
Sample size	298		212		84		29	
Mean	278		424[e,f,g]		497		543	
Standard deviation	44		32		39		49	

[a,b,c,d,e,f,g] Means with the same superscript are significantly different at the 0.05 level.

Table 5. Back-calculated total length (mm) at scale annuli of walleye, striped bass, and smallmouth bass from several lakes and reservoirs in the U.S.

Location	Reference	Age				
		1	2	3	4	5
	Smallmouth bass					
Claytor Lake, VA	Roseberry 1950	91	198	300	384	423
Norris Reservoir, TN	Stroud 1948	118	258	358	411	445
Lake Erie	Doan 1938	85	142	255	369	539
Oneida Lake, NY	Forney 1972	94	171	269	342	405
Dresden Pond		188	287	355	376	396[a]
	Striped bass					
Santee-Cooper, SC	Stevens 1958	216	399	503	583	
Keystone, OK	Mensinger 1971	258	455	541	606	
Herrington Lake, KY	Axon 1979	251	404	559	653	
Kerr Reservoir, VA/NC	Domrose 1963	129	280	415	561	
Collins Pond		182	451	604	653	
	Walleye					
Norris Reservoir, TN	Stroud 1949	264	416	474	505	533
Center Hill Reservoir, TN	Muench 1966	248	415	497	512	583
Claytor Lake, VA	Roseberry 1950	251	386	503	580	663
Canton Reservoir, OK	Lewis 1970	309	426	495	553	607
Collins Pond		279	431	499	538	608[b]

[a] Mean length in November of 3 smallmouth bass.

[b] Mean total length in November 1984 of 9 walleye collected.

DISCUSSION

Most investigators who have used the otolith to age fish attempted to verify the accuracy of the method without using known-age fish (Ambrose 1983). Erickson's (1983) use of 100, 3+ walleye in a Manitoban lake, and Taubert and Tranquilli's (1982) use of 9, 2+ largemouth bass from a Florida fish hatchery are notable exceptions. In both cases there was agreement between known-age and otolith-derived age.

In the present study, walleye, striped bass, and smallmouth bass of known age (0+ to age 4+), collected from power cooling ponds were aged correctly with the otolith. The rapid growth rate and the relative youth of these fish may have contributed to this accuracy. Regardless, we found that determining the age from the otolith is more precise and accurate than determining the age from the scales.

For all three species, back-calculated mean total lengths from the scales were not significantly different from the mean empirical total lengths. This suggests that valid back-calculated lengths can be obtained from the scales, if those fish whose number of otolith annuli do not agree with the number of scale annuli are discarded. In the present study, 20% of the striped bass, 25% of the walleye, and 29% of the smallmouth bass sampled would have been discarded under those conditions.

Back-calculated mean total lengths derived from the otolith using the log-log or assumed zero intercept also gave a good estimate of the empirical mean total lengths for all species at all ages except for the age 2 walleye. The back-calculated means for walleye age 2, using either the log-log derived intercept or the assumed zero intercept, did not agree with the empirical means. We could find no reason why so many of the age 2 walleye mean total lengths were significantly different.

Considering all three species, none of the various methods of calculating the intercept (log-log, zero, linear derivation) consistently affected whether or not the back-calculated mean total length and the corresponding empirical mean total length was significantly different.

Although the present study was limited to fish with a maximum age of 4+, the otolith aging method may be valid for older fish as well. Unmarked white bass Morone chrysops x striped bass hybrids were stocked into Collins Pond in 1978 and 1979. Otoliths from 53 of those fish were examined in 1984; all of the otoliths exhibited either 5 or 6 annuli.

ACKNOWLEDGMENTS

This research was funded by Commonwealth Edison Company, Chicago, Illinois, and the Electric Power Research Institute, Palo Alto, California (Project 1743).

REFERENCES

AMBROSE, J. 1983. Pages 301-324 in L. A. Nielsen and
    D. L. Johnson, editors, Fisheries techniques, American
    Fisheries Society, Bethesda, Maryland, U.S.A.

AXON, J. R. 1979. An evaluation of striped bass introduction
    in Herrington Lake. Kentucky Department of Fisheries
    and Wildlife Resources. Bulletin No. 63.

BEAMISH, R. J., and G. A. MCFARLANE. 1983. The forgotten
    requirement for age validation in fisheries biology.
    Transactions of the American Fisheries Society 112:735-743.

CARLANDER, K. D. 1982. Standard intercepts for calculating
    lengths from scale measurements from centrarchid and
    percid fishes. Transactions of the American Fisheries
    Society 111:332-336.

DOAN, K. H. 1938. Lake Erie smallmouths. Ohio Conservation
    Bulletin. 2:18-39

DOMROSE, R. J. 1963. Warmwater fisheries management
    investigations - striped bass study. Virginia Commission
    of Game and Inland Fisheries, Federal Aid Project in
    Fish Restoration, F-5-R-8, Richmond, Virginia, U.S.A.

ERICKSON, C. M. 1983. Age determination of Manitoban walleyes
    using otoliths, dorsal spines, and scales. North American
    Journal of Fisheries Management 3:176-181.

FORNEY, J. L. 1972. Biology and management of smallmouth
    bass in Oneida Lake, New York. New York Fish and Game
    Journal 19:132-154.

LEWIS, S. A. 1970. Age and growth of walleye, Stizostedion
    vitreum vitreum (Mitchill), in Canton Reservoir, Oklahoma.
    Proceedings of the Oklahoma Academy of Science 50:84-86.

MENSINGER, G. C. 1971. Observations of the striped bass,
    Morone saxatilis, in Keystone Reservoir, Oklahoma. Pro-
    ceedings of the Annual Conference of Southeastern
    Association of Game and Fish Commissioners 24:447-463.

MUENCH, K. A. 1966. Certain aspects of the life history
    of the walleye, Stizostedion vitreum vitreum (Mitchill),
    in Center Hill Reservoir, Tennessee. M.S. thesis.
    Tennessee Technological University, Cookeville,
    Tennessee. 66 p.

PRATHER, E. E. 1967. The accuracy of the scale method in
    determining the ages of largemouth bass and bluegills.
    Proceedings of the Annual Conference Southeastern Association
    of Game and Fish Commissioners 20:483-486.

ROSEBERRY, D. A. 1950. Game fisheries investigation of
    Claytor Lake, a main stream impoundment of New River,
    Pulaski County, Virginia, with emphasis on Micropterus
    punctulatus (Rafinesque). Ph.D. thesis. Virginia
    Polytechnic Institute and State University, Blacksburg,
    Virginia, U.S.A.

STEVENS, R. E. 1958. The striped bass of the Santee-Cooper
    Reservoir. Proceedings of the Annual Conference Southeastern
    Association of Game and Fish Commissioners 11:253-264.

STROUD, R. H. 1948. Growth of the basses and black crappie in Norris Reservoir, Tennessee. Journal of the Tennessee Academy of Science 23:31-99.

STROUD, R. H. 1949. Growth of Norris Reservoir walleye during the first twelve years of impoundment. Journal of Wildlife Management 13:157-177.

TAUBERT, B. D., AND J. A. TRANQUILLI. 1982. Verification of the formation of annuli in otoliths of largemouth bass. Transactions of the American Fisheries Society 11:531-534.

VAN OOSTEN, J. 1941. The age and growth of freshwater fishes. Pages 196-205 in A symposium on hydrobiology. University of Wisconsin Press, Madison, Wisconsin, U.S.A.

Age and validation of age from otoliths for warm water fishes from the Arabian Gulf

M. SAMUEL, C. P. MATHEWS, AND A. S. BAWAZEER

MARICULTURE AND FISHERIES DEPARTMENT
KUWAIT INSTITUTE FOR SCIENTIFIC RESEARCH
P.O. BOX 24885 SAFAT
STATE OF KUWAIT

ABSTRACT

Most marine fishes in Kuwait have conspicuous marks on their otoliths. For some species, e.g. newaiby Otolithes argenteus, hamoor Epinephelus tauvina, and hamra Lutjanus coccineus, validation shows that the marks are true annuli. Other species, e.g. nakroor Pomadasys argenteus, have clear marks but the validation procedure demonstrates that they are not formed annually; and for still other species, e.g. zobeidy Pampus argenteus, no clear marks can be seen.

The otoliths of tropical fish are very different from those of temperate water fish. The general appearance is hyaline with thin opaque zones, and there are no clear-cut differences between hyaline and opaque zones. Newaiby otoliths are typical of this condition; nevertheless, their age may easily be determined. In contrast, hamra otoliths need to be heated and the opaque and hyaline materials are then seen as whitish-grey bands in reflected light; thin, dark carbonaceous lines with some hyaline material alternate with these.

The advantages and drawbacks of the traditional aging techniques are examined for about 10 species, and the choice of technique is discussed. The need for aging by daily marks is also addressed. Further research on the timing of annual mark formation, body growth and otolith growth is needed in tropical fish.

Stock assessment and rational management of fin fisheries require knowledge of growth and mortality rates of fish stocks. Growth and mortality rates are usually estimated from age composition of either the landings or suitably large samples of fish. Considerable literature exists on methods of accurately aging fish

The Age and Growth of Fish, edited by Robert C. Summerfelt and Gordon E. Hall © 1987 The Iowa State University Press, Ames, Iowa 50010.

by means of annual marks, but this technology was developed
primarily on fishes of temperate waters.

Fishes in tropical and equatorial waters are generally considered difficult or impossible to age, because annual marks are
usually absent in their otoliths. A solution to this problem was
provided by the discovery of daily increments in otoliths (Panella
1971). Brothers et al. (1976) applied this technique to compare
ages of Merluccius sp. from the Gulf of California obtained from
counts of annuli and daily increments; they demonstrated that both
techniques gave similar results.

Routine stock assessents designed to produce estimates of
growth and mortality rates require aging of several hundred specimens of each species and which cover the full size and age ranges
found in the landings. For old fish, especially, the slicing and
counting of daily marks takes too much time. Therefore, for
routine assessments, the high cost and the great labour associated
with using daily marks makes the technique difficult to apply
in the developing tropical countries, where the need for it is
greatest. On the other hand, daily marks can be used easily and
cheaply to validate estimates of growth obtained from application
of other methods (Morgan, 1986). Although they can also be used to
obtain age-length keys, this has not been the practise.

An alternative approach was tried by Cassie (1954), who
obtained ages by analysis of size-frequency distributions. Mathews
(1974) used this same procedure on several tropical fish populations, and concluded that aging by means of otoliths was more
reliable.

Because many of the fish populations in the Arabian Gulf have
been heavily exploited, it was considered desirable to assess its
fishery resources and study the life history of individual species
to aid in preventing over exploitation, such as occurred in the
shrimp fishery (Mathews and Al-Ghaffar, 1986). Mathews et al.
(1979) attempted to age fish from this area by means of scales, but
concluded that it was too difficult. Bedford (1982) and Williams
(1986) applied established aging methodology with otoliths (Holden
and Raitt 1974; Williams and Bedford 1974) to Kuwaiti fish stocks.
After examination of the different kinds of otoliths, sagittae were
chosen for all species because they were larger and more easy to
break and polish.

The objective of this paper is to demonstrate that traditional
aging methods developed for fish of temperate waters, especially
those of the North Sea and northern Europe (Holden and Raitt 1974),
can be applied successfully to age various fishes from the Arabian
Gulf. Oceanographic surveys have been conducted regularly in
Kuwait waters since 1978 (Mathews and Lee in press), and they
provide information on oceanographic conditions during the period
covered by these ageing studies.

METHODS

In a preliminary examination of scales, vertebrae and otoliths of Kuwaiti fishes, otoliths were found to provide the clearest marks (Bedford 1982; Williams 1986). Three species, the newaiby, Otolithes argenteus, hamra Lutjanus coccineus and hamoor Epinephelus tauvina were chosen for age and growth studies and validation of the ageing methods.

Newaiby sagittae were ground on a wet grinder so that a polished edge passed through the otolith nucleus. The fine marks on the polished edge are seen more easily when the otolith is kept under water and observed against a black background. Successive hyaline and opaque marks can easily be distinguished in both reflected and transmitted light. Hamoor sagittae were also examined in this way for two to six year old fish, but marks on the otolith were more conspicuous after slight burning of the otolith in an alcohol flame; otoliths of older fish all had to be examined after burning. Hamra otoliths were also broken and burned in an alcohol flame; the amount of heat applied was determined carefully and varied with the thickness of the otolith.

Marks on otoliths were validated by examination of the marginal material over a 12 month cycle. For newaiby and hamoor, the presence of opaque or hyaline material was noted at the margin, but for hamra, it was not possible to distinguish these zones. Therefore, after heating of these otoliths, the presence of white or dark material at the margin was noted.

In addition to the detailed work validating ages in newaiby, hamoor and hamra, the otoliths of 10 other species were also examined to determine whether routine age determination based on otoliths could be carried out.

RESULTS

Newaiby

Newaiby have relatively large and thick sagittae (Williams, 1986). The large nucleus with its surrounding thin opaque zone can be seen clearly in the otolith section. The newaiby sagitta, which is mostly hyaline with a thin opaque zone, showed a marked contrast to the sagitta of many temperate species, e.g. cod (Blacker 1974), which typically tends to have a more prominent opaque zone.

The incidence of opaque and hyaline material at the margin of sagittae through a calendar year is shown in Fig. 1. Young-of-year newaiby were taken in July and August, and all individuals had an opaque margin; hyaline margins appeared in a few individuals in September, and by October all sagittae examined had hyaline margins. In February, some age I individuals showed opaque material at the margins; hyaline material appeared at the margins of some otoliths in July and occurred at the margins of all such otoliths in August. The appearance of opaque and hyaline material at the edge of sagittae of two-and three-year old fish was

similar. Only a few four-year old and one five-year old fish were
sampled, so the period of annulus formation was not determined for
older fish.

Annual marks appeared on the sagittae of newaiby with
sufficient regularity and precision for them to be used routinely
in age determination. Sexually mature, one-year old newaiby are
found in March and April and the first spawning appeared to occur
at this time (Fig. 1).

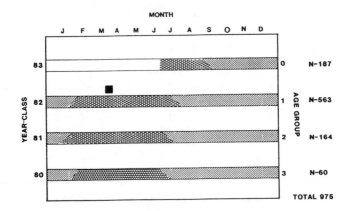

Figure 1. Validation diagram for newaiby; N = number of fish
    sampled in each age group.

Young-of-year newaiby grew rapidly between July, when they
first appeared in the catches, and December (Fig. 2). Growth rates
were slower for one-and two-year old fish than for young-of-year.
There was, however, no overlap of the growth curves for any age
newaiby (Fig. 2). Increase in length occurred mainly from about
May to December for all age groups. Mathews and Samuel (in press)
noted that different cohorts tended to show rather different
growth rates.

Hamoor
    The sagittae of hamoor were large and usually thin in younger
fish and rather thick and heavy in older fish. In transmitted
light, the sagittae appeared to be composed of more hyaline than
opaque material, and light zones predominated; in reflected light
the sagittae appeared darker with thin light zones. The opaque
zones appeared as whitish grey in reflected light or dark in
transmitted light; the hyaline zones were a yellowish-rust color
in transmitted light. There was no clear transition between the

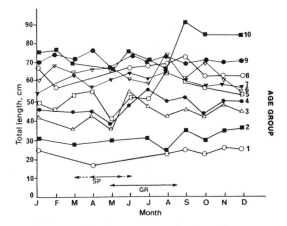

Figure 2. Mean size at age of newaiby during successive months.
Numbers indicate age group

hyaline and opaque zones, although sometimes it was possible to
discern a faint dark line of carbonaceous material between the
opaque and hyaline zone.

Opaque margins were found on hamoor sagittae from March to
September for 1-to 3-yr old fish and from March to July or August
for 5-to 10-yr old fish (Fig. 3). Fish older than 10 years were
caught too infrequently to validate ages, but opaque material was

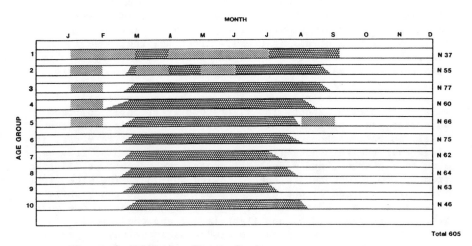

Figure 3. Validation diagram for hamoor.

found at the margins only from February–March to August–September; all 10-to 24-yr old fish showed the same type of marginal material as found in 0-to 9-yr old fish. Seasonal changes in average length of hamoor indicate that the growing season is from March to September (Fig. 4).

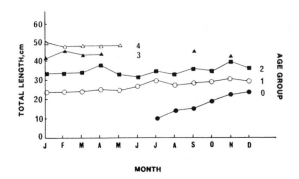

Figure 4. **Mean size at age** of hamoor during successive months. SP: Duration of the spawning season. The solid line indicates the period during which ripe fish are common; the broken line indicates the probable spawning range. GR: The period during which the otolith grows rapidly and has opaque material at the margin. Numbers indicate the age group.

Hamra

Similar to hamoor, the sagittae of hamra were thin in younger fish and increasingly thick and heavy in older fish. Hamra reached 45 years of age while growing slowly or not at all in length after 10 years (Mathews et al. in press). After heating, a series of distinct, broad white or grey zones were visible. They alternated with thin dark, carbonaceous layers, which sometimes contained a thin semihyaline layer on the inner side (i.e. on the side of the dark zone facing the nucleus). The dark zones were easily distinguished from light zones. The light zone is probably composed of an opaque and a hyaline zone combined, and the thin dark zone is likely to be a carbonaceous layer. For fish aged 1 to 9 yrs, it is clear that the thin dark lines occurred at the margins from May–June to September, and wide light zones were present at all times from September–October to the following May–June. The wide light zones are composed of one hyaline plus one opaque zone, while the dark zones separated successive light zones. Therefore one year's growth is represented by one light layer; and the dark layers separated the different growth periods. Sagittae may be used to age hamra up to 9 yr with precision (Fig.5). The dark zones were present from June to September, and the wide light zones were absent at the edges at this time. The wide light zones were always present from September–October to the following May–June.

The uniform thinness of the dark layer suggests that growth of these otoliths may slow down from June to September; however,

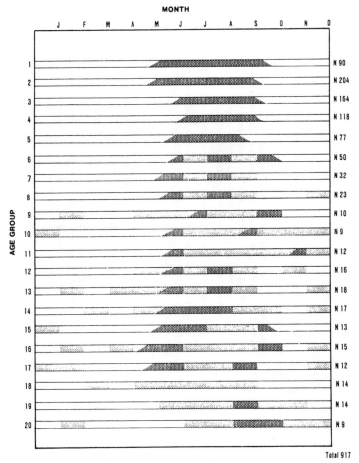

Figure 5. Validation diagram for hamra.

☐ thick light layer at otolith margin

▨ thin dark layer at otolith margin

▨ no samples

it is not known whether the otolith grew during the period when
the dark thin layer was forming at the margin.

There was little overlap in length of age-groups in the first
3 years of life, but considerable overlap from 4 to 6 years.
Overlap of length distributions of fish aged 7 years was so great
that it was difficult to distinguish different age-groups.

Other Species

The sagittae of a large number of other species of fish have
been screened for their suitability in age-determination work
(Table 1). Sheim Acanthopagrus latus and sobaity Acanthopagrus

Table 1. Notes on the type of otolith readibility for stock assessment purposes, age and size range of fish examined, and the best methods for age determination in each species. Readibility: (a) Difficult; (b) Easy; (c) Moderate; (d) Impossible.

Otolith Readability	Arabic Name	Latin Name/ English Name	Age Range (Years)	Length Range (total length, cm)	Shape and some notes on use of otolith for aging
(a)	Sheiry	Lethrinus spp.	1-10	16-61	Oval shaped and thick otoliths. The whole otolith can be read when illuminated from above and immersed in water against dark background – burning gives the same result. Otoliths are difficult to read especially for old and large fish owing to poor contrast between opaque and hyaline zones.
(a)	Wahar	Platycephalus indicus (Indian flathead)	1-7	13-67	Otoliths long and narrow. Whole otolith unreadable, burned sections improve clarity. They are difficult to interpret especially in young fish.
(a)	Nakroor	Pomadays argenteus (Silvery grunt)	-	18-74	Large and thick otolith. The burned cross section shows rings clearly but split rings and irregular ring formation through the year make aging impossible.
(b)	Sobaity	Acanthopagrus cuvieri (Silivery black porgy)	0-11	11-83	Oval shaped, slightly thick in older fishes. The whole otolith can be readwhen immersed in water and illuminated from above and viewed against a dark background. Burned sections confirm otolith readings especially in older fishes.
(b)	Chim	Arius thalassinus	0-18	11-79	Large, spherical otolith. Easy to read in unburned cross section with transmitted light. Split rings in otoliths of young fish near the nucleus may be confused with annual marks.

Table 1. (continued)

Otolith Readability	Arabic Name	Latin Name/English Name	Age Range (Years)	Length Range (total length, cm)	Shape and some notes on use of otolith for aging
(b)	Sheim	Acanthopagrus spp.	1–14	10–44	Thin, almost oval shaped sagitta. The whole sagittae can be read when immersed in water and illuminated from above and viewed against a dark background. Burned sections confirm otolith readings especially in older fishes. A white diffused opaque zone is present between the nucleus and the first opaque zone. In this diffused zone radial striations giving a halo effect occur around the nucleus; care must be taken not to count this as an annual mark.
(c)	Bassi	Nemipterus spp.	1–5	8–32	Oval shaped and thin sagitta. The whole sagitta can be read easily when illuminated from above against dark background immersed in water.
(c)	Khofaah	Pseudorhombus arsius	1–4	15–43	Hemispherical, thick and small sagittas. A burned cross section is possible to read, but large nuclear zone makes it difficult to interpret.
(c)	Mezlek	Brachyurus orientalis	1–15	18–37	Hemispherical and thick sagitta. Left hand sagitta may be used in burned cross section, right hand sagitta is usually unclear, showing diffused rings and nucleus.
(d)	Zobaidy	Pampus argenteus	–	10–32	Very thin and fragile otolith. Rings may be distinguished with difficulty in whole sagitta. Dark thick opaque and surrounding split hyaline zone make identifying the nature of rings impossible. It cannot be used for aging.

261

cuvieri are now aged on a routine basis, although the aging has
not yet been validated. Other species, nakroor Pomadasys argenteus
and zobeidy Pampus argenteus, cannot be aged at all.

Most of the sagittae examined in detail were from demersal
fish; they had a structure generally similar to those already
described i.e., they were predominantly composed of more hyaline
material than was commonly the case in European fish, and in some
species thin dark, apparently carbonaceous, layers appear.

Most pelagic fish sagittae could not be aged by annual marks.
Oom Sardinella spp., various species of boufchech Engraulis spp.
and other small pelagic fish have sagittae without clear annual
marks, possibly because of their short life cycles. Large pelagic
fish, such as Khubbat Scomberomorus guttatus and chenaid
Scomberomorus commersoni, could not be aged by means of annual
marks, while zobeidy showed only very irregular marks.

DISCUSSION

In general, sagittae of Arabian Gulf species have a hyaline
appearance with thin opaque zones, whereas sagittae of temperate
species (i.e. cod, Blacker 1974; cod, various flat fish, Williams
and Bedford 1974) have an opaque appearance with thin hyaline
marks. The distinction between the opaque and hyaline zones was
also less dramatic in Arabian Gulf fish than is usually the case
for temperate fish. Newaiby sagittae provided clear examples of
this.

Hamra sagittae appeared to be similar to those of European
flatfish (Christensen 1964) with a thin, carbonaceous layer
separating successive zones, each composed of an opaque and
hyaline zone. The thin dark marks in hamra sagittae aged 1 to 9
may definitely be used in aging, but because of the small samples
available, the annual occurrence of the marks in older hamra needs
further confirmation. Nevertheless, out of 159 fish aged 9 to 20
years (Fig. 6), thin dark lines appeared at the margins of all
fish sampled from May to September, and an opaque white layer
appeared at the margins of all fish sampled from October to April;
the only exception was in September and April, when some
overlapping of results was found. Therefore, even though further
validation work is necessary, annual marks were found in hamra
sagittae of all ages; and they may be used in routine age
determination and as a basis for stock assessments.

Unpublished data on the reproductive biology of newaiby,
hamra and hamoor are available (Mathews, personal communication).
These data show that a few newaiby spawn at 1 year, but the
majority do not spawn until they are 2-years old. Because
accretion of opaque material occurred in otoliths of young-of-year
fish prior to spawning, these marks are not spawning marks.
Spawning in hamoor peaked in April-May. The accretion of opaque
material on the otolith margin started slightly later than body
growth and appears to coincide with spawning. Nevertheless, the
marks are not spawning marks because they appear in one and two
year old fish, which are still immature. Spawning in hamra occurs

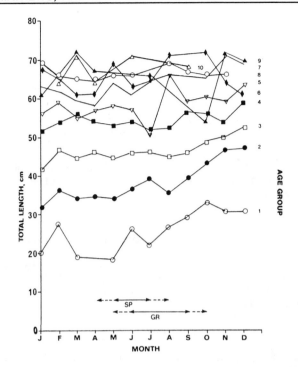

Figure 6. Mean size at age of hamra during successive months.
SP: Duration of the spawning season. The solid line indicates
the period during which ripe fish are common; the broken line
indicates the probable spawning range. GR: The period during
which the otolith grows rapidly and has opaque material at the
margin. Numbers indicate the age group.

mainly from May-June. The marks found in hamra otoliths are,
therefore, not spawning marks since they occurred in both 1-year
old fish and in older fish (11-45 yrs) which spawn less
frequently. Further investigation is necessary if the relations
between spawning, body growth and otolith growth are to be
understood.

    The incidence of annual mark formation may be related to
environmental conditions; in Kuwait, summer temperature usually
peaks in September (33°C; Mathews and Lee, in press) when opaque
zones appeared in sagittae of newaiby and hamoor and when otolith
growth started in all but young hamra. Hyaline material occurred
from about December to May. The lowest temperature (13-14°C)
usually occurred in January or February after deposition of
hyaline material at the margin. By May, temperatures may reach
25°C (Mathews and Lee, in press). It is clear that opaque
material does not simply represent "summer" growth and that
hyaline material does not simply represent "winter" growth in

Arabian Gulf fishes. The whole question of the relation between
growth of sagittae, body length and environmental changes needs to
be examined in detail for tropical and subtropical fishes. The
type of material laid down in tropical otoliths also needs to be
studied in further detail, especially in hamra. It is clear,
however that annual marks do appear in the sagittae of newaiby,
hamoor and hamra and that they may be used to estimate the age of
these fish reliably.

REFERENCES

BAGENAL, T.B., editor. 1974.   The ageing of fish: Proceedings of
    an Interational Symposium. Unwin Brothers Limited, Old Woking,
    Surrey, England.

BEDFORD, B. 1982.   The use of otoliths for ageing Kuwait's fish
    species. Kuwait Institute for Scientific Research, Kuwait

BLACKER, R.W. 1974. The ICNAF cod otolith. Photograph exchange
    scheme. Pages 108-113 in T.B. Bagenal, editor. Ageing of fish.
    Unwin Brothers Limited, Old Woking, Surrey, England.

BROTHERS, E.B., C.P. BROTHERS, AND R. LASKER 1976.   Daily growth
    increments in otoliths from larval and adult fishes. Fishery
    Bulletin 74:1-8.

CHRISTENSEN, J.M. 1964. Burning of otoliths, a technique for age
    determination of soles and other fish. Journal du Conseil
    29:73-81.

CASSIE, R.M. 1954. Some use of probability paper in the analysis
    of size frequency distributions. Australian Journal of Marine
    and Freshwater Research 5:513-22

HOLDEN, M.J., AND D.F.S. RAITT.   1974.   Manual of fisheries
    science. Part 2. Methods of resource investigation and their
    application.   FAO Fisheries Technical Paper No. 115 (Rev. 1).

MATHEWS, C.P. 1974.   An account of some methods of overcoming
    errors in ageing tropical and subtropical fish populations when
    hard tissue growth markings are unreliable and the data are
    sparse. Pages 158-166 in T.B. Bagenal editor, Ageing of fish.
    Unwin Brothers Limited, Old Woking, Surrey, England.

MATHEWS, C.P. 1986.   Fisheries Management in a Tropical Country:
    the most appropriate balance of size and age/length related
    methods for practical assessments. In D. Pauly and G.R. Morgan,
    editors, International Conference on the Theory and Application
    of Length based Fish Stock Assessment, 10-15 February 1985,
    Sicily, Italy (In Press).

MATHEWS, C.P. AND A.R. ABDUL-GHAFFAR.   1986.   A review of the
    present status of Kuwiat's shrimp fisheries with special
    reference to the need for effort limitation.   In A. Landry,
    editor.   First Shrimp Yield Workshop, 16-17 November 1983,
    Galveston, Texas (In press).

MATHEWS, C.P. AND J.W. LEE.   1986.   The Development of
    Oceanography in Kuwait. In the Symposium/Workshop on
    Oceanographic Modelling of the Kuwait Action Plan Region held

at the University of Petroleum and Minerals, Dharan, Saudi
Arabia, 15-17 October 1983. UNEP - Nairobi (In Press).

MATHEWS, C.P. AND M. SAMUEL. 1986. Stock assessment and
management of newaiby, hamoor and hamra in Kuwait. In C.P.
Mathews, editor, Proceedings of the 1984 Shrimp and Fin
Fisheries Management Workshop 15-17 December 1984. Kuwait
Institute for Scientific Research, Kuwait (In Press).

MATHEWS, C.P., M. SAMUEL, AND K. AL-ABDUL-ELAH. 1979. Fisheries
management studies in Kuwait. Kuwait Institute for Scientific
Research, Annual Research Report (1978):48-57

MORGAN, G.R. 1986. Incorporating age data into length based
stock assessment methods. In G.R. Morgan and D. Pauly, editors.
The International Conference on the Theory and Application of
Length Based Fish Stock Assessment, 10-15 February 1986,
Sicily, Italy (In Press).

PANNELLA, G. (1974). Otolith growth patterns: an aid in age
determination in temperate and tropical fishes. Pages 28-39 in
T.B. Bagenal, editor, Aging of fish. Unwin Brothers Limited,
Old Woking, Surrey, England.

SAMUEL, M. 1983. Preliminary stock assessment of newaiby
Otolithes argenteus in Kuwait waters. Pages 108-204 in C.P.
Mathews and A.R. Desai, editors, Proceedings of the Third
Shrimp and Fin Fisheries Management Workshop, 4-5 December
1982, Kuwait. Kuwait Institute for Scientific Research, Kuwait.

SAMUEL, M., A.S. BAWAZEER, AND M.K. BADDAR. 1985. Age
distribution and preliminary mortality estimation of Hamra
Lutjanus coccineus in Kuwait waters. Kuwait Institute for
Scientific Research, Annual Research, Kuwait (In Press).

SAMUEL, M., A.S. BAWAZEER, AND C.P. MATHEWS. 1985. Validation of
ring formation in the otoliths of newaiby and other fin fish in
Kuwait waters. In C.P. Mathews, editor, Proceedings of the
Shrimp and Fin Fisheries Management Workshop, 9-1 October 1983,
Kuwait. Kuwait Institute for Scientific Research, Kuwait (In
Press).

SAMUEL, M., AND C.P. MATHEWS. 1985. Validation and ageing of
newaiby, hamoor and hamra in Kuwait. In C.P. Mathews, editor,
Proceedings of 1984 Shrimp and Fin Fisheries Management
Project Workshop, 15-17 December 1984. Kuwait Institute for
Scientific Research, Kuwait (In Press).

SAMUEL, M., AND G.R. MORGAN. 1986. A comparision of length
related and age related stock assessment of newaiby, Otolithes
argenteus in Kuwait waters. Canadian Journal of Fisheries and
Aquatic Sciences (In Press).

WILLIAMS, T. 1986. Aging manual for Kuwaiti fish. Kuwait Institute
for Scientific Research, Annual Research Report (In Press).

WILLIAMS, T., AND B.C. BEDFORD. 1974. The use of otoliths for age
determination. Pages 114-123 in T.B. Bagenal, editor, Aging of
fish. Proceedings of an International Symposium, Unwin
Brothers, Old Woking, Surrey.

Verification and use of whole otoliths to age white crappie

MICHAEL J. MACEINA AND
ROBERT K. BETSILL
DEPARTMENT OF WILDLIFE AND FISHERIES SCIENCES
TEXAS A&M UNIVERSITY
COLLEGE STATION, TX 77843

ABSTRACT
    White crappie Pomoxis annularis were collected for one
year from two Texas reservoirs to determine if their age could
be interpreted from otoliths (sagittae). We found that (1) a
single opaque band was formed once a year, in late spring; (2)
fish length was highly correlated with otolith radius; and (3)
back-calculated lengths were similar to those derived from
length-frequency analysis. A sample of 257 otoliths, ages 1
to 9, were sectioned and annuli counted to provide a comparison
with whole-view age determination. Ages of only three slow-
growing fish, age 3 to 4, were underestimated by one year using
whole view. Computed coefficient of determination values, for
regressions predicting total length from otolith radius, were
lower for sectioned otoliths than for whole view. Whole-view
otolith examination was a reliable technique for determining
age in white crappie.

    Otoliths (sagittae) have been verified as valid aging
structures for largemouth bass Micropterus salmoides (Taubert
and Tranquilli 1982; Hoyer et al. 1985) and black crappie
Pomoxis nigromaculatus (Schramm and Doerzbacher 1982).
Otoliths have been used successfully to back-calculate lengths
for these fish to assess growth in a macrophyte-infested lake
(Maceina and Shireman 1982) and in a thermally altered
reservoir (Perry and Tranquilli 1984).
    Whole-view examination of otoliths may not allow accurate
age interpretation for fish species which have thick otoliths,
are slow growing or display considerable longevity. In these
cases, annuli may be indistinguishable and otoliths must be
viewed in thin transverse sections to correctly age fish
(Williams and Bedford 1974; Beamish 1979; Hoyer et al. 1985).

The Age and Growth of Fish, edited by Robert C. Summerfelt and Gordon E. Hall © 1987 The Iowa
State University Press, Ames, Iowa 50010.

However, preparation and sectioning of otoliths is more time
consuming than whole-view examination. In addition, Williams
and Bedford (1974) stated that caution should be used when
back-calculating lengths from sectioned otoliths.
Discrepancies in determining the exact otolith center and
inconsistent sectioning may alter back-calculated lengths.
Beamish (1979) and Hoyer et al. (1985) reported young fish
were accurately aged when otoliths were examined in whole view;
older fish, however, required sectioning to insure correct
aging. Schramm and Doerzbacher (1982) successfully aged black
crappie from ages 1 to 7 using whole view examination, but
overall sample size was small (N = 32).

Age determination from white crappie <u>Pomoxis</u> <u>annularis</u>
otoliths has not been verified. White crappie were collected
for one year from two Texas reservoirs to determine if age
could be interpreted from otoliths. Comparisons were made
between whole and transverse-sectioned views for annuli counts
and corresponding fish-length to otolith-radius relationships.
If whole-view otolith examination could be employed to age
white crappie, considerable savings in labor and materials
could be achieved.

MATERIALS AND METHODS

A total of 373 white crappie, 90 to 345 mm total length
(TL), were collected bimonthly from Lake Aquilla, TX, between
February 1984 and April 1985. The majority of fish (88%) were
captured with 46-m experimental multifilament gillnets with
bar mesh ranging from 25 to 89 mm. Remaining fish were caught
utilizing a pulsed DC electrofishing boat. In Lake Conroe, TX,
a total of 240 white crappie, 89 to 403 mm TL, were collected
between May 1984 and June 1985. In May of each year, fish were
collected by applying rotenone, 3 mg/l (5% active ingredient),
to six coves totaling 6.5 ha. From June 1984 to June 1985
white crappie (60%) were collected twice monthly with gillnets
similar to those used in Lake Aquilla. White crappie length-
frequency data was available from Lake Conroe cove rotenone
samples conducted from 1980 to 1982. Fish had been sorted and
enumerated into 25-mm size groups providing length-frequency
distributions that could be used to corroborate age estimates
and back-calculated lengths for fish collected in 1984 and 1985.

Fish TL was measured to the nearest mm, sex was determined
by dissection, and the largest otoliths, the sagittae, were
removed. Otoliths were placed in glycerin for one to two weeks
to clear the otolith and allow for improved reading. Otolith
size did not appear to be important in determining time required
for clearing. Whole otoliths exhibited alternating opaque and
hyaline (clear) bands. Otoliths were placed in a black dish
with water; opaque bands were counted and measurements were
made under a dissecting microscope (X10-X40) using reflected
light. Otoliths were read and measured in whole view for all

white crappie collected. Otolith radius (OR) was measured to
the nearest 0.05 mm from the kernel to the anterior tip or
rostrum of the otolith (Fig. 1) with an occular micrometer.
Distance to opaque bands was measured along the same axis. The
right otolith was used for both whole and sectioned views.

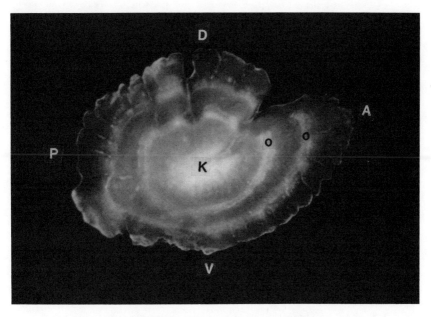

Figure 1. Whole view of an age-2 white crappie otolith
collected from Lake Aquilla in October 1984. o =
annulus; A = anterior end; D = dorsal end; P =
posterior end; V = ventral end; K = kernel.

A sample of 257 otoliths were sectioned transversely along
the dorsoventral plane (Beamish 1979) for comparison with
whole-view band counts. The posterior edge of the otolith was
ground to near the kernel with a 13-cm circular sander and fine
(#300) sand paper; otoliths were then mounted on glass
microscope slides using thermoplastic cement and the anterior
side was sanded to the kernel. The section was covered with
immersion oil and examined with transmitted light. Opaque
bands were counted at X40. OR was measured for 202 otoliths
along two different axes (Fig. 2) at X100. The first
measurement was from the kernel to the ventral edge of the
otolith, the second was from the kernel to the proximal edge,
just ventral to the central groove, as proposed by Hoyer et al.
(1985) for largemouth bass.

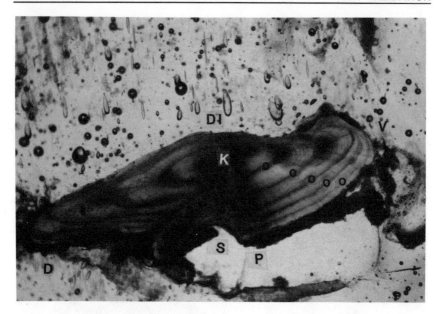

Figure 2. Transverse view of an age-5 white crappie
otolith collected from Lake Conroe in October 1984.
o = annulus; DI = distal surface; P = proximal
surface; D = dorsal end; S = sacculus groove; K =
kernel; V = ventral end.

Criteria modified from Everhart et al. (1975) were used to
determine if otoliths could be used to age and back-calculate
lengths for white crappie: an annulus must be formed once a
year and approximately at the same time each year for various
age fish, fish length should be proportional to or correlated
with OR, and back-calculated lengths should agree with those
determined from length-frequency analysis.

To determine if opaque bands were formed once a year,
marginal increment from the last band to the otolith edge was
calculated for otoliths displaying 1 to 5 bands. Total length
to otolith radius relationships (TL to OR) were analyzed with
linear regression using untransformed, single and double common
logarithmic (log to the base 10) transformed data and third-
degree polynomial equations. Residual plots were examined for
each model. When applicable, back-calculated lengths were
compared to corresponding age-group length-frequencies.

Data were analyzed using the statistical analysis programs
(primarily GLM and REG procedures) presented by the SAS
Institute Inc. (1985).

RESULTS AND DISCUSSION

Whole-view examination of white crappie otoliths indicated a single, white opaque band was formed once during the sampling year. In both reservoirs, a sharp decline in the mean marginal increment was evident in fish collected in May and June (Fig. 3). Marginal increment increased steadily throughout the fall and winter months for all fish displaying 1 to 5 bands. Based on these observations, opaque bands were considered to be annuli.

In Lake Aquilla, age-1 and age-2 white crappie were forming new annuli in April. By June, all age-1 fish and 98% of the age-2 fish had completed annulus formation. In Lake Conroe, all age-1 and age-2 fish had completed annulus formation in May. For age-5 white crappie, annulus formation

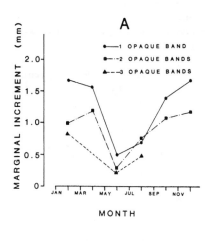

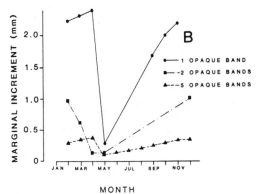

Figure 3. Yearly cycle of mean marginal increments since the last opaque band was formed on white crappie otoliths collected from Lakes Aquilla (A) and Conroe (B).

was complete for 25%, 83% and 100% of the fish examined in May,
June and July, respectively. These observations are similar to
results reported for other centrarchids (Schramm and Doerzbacher
1982; Taubert and Tranquilli 1982; Hoyer et al. 1985).

Analysis of covariance indicated no significant ($P > 0.05$)
differences between sexes for TL:OR relationships in either
reservoir. Therefore, data were pooled for sexes when
formulating regression equations. For each reservoir, TL:OR
regressions indicated a strong positive relationship between
these two parameters (Table 1). However, an increase in the
variability between TL and OR was evident with larger fish
(Fig. 4). In Aquilla Lake, the TL:OR relationship was linear,
and double log, transformed data stabilized the variance and
improved model predictability, as evidenced by an increase in
the coefficient of determination (r^2) value. Fish from Lake
Conroe displayed a slight S-curve response between TL and OR.
Highest r^2 values were computed for polynomial equations and
these models appeared to be the best expression of the TL:OR
relationship for Lake Conroe white crappie. Residual plots
showed single log transformation of the dependent variable, TL,
stabilized the variance for the polynomial equation. However,
with a few Lake Conroe fish, distance to the first annulus was
less than the minimum OR measured for fish collected during the
study period. With these fish, extrapolation was required to
compute back-calculated lengths. Extrapolation to determine
predicted values using polynomial models can give erroneous
results (Montgomery and Peck 1982), and Carlander (1981)
cautioned against the use of regression techniques to back-
calculate lengths from scale annuli when data extrapolation

Table 1. Regression equations[a] and corresponding coefficient
of determination (r^2) values describing white crappie total
lengths (TL) to otolith radius (OR) relationships from Lake
Aquilla and Conroe.

Lake	Regression equation	r^2
Aquilla	$TL = -35.4 + 60.4(OR)$	0.885
	$\log_{10}(TL) = 1.59 + 1.191 \log_{10}(OR)$	0.912
Conroe	$TL = -14.5 + 58.6(OR)$	0.955
	$\log_{10}(TL) = 1.63 + 1.161 \log_{10}(OR)$	0.967
	$TL = -26.4 + 33.4(OR) + 13.7(OR^2) - 1.5(OR^3)$	0.981
	$\log_{10}(TL) = 1.22 + 0.460(OR) - 0.049(OR^2)$ $+ 0.0016(OR^3)$	0.987

[a]All regression equations were significant ($P < 0.01$).

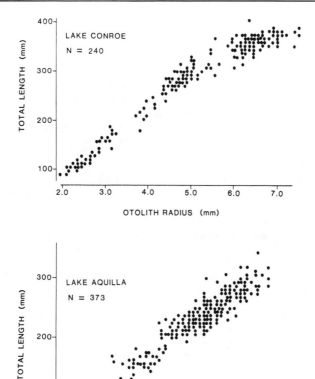

Figure 4. The total length to otolith radius
relationships for white crappie collected from Lakes
Aquilla and Conroe.

occurred. For simple linear regression, residual plots
indicated double log, transformed data computed a homogeneous
model error variance, helped to linearize the data, and
increased the r^2 value of the regression model. Linear
regressions using log transformed data computed a positive
intercept and resulted in a five-fold decline in intercept
variance when compared to regressions using untransformed data
in both reservoirs.

Incorporating log-transformed regression equations,
back-calculated lengths were derived for Lake Conroe and
Aquilla fish from a modified Fraser-Lee formula:

$$\log L_i = a + \log OD_i(\log L_c - a)/\log OR$$

where L_i is the estimated TL at age i (mm),
   OD_i is the otolith distance from the kernel to
      annulus i (mm),
   OR is the otolith radius measured from the kernel
      to the edge of the anterior rostrum (mm),
   L_i is the TL at capture (mm),
   a is the intercept of the log TL:log OR regression
      determined from the sample (mm).

The Fraser-Lee formula has been proposed as the best-suited equation for deriving back-calculated fish lengths using scales (Carlander 1981). Use of logarithmic transformation in the Fraser-Lee equation reduced the variance associated with mean back-calculated lengths.

For Lake Conroe white crappie, we attempted to compare back-calculated lengths for fish of the 1979 year-class, captured in 1984 and 1985, to observed length-frequency distributions from 1980 to 1982. We believe a relatively large white crappie year-class was formed in 1979, as 897 fish (140/ha), 64 to 140 mm TL, were obtained from cove rotenone samples in May 1980 (Fig. 5). Based on otolith examination of similar-sized fish in May 1984 and 1985, and the distinct length-frequency distribution observed in 1980, we presumed these fish were age 1. Because of the large size of the 1979 year-class and the relatively weak year-classes produced in 1980 and 1981 (only 4 fish from the 1980 and 1981 year classes were collected in 1984-85), this year-class was traceable to age-3. Back-calculated lengths for fish ages 1 to 3 from the 1979 year-class were similar to length-frequency data collected from rotenone samples (Fig. 5). Back-calculated lengths to specific ages were considered accurate with respect to time of annulus formation, since rotenone sampling was conducted in May when white crappie either had completed or were forming an annulus.

In Lake Aquilla, 22 white crappie from the 1982 year-class were collected in April 1984. At this time, all fish were forming their second annulus. One year later, 1982 year-class crappie were again collected and back-calculated lengths were computed to age-2. Length-frequency distributions for age-2 fish collected in 1984 were similar to back-calculated lengths to age-2 (Fig. 5). These results suggest that the methods we used derived accurate back-calculated lengths from whole-view otolith examination in Lakes Conroe and Aquilla.

Ages determined from sectioned otoliths were considered accurate and were used to evaluate the reliability of age determinations from whole otoliths. We compared whole and sectioned counts for 257 white crappie. Whole-view examination underestimated age by 1 year in only three age 3 to age 4 fish when compared to section views (Table 2). No differences

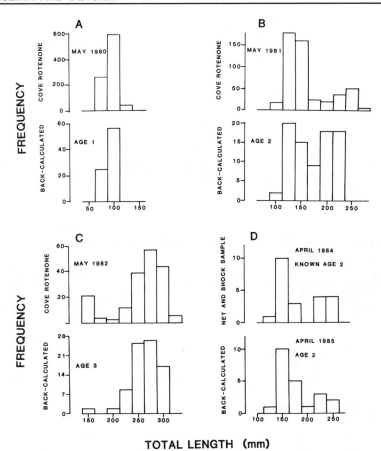

TOTAL LENGTH (mm)

Figure 5. White crappie length frequencies for fish
collected from Lake Conroe in May 1980 to 1982 (A-C)
in cove rotenone samples compared to corresponding age
1 to 3 back-calculated lengths derived for the 1979
year-class. Length-frequency of the 1982 year-class
collected in April 1984 from Lake Aquilla (D) compared
to the age 2 back-calculated lengths of the same year-
class in April 1985.

occurred between whole and sectioned counts for age-5 to age-9
fish (N = 74).
     The two Lake Conroe otoliths that differed in whole-view
were slow-growing fish. These fish were 312 and 314 mm TL,
respectively, while the average size of fish in that year-class
was 350 mm TL. Since the right otoliths were sectioned and
could not be read again in whole view, the left otoliths of the
3 fish that differed were allowed to resoak in glycerin and

Table 2. Numbers of white crappie in each age group
          determined from section view counts, with indications
          of those which were aged differently in whole view.

Lake	1	2	3	4	5	6	7	8	9
				Age					
Aquilla	45	43	3[a]	2	2				
Conroe	46	19	5	20[a]	58	4	3	4	3

[a]One fish from Lake Aquilla and 2 fish from Lake Conroe were
interpreted to be one year younger in whole view otolith
examination.

reexamined to determine why discrepancies in band counts
occurred. In two otoliths, very light opaque bands not
considered to be annuli in whole view appeared to be true
annuli when sectioned. The other otolith, from Lake Aquilla,
displayed extremely slow first-year growth and the kernel had
merged with the first annulus, so this annulus was not
enumerated. Since there was only a 1% discrepancy rate between
whole-view and sectioned counts, we believe this method
probably described accurate population age in Lakes Aquilla and
Conroe. White crappie otoliths are relatively thin and show
distinct opaque and hyaline bands. Our data suggests that
crappie age determination from whole-view otolith examination
may be easier than for other fish species with thicker otoliths.
Regression equations were fitted to sectioned TL:OR
relationships and compared to corresponding whole-view TL:OR
relationships. All regression equations were significant ($P <$
0.01). Coefficient of determination values indicated that OR
measured in whole view was a better predictor of fish TL than
sectioned OR (Table 3). When measuring sectioned OR, we
observed difficulty in determining the precise center of the
kernel, which probably caused some error in describing
white crappie TL. In addition, obtaining perpendicular

Table 3. Coefficient of determination (r^2) values describing
          best fitting regression equations[a] for crappie TL:OR
          relationships for whole-and sectioned-view measurements.

Lake	N	Size range (mm TL)	Whole view	Sectioned view	
				Ventral	Proximal
Aquilla	93	90–320	0.91	0.79	0.72
Conroe	109	110–392	0.97	0.84	0.89

[a]All regression equations were significant ($P < 0.01$).

transverse sections through the kernel center was difficult, and inconsistencies probably led to additional error.

In conclusion, whole-view otolith examination can provide a precise method for determining age and back-calculated lengths of white crappie. For slow-growing white crappie populations, otolith sectioning may be necessary to determine age. When estimating back-calculated lengths, however, some precision may be sacrificed as the correlation between TL and OR may be lower for sectioned otoliths than for whole-view.

ACKNOWLEDGMENTS

Primary support for this study was provided by the Fort Worth District Corps of Engineers, Contract Number DACW-79-C-0410. Additional funding was provided by the Texas Agricultural Experiment Station Project H-6634. Phil Bettoli, Cindy Chritton, Mark Luedke, Alan Rudd, and Scott Smith assisted in field collections and laboratory analysis. This paper is TA-20359 of the Texas Agricultural Experiment Station.

REFERENCES

BEAMISH, R. J. 1979. Differences in age of Pacific hake (Merluccius productus) using whole otoliths and sections of otoliths. Journal of the Fisheries Research Board of Canada 36:141-151.

CARLANDER, K. D. 1981. Caution on the use of the regression method of back-calculating lengths from scale measurements. Fisheries 6:2-4.

EVERHART, W. H, A. W. EIPPER, and W. D. YOUNGS. 1975. Principles of fishery science. Cornell University Press, Ithaca, NY, USA.

HOYER, M. V., J. V. SHIREMAN, and M. J. MACEINA. 1985. Use of otoliths to determine age and growth of largemouth bass in Florida. Transactions of the American Fisheries Society 114:307-309.

MACEINA, M. J., and J. V. SHIREMAN. 1982. Influence of dense hydrilla on black crappie growth. Proceedings of the 36th Annual Conference Southeastern Association of Fish and Wildlife Agencies. 36:394-402.

MONTGOMERY, D. C. and E. A. PECK. 1982. Introduction to linear regression analysis. John Wiley & Sons, New York, NY, USA.

PERRY, L. G., and J. A. TRANQUILLI. 1984. Age and growth of largemouth bass in a thermally altered reservoir, as determined from otoliths. North American Journal of Fisheries Management 4:321-330.

SAS Institute Inc. 1985. SAS users guide: Statistics, version 5 edition. SAS Institute Inc. Cary, NC, USA.

SCHRAMM, H. L., Jr., and J. F. DOERZBACHER. 1982. Use of otoliths to age black crappie from Florida. Proceedings of the 36th Annual Conference Southeastern Association of Fish and Wildlife Agencies. 36:95-105.

TAUBERT, B. D., and J. A. TRANQUILLI. 1982. Verification
    of the formation of annuli in otoliths of largemouth
    bass. Transactions of the American Fisheries Society
    111:531-534.
WILLIAMS, T., and B. C. BEDFORD. 1974. The use of otoliths
    for age determination. Pages 114-223 in T. B. Bagenal,
    editor, The Aging of Fish. Gresham Press, Old Working,
    England.

Use of on-going tagging programs to validate scale readings

GARY C. MATLOCK, ROBERT L. COLURA,
ANTHONY F. MACIOROWSKI, AND
LAWRENCE W. McEACHRON
TEXAS PARKS AND WILDLIFE DEPARTMENT
4200 SMITH SCHOOL ROAD
AUSTIN, TX 78744

ABSTRACT
Red drum Sciaenops ocellatus is an economically important
sport and commercial fish along the Atlantic and Gulf coasts
of the United States. Continuation of these fisheries depends
on reliable age and growth information for use in yield models.
Red drum have been aged using length-frequency distributions,
scales, otoliths, and mark-recapture methods, but none of the
methods have been validated for wild populations. An on-going
tagging program for red drum in Texas bays was used to determine
if valid annuli are formed on scales. Impressions of scales
collected from tagged fish at release and recapture were compared
to determine the time and number of growth checks formed.
Scales of 22 recaptured fish indicated Texas red drum formed
valid annuli as early as January. The relationship between
red drum total length and scale radius at release and recapture
was statistically similar, which suggested that tagging did
not influence the body-scale relationship or annulus formation.
Using this procedure the scale method was validated for aging
of red drum to 4.5 years.

Red drum Sciaenops ocellatus supports economically important
sport or commercial fisheries throughout its range (from Massachusetts to Tuxpan, Mexico). Sport fishermen caught 4.5 million
red drum in the United States in 1981 (National Marine Fisheries
Service 1985); commercial landings were 1.7 million kg in 1977
(Thompson 1984). Reliable age and growth information for use
in yield models will be useful in management efforts to maintain
these fisheries. Length-frequency distributions, scales, otoliths,
and mark-recapture methods have been used to determine ages
of red drum (Pearson 1929; Theiling and Loyacano 1976; Wakefield
and Colura 1983; Matlock 1985). Although otoliths have been

The Age and Growth of Fish, edited by Robert C. Summerfelt and Gordon E. Hall © 1987 The Iowa
State University Press, Ames, Iowa 50010.

validated for fish trapped in a South Carolina impoundment (Theiling and Loyacano 1976), none of the methods have been validated for free-ranging populations. The most direct validation approach is the use of structures from known-age fish to verify that fish are not older or younger than estimated; and that each annulus forms once a year (Casselman 1983; Beamish and McFarlane 1983). However, the direct comparison technique has not been widely employed (Casselman 1983). The objective of this study was to determine the feasibility of using an on-going tagging program to validate annulus formation on red drum scales.

MATERIALS AND METHODS

Red drum were caught in gill nets at randomly selected sites in eight Texas bays (McEachron and Green 1985) during spring (Apr-Jun) and fall (Sep-Nov) 1983-1984. Fish were also caught with hook and line at the cooling lake outfall of Houston Lighting and Power Company's Cedar Bayou Electric Generating Station on the Galveston Bay system. Prior to release of each fish two or more scales were removed from beneath the distal end of the left pectoral fin immediately ventral to the lateral line. Captured fish were tagged with internal abdominal tags with external plastic streamers (Osburn et al. 1980) and released. Fishermen were requested through posters and news-media advertisements to return at least one scale from the same area from each recaptured tagged fish. Each scale was washed in soapy water and impressed on a cellulose acetate slide with a roller press (Smith 1954); the impression was examined at 16.5 diameters magnification with a microprojector. Annuli were identified following Pearson's (1929) description and separately counted by two examiners without collaboration until agreement was achieved. Scale radii and distances from the focus to successive annuli were measured along a diagonal line (Figure 1) to the right antero-lateral scale corner (Klima and Tabb 1959).

The number of growth checks at release and recapture were compared to determine if only one check was formed annually, since the number of days following release was known. The effect of tagging on annulus formation and growth was examined using standard, least-squares linear regression and analysis of covariance (Sokal and Rohlf 1969). The distance to each annulus (Y) was regressed on the number of annuli (X) using the technique for multiple Y values at each X; these regressions were compared statistically. Effects on growth were examined by comparing the relationship between body length (Y) and scale radius (X) at release with that at recapture. The regression-fitting technique for single Y at each X was used to estimate the relationships.

RESULTS AND DISCUSSION

The on-going tagging program for Texas red drum was useable to validate scale readings because: 1) tagging did not appear to influence scale growth or annulus formation; and 2) sufficient scales were returned at recapture (22 useable scales) to give

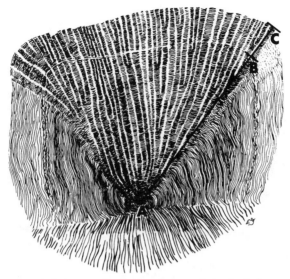

Figure 1. Red drum scale indicating morphological features
used to measure radii and annuli: A = focus, B = annulus,
and C = scale margin.

evidence of annulus formation at expected times. The relationship
between fish total length and scale radius did not change from
release to recapture (Figure 2). The relationship at release
was statistically similar to that obtained at recapture (for
slopes: F = 0.530; d.f. = 1, 40; P > 0.05; for y-intercepts:
F = 0.002, d.f. = 1, 41; P > 0.05). The regression for all
data combined indicated that total length increased two units
for each unit of scale increase (Table 1). The relationship
between scale radius and each annulus (Table 2) at release was

Table 1. Regression statistics for total length (mm) - scale radius
(mm x 16.5X) relationships for scales taken from tagged and recaptured
red drum.

Time measurements recorded	N	Y-intercept	Slope	95% Confidence interval around slope	r
Release	22	61.87	2.18	1.70-2.66	0.906
Recapture	22	124.08	1.92	1.36-2.48	0.846
Combined	44	89.71	2.05	1.72-2.38	0.894

Table 2. Regression statistics for the scale radius (mm x 16.5X) at each
annulus relationship for scales removed from tagged and recaptured red
drum.

Time measurements recorded	N	Y-intercept	Slope	95% Confidence interval around slope	r
Release	30	118	52	0-179	0.605
Recapture	40	117	58	51- 65	0.732
Combined	70	117	56	30- 82	0.691

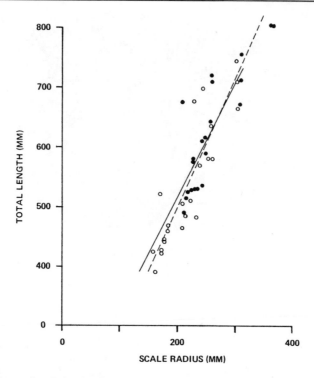

Figure 2. Relationship between fish total length and
    scale radius (16.5X magnification) for tagged red drum
    at release (open circles and dashed line) and at recapture
    (closed circles and solid line).

also similar to that obtained at recapture (Figure 3). The
regression for all data combined indicated that annuli were
formed when the scale radius was 173 mm, 229 mm, 285 mm, and
341 mm (at 16.5X magnification).

The scale method for aging red drum <4.5 years old has
been validated in this study. Red drum spawn in fall (Sep–Oct)
and do not form a scale annulus during their first year of life
(Matlock 1984, Hysmith et al. 1983), but form one as early as
January thereafter. This conclusion was reached by a process
of elimination. Nine of 22 fish examined had scales on which
the number of opaque areas (annuli) increased between release
and recapture, and all were free during November through January
(Figure 4). Scales of four fish that were free during November
or December had no annulus increase. This is further supported
by fish free during months other than November through January;
none of these fish formed an additional annulus. However, of
the two fish free during two January's, one formed two annuli
and the other formed one annulus. Some annulus formation may
occur as late as May. Pearson (1929) reported scale annulus

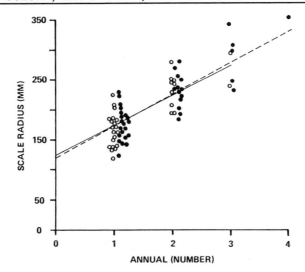

Figure 3. Scale radius distance (16.5X magnification)
to each annulus on each of 22 tagged red drum at release
(open circles and dashed line) and at recapture (closed
circles and solid line). Scales from one recaptured
fish had no annulus at release or recapture.

formation for Texas red drum occurred in winter. However, Matagorda
Bay red drum were reported to have formed scale annuli by March
or April (Wakefield and Colura 1983). South Carolina fish formed
otolith annuli in April and May (Theiling and Loyacano 1976).

Additional research is needed to validate annulus formation
on fish < 4.5 years old. Fish caught in the Gulf of Mexico
could be handled as in this study. However, more effort to
obtain scales from recaptured fish would be needed than was
exerted here to obtain at least 20 readable scales. During
1.5 years of tagging, 3,450 fish had scales removed at tagging
and 46 at recapture. Of the latter, 22 contained readable,
non-regenerated scales. Fewer released tagged fish would be
obtainable in the Gulf because older fish would be involved.
Fewer recaptured fish could be expected because fishing pressure
in the Gulf is less than in Texas bays (Ferguson and Green
1986), and there is currently a maximum size limit (762 mm)
in effect. However, an intense advertising campaign through
Gulf fishing piers, Gulf beach parks and businesses, and on-site
interviews of anglers along Gulf beaches could increase the
percent of returned scales, if the size limit was relaxed or
eliminated. Removing scales from only one side at tagging
and from the other at recapture should reduce the number of
regenerated scales returned.

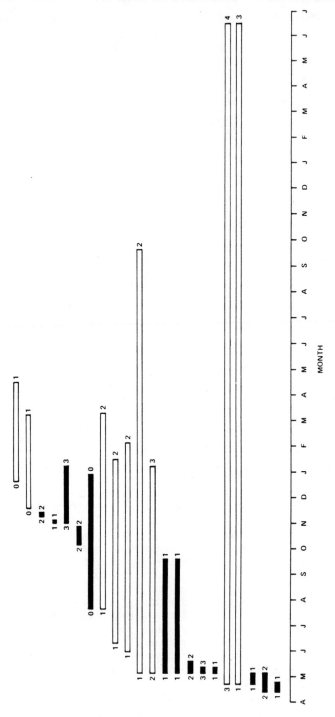

Figure 4. Time of release and recapture for each of 22 tagged red drum with number of growth checks (annuli) present on scales removed at release (to the left of each bar) and recapture (to the right of each bar). Shaded bars indicate scales with no increase in annuli. Open bars indicate scales with an increase in annuli.

REFERENCES

BEAMISH, R. J., AND G. A. MCFARLANE. 1983. The forgotten requirement for age validation in fisheries biology. Transactions of the American Fisheries Society 112:735-743.

CASSELMAN, J. M. 1983. Age and growth assessment of fish from their calcified structures-techniques and tools. Pages 1-18 In E. D. Prince and L. M. Pulos, editors. Proceedings of the International Workshop on age determination of oceanic pelagic fishes: tunas, billfishes, and sharks. United States Department of Commerce, National Oceanic and Atmospheric Administration, National Marine Fisheries Service, NOAA Technical Report 8.

FERGUSON, M. O., AND A. W. GREEN. 1986. An estimate of unsurveyed recreational boat fishing activity on the Texas coast and its effects on estimated landings. Marine Fisheries Review (in review).

HYSMITH, B. T., G. C. MATLOCK, AND R. L. COLURA. 1983. Effects of stocking rate and food type on growth and survival of fingerling red drum. Proceedings of the Warmwater Fish Culture Workshop. World Mariculture Society Special Publication Number 3:133-141.

KLIMA, E., AND D. C. TABB. 1959. A contribution to the biology of the spotted weakfish, Cynoscion nebulosus (Cuvier), from northwest Florida with a description of the fishery. Florida State Board of Conservation Technical Series Number 30.

MATLOCK, G. C. 1984. A basis for the development of a management plan for red drum in Texas. Ph.D. Dissertation, Texas A & M University, College Station, Texas.

MCEACHRON, L. W., AND A. W. GREEN. 1985. Trends in relative abundance of selected finfishes along the Texas coast: November 1975-June 1984. Texas Parks and Wildlife Department, Coastal Fisheries Branch. Management Data Series Number 79.

NATIONAL MARINE FISHERIES SERVICE. 1985. Marine recreational fishery statistics survey, Atlantic and Gulf coasts, 1981-1982. United States Department of Commerce, National Oceanic and Atmospheric Administration, Current Fishery Statistics Number 8324.

OSBURN, H. R., G. C. MATLOCK, AND H. E. HEGEN. 1980. Description of multiple census tagging program for marine fisheries management. Annual Proceedings of the Texas Chapter of the American Fisheries Society 2:9-25.

PEARSON, J. C. 1929. Natural history and conservation of redfish and other sciaenids on the Texas coast. Bulletin of the United States Bureau of Fisheries 44:129-214.

SMITH, S. H. 1954. Method for producing plastic impressions of fish scales without using heat. The Progressive Fish-Culturist. 16:75-78.

SOKAL, R. R., AND F. J. ROHLF. 1969. Biometry. W. H. Freeman and Company, San Francisco.

THEILING, D. O., AND H. A. LOYACANO, JR.   1976.   Age and
    growth of red drum from a saltmarsh impoundment in South
    Carolina.   Transactions of the American Fisheries Society
    105:41-44.
THOMPSON, B. G.   1984.   Fishery statistics of the United
    States, 1977.   United States Department of Commerce, National
    Oceanic and Atmospheric Administration, National Marine
    Fisheries Service Statistical Digest Number 71.
WAKEFIELD, C. A., AND R. L. COLURA.   1983.   Age and growth
    of red drum in three Texas bay systems.   Annual Proceedings
    of the Texas Chapter of the American Fisheries Society
    5:77-87.

Validation of the dorsal spine method of age determination for spiny dogfish

GORDON A. McFARLANE AND
RICHARD J. BEAMISH
DEPARTMENT OF FISHERIES AND OCEANS
FISHERIES RESEARCH BRANCH
PACIFIC BIOLOGICAL STATION
NANAIMO, BRITISH COLUMBIA V9R 5K6
CANADA

ABSTRACT
    The annuli that formed on the spines of 68 spiny dogfish
Squalus acanthias, at liberty for 2 years or more after marking
with oxytetracycline, corresponded to the years at liberty.
Because the recovered fish ranged in age from 17 to 70 years we
consider that this age determination method is valid. The
grouping of marks or the exclusion of ages because annuli were
difficult to interpret has led to a misunderstanding of life
history parameters, particularly the over-estimation of growth
rate. A revised method of age determination based on the
validation study is proposed.

    Spiny dogfish Squalus acanthias have been exploited for
their oil since the 1870s. By the early 1940s, dogfish became
the single most important groundfish species in the commercial
fishery off the west coast of Canada (Ketchen, 1986). This
fishery collapsed in the late 1940s as a consequence of
declining stocks and decreasing demand. The stocks recovered
more quickly than anticipated and by the late 1950s spiny
dogfish were considered a nuisance. The relatively rapid
increase in stock size resulted from the recruitment of the
unexploited juvenile biomass (Wood et al. 1979).
    At present there is a small commercial fishery for dogfish
that does not reduce its nuisance to other fisheries. Effective
management of this species must consider the value of a
commercial fishery, as well as the nuisance caused by incidental
catches and the importance of this predator in the ecosystem.
The basis for these management decisions requires accurate
estimates of age.
    Ketchen (1975) estimated age using growth zones on the
second dorsal spine, accepting only very clear zones, and

The Age and Growth of Fish, edited by Robert C. Summerfelt and Gordon E. Hall © 1987 The Iowa
State University Press, Ames, Iowa 50010.

excluding ages that differed +2 yr. Similarly, Holden and Meadows (1962) rejected all readings from spines which appeared to be worn down below the first annual ring. Bonham et al. (1949) selected spines for which there was little disagreement. The consequences of accepting or rejecting ages in these studies were not evaluated.

The present study validated the method of age determination using the second dorsal spine. This resulted in a modification of the method previously described (Bonham et al. 1949; Ketchen 1975). This paper also provides the detailed methodology on the age validation studies that were not included in a previous paper (Beamish and McFarlane 1985) on the histology and growth of the spine in relation to annulus formation.

METHODS

Dogfish were tagged with a modified Petersen disc in the Strait of Georgia (Fig. 1) between 1980 and 1983 (McFarlane and Beamish 1986). Approximately 52% of the fish tagged received either an intermuscular or intraperitoneal injection of 25 mg/kg of oxytetracycline (OTC) (Table 1). The dosage was similar to that recommended by Beamish et al. (1983) for sablefish Anoplopoma fimbria; laboratory tests found this dosage to produce a strong mark in the spine without causing mortality.

All dogfish receiving OTC injections were caught on a longline anchored to the bottom at depths from 20 m to 220 m (79%), or in a bottom trawl (21%), and held in a 3000 L holding tank which received a constant flow of seawater. Fish to be tagged and injected were transferred into a smaller (150 L) fiberglass tank and anaesthesized using tricaine methane sulfonate (MS222). Each fish was measured from the tip of the snout to the tip of the upper lobe of the caudal fin when depressed in a line horizontal with the body. After length and sex were determined, fish were tagged and injected with OTC. Some fish were held in recovery tanks prior to release if they had not recovered from the anaesthetic by the end of the tagging operation.

Table 1. Number of spiny dogfish tagged, tagged and injected, and released from 1980–1983, and number of fish recovered by year as of December 31, 1984.

Release year	Number released total	Number[a] released injected	Number of injected fish recovered by year (spine recovered)					
			1980	1981	1982	1983	1984	Total
1980	7482(2721)[b]	1460	3(2)	9(5)	6(4)	7(5)	6(27)	31(18)
1981	6968(3574)[b]	3572	–	13(8)	83(54)	82(57)	47(29)	225(148)
1982[c]	10502(10502)[b]	7981	–	–	67(50)	92(70)	79(58)	238(178)
1983	1613(1613)[b]	812	–	–	–	7(6)	7(4)	14(10)
Total	26565(18410)	13825	3(2)	22(13)	156(108)	188(138)	129(93)	508(354)

[a]Both intermuscular and intraperitoneal.
[b]Number released from cruises where fish were injected.
[c]3965 (2938 injected) were trawl caught, all others longline caught.

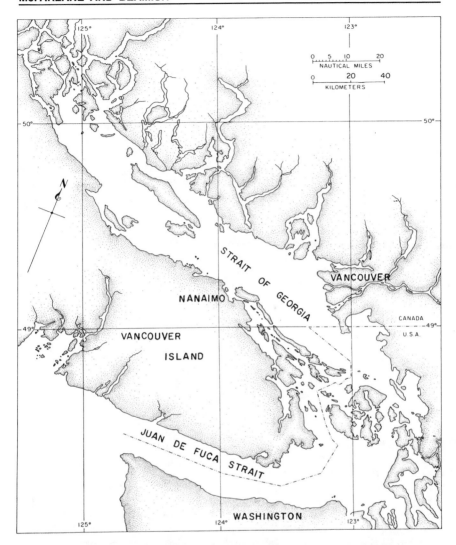

Figure 1. Study area, Strait of Georgia, for spiny dogfish
tagged and injected with OTC, 1980-1983.

Recaptured fish were measured and the second dorsal spine
(Fig. 2) removed by cutting horizontally just above the
notochord to ensure that the spine base and stem were intact.
These spines were frozen prior to processing according to the
procedure outlined by Chilton and Beamish (1982). Three readers
identified the annuli on each spine using UV light and a
dissecting microscope. The distance from the OTC mark on the

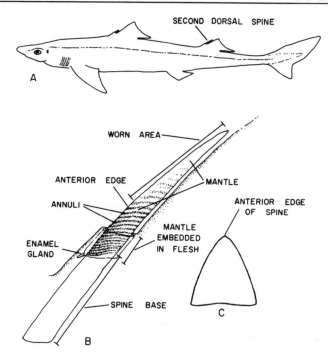

Figure 2. (A) Diagram of a spiny dogfish showing position
of second dorsal spine. (B) Second dorsal spine with
structural components identified. (C) Cross section of
second dorsal spine.

leading edge of the spine to the base of the enamel was
measured. Using this measurement the position of the OTC mark
was identified under reflected light and the number of annuli
that formed after the mark were counted. The annulus was
defined according to Beamish and McFarlane (1985) as a darkened
band or ridge or both, present on the enamelled portion of the
spine.

Estimates of growth in length of spiny dogfish were
obtained using the von Bertalanffy growth equation. The median
length at 50% maturity for female dogfish was identified using
probit analyses (Leslie et al. 1945), and the inflection of the
growth curve.

RESULTS

A total of 26,565 spiny dogfish were tagged and released
between 1980 and 1983, of which 13,825 received an injection of
OTC (Table 1). Of these, 11,211 were injected intermuscularly
and 2616 intraperitoneally. Recoveries as of December 31, 1984,
were used in this report. The recovery percentage for fish

injected intermuscularly (3.7%) was not significantly different
(t-test P > 0.05) for fish which received an intraperitoneal
injection (3.6%). Thus recovery results of injected fish have
been pooled.

The recovery percent (6.3%) of fish tagged and injected
with OTC in the fall (November 1981) was significantly greater
(t-test P < 0.05) than for fish tagged and injected in the
spring (March-July, 1980-1983) (2.8%) (Table 2). The higher
recovery percentage from fish released in the fall was probably
a result of tagging larger fish which would be more susceptible
to capture in the commercial fishery. The percentage of fish
(83%) which had an identifiable OTC mark on the spine was also
greater for fish tagged and injected in the fall than in the
spring (53%).

Table 2. Recovery percentage of spiny dogfish from fall
(1981) and spring (1980-1983) releases and percentage
of dorsal spines recovered which show an OTC mark.

	Fall release (Nov 1981)	Spring release (1980-1983)
No. released	3573	10254
No. recaptures	225	283
Percent recaptured	6.3	2.8
No. spines recovered	148	206
Percent spines with OTC mark	83	53

In the spring cruises, where approximately 75% of the fish
received OTC injections, the recovery rate of injected (2.8%)
and uninjected (2.9%) fish was similar, indicating that there
was no mortality introduced with the injection of OTC.

A total of 508 fish have been recaptured and spines were
recovered from 354 of these fish. Sixty-six percent (234) of
the recovered spines had an OTC mark. Approximately 20% of the
spines were not used in the validation study because they were
damaged or the wrong spine was collected.

Validation
Preliminary results are summarized in Beamish and McFarlane
(1985) and have been updated for this report (Table 3). In the
present study we used fish that had been recovered after being
at liberty for two years or more; those at liberty less than two
years were excluded because the slow growth of the spine made it
difficult to separate the OTC mark from the annulus. Of the
fish examined that had an OTC mark, 41 fish had been at liberty
for two years, 23 for three years, 2 for four years, and 2 for
five years (Table 3). Fish at liberty for two, four, and five
years developed 2, 4, and 5 annuli, respectively, after the OTC
mark formed (Fig. 3). Of the 23 fish at liberty for three
years, 20 had developed 3 annuli, and in 3 spines only 2 annuli
could be identified.

Table 3. Expected number of annuli that formed on
spines from fish tagged and injected with
oxytetracycline that had been at liberty for 2 or
more years.

Expected no. of annuli (yr) beyond OTC mark[a]	n	Estimated no. of annuli (n)	Age range (yr) at time of recapture (n)[b]
2	41	2(41)	17–58(39)
3	23	3(20), 2(3)	19–70(17)
4	2	4(2)	26–43(2)
5	2	5(2)	40(1)

[a]Annulus formation was assumed to occur from November
to April. A fish that was expected to form 2 or
more annuli had to be at liberty for at least two
consecutive November to April periods.
[b]Ages could not be estimated on all spines because
portions of the spine were damaged.

The OTC mark frequently appeared as a double line
(Fig. 3B). Because spine growth is slow and OTC is incorporated
into the tissue over a short time (Milch et al. 1958), the 2 OTC
marks must have formed simultaneously. This means that one mark
was deposited in the enamel and the other in the mantle dentine
(Beamish and McFarlane 1985).

In most samples the annulus was easily identified as a
ridge that was darker than the areas above and below it
(Fig. 4A). In some older fish, annuli that formed after the OTC
mark were quite closely spaced (Fig. 4B), but under reflected
light they were very distinct, forming prominent darkened
ridges. In others, annuli could be identified only on the
anterior edge of the spine (Fig. 3A). In younger or faster-
growing fish the annuli were more widely spaced and ridges were
less prominent (Fig. 4C). Spines showing abnormal enamel
deposition (Fig. 4D) or extreme wear (Fig. 4E) were common,
however, annuli were still readily identified on the enamelled
portion of the mantle. Spines displaying variable growth zones
(Fig. 4F) were quite common and easily aged using our definition
of an annulus. No checks were observed in any of the tagged and
injected mature fish examined.

Fish that had been at liberty for 2 or more years ranged in
age from 17 to 70 years. All spines showed some wear at the
tip. The number of missing annuli were estimated using the
proceedure of Ketchen (1975) and added to the annuli count to
produce an estimated age. For example, the youngest fish
recaptured in this study (17 yr) had an estimated 4 annuli worn
from the tip of the spine.

Growth and Maturity

The Strait of Georgia growth curve developed using the
validated age determination techniques indicated slower growth
than reported previously (Bonham et al. 1949; Ketchen 1975)

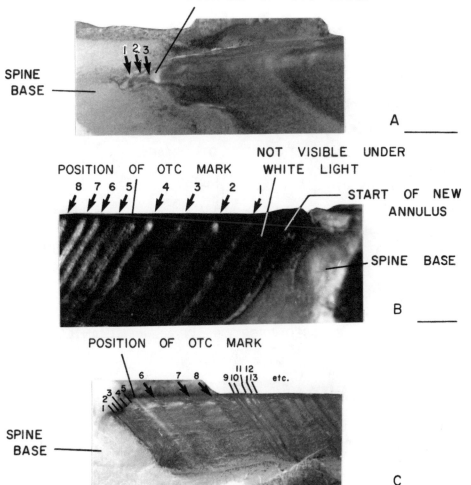

Figure 3. Dorsal spines from recaptured tagged and injected spiny dogfish. All spines photographed using ultraviolet light. (A) Spine from 87 cm male recaptured in April 1984 after 3 yr at liberty. Note that annuli formed after the mark appeared on the anterior edge of the spine only. (B) Spine from 75 cm male, recaptured in late 1984 after 4 yr at liberty. (C) Spine from 91 cm female, recaptured in January 1985 after 5 yr at liberty, showing variable spacing between annuli and the potential for grouping annuli. Annuli are numbered beginning with the most recent year. (Scale bar = 2mm).

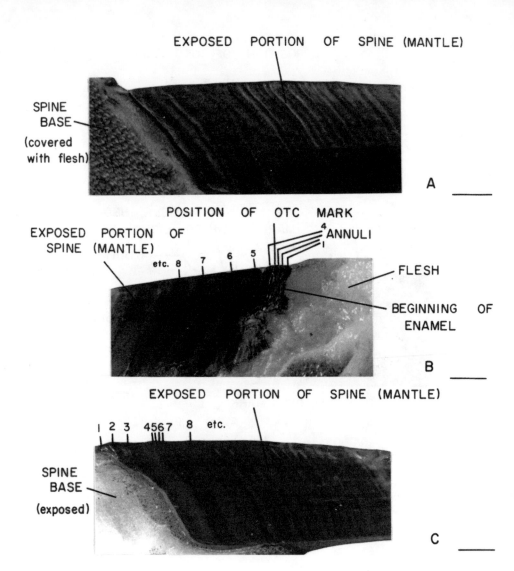

EXPOSED PORTION OF SPINE (MANTLE)

SPINE BASE (covered with flesh)

A

POSITION OF OTC MARK

EXPOSED PORTION OF SPINE (MANTLE)

4 ANNULI

etc. 8 7 6 5

FLESH

BEGINNING OF ENAMEL

B

EXPOSED PORTION OF SPINE (MANTLE)

1 2 3 4567 8 etc.

SPINE BASE (exposed)

C

Figure 4. Examples of dorsal spines from recaptured tagged and injected spiny dogfish. All spines photographed using white light. (A) Annulus easily identified as a ridge that is darker than areas above or below it. (B) Spine from a slow-growing fish at liberty 3 yr showing crowded ridge-like annuli. Position of OTC mark is indicated. (C) Spine from a faster-growing fish. Note annuli are wider spaced and ridges less prominent than A and B (above). (D) Spine showing abnormal enamel deposition in recent yr. (E) Spine showing extreme wear. (F) Spine displaying variable growth zones.

294

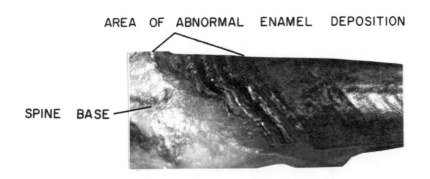

AREA OF ABNORMAL ENAMEL DEPOSITION

SPINE BASE

D

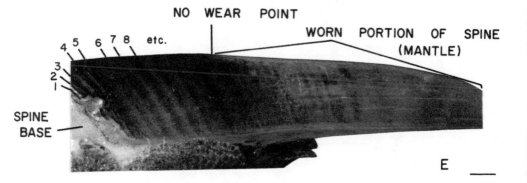

NO WEAR POINT

WORN PORTION OF SPINE (MANTLE)

4 5 6 7 8 etc.
3
2
1

SPINE BASE

E

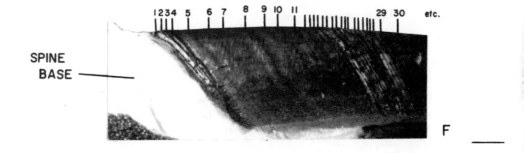

1234 5 6 7 8 9 10 11 29 30 etc.

SPINE BASE

F

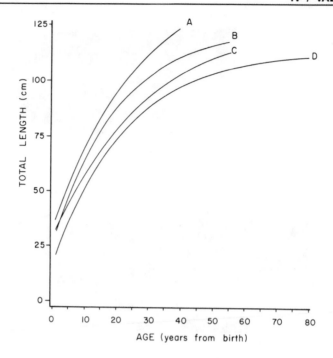

Figure 5.   Comparison of von Bertalanffy growth curves for
female spiny dogfish from the northeast Pacific Ocean.
A.   Bonham et al. 1949 (Washington coast from Ketchen
1975); B.   Ketchen 1975 (composite of Strait of Georgia
and Washington coast); C.   Ketchen 1975 (Strait of
Georgia); D.   Present study (Strait of Georgia).   (Data
from Ketchen 1975 reproduced with permission of the
author and editors of the Journal of the Fisheries
Research Board, publisher.)

(Fig. 5).   In addition, maximum ages obtained in this study
(80+) were twice that reported by Bonham (as reported in Ketchen
1975) and 45% greater than those reported by Ketchen (1975).

Previous estimates of the age at 50% maturity for female
spiny dogfish in the northeast Pacific Ocean range from 20 yr
(west coast of Washington) to 31 yr (Strait of Georgia) (Ketchen
1975).   Ketchen (1975) concluded that the age that best
accommodates the data from Bonham et al. (1949) and Ketchen
(1975) is 23 yr.   In our study the age at 50% maturity for
female dogfish for the Strait of Georgia is 35 yr (Fig. 6),
similar to the maximum value reported by Ketchen (1975).
Despite the difference in time and location of these previous
studies, there can be little doubt that the difference in
reported age at maturity is attributable, to a large degree, to
the rejection of difficult-to-age spines and the grouping of
annuli.

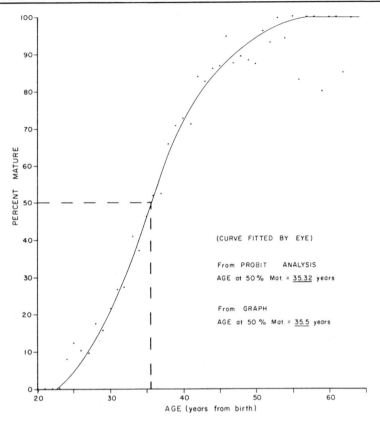

Figure 6. Age at 50% maturity for female spiny dogfish
captured in the Strait of Georgia, fall of 1980 and
1981.

DISCUSSION
      We can only assume that injected and tagged fish behaved in
the same manner as untagged fish. This is a standard assumption
made in tagging studies that is almost impossible to test.
Accepting these assumptions, our study demonstrates that the
method of ageing spiny dogfish using marks on the spine (Bonham
et al. 1949; Holden and Meadows 1962; Ketchen 1975) does produce
valid estimates of age, provided an annulus is defined as a
darkened band or ridge or both. Darkened bands or ridges that
are closely spaced are distinct annuli and must not be grouped.
Once it is known that these zones must be counted the number of
spines rejected as being difficult to age is greatly reduced.
      Other marks or zones form annually in the spine (Beamish
and McFarlane 1985) and these marks have been used by others to
estimate age (Holden and Meadows 1962; Soldat 1982). A mark

forms in the stem dentine that is probably annual, but it is difficult to detect and requires the careful removal of enamel, pigment, and mantle dentine in order to identify marks in the exposed area of the spine (Beamish and McFarlane 1985). An annual mark also appears in cross section which results from the growth of the stem dentine. Once again the mark (zone) is indistinct, and because of the upward growth of the spine it is not possible to produce sections for older fish that have all the annuli. Beamish and McFarlane (1985) indicated that the annulus on the mantle is much clearer than the other two areas and that it be used exclusively.

Using our definition of the annulus we conclude that spiny dogfish are older, slower growing, and later maturing than previously thought. Previous investigations have tended to group annuli or reject difficult-to-read or worn spines (Bonham et al. 1949; Holden and Meadows 1962; Ketchen 1975). While it is not possible to compare directly the methodologies of these other investigations with this study, it is probable that these procedures did result in the exclusion of the older and slower-growing component of the population, which can affect understanding of the biology of this species.

As a result of the present study, we recommend that the method for ageing spiny dogfish be modified as follows:
1. Remove spine by cutting close to the notochord; the spine is then air dried and cleaned.
2. Count annuli using reflected light and a dissecting microscope using sufficient magnification (120-250x).
3. All zones that occur on the enamelled surfaces that are darkened bands, ridges, or both must be counted; spacing should not be an overriding factor in the decision to count an annulus.
4. If the tip of the spine is worn then the number of missing annuli can be estimated. The "no wear point" occurs at the first complete annulus below the worn spine. The diameter of the spine is measured at this point, and the number of missing annuli are estimated using the procedure of Ketchen (1975) and added to the number of zones counted on the spine. We used Ketchen's original data to estimate the 95% confidence interval (Fig. 7). The interval increases with the spine base diameter (number of missing annuli). For example, an estimate of 20 missing annuli can range from 10 to 30 annuli.

ACKNOWLEDGMENTS
The assistance of M. Smith and M. Saunders in all aspects of the study is greatly appreciated. A number of fishermen and staff of the Fisheries Research Board participated in the tagging program. Students and staff of the Campbell River and Southgate Secondary schools also assisted in the study and we appreciate the continued cooperation of V. Egan and D. Brown.

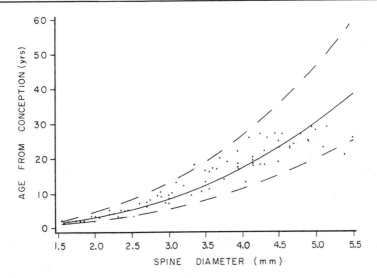

Figure 7. Relationship of age of spiny dogfish to the
diameter of the base of the second dorsal spine, y =
0.5097 x 2.5469 (Ketchen 1975). (Original data from
Ketchen 1975 used with permission of the author.)

REFERENCES
BEAMISH, R. J., G. A. MCFARLANE, AND D. E. CHILTON. 1983. Use
    of oxytetracycline and other methods to validate a method
    of age determination for sablefish (Anoplopoma fimbria).
    Pages 95-116 in: Proceedings of the Second Lowell
    Wakefield Fisheries Symposium. Alaska Sea Grant Report
    83-8.
BEAMISH, R. J. AND G. A. MCFARLANE. 1985. Annulus
    development on the second dorsal spine of the spiny dogfish
    (Squalus acanthias) and its validity for age
    determination. Canadian Journal of Fisheries and Aquatic
    Sciences 42:1799-1805.
BONHAM, K., F. B. SANFORD, W. CLEGG, AND G. C. BUCHER. 1949.
    Biological and vitamin A studies of dogfish (Squalus
    suckleye) in the State of Washington. Washington State
    Department of Fisheries. Biological Report 49A:83-114.
CHILTON, D. E., AND R. J. BEAMISH. 1982. Age determination
    methods for fishes studied by the groundfish program at the
    Pacific Biological Station. Canada Special Publications of
    Fisheries and Aquatic Sciences 60.
HOLDEN, M. J. AND P. S. MEADOWS. 1962. The structure of the
    spine of the spur dogfish (Squalus acanthias L.), and its
    use for age determination. Journal of Marine Biological
    Association of the United Kingdom 42:179-197.

KETCHEN, K. S. 1975. Age and growth of dogfish, (Squalus acanthias), in British Columbia waters. Journal of the Fisheries Research Board of Canada 32:13-59.

KETCHEN, K. S. 1986. The spiny dogfish (Squalus acanthias) in the northeast Pacific and a history of its utilization. Canada Special Publication of Fisheries and Aquatic Sciences. (In press).

LESLIE, P. H., J. S. PERRY, AND J. S. WATSON. 1945. The determination of the median body-weight at which female rats reach maturity. Proceedings of the Zoological Society of London 115:473-488.

MCFARLANE, G. A. AND R. J. BEAMISH. 1986. A tag suitable for assessing long-term movements of spiny dogfish and preliminary results using this tag. North American Journal of Fisheries Management. (In press).

MILCH, R. A., D. P. RALL, AND J. E. TOBIE. 1958. Fluorescence of tetracycline antibiotics in bone. Journal of Bone and Joint Surgery, Volume A4:897-909.

SOLDAT, V. T. 1982. Age and size of spiny dogfish, (Squalus acanthias), in the Northwest Atlantic. Atlantic Fisheries Organization, Science Council Studies 3:47-52.

WOOD, C. C., K. S. KETCHEN, AND R. J. BEAMISH. 1979. Population dynamics of spiny dogfish (Squalus acanthias) in British Columbia waters. Journal of the Fisheries Research Board of Canada 36:647-656.

Comparison of radiometric with vertebral band age estimates in four California elasmobranchs

BRUCE A. WELDEN,[1] **GREGOR M. CAILLIET,
AND A. RUSSELL FLEGAL**[2]
MOSS LANDING MARINE LABORATORIES
P.O. BOX 450
MOSS LANDING, CA 95039

ABSTRACT
A radiometric dating technique utilizing the naturally occurring
radionuclide lead-210 (half-life = 22.3 years) was used to
determine ages in 4 species of elasmobranchs from California
waters. The applicability and effectiveness of this technique
were assessed by comparing radiometric age estimates with ages
determined from counts of vertebral growth bands. Radiometric
age estimates in 2 species, the leopard shark <u>Triakis</u> <u>semi-</u>
<u>fasciata</u> and common thresher shark <u>Alopias</u> <u>vulpinus</u> were too
variable to be used; in 2 other species, the Pacific angel shark
<u>Squatina</u> <u>californica</u> and white shark <u>Carcharodon</u> <u>carcharias</u>,
radiometric age estimates were less variable and roughly agreed
with estimates from other studies. The high variability of these
radiometric age estimates is a function of both the analytical
limitations of low-level radioactive analyses and the physiological processes within the calcified vertebral centra, which
may invalidate the "closed system" and "constant uptake" assumptions necessary for valid radiometric dating. Violations of
these assumptions occurred in all 4 species; thus, present
radiometric age estimates are not completely reliable indicators
of age. Ultimately, a more comprehensive understanding of the
calcium physiology of elasmobranchs is needed to produce confident radiometric age estimates of their calcified structures.

---

[1] Present address: P.G.& E. Biological Laboratory, P.O. Box 117,
Avila Beach, California 93424.

[2] Present address: Marine Sciences Institute, University of
California, Santa Cruz, California 95064.

The Age and Growth of Fish, edited by Robert C. Summerfelt and Gordon E. Hall © 1987 The Iowa
State University Press, Ames, Iowa 50010.

The recent, rapid increase in commercial and recreational exploitation of elasmobranchs in California (Cailliet and Bedford 1983) has created an urgent need for accurate and reliable methods of elasmobranch age determination and validation. The development of such methods has lagged behind that of teleosts because, historically, elasmobranch fisheries have been of minor economic importance, and because elasmobranchs lack the calcareous hardparts traditionally used to age teleosts.

Several techniques have been used to estimate ages of elasmobranchs (Cailliet et al. 1983b); however, the enumeration of growth zones deposited in vertebral centra is the most effective method because of its wide applicability, consistent results, low cost, and simplicity. Several authors have postulated that growth zones in vertebral centra are deposited on an annual basis; but conclusive validation has only been accomplished for the thornback ray Raja clavata (Holden and Vince 1973), the leopard shark Triakis semifasciata (Smith 1984), and the lemon shark Negaprion brevirostris (Gruber and Stout 1983).

In the above studies, animals were tagged and injected with tetracycline, and either kept alive in the laboratory or released for later recapture. Such techniques are technically difficult, time consuming, and costly, and may not be applicable to some species. For example, the white shark Carcharodon carcharias has never survived in captivity for more than several days, and its rarity and large size make field tagging difficult. Several other approaches to age validation have been used (Brothers 1983), most of which require large numbers of random samples over a long sampling period and have poor resolution when applied to slow-growing or long-lived species.

A possible alternative to these validation methods is the application of radiometric dating techniques (Goldberg and Bruland 1974) to produce independent age estimates. Only a few samples of different selected size and age classes would be needed for radiometric analysis to estimate ages. If vertebral band pairs are shown to be annual, then age determination of most of the specimens could be accomplished using conventional vertebral growth-zone enhancement techniques.

Radiometric dating techniques have been used to estimate the growth rates of clams (Turekian et al. 1983), corals (Dodge and Thompson 1974), and chambered nautilus (Cochran et al. 1981), and to confirm the longevity of the splitnose rockfish Sebastes diploproa (Bennett et al. 1982). Radiometric dating techniques rely on the constant and known rate of decay of radioactive material over time. The present (or end) state of the system is experimentally determined, and the initial state of the system is estimated; the difference between these 2 states can then be used to calculate age.

Two assumptions are necessary for radiometric dating to produce valid age estimates of organisms. First, the radionuclide must be incorporated at a constant or known rate over the lifespan of the organism, so that the initial activity (or initial state) of the system can be estimated. Second, the structure which incorporates the radionuclide (calcified cartilage of the vertebral centra) must act as a closed system with respect to that radionuclide. Once the radionuclide is incorporated into the structure, there must be no loss or gain except by radioactive decay.

Lead-210 was selected as the most appropriate radionuclide to be analyzed in the calcified cartilage of the vertebral centra. Lead-210 is a suitable radionuclide for use in age determination for several reasons: it has a short half-life (22.3 years); it is biochemically similar to calcium and is therefore incorporated into calcified structures (Burnett and Patterson 1980); it forms a very stable complex with the calcified cartilage mineral apatite; it is ubiquitous in the marine environment; and it is rather easily measured from the ingrowth of its alpha-emitting granddaughter polonium-210. The calcified cartilage of the vertebral centra was chosen as the most suitable storage structure because the concentric calcified growth zones are easily separated using saws and drills, and because it appeared to best satisfy the "closed system" assumption due to its lack of cellular ability to resorb or mobilize elements stored in its matrix (Urist 1964).

In this study, a radiometric age determination technique utilizing the naturally occurring radionuclide lead-210 was developed and applied to vertebrae from 4 species of elasmobranchs from California. Radiometric age estimates were compared with ages determined from counts of calcified vertebral bands and the results were used to assess the applicability and effectiveness of radiometric age determination in these species.

The Pacific angel shark _Squatina californica_ and leopard shark are nearshore, demersal species found in shallow, temperate waters off the west coast of North America (Castro 1983). The common thresher shark _Alopias vulpinus_ and white shark are pelagic species found world wide in temperate, sub-tropical and tropical waters (Castro 1983). Three of these species, the Pacific angel, leopard, and common thresher sharks, are targets of an expanding commercial fishery off California (Cailliet and Bedford 1983). Age estimation has been attempted for all 4 species (Cailliet et al. 1983a; Cailliet et al. 1985; Natanson 1984; Smith 1984); but these age estimates have been validated only in the leopard shark (Smith 1984).

METHODS

Radiometric age determination was investigated by analyzing for lead-210 activities in inner and peripheral vertebral growth bands. Comparison of inner and peripheral band activities yielded an estimated age for the individual specimen, which was then compared with the number of vertebral growth bands counted from x-radiographs and resin-embedded thin sections.

Collection and Processing

For each species investigated, 2 to 9 individuals were selected from the collection of elasmobranch vertebral centra accumulated during 1978–1984 at Moss Landing Marine Laboratories, California (Cailliet et al. 1983b). Measurements of total length were taken on whole specimens of all 4 species, and alternate lengths (distance between origin of first and second dorsal fins) were taken from dressed specimens of the common thresher shark and converted to total lengths using a regression developed from measurements of whole animals (Cailliet and Bedford 1983).

A 6 to 12 inch section of the vertebral column was removed from each individual, and either frozen in a plastic bag or, for 2 specimens of the white shark, preserved in formalin and later stored in isopropyl alcohol. Vertebral sections were taken from the area anterior to the first dorsal fin on whole specimens of the leopard, common thresher and white sharks, and posterior to the precaudal pit on dressed specimens of the common thresher shark. Pacific angel shark vertebrae were taken from an area just posterior to the head, in the region of maximum centrum diameter and band counts (Natanson 1984).

Six to 20 individual centra were removed from each vertebral section. The centra were cleaned of all excess tissue, immersed in a concentrated solution of sodium hypochlorite to remove connective tissue from the centrum face, rinsed in distilled water for 20 minutes, and allowed to dry.

Vertebral Band Counts

X-radiographs of 2 to 4 centra per individual of the Pacific angel, common thresher and white shark were taken using an x-ray machine with industrial-grade film (Cailliet et al. 1983b). Growth zones were counted by viewing these radiographs through a dissecting microscope on low power using transmitted light.

Cleaned leopard shark centra were embedded in clear poly-ester casting resin after dehydration in 70%, 90%, and absolute ethanol and acetone. These centra were then sectioned longitu-dinally at their widest diameter into 0.5 to 0.6 mm slices with a double blade isomet saw (Smith 1984). Sections were mounted on slides with clear enamel, and growth zones were counted under a dissecting microscope using transmitted light.

For all species, at least 2 observers made independent counts of 2 replicates from each specimen. In the event obser-vers could not reach a consensus about the number of growth zones present, the specimen was not used for age analysis. Measure-ments of growth-zone location, thickness, and diameter were made from these x-radiographs and sections with vernier calipers; they were used to determine the location and size of bands to be isolated for radiometric analysis.

Growth Curves

Vertebral band counts from 21 specimens of the white shark were used to generate a growth curve based on the von Bertalanffy

growth equation (Cailliet et al. 1985). Growth data from laboratory-reared (Natanson 1984) and field-tagged (Pittenger 1984)
Pacific angel sharks were used to generate an approximation of
growth from juvenile through adult size-classes. Juvenile growth
was estimated from laboratory grow-out studies on individuals
250-600 mm total length; adult growth was estimated from field-
tagged individuals greater than 900 mm.

Isolation of Bands

Inner and peripheral vertebral band pairs representing
summer and winter growth (1 year) were mechanically isolated from
cleaned centra using a jeweler's saw and dental grinding equipment. These ranged from 0.6 mm thick and 5.6 mm in diameter for
inner band pairs of small individuals of the leopard shark, to
2.2 mm thick and 73 mm in diameter for peripheral band pairs of
large individuals of the white shark. Care was taken to isolate
only 1 pair of growth bands; however, on smaller specimens more
than 1 band pair was taken in order to get sufficient material
(1 to 2 grams) for analysis. Several centra were needed from each
animal and up to 20 centra were used from small individuals. The
separated band pairs were cleaned in a water-filled ultrasonic
bath, rinsed in distilled water, and dried to a constant weight
in an oven at 30°C.

Lead-210 Analysis

Lead-210 activities were determined through the measurement
of the ingrowth of its granddaughter polonium-210 (half-life =
138 days). Polonium-210 is advantageous for analyses of natural,
low level radioactivity, because it emits an alpha particle that
can be detected in counters with very low backgrounds (0.01 -
0.03 cpm). In addition, the chemical separation of polonium is
simple and quantitative, and the polonium-208 tracer (energy =
5.1 Mev) may be readily identified using solid state, silicon-
barrier detectors and pulse height analyzers.

In specimens collected at least 1 year prior to analysis,
and at least 1 year old, polonium-210 activities were measured
directly because of the relatively complete attainment of secular
equilibrium with lead-210. In samples collected more recently,
lead-210 was determined by measuring the ingrowth of new polon-
ium-210 over a 3-month period following removal of all polonium-
210 from the original solution. Lead-210 activities were then
calculated using equations for growth of daughter isotopes from
initially pure parent isotopes (Friedlander et al. 1949).

Following the methods of Flynn (1968), the samples were
digested in acid, and polonium-210 was plated out onto silver
discs and counted in an alpha spectrometer. To determine ingrowth of polonium-210, a second plating was done to ensure
complete removal of all polonium-210, the solutions were stored
for 3 months and were then replated and counted.

Calcium and barium concentrations were determined using
atomic absorption spectrophotometry; and they were used as addi-

tional normalizing factors to compensate for possible differential rates of lead-210 uptake relative to mass over the lifespan of the animal.

Calculation of Lead-210 Activities

Lead-210 activities were normalized to grams dry weight, and grams calcium for all species and, in addition, to grams barium for the white shark. These values were then used to calculate estimated ages of individual specimens using the decay equation:

$$A(t) = A(o) * e^{-(\lambda t)}$$

where $A(t)$ = lead-210 activity of inner band
       $A(o)$ = lead-210 activity of peripheral band
       λ = decay constant for lead-210
       t = time elapsed between formation of inner and peripheral bands

Because of the small number of samples analyzed, statistical analyses of the results were not necessary to make preliminary conclusions about the relative success of this technique.

RESULTS

Inner and peripheral growth bands were analyzed from 2 adult specimens of leopard shark 1240 and 1342 mm in total length. Radiometric age estimates, normalized to grams dry weight and grams calcium, ranged from 9 to 13 yr in the 1290 mm specimen and from 47 to 48 yr in the 1342 mm specimen. Age estimates for the same specimens determined from resin-embedded thin sections were 20 yr in the 1290 mm specimen and 15 yr in the 1342 mm specimen) (Table 1).

Inner and peripheral growth bands were analyzed from 2 adult specimens of the Pacific angel shark 1016 and 1110 mm. Radiometric age estimates ranged from 1 to 6 yr in the 1110 mm specimen and from 6 to 8 yr in the 1016 mm specimen (Table 2). These estimates were much lower than the number of growth zones counted using x-radiography (33 and 31); but they more closely

Table 1. Age estimates of leopard shark determined from resin-embedded thin sectioning and lead-210 radiometric ages normalized to cartilage dry weight and calcium.

Total Length (mm)	Resin Section Age	Radiometric Age ($\pm$ 1 SD) a. Dry Weight	Calcium
1290	20	9	13
		(14)	(18)
1342	15	48	47
		(b)	(b)

a. Errors ($\pm$ 1 SD) represent the sum of radiometric counting and elemental analysis variabilities.
b. Radiometric counting errors exceeded sample activities

Table 2. Age estimates of Pacific angel shark determined from
growth data (Natanson 1984, Pittenger 1984) and lead-210
radiometric ages normalized to cartilage dry weight and
calcium.

Total Length (mm)	Number of Bands (x-ray)	Age From Growth Data	Radiometric Age (± 1 SD) a. Dry Weight	Calcium
1110	33	12	1 (7)	6 (10)
1016	31	9	6 (10)	8 (13)

a. Errors (± 1 SD) represent the sum of radiometric counting and
elemental analysis variabilities.

approximated age and growth as determined from a synthesis of
growth information from field (Pittenger 1984) and laboratory
(Natanson 1984) growth studies (Fig. 1).

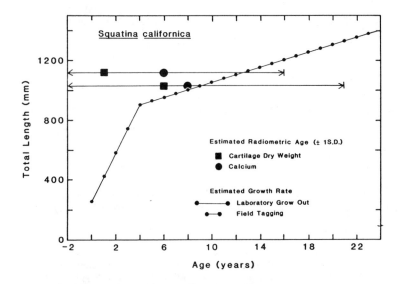

Figure 1: Comparison of Pacific angel shark growth estimated from: 1) laboratory and field measurements in juveniles (Natanson 1984) and adults (Pittenger 1984); and 2)
radiometric age estimates.

Inner and peripheral growth bands were analyzed from 3 juvenile and 6 adult common thresher sharks ranging in size from 3441
to 5733 mm. Radiometric age estimates were highly variable, with
a number of anomolously low (negative) ages (Table 3).

Table 3. Age estimates of common thresher shark determined from x-radiography and lead-210 radiometric ages normalized to cartilage dry weight and calcium.

Total Length (mm)	Age (x-ray)	Radiometric Age (± 1 SD) a.	
		Dry Weight	Calcium
3441	5	-4 (7)	-6 (10)
3441	6	8 (b)	NA
3613	6	-36 (b)	-60 (b)
4530	11	9 (12)	10 (16)
5102	15	64 (32)	71 (35)
5332	11	10 (3)	19 (6)
5389	15	-6 (14)	2 (17)
5618	11	-18 (14)	-7 (17)
5733	12	19 (21)	21 (24)

a. Errors (± 1 SD) represent the sum of radiometric counting and elemental analysis variabilities.
b. Radiometric counting errors exceeded sample activities.
NA Sample not analyzed.

Inner and peripheral growth bands were analyzed from 1 juvenile (2340 mm) and 3 adult (3930, 4609, and 5079 mm) white sharks. Radiometric ages generally increased with size from the smallest (2 to 7 years) to the largest (26 to 41 years) shark (Table 4). Radiometric ages were older and produced lower growth rates than those estimated from counts of vertebral bands (Fig. 2).

Uptake of lead-210 increased in larger individuals of all 4 species. Whole centra lead-210 concentrations (dpm/gm) increased

Table 4. Age estimates of white shark determined from x-radiography and lead-210 radiometric ages normalized to cartilage dry weight, calcium, and barium; mean growth rates determined from means of radiometric ages (dry weight, calcium, barium), and size at birth of 1300 mm.

Total Length (mm)	Age (x-ray)	Radiometric Age (± 1 SD) a.			Mean Growth Rate (mm/year)
		Dry Wt.	Calcium	Barium	
2340	2	2 (13)	7 (19)	3 (19)	260
3930	9	41 (4)	53 (7)	49 (7)	55
4609	13	19 (3)	20 (6)	16 (6)	180
5079	14	26 (4)	35 (7)	41 (7)	110

a. Errors (± 1 SD) represent the sum of radiometric counting and elemental analysis variabilities.

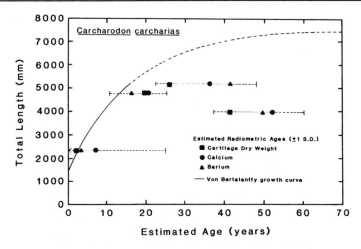

Figure 2:   Comparison of white shark growth estimated from:
1) vertebral band counts (von Bertalanffy growth curve
from 21 specimens aged using x-radiography, Cailliet et
al. 1985); and 2) radiometric age estimates.   Dashed line
is extrapolation of von Bertalanffy curve beyond actual
data points.

in larger specimens of the leopard and Pacific angel sharks (Fig.
3).   Lead-210 concentrations in peripheral bands increased in
larger specimens of the common thresher shark (Fig. 4); increased
and then decreased with increasing size in the white shark (Fig.
4).

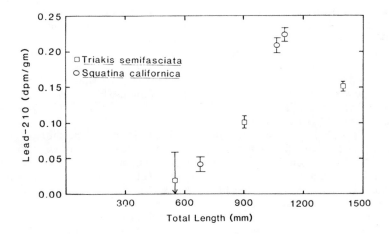

Figure 3:   Whole centrum lead-210 concentrations versus
total length of small and large leopard and Pacific angel
sharks.

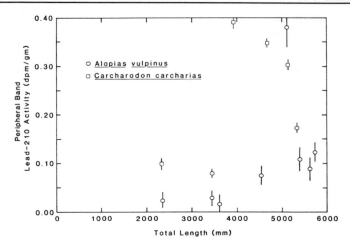

Figure 4: Peripheral band lead-210 concentrations versus
total length of small and large common thresher and white
sharks.

DISCUSSION
    The radiometric age determination technique was not uni-
formly successful for the 4 species tested. Estimates of age of
the Pacific angel and white sharks roughly agreed with other
ageing studies. But in the leopard and common thresher sharks,
age estimates were too variable to be used, and have limited
value for verification of existing age estimates.
    Age estimates of the leopard shark determined from resin-
embedded thin sections of vertebrae have recently been validated
from field-tagged and tetracycline-injected animals (Smith 1984).
Smith found that the band pairs were deposited on an annual
basis, providing evidence that the leopard shark matures at
between 7 and 10 yrs of age and lives to ages of 25 to 30 yrs.
Radiometric age estimates were both higher and lower than the
validated resin-embedded, thin-section age estimates, which suggests that systematic errors were not responsible for the high
variability but that violations of the "closed system" and "constant uptake" assumptions probably occurred.
    Radiometric age estimates of the Pacific angel shark agreed
roughly with ages determined from a synthesis of growth data from
juvenile (Natanson 1984) and adult (Pittenger 1984) sharks (Fig.
1). This synthesis assumes a size at birth of 260 mm, a growth
rate of 160 mm/yr for juvenile sharks (Natanson 1984) up to 900
mm, and a growth rate of 25 mm/yr for mature adult sharks greater
than 900 mm (Pittenger 1984). This combination of growth rates
is reasonable because younger animals typically exhibit much
faster growth rates than older ones, and growth greatly dimin-
ishes when animals reach maturity and energy is channeled into
reproductive functions.

Radiometric age estimates and the above growth data support the hypothesis that vertebral growth bands in the Pacific angel shark are deposited in relation to growth and not in relation to time. Assuming annual band formation, angel shark gestation would have to last 5 to 6 years, and adult sharks would range in age from 30 to 40 years. However, angel shark gestation has been estimated at 9 to 11 months based on embryonic growth rates, and laboratory studies of tetracycline-injected juveniles found deposition rates of up to 7 bands per year (Natanson 1984). Evidently, angel sharks lay down vertebral bands in direct relation to their growth, depositing several bands per year as fast-growing juveniles, and as few as 0 or 1 band per year as slow- (or non-) growing adults.

Age determination of the common thresher shark has been accomplished using counts of vertebral growth bands (Cailliet et al. 1983a, Cailliet and Bedford 1983). Those age estimates indicate the thresher shark matures at between 6 and 8 yrs of age, and may live to be 30 to 50 yrs old. These ages, however, have not yet been conclusively validated because of the difficulties involved in field and laboratory studies of these large, pelagic animals. Radiometric age estimates were too variable to support or disprove the hypothesis of annual band formation. The highly variable, and, in some cases, negative ages, suggest that the essential radiometric dating assumptions have been grossly violated in this species (Table 3).

Age determination of the white shark has been accomplished using counts of vertebral growth bands (Cailliet et al. 1985). Those age estimates indicate that the white shark matures at an age of 10 to 12 yrs and suggest that it lives 25 to 30 yrs. Radiometric age estimates (Fig. 2) indicate that the white shark grows more slowly, maturing at a later age (16 to 20 yrs) and living to older maximum ages (50 to 60 yrs).

The average growth rates derived from these radiometric ages ranged from 55 mm/yr in the white shark with the oldest radiometric age to 110-260 mm/yr in the other sharks. The latter growth rates overlap, but are generally lower than, those estimated using vertebral bands (218-300 mm/yr, Cailliet et al. 1985). Also, they would produce a lower estimated maximum size than was previously determined using the Von Bertalanffy growth equation (Fig. 2).

This radiometric dating technique was not completely successful in estimating age in elasmobranchs, because at least 1 and possibly both of the essential assumptions, were violated by all 4 species. Uptake of lead-210 was not constant and increased in larger individuals of all 4 species (Figs. 3 and 4). Two possible reasons for this observed increase are shifts in habitat and diet with increasing age.

As elasmobranchs age, many species move from shallow, nearshore pupping grounds, with relatively low levels of lead-210 in seawater (7-10 dpm/100kg), to deeper habitats with higher lead-210 levels (10-25 dpm/100 kg; Nozaki et al. 1980). Shifts in diet with increasing age are known to occur in several shark species. For example, the white shark is believed to shift from predominantly fish to marine mammals upon reaching approximately 3 m in length (Compagno 1984); the leopard shark begins life feeding primarily upon benthic invertebrates and switches to small fishes upon reaching a larger size (Russo 1975). Since different taxonomic groups tend to have widely different lead-210 concentrations (Holtzman 1969), such prey shifts could alter rates of uptake of lead-210.

Further, it is likely that the "closed system" assumption may have been violated in these organisms. Hoenig and Walsh (1982) noticed the occurrence of vascularized canals in the vertebrae of several elasmobranchs, including the white shark. These canals may serve in the regulation of serum calcium, and they offer evidence that the elasmobranch vertebral centra, composed primarily of calcified cartilage, may not be as static a structure as proposed by Urist (1964) and Simkiss (1974). If exchange of calcium (and the biochemically similar lead-210) is occurring within the calcified cartilage of the centra, it would invalidate the "closed system" assumption and produce erroneous radiometric age estimates. Although the lead-210 /apatite complex is very stable, ion exchange can and obviously does occur, very probably with the exchange of stable lead for lead-210. Because of the relatively low natural levels of lead-210 found in calcified cartilage, and the increasingly high and yet variable concentrations of anthropogenic stable lead found in seawater and in marine organisms (Burnett and Patterson 1980), stable lead/lead-210 ion-exchange could be responsible for a large amount of the observed variability.

There are several approaches which may serve to reduce some of the variability and increase the accuracy and precision of this radiometric age determination technique. These include the investigation of other natural and anthropogenic radionuclides, the use of parent-daughter decay couples, radiometric analysis of the centrum edge using short-lived radionuclides, investigation of other calcified storage structures such as spines and teeth, and whole-structure analysis (Welden 1984).

Ultimately, the confident use of radiometric dating techniques for age determination is dependent upon a better understanding of the physiology of the calcified structures analyzed. Elasmobranch calcified structures lack the stability of clam shells (Turekian et al. 1983), coral skeletons (Dodge and Thompson 1974), and rockfish otoliths (Bennett et al. 1982) which have been used successfully in radiometric growth studies. Information is needed about the rates of uptake, incorporation and loss of calcium, and biochemically similar radionuclides such as strontium-90, radium-226 and 228, and lead-210. For example, how do these rates change in response to environmental fluctuations, increasing age of the animal, or reproductive cycles?

Finally, a complete evaluation of this radiometric ageing technique will only be possible when it is applied to animals of known age. In this study, only the leopard shark ages have been conclusively validated; yet radiometric age estimates were so variable that comparisons yielded no useful information, except to indicate that the essential assumptions had been violated. Further refinement of this technique coupled with its application to known-age animals is essential before radiometric ageing can be confidently used to determine age in elasmobranchs.

ACKNOWLEDGMENTS

G. Boehlert first suggested the use of radiometric dating techniques on elasmobranchs. V. Noshkin, K. Wong, and T. Jokela from the Lawrence Livermore Laboratories provided advice and the use of their facilities for initial testing and development. This research was sponsored by the California Sea Grant College Program and California State Resources Agency (R/F-81 and R/F-84) and the National Science Foundation (OCE-8309392).

REFERENCES

BENNETT, J. T., G. W. BOEHLERT, and K. K. TUREKIAN. 1982. Confirmation of longevity in Sebastes diploproa (Pisces: Scorpaenidae) from Ra-226/ Pb-210 measurements in otoliths. Marine Biology 71:209-215.

BROTHERS, E. B. 1983. Summary of round table discussions on age validation. Pages 35-44 in E. Prince and L. Pulos, editors. Proceedings of the international workshop on age determination of oceanic pelagic fishes: tunas, billfishes, sharks. National Oceanographic and Atmospheric Administration Technical Report. National Marine Fisheries Service 8. Washington, DC, USA.

BURNETT, M. W., and C. C. PATTERSON. 1980. Perturbation of natural lead transport in nutrient calcium pathways of marine ecosystems by industrial lead. Pages 413-438 in E. Goldberg, Y. Horibe and K. Saruhashi, editors. Isotope marine chemistry. University of Rokakuho Publications, Tokyo, Japan.

CAILLIET, G. M., and D. BEDFORD. 1983. The biology of three pelagic sharks from California waters and their emerging fisheries: a review. California Cooperative Oceanic Fisheries Investigations Reports 24:57-69.

CAILLIET, G. M., L. K. MARTIN, J. T. HARVEY, D. KUSHER, and B. A. WELDEN. 1983a. Preliminary studies on the age and growth of blue, common thresher, and shortfin mako sharks from California waters. Pages 157-165 in E. Prince and L. Pulos, editors. Proceedings of the international workshop on age determination of oceanic pelagic fishes: tunas, billfishes, sharks. National Oceanographic and Atmospheric Administration Technical Report. National Marine Fisheries Service 8. Washington, DC, USA.

CAILLIET, G. M., L. K. MARTIN, D. KUSHER, P. WOLF, and B. A. WELDEN. 1983b. Techniques for enhancing vertebral bands in age estimation of California elasmobranchs. Pages 179–188 in E. Prince and L. Pulos, editors. Proceedings of the international workshop on age determination of oceanic pelagic fishes: tunas, billfishes, sharks. National Oceanographic and Atmospheric Administration Technical Report. National Marine Fisheries Service 8. Washington, DC, USA.

CAILLIET, G. M., L. J. NATANSON, B. A. WELDEN, and D. A. EBERT. 1985. Preliminary studies on the age and growth of the white shark Carcharodon carcharias, using vertebral bands. Southern California Academy of Sciences Memoirs 9:49–60.

CASTRO, J. I. 1983. The Sharks of North American Waters. Texas A & M University Press. College Station, Texas, USA.

COCHRAN, J. K., D. M. RYE, and N. H. LANDMAN. 1981. Growth rate and habitat of Nautilus pompilius inferred from radioactive and stable isotope studies. Paleobiology 7:469–480.

COMPAGNO, L. J. V. 1984. Sharks of the World. Food and Agriculture Organization of the United Nations. Fisheries Synopsis Number 125, Volume 4, Part 1. Rome, Italy.

DODGE, R. E., and J. THOMPSON. 1974. The natural radiochemical and growth records in contemporary hermatypic corals from the Atlantic and Caribbean. Earth and Planetary Science Letters 23:313–322.

FLYNN, W. W. 1968. The determination of low levels of polonium-210 in environmental materials. Analytica Chimica Acta 43:221–227.

FRIEDLANDER, G., J. W. KENNEDY, and J. M. MILLER. 1949. Nuclear and Radiochemistry. John Wiley and Sons, New York, NY, USA.

GOLDBERG, E., and K. BRULAND. 1974. Radioactive Geochronologies. Pages 451–489 in E.D. Goldberg, editor. The Sea. Vol 5. John Wiley and Sons, New York, NY, USA.

GRUBER, S. H., and R. G. STOUT. 1983. Biological materials for the study of age and growth in a tropical marine elasmobranch, the lemon shark Negaprion brevirostris Poey. Pages 193–205 in E. Prince and L. Pulos, editors. Proceedings of the international workshop on age determination of oceanic pelagic fishes: tunas, billfishes, sharks. National Oceanographic and Atmospheric Administration Technical Report. National Marine Fisheries Service 8. Washington, DC, USA.

HOENIG, J. M., and A. H. WALSH. 1982. The occurrence of cartilage canals in shark vertebrae. Canadian Journal of Zoology 60: 483–485.

HOLDEN, M. J., and M. R. VINCE. 1973. Age validation studies on the centra of Raja clavata using tetracycline. Conseil Permanent International Pour L'exploration de la Mer 35(1):13–17.

HOLTZMAN, R. B. 1969. Concentrations of the naturally occurring
    radionuclides Ra-226, Pb-210 and Po-210 in aquatic fauna.
    Pages 535-546 in Proceedings of the 2nd national symposium
    on radioecology. Ann Arbor, Michigan, USA.
NATANSON, L. J. 1984. Aspects of the age, growth and repro-
    duction of the Pacific angel shark Squatina californica, off
    Santa Barbara, California. M.A. Thesis. San Jose State
    University, Moss Landing Marine Laboratories, Moss Landing,
    California, USA.
NOZAKI, Y., K. K. TUREKIAN, and K. VON DAMM. 1980. Pb-210 in
    GEOSECS water profiles from the North Pacific. Earth and
    Planetary Science Letters 49:393-400.
PITTENGER, G. 1984. Movements, distribution, feeding, and
    growth of the Pacific angel shark Squatina californica at
    Catalina Island, California. M.A. Thesis. California State
    University - Long Beach, Long Beach, California, USA.
RUSSO, R. A. 1975. Observations on the food habits of leopard
    sharks Triakis semifasciata and brown smoothhounds Mustelus
    henlei. California Fish and Game 61:95-103.
SIMKISS, K. 1974. Calcium metabolism of fish in relation to
    ageing. Pages 1-12 in T.B. Bagenal, editor. Ageing of
    fish. Unwin Brothers Ltd., Surrey, England.
SMITH, S. E. 1984. Timing of vertebral band deposition in
    tetracycline injected leopard sharks. Transactions of the
    American Fisheries Society 113:308-313.
TUREKIAN, K. K., J. K. COCHRAN, and J. T. BENNETT. 1983. Growth
    rate of a vesicomyid clam from the 21°N East Pacific Rise
    hydrothermal area. Nature 303:55-56.
URIST, M. R. 1964. Further observations bearing on the bone-
    body fluid continuum: composition of the skeleton and serums
    of the cyclostomes, elasmobranchs and bony vertebrates.
    Pages 151-179 in H.M. Frost, editor. Bone biodynamics.
    Little Brown, Boston, Massachusetts, USA.
WELDEN, B. A. 1984. Radiometric verification of age
    determination in elasmobranch fishes. M.S. Thesis.
    California State University, Hayward. Moss Landing Marine
    Laboratories, Moss Landing, California, USA.

V
Methods

Methodological approaches to the examination of otoliths in aging studies

EDWARD B. BROTHERS

EFS CONSULTANTS
3 SUNSET WEST
ITHACA, NY 14850

ABSTRACT

Otolith analysis in aging studies may proceed on two levels; macrostructural, involving the enumeration of annual marks; or microstructural, where daily growth increments are usually the basic counting unit. Research protocols for different studies will ultimately vary and diverge, however the initial stages will have common elements. Collection, storage, fixation, extraction and measurement procedures should be carefully designed to insure adequate samples for in-depth analysis. Preparation and examination methods should be developed after experimentation with a wide variety of approaches. Data collection (e.g. counting) and documentation (statistical analysis and photography) and validation are also essential components of otolith aging studies. Microstructural analysis for both larval and adult fishes is emphasized, with particular attention to the need for maintaining flexibility and innovation in dealing with the many species which have not yet been examined.

Because of the vast diversity of bony fishes, their highly variable growth patterns, otolith sizes and morphologies, it is not possible to describe a general methodological procedure that will be applicable for all aging studies. Approaches to otolith aging have undergone profound changes in the last fifteen years, primarily due to the introduction of microstructural and chemical analyses. In addition, the more traditional "annual aging" techniques are constantly being modified and successfully applied to many more species, some of which were incorrectly aged in the past, by otolith or other methods.

This paper is not intended to give a comprehensive review of a rather diffuse and often poorly documented literature, but rather to provide some personal insights to otolith aging,

The Age and Growth of Fish, edited by Robert C. Summerfelt and Gordon E. Hall © 1987 The Iowa State University Press, Ames, Iowa 50010.

including some of the alternative techniques, their advantages
and disadvantages, and considerations when embarking on a new
otolith-aging project. Otoliths are often identified as a
subject for study, either as the primary aging structure, or as
an alternative for verifying some other approach. The potential
power of otolith aging is great for bony fishes and therefore the
method deserves exploration; however, sometimes practical
considerations or structural peculiarities favor use of fin
spines, scales, other hard parts or even length frequency
analyses. Techniques relying solely on chemical analyses, such
as radioactive isotopic ratios (Bennett et al. 1983) or elemental
analysis (Radtke and Targett 1984), will not be considered in
this paper.

The most important consideration when initiating an otolith
study is to come into it with an open mind, i.e., without
unjustified preconceived notions which would limit the options
explored. Techniques which may have appeared to have been
successful have often found to be incorrect, inadequate, badly
biased and unvalidated (Beamish and McFarlane 1983). We should
be careful not to perpetuate the errors of the past by taking a
myopic view based on limited experience (taxonomic and
geographic), but, at the same time we should avoid being seduced
by the apparently unbounded (and often unjustified) promise of
new techniques described as "state of the art." The fact remains
that all otolith-aging methods require some level of subjective
decision beyond careful technique or sophisticated
instrumentation.

Use of high technology may give the outward appearance of
greater objectivity, but setting a threshold level for
automatically counting aging marks still involves making a
somewhat arbitrary decision. The consequences of errors in
judgement may be relatively minor if the units of age
determination measurement are small (e.g. days), or substantially
larger if annual marks are being counted. Improvements in the
accuracy and precision of otolith aging can be achieved when the
reliability of the decision-making process is enhanced by an
extensive knowledge of, and experience with, a wide variety of
otolith studies, including familiarity with many techniques and
many kinds of fish; and a study plan which starts with a broad
attack and considers, tests, and gradually narrows to the optimal
methodology for a particular species and the data needs of the
research. Neglecting experimentation and blind reliance on
"standard" or established techniques will impede the flexibility
and ingenuity necessary to develop reliable otolith-aging methods
for the remarkable diversity of living fishes which will require
age and growth studies in the years to come. Recent references
which describe the current state of otolith-aging methods are:
Blackler 1974; Bagenal and Tesch 1978; Beamish 1979; Pannella
1980; Bennett et al. 1982; Brothers 1982; Gjosaeter et al 1984;
and Campana and Neilson 1985.

COLLECTION OF FISH

Otolith material may come from fish representing all life history stages; larvae, juveniles, adults and even unhatched embryos. Extensive collections of fishes are most desirable to develop the most complete series of ages and growth curves, to facilitate validation studies, and to allow for highly instructive comparisons of populations. These should include: efforts to obtain samples with good representation over the whole size range and life history; notation of sex, state of maturity or reproductive activity; collections drawn over all seasons of the year and coverage of a wide geographic range.

STORAGE, FIXATION AND PRESERVATION OF FISH

Otoliths should be extracted from freshly killed fish. When this is not possible, then the fish may be frozen or preserved in ethanol. Extreme care should be afforded to larval and juvenile fishes. Because of the small absolute size of the otoliths of these life stages and the attendant relatively high otolith surface to volume ratio, these samples are most sensitive to degradation and dissolution. The safest methods of preservation are fixation and storage in 95% ethanol or freezing. Quick drying of whole fish or heads is also a possibility, especially for smaller specimens. Each method has associated advantages and disadvantages, primarily related to the ease of later extraction but also to the distortion and shrinkage of the fish, which makes sorting and identification difficult and biases measurements. To avoid the latter, it is necessary to make fish length (and mass) measurements both before and after fixation and storage so appropriate correction factors can be calculated. In general, it is best to avoid any sort of formalin fixation. Even buffered formalin can change pH over time, with disastrous results to the otoliths of small fish. It is equally important to fix and preserve fish in adequate volumes of ethanol, since packing of a large mass of fish into a small volume of alcohol results in a substantial dilution of the effective concentration. Adding marble chips or calcium carbonate to alcohol-stored specimens is an added safety measure for stabilizing pH. Occasional checking of otolith integrity and changing of alcohol is also recommended, if long-term storage is required. Cool, dark storage areas are probably best.

OTOLITH EXTRACTION

The removal of otoliths from adult fishes is a general practice in fisheries science and needs little comment here, except that locating and removing the smaller lapilli requires extra care and should not be overlooked. Several dissection approaches are possible; I usually favor a frontal head cut, at approximately the level of the top of the eye. Special equipment

and methods may be required for very large fish, but generally
only a sharp knife, bone saw or scalpel is necessary. The
relative and absolute size of the otoliths of adult fishes is
highly variable, resulting in varying degrees of difficulty in
locating the otoliths. Dissection under a binocular microscope
may be necessary, even for moderately large fish. If the
otoliths are small, it is best to remove the entire membraneous
labyrinth (especially the utriculus, sacculus and lagena),
including any pieces of the semicircular canals. These tissues
can then be teased apart and searched with the aid of a
microscope, or fixed in alcohol for later examination. I
routinely remove all otoliths, especially the sagittae and
lapilli, and often the asterisci. As will be discussed later,
the lapilli often prove to be as or more valuable than the
traditionally studied sagittae.

     During dissection it is important to take careful notes on
the orientation of the otoliths so that their morphological
features can be related to their positions in situ.
Separation of left and right otoliths is advisable first, until
their structures are well enough known to allow for sorting based
on morphology. Otoliths should be cleaned of attached tissue
before storage or examination; this is best accomplished at the
time of removal, especially if samples are to be stored dry.
Cleaning is simply a matter of gentle mechanical scrubbing,
rubbing and teasing. Immersion in a weak solution of sodium
hypochlorite or hydrogen peroxide for several minutes will help
to dissolve or loosen resistant material, as will the use of an
ultrasonic cleaner.

     Otolith removal from embryonic and larval fishes obviously
requires special care. Otoliths may be only 10 or 20 um in
diameter. If the specimen is relatively clear and unpigmented,
the otoliths can often be seen in situ in the otic vesicles
or capsules by examination with transmitted light and a binocular
dissecting microscope (at 30-50x). Otoliths are birefringent and
show up as bright points of light, if double (crossed) polarized
illumination is used. Alcohol- fixed larvae can be cleared by
brief (1-3 min.) immersion in a 1% KOH solution. Dissection must
be performed with the finest needles obtainable (ground to sharp
points!) and fine, sharp forceps. Fish are placed in a drop of
water on a clean glass slide; once the otoliths are located, they
can be gently teased out from the head of the larva, pushed out
of the surrounding water and allowed to air dry for a few seconds
on the slide. Needles, not forceps should be used to pick up
very small otoliths; the dried otoliths can be picked up and
transferred by simply touching them with the needle tip, after
first oil-wetting the tip and then wiping it off with the
fingers; enough remains to produce a "sticky" surface.

     For embryos and very small larvae with fairly transparent
tissues otoliths may be examined in situ. This is a rapid
procedure, but not recommended because of the generally poorer
image quality obtained. Alcohol-fixed larvae can be dehydrated

in absolute ethanol, air dried and then soaked in immersion or mineral oil. Otoliths can be examined with the compound light microscope, if the fish are covered with a coverslip. It is even possible to view the otoliths of freshly killed or anesthetized larvae by placing the fish under a coverslip in a water mount.

STORAGE

Whole otoliths, particularly of adults, may be stored dry in envelopes, capsules, vials, etc. Alcohol storage is also possible and is more appropriate for the smaller, more fragile otoliths of juveniles. Otoliths of larvae, being very difficult to handle once they are removed from the fish, are best slide-mounted for storage. They may be covered with a neutral, clear medium (e.g. Permount), topped with a coverslip and permanently preserved in this manner. Before permanent mounting, decisions must be made as to how the otoliths are to be oriented to give the best view of internal structures (i.e. which side up), and that grinding and/or preparation for the scanning electron microscope (SEM)is not required. Small otoliths have a tendency to be crushed by the coverslip as the mounting medium dries and shrinks. Very fine nylon monofilament, or pieces of glass fiber, may be used as a spacer to prevent this; however, care must be taken not to make the mount too thick, because working distances may be very small at higher magnifications. After complete drying, the slides can be stored in typical vertical fashion.

Use of oil instead of a hardening medium is a simpler but still a semi-permanent (good for at least five years and probably much longer) mounting procedure for larval otoliths and ground sections. It has the advantage of allowing easy removal of the otoliths for further preparation and turning over. Immersion oil has been used for this purpose; however, I recently discovered that some oils are (or become) slightly acidic over time and may cause degrading of small otoliths. Mineral oil (to which some marble or $CaCO_3$ chips have been added) appears to have a more neutral to basic pH. Otoliths on slides may be stored in oil with or without (not recommended for very small otoliths) coverslips. The slides must be stored flat on trays and protected from dust.

OTOLITH MEASUREMENT

Otoliths are often weighed or measured to maximum total length along specified axes, or radii, to establish fish/otolith relationships for use in backcalculations, or to define population morphometric characteristics. Obviously, otolith and fish measurements must be made before treatments or preparations which might alter these features. Once in situ orientation of the otoliths is known and established, standard terminology (e.g. Hecht 1978; Jensen 1965) should be applied to describe the precise location and orientation of all measurement and counting paths. Special terms should be defined carefully. Measurements can be made with calibrated reticles in dissecting or compound light microscopes, or from photomicrographs taken with optical or SEM instruments. If measurements are taken from a video monitor, it is particularly important to check for distortion and non-linearity at the edges of the field of view.

PREPARATION AND EXAMINATION

The data requirements of the research, the life stage of the fish, and the size and morphological characteristics of the otoliths will determine which of two basic approaches to otolith aging is most appropriate: aging by annual or seasonal zones or aging by microstructural features (daily growth increments). Both approaches may be used to help in validating age determinations. Aging by annual and seasonal zonations in the otoliths has a well-established set of methods, Jearld (1983); Blackler (1974); Bagenal and Tesch (1978). Attention almost always centered on sagittae, although lapilli and even asterisci often show comparable translucent and opaque zones. Otoliths may be examined whole, ground, cut, broken, burned, stained and immersed in a variety of clearing agents. Without discussing the relative merits of each of the above approaches, I would emphasize that one of the most important considerations is the choice of orientation for any "sectioning". Many sagittae grow in a very asymmetrical fashion, with later growth additions appearing only on the interior (sulcal) face. Lateral views will not reveal many of the annual growth zones and ages may be significantly biased downwards. Transverse or oblique sections are most appropriate for sagittae showing this sort of growth pattern. Boehlert (1985) discusses this phenomenon and also presents a new method of aging which relies on multiple regressions incorporating otolith mass and linear measurements. Lighting is another important variable which can substantially improve a reader's ability to discriminate annuli. Reflected, transmitted, oblique, "glancing", and polarized lighting should all be tried. Annual zones in otoliths can also be enhanced by image analysis instrumentation or even electron microprobe analysis (Casselman 1983; Radtke and Targett 1984; McGowan et al. 1986).

Examination of otoliths for presumptive daily growth increments can be as fast and simple as looking at an oil-mounted whole otolith with the compound microscope, usually at magnifications of 500X or greater. This is all that is required for the otoliths of many small larvae. Even if the sagittae is too large (thick) to examine in this manner, the smaller lapilli may be perfectly readable without further preparation. Try both "sides" and rotating a polarized light source to enhance contrast of increments and toreduce background "noise". Experimentation with lighting (brightness, condenser focus, field and condenser diaphrams, filters etc.) and manipulation of the orientation of the otoliths are important during preliminary examinations. As discussed for annual zones, microstructural growth increments are generally not deposited in a symmetrical and uniform fashion over the entire otolith. Certain axes exhibit much faster growth and wider increments, while others may show very fine increments, occasional growth interruptions, or a complete, long-term cessation of otolith growth and deposition. Determination of the optimal plane for grinding or sectioning is critical. For moderately small otoliths of "typical" flattened spheriod shape, it is often sufficient to thin the otolith by grinding to a para-sagittal plane on one lateral face, flipping the otolith, and then grinding the other side to a near mid-sagittal plane. Thinning to a wafer less than 1 mm thick is adequate to allow sufficient light passage for examination with the compound microscope; more opaque otoliths require more grinding.

Otoliths larger than a couple of mm long can be ground by slight finger pressure on a glass plate covered with a mixture of carborundum grit (about 600) and mineral oil. Grinding papers and coarser grits can also be used with practice, even smaller otoliths can be worked on this way. Grinding is alternated with microscopic inspection until the appropriate levels are reached. Excess grit can be removed with solvent in an ultrasonic cleaner, and the otoliths are then placed in oil on a glass microscope slide. Inspection will also reveal whether slightly tilted, or oblique, "sagittal" planes expose a more complete growth record, ie. without "grinding through" peripheral or central increments. Slight adjustments in finger pressure can be used to make necessary corrections in grinding.

If otolith breakage is a major problem (otoliths from some fish are more prone to break, they can be mounted in a small block of hard material (e.g. epoxy, polyester resin, and various metalurgical or histological embedding compounds), or cemented onto a slide or other surface to facilitate handling (see ref. cited in Campana and Neilson 1985). Larger otoliths can be sectioned for microstructural or annual zonation examination with special, low speed saws.

Examination of several sectioning planes, for both sagittae and lapilli, should be a major effort at the beginning of a study. There are peculiarities of the growth pattern of the sagittae of many fish which make counting daily growth increments very difficult. At certain periods in their life history, discontinuities, disruptions and subdaily growth increments may prevail. These problems may be circumvented by analysis of lapilli, especially for growth during the first few hundred days of life. Lapilli may present other problems however: smaller size, more difficult dissection handling and sectioning; finer increments; slower growth; and earlier onset of growth interruptions. Nevertheless, lapilli may offer distinct advantages in certain aging studies which do not deal with old, slow growing fish.

Acid etching of otoliths is necessary for SEM examination.

It has also been recommended for light-microscope studies on whole otoliths. Despite claims to the contrary, for "typical" otoliths such etching should not provide any enhancement of internal increment visibility, because only the outer surface should be affected by the acid. The only possible positive effect could be a slight smoothing of surface irregularities; this is also functionally performed by oil immersion. Water immersion does not have the desired effect because water's refractive index is not as close as oil to that of the otolith material. On the negative side, otoliths which are treated with an acidic solution will lose peripheral material and increments, and, unless thoroughly washed, small otoliths will continue to degrade in storage. Etching does help to enhance increment visibility in whole otoliths of certain species, such as many scombrids, because of the atypical nature of their increment-deposition pattern. Etched otoliths of tuna may be examined whole, or an acetate replica may be prepared (Wild and Foreman 1980).

Preparation of acetate replicas from ground, polished and etched otoliths (Pannella 1980) is often cited in the literature; however, this methodology has several disadvantages when compared with direct observation of ground and polished specimens. Replicas only reproduce structures on the surface, and, unless grinding intersects increments at the right level so all growth surfaces are perpendicular to the section plane, increments will be completely missed, or their widths will be distorted. It is often impossible to find a single section plane which will be "perfect" for all increments from the center of the otolith to the margin. Direct microscopy on the otolith itself allows for optical sectioning to reveal increments not properly exposed on the ground surface. Further, the appearance of increments changes radically (to the point where they may become unreadable), depending on etching times and the strength of the etching solution. Sequential or differential etching may also be required (Wild and Foreman, 1980).

Otolith microstructure may be viewed by two primary methods; light microscopy or SEM. Each technique has its own strengths and weaknesses. Light microscopy is generally simpler, requiring readily available equipment and easier preparation. Grinding, if necessary, has wider tolerances than for the SEM, since light images can be formed of internal features as well as those directly on the surface. Disadvantages of light microscopy include the appearance of confusing visual artifacts and the limited resolving power, the practical lower level being about 0.5 to 1 um. A bright light source, oil immersion objectives (100X), rotatable polarized light source, and video imaging are all important options for light microscopy. Video imaging can assist in increasing magnification to obtain measurements (but does not improve resolution) and in electronically enhancing image contrast and brightness. It is also a component of automatic and semi-automatic image analysis. Video images can also be easily recorded on tape for permanent documentation or exchange with other readers.

SEM has the advantage of providing much higher magnifications (e.g. to over 100,000X) and resolving power. Growth increments below 1 um thick are not uncommon in some fish or life stages, and they may be essentially undetectable with the light microscope. Optical artifacts are not a problem with SEM, although preparation artifacts due to the etching process may be introduced. Preparation for SEM is somewhat more involved and requires grinding or sectioning, polishing and etching. As noted for acetate replicas, grinding to the proper plane is critical to reveal an undistorted view of increments for an entire otolith. Grinding and polishing is greatly facilitated by the use of mechanically operated wheels, as used in metallurgical laboratories. Polishing with diamond compounds (e.g. 9, 3, 1, 1/4 um) is followed by etching with weak acidic solutions (e.g. 0.1 or .001 N HCl, or various strengths of EDTA or gluteraldehyde (see Haake et al. 1982). Recently I had good success with a combination of the following - 5.5% EGTA, 2% SDS, 0.4M NaOH and 0.1M phosphate buffer. For this solution, optimal etching times varied between 1/2 and 2 minutes. After etching, the sample should be coated with a 100-200 A layer of a conductive film, such as a gold/palladium alloy. The SEM is an expensive instrument and it requires a skilled operator to give good results. Variables such as tilt, accelerating voltage, and spot size can greatly alter SEM results. Typically it is used in the secondary electron image mode; however, I have found that the backscattered electron image mode can substantially enhance the visibility of increments. Work sessions may be recorded on video tape if the SEM has a "TV" mode.

Electron image enhancement and automatic or semi-automatic analysis are applicable to both light and electron microscopy; McGowan et al. (1986) in this symposium deal with the subject.

PHOTOGRAPHY

Photomicrography documentation is an important requirement for otolith-aging studies. Photographs should be of the highest quality obtainable and include magnification or scale bars, and a careful description of preparation, orientation, and source of material (e.g. species, size, sex,). Photographs illustrate exactly what structures have been counted for aging; they provide an opportunity to make a comparison between readers; and they illustrate the actual image on which counts and/or measurements are made. The last use must be employed with great caution, especially for light micrographs where artifacts cannot be readily identified in a frozen image, and where the ability to focus through a specimen is also lost.

COUNTING

Counts on otoliths, whether they are microstructural increments or annual zones, are best made in a "blind" fashion, i.e. without the reader having knowledge of the size, sex, etc. of the specimen. Comparison and verification between readers is also advisable, after a set of counting criteria is established. The path of counting should also be clearly specified or recorded.

VALIDATION AND MARKING

Validation is required to ensure the accuracy of any aging methodology. Approaches to validation for annual aging and microstructural aging are similar (Beamish and McFarlane 1980; Brothers 1981; Geffen 1985). Researchers must strive for the most rigorous validation achievable within the biological and practical limitations of the study. An important tool in otolith-validation studies is the use of marked otoliths. Marks may be natural, i.e. caused by environmental phenomena in the wild, or produced by a variety of manipulations in the field or laboratory, including modification of food available, temperatures and light, and the exposure to a variety of chemicals such as tetracycline, acetozolamide, or strontium. Recent references to these phenomena and techniques are provided by Campana and Neilson (1985), Tsukamoto (1985) and Hurley et al. (1985).

In summary, using otoliths to age fish requires that the investigator be flexible in applying a wide variety of methods to fit the specific characteristics of the fish being studied. Many options are possible and arriving at the optimal approach can only be achieved by careful testing and evaluation of the full range of techniques available. The most common errors introduced into otolith studies are the result of inadequate technique, compounded by inappropriate interpretation. Other deficiencies include lack of validation and insufficient sampling (numbers of fish, geographic and seasonal coverage, etc.). Better methodology can substantially improve the accuracy and precision of otolith-aging studies.

REFERENCES

BAGENAL, T.B. and F.W. TESCH. 1978. Age and growth. Pages 101-136 in T.B. Bagenal, editor, Methods for the assessment of fish production in freshwater, 3rd edition. Blackwell Science Publications, Oxford, England.

BEAMISH, R.J. 1979. New information on the longevity of Pacific ocean perch (Sebastes alutus). Journal of the Fisheries Research Board of Canada 36:1395-1400.

BEAMISH, R.J. and G.A. McFARLANE. 1983. The forgotten requirement for age validation in fisheries biology. Transactions of the American Fisheries Society 112:735-743.

BENNETT, J.T., G.W. BOEHLERT, and K.K. TUREKIAN. 1982.
Confirmation of longevity in Sebastes diploroa
(Pisces: Scorpaenidae) using Ra measurements in otoliths.
Marine Biology 71:209-215.

BLACKLER, R.W. 1974. Recent advances in otolith studies.
Pages 67-90 in F.R. Harden Jones, editors. Sea Fisheries
Research. John Wiley, New York, USA.

BOEHLERT, G.W. 1985. Using objective criteria and multiple
regression models for age determination in fishes. U.S.
National Marine Fisheries Service Fishery Bulletin
83:103-117.

BROTHERS, E.B. 1982. Aging reef fishes. Pages 3-22 in G.R.
Huntsman, W.R. Nicholson and W.W. Fox, Jr., editors. The
biological bases for reef fishery management. National
Marine Fisheries Service, NOAA Techincal Memorandum,
Southeastern Fisheries Center, Miami, Florida, USA.

CAMPANA, S.E. and J.D. NEILSON. 1985. Microstructure of fish
otoliths. Canadian Journal of Fisheries and Aquatic Science
42:1014-1032.

CASSELMAN, J.M. 1983. Age and growth assessment of fish from
their calcified tissues - techniques and tools. Pages 1-17
in E.D. Prince and L.M. Pulos, editors. Proceedings of the
international workshop on age determination of oceanic
pelagic fishes: tunas, bullfishes, and sharks. NOAA
Technical Report 8, National Marine Fisheries Service.

GEFEN, A.J. 1986. The validation of daily otolith ring
deposition in larval fish. In Proceedings of the fifth
larval fish conference. Louisiana Cooperative Fisheries
Unit, Baton Rouge, Louisiana, USA.

HAAKE, P.W., C.A. WILSON and J.M. DEAN. 1981. A technique for
the examination of otoliths by SEM with application to
larval fishes. In Proceedings of the fifth larval fish
conference. Louisiana Cooperative Fisheries Unit, Baton
Rouge, Louisiana, USA.

HECHT, T. 1978. A descriptive systematic study of the otoliths
of the neopterygean marine fishes of South Africa. Part I.
Introduction. Transactions of the Royal Society of South
Afraica 43(2):191.

HURLEY, G.V., P.H. ODENSE, R.K. O'DOR and E.G. DAWE. 1985.
Strontium labeling for verifying daily growth increments in
the statolith of the short-finned squid (Illex
illecebrosus). Canadian Journal of Fisheries and
Aquatic Science 42:380-383.

JERALD, A., JR. 1983. Age determination. Pages 301-324 in
L.A. Neilsen, D.L. Johnson and S.S. Lampton editors,
Fisheries Techniques. American Fisheries Society, Bethesda,
Maryland, USA.

JENSEN, A.C. 1965. A standard terminology and notation for
otolith readers. ICNAF Research Bulletin (2):5-7.

McGOWAN, M.F., E.D. PRINCE and D.W. Lee. 1986. An inexpensive
microcomputer-based system for making rapid and precise

counts and measurements of video displayed zonations in
skeletal structures of fish. In R.C. Summerfelt and G.E.
Hall, editors, Age and Growth of Fish. Iowa State
University, Ames, Iowa, USA.

PANNELLA, G. 1980. Methods of preparing fish sagittae for the
study of growth patterns. Pages 619-624 in D.C. Rhoades and
R.A. Lutz, editors, Skeletal growth of aquatic organisms:
biological records of environmental change. Plenum Press,
New York, New York, USA.

RADTKE, R.L. and T.E. TARGETT. 1984. Rhythmic structural and
chemical patterns in otoliths of the Antarctic fish
Notothenia larseni: their application to age
determination. Polar Biology 3:203-210.

TSUKMATO, K. 1985. Mass-marking of ayu eggs and larvae by
tetracycline-tagging of otoliths. Bulletin of the Japanese
Society of Scientific Fisheries 51:903-911.

WILD, A. and T.J. FOREMAN. 1980. The relationship between
otolith increments and time for yellowfin and skipjack tuna
marked with tetracycline. Inter-American Tropical Tuna
Commission Bulletin 17:509-541.

Ultrastructure of the organic and inorganic constituents of the otoliths of the sea bass

BEATRIZ MORALES-NIN
PASEO NACIONAL S/N
08003 BARCELONA, SPAIN

ABSTRACT

The ultrastructure of the organic and inorganic components of sea bass Dicentrarchus labrax otoliths was studied using sections taken through the nucleus along different planes and observed with scanning electron microscope.

The otoliths were composed of prismatic structures of crystalline material running radially from the centre out to the edge. Each crystalline prism was composed of aggregates of crystalline fibres, which were in turn made up of numerous acicular microcrystals arranged along the axis of growth.

Rhythmic variations in the deposition of microcrystals and the surrounding organic matrix over a daily cycle lead to the formation of daily growth increments. Each increment consisted of a continuous and a discontinuous unit. The size and continuity of the microcrystals making up such units depended on the rate of growth, but the method used to prepare the otoliths for examination also affected the results.

Structural discontinuities were relatively frequent in this species, depending on age, growth rate and reproductive stage.

The otoliths of teleost fishes are complex polycristalline bodies which serve as statoliths in the inner ear (Carlstrom 1963). They are composed chiefly of crystallized calcium carbonate in the form of aragonite and of a collagenous, fibrous protein (Degens et al. 1969). Otolith growth occurs through the apposition of new material over its surface. This apposition is cyclic and is a function of daily calcium metabolism rhythms and amino acid synthesis cycles that result in the formation of daily growth units or increments. Each such unit is made up of a continuous or incremental zone, deposited during the active period of the calcium metabolism, and a discontinuous zone (Pannella 1974).

The Age and Growth of Fish, edited by Robert C. Summerfelt and Gordon E. Hall © 1987 The Iowa State University Press, Ames, Iowa 50010.

Daily growth increments can be used to determine the age and growth rate of larvae and juveniles (Brothers et al. 1976, 1981), to age species that do not lay down seasonal rings and to verify the periodicity of such seasonal rings (Pannella 1980; Victor and Brothers 1982).

Recently, an extensive literature has appeared on the structure of the organic matrix (Dunkelberger et al. 1980; Watabe et al. 1982) and otolith formation (Irie 1960; Mugiya 1981). Little attention has been focused on the nature and relationship of the organic matrix and the crystals in otoliths from a single species, and little is known about the integrated structure of these complex bodies.

This paper describes a scanning electron microscope study of the organic and inorganic components of the sagittae of the sea bass Dicentrarchus labrax and discusses otolith growth of that species.

MATERIAL AND METHODS

The sagittal otoliths used in this study were from 50 wild sea bass collected with bottom trawl off the coast of Catalonia, Spain, and from 20, 1-year old fish spawned in January 1979 at the Torre la Sal aquaculture facility in Castellón, Spain and reared at the Instituto de Investigaciones Pesqueras in Barcelona until June 1980 (Table 1).

Otoliths were removed from freshly caught fish, cleaned, and stored dry in vials. The crystalline structure of the otoliths was examined from cross-sections and sagittal sections taken through the nucleus. Sections were obtained by breaking and grinding. Breaking was done along the transverse plane by pressing with the edge of a scalpel blade over a semirigid surface. Some of the broken surfaces were polished by grinding. The rest were washed in distilled water and dehydrated in a graded alcohol series.

Grinding was done by hand on a glass plate coated with a graded series of aluminum oxide compounds (400, 600 and 900 grit), followed by polishing with 1 μ diamond paste. The polished surfaces were washed several times in distilled water and decalcified with 0.1 N HCl for 3-4 seconds.

Some recently removed otoliths were used to study the protein matrix. Sagittae were briefly rinsed in 0.1 M sodium cacodylate buffer and split in half horizontally along the transversal plane with a razor blade. They were fixed in 2.5 % glutaraldehyde-2 % paraformaldehyde in 0.1 M sodium cacodylate buffer containing 4 % cetylpyridinium chloride (CPC) for 30 minutes. After etching with 0.2 M EDTA (pH 7.4) for 5 minutes, each otolith half was critical point dehydrated.

The otoliths sections were mounted and coated with gold for observation under a scanning electron microscope.

Table 1. Summary of samples and techniques used to study otolith structure in sea bass.

	Wild fish (n = 50)	Laboratory reared fish (n = 20)
Total length range	20-63 cm	13-17 cm
Age range	2-6 years	1 year
Date of catch or sacrifice	January 1980 to November 1981. Freshly caught fish for organic matrix studies: July 1983.	January and June 1980.
Techniques and number of samples	Otoliths cut along sagittal plane, fixed, etched and critical-point dehydrated (n = 15). Cross-sections by breaking (n = 7). Cross-sections by breaking and acid etched (n = 8). Sagittal-sections, ground and acid-etched (n = 20).	Cross-sections by breaking (n = 6). Cross-sections, ground and acid-etched (n = 4). Sagittal-sections, ground and acid-etched (n = 10).

RESULTS
Daily growth
        The crystalline structure of otolith sections prepared by
breaking alone, remained unaltered and consisted of many prismatic structures of crystalline material running radially from
the center out to the edges of the otolith (Figs. 1a, b). Each
prism was composed of a large number of growth units (Fig. 1a,
black arrows) and narrow discontinuities running transversely
to the prism's axis of growth (Fig.1a, white arrows). Prism
formation was sometimes interrupted by halts in growth, appearing
as discontinuous surfaces within the structure. The direction
of crystalline growth was sometimes altered after such a
discontinuity (Figs.1b, c, circle). In certain otolith sections
prism edges were not clearly defined, depending on the plane of
exfoliation followed by the crystalline material on separation
(Figs.1d, e).
        Examination of prism structure at high magnification revealed
that each prism was composed of a number of parallel, acicular
microcrystals arranged along the axis of crystalline growth (Figs.
1d to f). The microcrystals were variable in length, and they
generally grew across several discontinuities. They also tended
to grow along the same axis, giving rise to the crystalline fibers
that formed the prisms, and which were visible in the ground sections, and in certain of the broken sections, depending on the
plane of fracture.
        The intergrowth of the crystalline fibers produced a zig-zag
pattern (Figs.1d, e, circles). In sections taken across the long
axis of the prisms, the acicular-shaped microcrystals were distinguishable, together with aggregations corresponding to the bundles
of microcrystals making up each prism (Fig.1f). The diameter of
the microcrystals varied around 0.3 µm.
        The organic fraction of the otoliths, as it was seen in the
EDTA-decalcified and critical-point dehydrated preparations, comprised a reticular matrix, which was dense in the region of the
nucleus (Fig.2a) and otolith center (Fig.2b) and less dense near
the edges, where the matrix usually followed the spatial orientation
of the aragonite prism structures (Fig.2c). The matrix was composed
of fibers 100 A in diameter forming a three-dimensional reticulum,

Figure 1. Microstructural features of the crystalline component
        of the sea bass otoliths. Scanning electron micrographs
        of broken sections, a and c acid–etched. a: Prismatic
        structures running radially to the edge, black arrows =
        incremental units, white arrows = discontinuous units.
        The zone near the edge was formed during a winter period.
        b, c: Prisms, circles = changes in the direction of
        crystalline growth. d: Edge of a section showing the
        prisms and their components, circle = intergrowth of
        crystalline fibers. e: Crystalline fibers, circle =
        intergrowth pattern. f: Section across the longitudinal
        axis of prismatic growth.
        (White line: 10 µm: a to e; 3 µm: f)

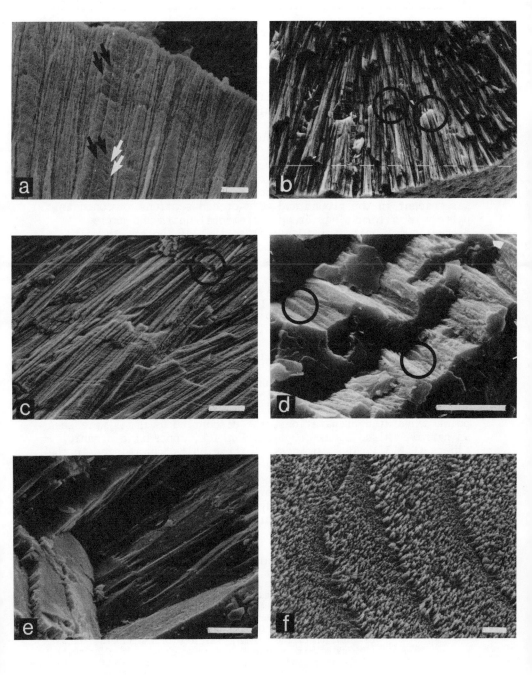

in which the size of the openings varied depending on the otolith region involved (Figs. 2a to 2f). At regular intervals of around 2 μm, the fibers combined to form bands of 300 A-thick fibers laid down transversely to the direction of otolith growth (Figs.2b and e, arrows). The spacing and orientation of these concentric fibers corresponded to the discontinuities in the crystalline prismatic structures. In addition to forming these concentric structures, the 300 A-thick fibers were also arranged obliquely, and together they formed a thick framework superimposed on and continuous with that formed by the 100 A thick fibers (Figs.2c to f, white arrows). Broken fibers were frequently present and formed globular knobs (Figs.2a to 2c and 2e, circles).

The rhythmic pattern of otolith formation in teleost fishes was evident in the ground and acid-etched sections. A concentric pattern of alternating, broad incremental units and narrow discontinuous units was identical to that of the growth units found in the daily growth increments; each was composed of an incremental and a discontinuous unit (Pannella 1980; Mugiya et al. 1981; Watabe et al. 1982).

The broad incremental units were composed of many acicular, branched microcrystals running perpendicularly to the bands (Figs. 3a to c). Most microcrystals were continuous from one band to the next (Figs.3b and c, circles). These incremental units corresponded to the incremental units described in the prisms. The relationship between the components of the otoliths is shown diagramatically in Fig.4.

The narrow discontinuous units formed sutures between the incremental units in which the growth of a part of the microcrystals was interrupted. In the fixed, EDTA-decalcified, and critical-point dehydrated preparations, the discontinuous units were seen as narrow, raised bands composed of thick, concentric organic fibers (Fig.3d). The thickness of these units was related to the technique employed in preparing the otolith for examination. In the cut and broken sections the discontinuous units were narrow (1.5 μm) (Fig.1c), while in the ground sections they were wider (2.1 μm) (Fig.3b). Polishing may destroy the faint edges of the microcrystals, and hence the acid etching might dissolve more

Figure 2. Microstructural features of the organic component of sea bass otoliths. Scanning electron micrographs of transversal sections, fixed etched and critical-point dehydrated. a: Organic matrix of the nuclear zone, circles = broken fibers. b: Regularly spaced thick fibers (arrows) superposed on the thin fibers, circle = broken fibers. c: Organic matrix following the spatial orientation of the prisms, circle = broken fibers, arrows = thick matrix growing across prism edges. d: Continuity of the thick and thin organic fibers (arrows). e,f: Organic 300 A-thick fibers forming concentric bands (thin arrows) and transverse fibers (thick arrows) and growing across a crystalline discontinuity (open arrow). (White line: 3 μm a to f)

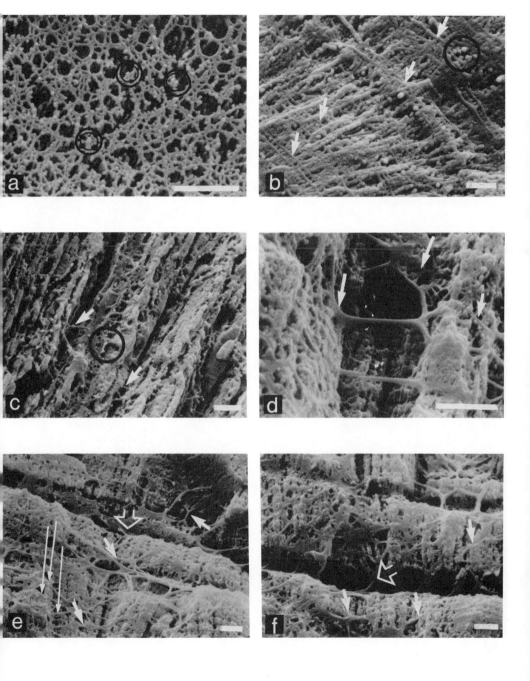

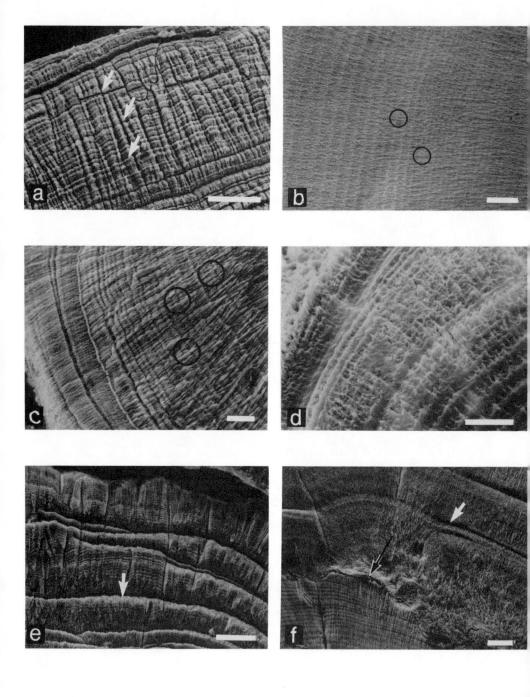

Figure 3. Daily growth increments and growth discontinuities
    on the otoliths of sea bass. Sagittal sections ground and
    acid-etched, except d which is prepared as in Fig.2.
    a: Rhythmic deposition of the daily growth increments,
    twoo-three subdaily units are visible, arrow = multidaily
    pattern. b, c: Incremental and discontinuous units in
    the daily growth increments, circle = microcrystals
    growing across discontinuous units. d: Discontinuous units
    composed by the concentric organic bands of 300 A-thick
    fibers. e, f: Structural discontinuities, arrow = growth
    disturbance, black arrow = laboratory-induced reabsorption.
    (White line: 10 μm b to d, 30 μm a, e and f)

of the crystalline material. In broken but not acid-etched
sections, the discontinuous units were hardly noticeable (Figs.
1d and e).

The marginal structures on the otoliths from fish collected
in different seasons of the year displayed differences in the
thickness and definition of the increments. During winter and
early spring the growth units were thin (x=1.5 μm, range 1.1 to
1.9 μm), whit the edges between prisms poorly defined (Figs.1a
and 3b); during summer and autumn they were thick (x=3.1 μm,
range 2.0 to 4.28 μm), with quite well-defined prisms (Fig.1a).

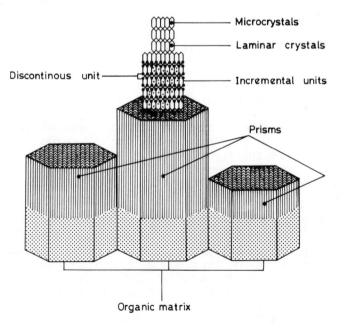

Figure 4. Diagram showing the relationships between otolith
    components.

To determine thickness of daily growth increments in relation to the growth rate, the structure of otoliths from slower-growing specimens among the 1-year old reared fish was compared to the structure in faster-growing fish; the mean length of slower-growing fish was 11.9 cm, of faster-growing specimens 15.7 cm. Fast growth rates were associated with thick microcrystals and wide growth units, while slow growth rates were related to narrow microcrystals and thinner growth units. Microcrystals were relatively longer and more closely spaced in the slower-growing specimens.

Two faint subunits were common in the daily growth increments in the wild fish, but less frequent in the laboratory-reared specimens (Fig.3a). A multiday pattern of groups of 7 and 14 daily growth increments was also more frequent in wild than in laboratory-reared fish (Fig.3a, arrows).

Structural Discontinuities

Structural discontinuities altering the otolith's growth pattern occurred often. When crystal development ceased and later resumed, sutures traversed by the growth of the 300 A-thick organic fibers were formed (Fig.2e). The density of the fibers in the discontinuities can be similar to that of those laid down earlier (Fig.2e, open arrow). After such a discontinuity, the direction of crystalline growth can also be altered (Fig.3e).

Discontinuities were more frequent in laboratory-reared than in wild fish, which was probably related to laboratory-induced stress. Several types of discontinuities were observed, ranging from small gaps (Fig.3f, white arrow) to major unconformities that could be traced continuously along the entire sagitta (Fig.3e, arrow). Discontinuities were more frequent near the sulcus acusticus and in the dorsal area. In wild fish the discontinuities were related to winter growth periods and spawing (Fig.3a).

The otoliths from a number of laboratory-reared fish exhibited a severe disturbance, which seemed to have been caused by reabsorption of the sagitta (Fig.3f, black arrow). Such discontinuities, probably a result of the stress caused by moving the fish from one aquarium to another, affected 60 % of the specimens.

DISCUSSION

The structure of sea bass otoliths is complex and exhibits closed interaction among its components. Otolith growth over a daily cycle can be divided into two distinct periods: a period of active calcification during which the microcrystals are laid down in the 100 A-thick matrix; this is followed by a period during which the calcification rate is lower, when the fibers join together to form concentric layers of fibers 300 A wide. The result of this repeated process is the formation of daily growth increments, consisting of an incremental and a discontinuous unit, laid down during the active calcification period and the low calcification period, respectively. The discontinuous unit is deposited around daybreak (Morales Nin 1984).

The growth of the microcrystals along a single axis of growth leads to the formation of crystalline fibers which, together with the surrounding organic matrix in which they are embedded, form prisms of aragonite. The continuity of the microcrystals, which do not undergo interruption simultaneously, gives rise to a high level of crystalline homogeinity in sea bass otoliths. No organic structures surrounding and delimiting each of the prisms could be identified.

Differences in the thickness of the discontinuous units, as well as in the continuity of the microcrystals passing through them, were observed, but these differences were in - fluenced by the method used to prepare the sagittae. In sections obtained by breaking and without demineralization, the discontinuous units were thin, with mostly uninterrupted microcrystals. The units were much broader on otoliths treated with HCl; they also had a larger number of interrupted microcrystals. These features were still more evident in polished, demineralized otoliths, the cristalline structure is altered by how much carbonate was dissolved. These results suggest that the reduction in the calcification rate along the daily cicle, may be slight. It appears that, because of their lower density, the discontinuous units have to be considered in the light of the technique used to prepare the otolith for examination.

The size of the microcrystals and the width of the incremental units in the ground sections appeared to be linked to the stage of growth. During active growth, thick microcrystals and broad incremental units with well-defined prisms are laid down; during periods of slower growth, the microcrystals are shorter and more compactly spaced, given an impression of greater density. Such a result agrees with observations on hypermineralization of tissues during periods of slow growth (Casselman 1974).

The organic matrix is much denser in the nucleus of the otolith than outside of it, although the latter density seems to increase during periods of high growth.

Structural discontinuities of varying importance are frequent in this species, particularly in long-lived indiciduals. The effect of discontinuities on the process of minaralization seems to be significant,since organic material was found to be present in all of them. Calcium carbonate plays an important role in maintaining the organism's osmotic balance (Berg 1974). As a result, carbonate secretion ceases during periods of stress or change. High blood acidity levels of fish held under anoxic conditions also cause reabsortion, as certain laboratory experiments carried out on sea bass would seem to indicate. Pannella (1980) reported similar results for Tilapia subjected to experimentally induced anoxia.

Environmental conditions where the juvenile are reared

seem to have had a large effect on otolith structure. Thus, subunits and multiday groupings of increments were much less frequent in laboratory-reared specimens. The relative stability of the experimental environment (food, temperature) may be the factor, which contrast to the variability of marine environment. On the other hand, in the experimental medium the specimens were subjected to conditions which may be stressful (cleaning of the aquarium, handling) leading to a larger number of discontiuities in the otolith structure.

ACKNOWLEDGMENTS
      The author wishes to thank Dr.M.Carrillo Director of the To-rre la Sal aquaculture facility for suplying the laboratory-reared sea bass specimens. Thanks also go to Dr.E.Macpherson and Dr.P.Arté for their support, Mr.A.Fauquet for mounting the samples and the long hours spent making observations, Mr.J.Biosca for developing the photographs and Mr.R.Sacks for translating the Spanish text.

REFERENCES
Brothers, E.B., C.P.Mathews and R.Lasker. 1976. Daily growth increments in otoliths from larval and adult fishes. Fisheries Bulletin U. S. 74:1-8.
Brothers, E.B. and W.N.MacFarland. 1981. Correlations between otolith microstructure, growth and life history transitions in newly recruited French grunts (Haemulon flavolineatum). Rapports Proces Verbaux Reunions Conseil International Explora-tion Mer 178:369-374.
Carlstrom, D. 1963. A cristallographic study of vertebrate oto-liths. Biological Bulletin, Marine Biological Laboratory Woods Hole 125:441-463.
Casselman, J.M. 1974. Analysis of hard tissue of pike Esox lu-cius L. with special reference to age and growth. Pages 13-27 in T.B.Bagenal, editor. Ageing of fish. Unwin Brothers Ltd, Surrey England.
Degens, E.T., W.G.Deuser and R.L.Haedrich. 1969. Molecular structure and composition of fish otoliths. Marine Biology 2: 105-113.
Dunkelberger, D.G., J.M.Dean and N.Watabe. 1980. The ultraestruc-ture of the otolith membrane and otolith in the juvenile mummi-chog Fundulus heteroclitus. Journal Morphology 163:367-377.
Irie, T. 1960. The growth of the fish otolith. Journal Faculty Animal Husbandry Hiroshima University 3:203-221.
Morales-Nin, B. 1984. Microarquitectura, características y compo-sición de los otolitos de los peces Teleósteos. Doctoral disser-tation. Barcelona University, Barcelona, Spain.
Mugiya, Y. 1981. Diurnal rhythm in otolith formation in the gold-fish Carassius auratus. Compendium Biochemistry and Physiology 68: 659-662.

Pannella, G. 1974. Otolith growth patterns: An aid in age deter-mination in temperature and tropical fishes. Pages 28-39 in T. B.Bagenal, editor. Ageing of fish. Unwin Brothers Ltd., Surrey, England.

Pannella, G. 1980. Growth patterns of fish sagittae. Pages 519-560 in D.C. Rhoads and R.A. Lutz, editors. Skeletal growth of aquatic organisms. Plenum Press, New York, U.S.A.

Simkiss, K. 1974. Calcium metabolism of fish in relation to ageing. Pages 1-12 in T.B. Bagenal editor. Ageing of fish. Unwin Brothers Ltd., Surrey, England.

Victor, B.C. and E.B. Brothers. 1982. Age and growth of the fal-lfish Semotilus corporalis with daily otolith increments as a method of annulus verification. Canada Journal Zoology 60:2543-2550.

Watabe, N., K.Tanaka, J.Yamada and J.M.Dean. 1982. Scanning electron microscope observations of the organic matrix in the otolith of the teleost fish Fundulus heteroclitus and Tilapia nilotica. Journal Experimental Marine Biology and Ecology 58: 127-134.

Techniques for the estimation of RNA, DNA, and protein in fish

LAWRENCE J. BUCKLEY
NATIONAL MARINE FISHERIES SERVICE
NORTHEAST FISHERIES CENTER
NARRAGANSETT LABORATORY
NARRAGANSETT, RI 02882

FRANK J. BULOW
DEPARTMENT OF BIOLOGY
TENNESSEE TECHNOLOGICAL UNIVERSITY
COOKEVILLE, TN 38505

ABSTRACT
     This paper presents a description of techniques used for the extraction and quantification of RNA, DNA, and protein in fish. The specific techniques employed in fish growth studies have varied with the life history stage of the fish and the nature of the investigation. A selection of the techniques most commonly used is presented so that the researcher may choose those that meet specific experimental needs.

The following description of techniques is intended to provide a guide to researchers new to the area of fish nucleic acid studies. It is recognized that alternative techniques may be favored by other researchers and that experimentation with techniques is a continuing effort. Munro and Fleck (1966, 1969) should be consulted for a review of the principles and techniques involved in measuring nucleic acids in tissues.

ESTIMATION OF RNA, DNA, AND PROTEIN IN LARVAL FISH
     Nucleic acids are extracted and partially purified using a modification of the Schmidt and Thannhauser method (Munro and Fleck 1966) as adapted for larval fish (Buckley 1979). The procedure is illustrated in Table 1. Preliminary treatment with cold acid removes free nucleotides, amino acids, and other low molecular weight compounds, and precipitates RNA, DNA, and

protein. Separation of RNA from DNA and protein is achieved by hydrolysis of RNA with dilute alkali; separation of DNA from protein by hydrolysis of DNA with hot dilute acid. Both RNA and DNA are estimated from the absorbancy of the appropriate hydrolysate at 260 nm (A_{260}).

Precautions

Perchloric acid ($HClO_4$) is a strong oxidizer and extreme care should be exercised, particularly with the concentrated acid. Nucleic acids are subject to enzymatic and heat degradation. Homogenization and extraction should be carried out at $0-4°$ C.

Collection

Larvae are removed from the net as soon as possible, placed individually on a labeled petri dish, and all excess water removed. If dry weights are desired, marine larvae should be first rinsed in distilled water or isotonic ammonium formate to remove salt. The petri dishes are immediately placed in a freezer. Alternatively, the samples can be frozen in liquid nitrogen or dry ice in alcohol.

Storage

Long-term storage (months) is best accomplished in an ultra-cold freezer ($-75°$ C). Samples can be stored for shorter periods of time in a conventional freezer ($-15°$ C).

Dry Weight Determination

If dry weights are required, the samples can be freeze dried and weighed. Care should be taken to minimize exposure of freeze-dried samples to room temperature. Samples should be kept in a dessicator in the freezer until weighed.

Sample Size and Homogenization

The sensitivity of the method is limited by the concentration of DNA in the tissue. The minimum DNA content per sample is 6 µg, which gives an A_{260} of 0.06 in the final DNA fraction. The maximum sample size is limited by the RNA content. The maximum RNA content per sample is 118 µg, which gives an A_{260} of about 1.0 for the RNA fraction. For larval fish where DNA and RNA make up about 1% and 5% of the dry weight, respectively, the optimum sample size is between 0.6 and 2.5 mg dry weight. Individuals may be pooled or homogenates diluted to achieve this optimum sample size. Whole larvae are homogenized in 2 ml of ice-cold, distilled water. In our laboratory, we have used motorized, all glass tissue grinders and a SDT tissue

mixer. Reference to trade names does not imply
endorsement by the National Marine Fisheries Service,
nor by Tennessee Technological University. A 1.4-ml
portion is used for estimation of nucleic acids.
Duplicate 0.1-ml or 0.05-ml portions can be used for
estimation of protein using a modification of the Lowry
method (Hartree 1972).

Centrifugation, Washes, and Removal of Supernatant

All centrifugations are for 10 min at 4000-6000
xg. A refrigerated centrifuge with a large capacity
head is most desirable. We used one head which
takes ninety-six 12 x 75 mm tubes, facilitating the
analysis of 60-90 samples in 1.5-2 days (Sorrall makes
this type of head). Supernatants are drawn off using a
drawn-out Pasteur pipet. Care must be taken not to
remove any of the pellet. Wash the pellet by adding
cold 0.2 N $HClO_4$ and stirring on a Vortex mixer or
breaking up the pellet with a glass rod. In the latter
case, the glass rod should be rinsed into the tube with
an added portion of $HClO_4$. After centrifugation, the
supernatant is carefully drawn off.

Quantitation

Quantitation is based on the relation between the
concentration of nucleic acid and the absorbancy at 260
nm (A_{260} of a 1 μg/ml solution of hydrolyzed RNA or DNA
is 0.030). If the volumes given in Table 1 are
followed exactly, then the total ug of nucleic acid in
the original 2 ml homogenate are:

$$RNA = \frac{(A_{260} \text{ RNA fraction} - A_{260} \text{ blank}) \text{ times } 2.52 \text{ times } 2}{0.030 \text{ times } 1.4}$$

$$= (A_{260} \text{ DNA fraction} - A_{260} \text{ blank}) \text{ times } 120$$

$$DNA = \frac{(A_{260} \text{ DNA fraction} - A_{260} \text{ blank}) \text{ times } 2.2 \text{ times } 2}{0.03 \text{ times } 1.4}$$

$$= (A_{260} \text{ DNA fraction} - A_{260} \text{ blank}) \text{ times } 104.8$$

Where 2.52 and 2.2 are the volumes (ml) of the RNA and
DNA fractions, 2 is the homogenate volume, 1.4 is the
volume of the homogenate actually used in the assay,
and 0.03 is the A_{260} of a 1.0 μg/ml solution of
hydrolized nucleic acid.

Blanks

For larval fish, a 1.4 ml distilled water blank
carried through the procedure is suitable.

Table 1.   Outline of procedure for estimation of RNA
   and DNA in fish
```
                        1.4 ml homogenate
                      | +0.7 ml cold 0.6 N HClO₄
                      | 15 min on ice
                      | centrifuge 6000 xg for 10 min.
                      | (less than 4° C)
```

Sediment	Supernatant
wash twice with	(discard)
2.0 ml cold 0.2 N HClO₄	
centrifuge after each wash	

Sediment	Supernatant
+ 1.12 ml 0.3 N KOH	(discard)
incubate at 37° C for 1 hr	
+ 1.4 ml 0.6 N HClO₄	
30 min on ice	
centrifuge	

Sediment	Supernatant
wash with 2.0 ml cold 0.2 N HClO₄	RNA fraction
centrifuge	read absor-
	bance at
	260 nm

Sediment	Supernatant
+ 2.2 ml 0.6 N HClO₄	(discard)
incubate at 85° C for 15 min.	
15 min on ice	
centrifuge	

Sediment	Supernatant
(Protein)	DNA frac-
	tion
	read absor-
	bance at
	260 nm

Standards and Quality Assurance

High and variable percentages of most commercial
preparations of RNA and DNA are soluble in 0.2 N HClO₄,
presumably due to low molecular weight and partial
degradation during preparation. This results in poor
recoveries in the RNA and DNA fractions of spikes added
to the homogenate with the balance going to the first
0.2 N HClO₄ wash. When standards are added just prior
to the hydrolysis steps, recoveries are very good.
Reproducibility of duplicate samples is also very good
(Buckley 1979, Barron and Adelman 1984). As a
quality-control measure, several hundred replicate

samples of larvae taken from a single spawn can be
stored in a freezer, then two of these standard samples
run along with each regular set of samples. The
absorbancy of the nucleic-acid fractions can be read at
more than one wavelength (i.e., 234, 260, and 280 nm),
and the absorption ratios compared with pure hydrolized
nucleic acids to insure against contamination.
Although generally not necessary for larval fish,
errors due to contamination by polypeptide material can
be minimized by using a two-wave length method (Fleck
and Begg 1965, Wilder and Stanley 1983).

Alternative Techniques

One limitation of the UV technique described is
that smaller larvae must be pooled to achieve the
minimum sample size, resulting in loss of information
on individual variability and necessitating collection
of a larger number of individuals. Several
fluorescence techniques for estimation of RNA and DNA
appear sensitive enough to run even the smallest larval
fish individually; however, none of these techniques
has been used extensively with fish. Many of these
techniques that involve the binding of a dye to RNA or
DNA, to produce a fluorescent product, do not require
preliminary extraction or clean-up of the tissue
homogenate (Morgan et al. 1979, Labarea and Paigen
1980).

ESTIMATION OF RNA, DNA, AND PROTEIN IN
JUVENILE AND ADULT FISH

Two problems arise in analysis of older fish. The
first is the selection of a suitable tissue, and the
second is the lower concentrations of nucleic acids and
the greater potential for contamination of the nucleic
acid fractions. Bulow (1971) has reviewed the
selection of tissue. Muscle and liver appear to be the
most closely related to growth. White muscle tissue is
recommended for most growth studies. Nucleic acid
levels in muscle can be quite low, necessitating a
larger tissue sample. Love (1970) considered, in
detail, the problem of obtaining a reproducible sample
of muscle tissue due to its heterogenous nature.
Contamination of the nucleic-acid fractions, if a
problem, can be overcome by several different
approaches. Tissue blanks should be run.

Lipid can be removed by prior extraction with
organic solvents (Bulow 1970). The A_{260} of the nucleic
acid fractions can be corrected for protein
contamination by first determining their protein
concentration using the Lowry method (Hartree 1972),

and then subtracting 0.001 O.D. units for each 1 µg of protein found per ml of the RNA or DNA fraction (Munro and Gray 1969). Wilder and Stanley (1983) estimated the nucleic acid levels in salmonid tissue from their UV absorbance using a two-wavelength method to correct for the presence of dissolved protein in the nucleic acid fractions.

More specific or sensitive methods can be used for analysis of the RNA and DNA fractions; however, these are generally more time consuming than the UV absorbance methods. RNA can be determined using the orcinol method (Schneider 1957). DNA can be determined using the indole method (Ceriotti 1952) or the 3,5-diaminobenzoic acid dihydrochloride fluorometric assay (Holm-Hansen et al. 1968, Hinegardner 1971).

Bulow (1970, 1971) and Bulow et al. (1981) used the technique outlined in Table 2 for juvenile and adult fish. While somewhat more involved and time consuming than the UV method, it gave good results with a variety of fish tissues and species.

Table 2. Outline of procedure for estimation of RNA
    and DNA in larger juvenile and adult fish

Preparation of Reagents

1. RNA standard stock solution. Warm mixture of 100 mg RNA powder and 5 ml distilled water until dissolved, add 10 ml 5% trichloroacetic acid (TCA), heat at 90° C for 60 min, cool, and bring volume to 200 ml with 5% TCA. Store under refrigeration.

2. DNA standard stock solution. Prepare in same way as RNA.

3. To accurately establish the concentrations of standard solutions, determine total phosphorus content by the procedure of Fiske and Subbarow (1925). An RNA standard stock solution typically may be 44 µg RNA P/ml. This is then diluted to 22, 11, and 5.5 µg RNA P/ml and standards of 22, 11, and 5.5 µg RNA P/ml are run with each batch of samples. The mean O.D./µg RNA P/ml is used in calculations.

4. Orcinol reagent. Immediately before use, dissolve 1 g purified orcinol in 100 ml concentrated HCl containing 0.5 g $FeCl_3$.

5. Diphenylamine reagent. Dissolve 1.5 g indicator grade diphenylamine in 100 ml glacial acetic acid and 1.5 ml concentrated H_2SO_4. Just before use, add 0.5 ml aqueous acetaldehyde (2.1 ml acetaldehyde per 100 ml of acetic acid) to the reagent. This reagent should be prepared immediately before each use.

Extraction

1. Mix 100 mg dry, fat-free (Bulow 1970) muscle tissue powder in 3 ml cold 5% TCA in thick-walled 15 ml centrifuge tubes.

2. Refrigerate 12 hr, mix, centrifuge, and discard supernatant. Add additional 3 ml cold 5% TCA mix, centrifuge, and discard supernatant.

3. Add 3 ml 5% TCA, mix, cover tubes with large marbles, incubate at 90° C for 30 min, and place tubes in cold-water bath.

4. Add 3 ml 5% TCA, mix, centrifuge for 15 min, draw off supernatant, and store supernatant (containing the combined nucleic acids) under refrigeration in capped glass vials.

RNA Analysis (modified from Schneider 1957)

1. Transfer duplicate (two tubes per sample) 0.5 ml aliquots of nucleic acid extract and 0.5 ml samples of RNA P and DNA P standard solutions to centrifuge tubes.

2. Add 1.5 ml distilled water and 2 ml orcinol reagent, mix, cover tubes, incubate at 90° C for 60 min, and cool in cold-water bath. (Blanks of 2 ml distilled water and 2 ml orcinol reagent are incubated with samples).

3. Read optical density against blank at 660 nm.

    (Since orcinol reacts with DNA as well as RNA, a correction should be made for the fraction of O.D. resulting from DNA in the sample. This is done through use of DNA P standards in the orcinol reaction).

DNA Analysis (After Burton 1956)

1. Combine duplicate 1 ml aliquots of nucleic acid extract with 1 ml 1 N $HClO_4$ and 2 ml diphenylamine reagent. Similarly prepare DNA P standards and prepare blanks by substituting 1 ml 5% TCA for the nucleic acid extract.

2. Mix and incubate at 30° C for 18 hr.

3. Read against blank at 600 nm.

Calculations for DNA

1. μg DNA P/ml = $\dfrac{\text{O.D. at 600 nm}}{\text{O.D./}\mu\text{g DNA P/ml}}$ (mean from series of standards run with each batch of samples)

2. µg DNA P/ml times 6 = µg DNA P/100 mg dry tissue.

3. DNA = DNA P times 10

Calculations for RNA

1. µg RNA P/ml =

$$\frac{\text{O.D. at 660 nm} - (\text{µg DNA P/0.5 ml times O.D./µg DNA P/0.5 ml})}{\text{O.D./µg RNA P/ml}}$$

where:

µg DNA P/0.5 ml = concentration of DNA P/ml of
                  extract as indicated by the
                  diphenylamine reaction times
                  0.5.

O.D./µg DNA P/0.5 ml = O.D./0.5 ml of DNA P
                       standard solution in orcinol
                       reaction.

O.D./µg RNA P/ml = mean from RNA P standards
                   in orcinol reaction

2. µg RNA P/ml times 6 = µg RNA P/100 mg dry tissue.

3. RNA = RNA P times 10.

Collection and Storage

Tissue samples should be removed from fish within
two hours of capture and immediately upon sacrifice of
the fish (Bulow et al. 1981). A 0.5-1.0 g wet weight
block of the left epaxial muscle is dissected from an
area just posterior to the head, quickly trimmed and
weighed, immediately submerged in liquid nitrogen or a
dry ice-isopropanol preparation, and then stored in a
capped vial below -15° C. The quick-freezing and cold
storage prevents enzymatic breakdown of nucleic acids
(Munro and Fleck 1969).

Extraction and Analysis of Nucleic Acids

Many of the basic techniques and instruments are
similar to those described for larval fish. Techniques
for centrifugation, washes, and removal of supernatant,
for example, are essentially the same.

The amount of tissue used for each sample should
be kept constant to avoid possible changes in recovery
associated with different amounts of tissue. Each
muscle block is minced with a sharp blade while still
frozen and then homogenized in ice-cold distilled

water. The procedure outlined in Table 1 can be
followed starting with a tissue concentration of 10 mg
dry weight/ml in the diluted homogenate. The UV
absorption spectrum of the final RNA and DNA fractions
should be compared with an authentic sample of
hydrolyzed RNA or DNA as an initial check for
contamination. Also, the protein content of the
nucleic acid fractions can be determined using a
modification of the Lowry method (Hartree 1972). If
contamination appears to be a problem, a two-wavelength
method is recommended. Alternatively, the
technique outlined in Table 2 or one of the
fluorometric techniques can be used.

REFERENCES
BARRON, M. G., AND I. R. ADELMAN. 1984. Nucleic acid,
    protein content, and growth of larval fish
    sublethally exposed to various toxicants.
    Canadian Journal of Fisheries and Aquatic Sciences
    41:141-150.
BUCKLEY, L. J. 1979. Relations between RNA-DNA
    ratio, prey density, and growth rate in Atlantic
    cod (Gadus morhua) larvae. Journal of the
    Fisheries Research Board of Canada 36:1497-1502.
BULOW, F. J. 1970. RNA-DNA ratios as indicators of
    recent growth rates of a fish. Journal of
    the Fisheries Research Board of Canada 27:2343-
    2349.
BULOW, F. J. 1971. Selection of suitable tissues for
    use in the RNA-DNA ratio technique of assessing
    recent growth rate of a fish. Iowa State Journal
    of Science 46:71-78.
BULOW, F. J., M. E. ZEMAN, J. R. WINNINGHAM, AND W. F.
    HUDSON. 1981. Seasonal variations in RNA-DNA
    ratios as indicators of feeding reproduction,
    energy storage, and condition in a population of
    bluegill, Lepomis macrochirus Rafinesque.
    Journal of Fish Biology 18:237-244.
BURTON, K. 1956. A study of the conditions and
    mechanism of the diphenylamine reaction for
    the colorimetric estimation of
    deoxyribonucleic acid. Biochemical Journal
    62:315-323.
CERIOTTI, G. 1952. A microchemical determination of
    deoxyribonucleic acid. Journal of Biological
    Chemistry 198:297-303.
FISKE, C. H., AND Y. SUBBAROW. 1925. The colorimetric
    determination of phosphorus. Journal of
    Biological Chemistry. 66:375-400.
FLECK, A., AND D. BEGG. 1965. The estimation of

ribonucleic acid using ultraviolet absorption measurements. Biochemica et Biophysica Acta. 108:333-339.

HARTREE, E. F. 1972. Determination of protein: A modification of the Lowry method that gives a linear photometric response. Analytical Biochemistry 48:422-427.

HINEGARDNER, R. T. 1971. An improved fluorometric assay for DNA. Analytical Biochemistry 39:197-201.

HOLM-HANSEN, O. W., H. SUTCLIFFE, JR., AND J. SHARP. 1968. Measurement of deoxyribonucleic acid in the ocean and its ecological significance. Limnology and Oceanography 13:507-514.

LABAREA, C., AND K. PAIGEN. 1980. A simple, rapid, and sensitive DNA assay procedure. Analytical Biochemistry 102:344-352.

LOVE, R. M. 1970. The chemical biology of fishes. Academic Press, New York, New York, USA.

MORGAN, A. R., J. S. LEE, D. E. PULLEYBLANK, N. L. MURRAY, AND H. EVANS. 1979. Ethidium fluoresence assays. Part I. Physiochemical studies. Nucleic Acids Research 7:547-569.

MUNRO, H. N., AND A. FLECK. 1966. The determination of nucleic acids. Pages 113-176 in D. Glick, editor. Methods of biochemical analysis. Volume 14. Interscience Publishers, New York, New York, USA.

MUNRO, H. N., AND A. FLECK. 1969. Analysis of tissue and body fluids for nitrogenous constituents. Pages 423-525 in H. N. Munro, editor. Mammalian protein metabolism. Volume 3. Academic Press, New York, New York, USA.

MUNRO, H. N., AND J. A. M. GRAY. 1969. The nucleic acid content of skeletal muscle and liver in mammals of different body size. Comparative Biochemistry and Physiology 28:897-905.

SCHNEIDER, W. C. 1957. Determination of nucleic acids in tissues by pentose analysis. Pages 680-684 in S. P. Colowick and N. O. Kaplan, editors. Methods in enzymology. Volume 3. Academic Press, New York, New York, USA.

WILDER, I. B., AND J. G. STANLEY. 1983. RNA-DNA ratio as an index to growth in salmonid fishes in the laboratory and in streams contaminated by carbaryl. Journal of Fish Biology 22:165-172.

Uptake of ^{14}C-glycine by fish scales (in vitro) as an index of current growth rate

GREG P. BUSACKER AND IRA R. ADELMAN

DEPARTMENT OF FISHERIES AND WILDLIFE
UNIVERSITY OF MINNESOTA
ST. PAUL, MN 55108

ABSTRACT

An index of current fish growth rate may be determined by measuring the uptake of ^{14}C-glycine by fish scales incubated in vitro. Procedures are given for solution preparations; selecting fish and scales; the choice of incubation methods and temperatures; and for radioactivity calculations, determinations, and expressions. The method is non sacrificial, logistically uncomplicated, and easily adapted for large numbers of scales.

HANDLING OF SCALES

Measurement of glycine uptake by fish scales in vitro may be used as an index of fish growth rate (Adelman 1986). Three or more scales of uniform size are removed from the mid-body region of the test fish and immediately incubated in a medium containing radioactively labeled ^{14}C-glycine, unlabeled glycine, and buffered physiological saline. The volume of incubation medium may be as little as 0.4 ml in small vials or greater volumes in suitable beakers. The general constraints are scale size and number, and the desire to use as little of the radioactively labeled glycine as possible. If possible, examine scales before incubation and discard any that are regenerated.

At the end of the incubation period (2 h is recommended) the scales are removed from the incubation medium using a fine forceps, rinsed in a large volume of cold saline, blotted of excess fluid, and placed in a liquid scintillation vial for drying and counting of radioactive emissions. The scales are dried at 60 C to a constant weight, weighed on a suitable balance and returned to the original scintillation vial.

After weighing, the scales are prepared for the scintillation counter by adding a tissue solubilizer to the scintillation vial and returning the vials to the oven for approximately 24 h. A suitable liquid scintillation fluor solution is then added to the vials, and the amount of radioactivity is determined. Radioactivity is expressed in terms of the area of the scale (pmol/mm^2). A relationship between scale area and weight can be determined for the size range of fish in question and a conversion factor applied to the scale weight.

When large numbers of scales are to be incubated, the incubation medium may be pipetted into vials, sealed, and kept frozen until use. These vials can be taken to the field in an insulated container with dry ice and be ready for use by quickly warming to the incubation temperature, typically the temperature of the water from which the fish are taken.

The choice of incubation temperature is a complex issue, but incubating the scales at the acclimation temperature of the fish appears to be most appropriate (Goolish and Adelman 1983). Incubation temperatures are easily controlled in a water bath.

PREPARATION OF PHYSIOLOGICAL SALINE

The physiological saline is a modification of Shuttleworth (1972) and is prepared as follows: 14.3 mM NaCl, 2.68 mM KCl, 1.53 mM $CaCl_2$, 15.0 mM $NaHCO_3$, 25.0 mM HEPES buffer, adjusted to a final pH of 7.5 using 0.1–1.0 N HCl. Add the $CaCl_2$ during the pH adjustment procedure when the pH is approximately 8.0 to avoid forming an insoluble carbonate.

In practice, it is convenient to make stock solutions for preparation of small volumes of saline. The following concentrations are useful: 0.1 g/ml KCl, 0.1 g/ml $CaCl_2$, 0.075 g/ml $NaHCO_3$; NaCl and HEPES can be weighed and added during solution preparation. For example, to make 100 ml of physiological saline (solution 1), weigh 837 mg NaCl and 651 mg HEPES and disolve in approximately 80 ml of distilled water; add 200 ul stock KCl and 1680 ul stock $NaHCO_3$; add 170 ul stock $CaCl_2$ during pH adjustment. Add HCl to adjust the final pH to 7.5 and add distilled water to bring the solution to final volume.

As with any bicarbonate buffer, the pH will shift as CO_2 is lost to the atmosphere. HEPES buffer and cold storage temperature reduce this problem but do not eliminate it. All stock solutions should be stored at 4 C.

PREPARATION OF GLYCINE SOLUTIONS

The incubation medium contains a mixture of radioactively labeled and unlabeled glycine at a total concentration of approximately 420 uM glycine, which is a saturation concentration for uptake by carp Cyprinus carpio scales. We have not determined a saturation concentration for other species but suspect that, in general, 420 uM will provide satisfactory results.

The unlabeled glycine is prepared by making a stock solution of 31.54 mg/ml in distilled water (solution 2). To prepare the solution of unlabeled glycine in physiological saline (solution 3), add 100 ul of solution 2 to 99.9 ml of solution 1. Solutions 2 and 3 should be prepared on the same day that the final incubation medium (solution 4) is made to avoid bacterial degradation of the glycine during storage.

Radioactively labeled ^{14}C-glycine is available with the label in the first or second carbon position. Either label is acceptable but use should remain consistent. Uniformly labeled glycine may also be used, but it is more expensive. Radioactively labeled glycine is typically purchased by activity. For example, assume 0.25 mCi of ^{14}C-glycine is purchased. It arrives in 2.5 ml of acidified water with a specific activity of approximately 45 mCi/mmol glycine. To calculate the amount of glycine present: 0.25 mCi glycine/45 mCi/mmol = 5.556 X 10^{-3} mmol glycine. In solution, this is 5.556 umol/2.5 ml = 2.222 umol/ml (2.222 mM) or 166.9 mg/l.

The incubation medium with radioactive glycine (solution 4) is prepared by adding 4 ul of labeled glycine (0.1 uCi/ul) per ml of the solution of unlabeled glycine (solution 3) to result in a final concentration of 0.4 uCi/ml. The concentration of labeled glycine can be increased if greater counting efficiency is required. The small change in concentration of total glycine in the incubation medium (solution 4), caused by a two to three fold increase in amount of radiolabeled glycine added, should not affect results.

ACKNOWLEDGMENT

This work was supported in part by the University of Minnesota Agricultural Experiment Station, project number 75, Scientific Journal Series Number 14,720.

REFERENCES

ADELMAN, I. R. 1986. Uptake of radioactive amino acids as indices of current growth rate of fish: a review. In R. C. Summerfelt and G. E. Hall, editors. Age and growth of fish. Iowa State University Press, Ames, Iowa, USA.

GOOLISH, E. M., AND I. R. ADELMAN. 1983. Effects of fish growth rate, acclimation temperature and incubation temperature on in vitro glycine uptake by fish scales. Comparative Biochemistry and Physiology 76A:127-134.

SHUTTLEWORTH, T. J. 1972. A new isolated perfused gill preparation for the study of the mechanisms of ionic regulation in teleosts. Comparative Biochemistry and Physiology 43A:59-64.

A progress report on the electron microprobe analysis technique for age determination and verification in elasmobranchs

GREGOR M. CAILLIET
MOSS LANDING MARINE LABORATORIES
P.O. BOX 450
MOSS LANDING, CA 95039

RICHARD L. RADTKE
HAWAII INSTITUTE OF MARINE BIOLOGY
UNIVERSITY OF HAWAII
P.O. BOX 1346
KANEOHE, HI 97644

ABSTRACT
Elasmobranchs have been aged using opaque and translucent bands
in their spines and vertebral centra, but few reports have
verified the temporal deposition patterns of these bands. We
present information from electron microprobe analyses for calcium
and phosphorus across vertebral centra of the tropical, nearshore
gray reef shark Carcharhinus amblyrhynchos and the temperate,
pelagic common thresher shark Alopias vulpinus. In both species,
the number of peaks in calcium and phosphorus concentrations was
equal to the number of opaque growth bands counted in either
x-radiographs or sections of the vertebrae. Also, those caught
in warm water, summer months had peaks, while those caught in
cold water, winter-spring months had valleys in these elemental
concentrations. Thus, electron microprobe analyses proved
helpful in verifying the annual periodicity of band formation for
two elasmobranchs from different habitats and geographical
regions.

The exploitation of elasmobranchs has been rapidly increasing in recent years, and this has stimulated research on their
life histories to provide information essential for understanding
and managing their populations (Holden 1974, 1977).

Most of the conventional structures used for age determination in bony fishes are not applicable to elasmobranchs, which

The Age and Growth of Fish, edited by Robert C. Summerfelt and Gordon E. Hall © 1987 The Iowa
State University Press, Ames, Iowa 50010.

lack calcareous otoliths and other usable skeletal parts. However, alternative methods have been developed for elasmobranchs, including size-mode analysis, tooth-replacement rates (Moss 1972), extrapolation from embryonic growth rates (Francis 1981), and growth-zone deposition in spines and vertebrae (reviewed in Cailliet et al. 1983a, 1986).

Because the amount and pattern of calcification in spines and vertebrae varies considerably among species (Ridewood 1921; Urist 1961; La Marca 1966; Applegate 1967), it is essential to evaluate the temporal periodicity with which the growth zones are deposited. Many authors have hypothesized that these growth zones are deposited annually, but only a few have actually tested this hypothesis (Beamish and McFarlane 1983).

Recently, terms commonly used in age determination of fishes have been carefully defined (Casselman 1983; Jearld 1983; Wilson et al. 1983), and the procedures necessary to adequately evaluate precision (verify) and/or evaluate accuracy (validate) of the temporal nature of band deposition in the calcified structures of fishes have been outlined (Brothers 1983).

Many of these verification and validation procedures have also been used in elasmobranch growth studies (Cailliet et al. 1986); they include size-frequency analysis, growth-model parameters, back-calculation, centrum-edge dimensions and histological characteristics, radiometric dating (Welden 1986), laboratory growth studies, tag-recapture results from the field, and tetracycline marking in both laboratory and field studies.

The difficulty is that discerning growth bands, especially those that are irregular in width or density and those at the outer edge of elasmobranch vertebral centra, requires objective techniques to establish solid criteria for detecting seasonal growth patterns and to validate the temporal periodicity of growth-zone formation. A relatively new technique involves chemical microanalysis. This method has only been used to evaluate growth zones in the centra of the spiny dogfish Squalus acanthias (Jones and Geen 1977) and in the otoliths of freshwater fish (Casselman 1983) by correlating seasons with peaks in calcium and phosphorus levels and opaque band dimensions. Microanalysis can further be used to characterize the width and chemical composition of the centrum edge from elasmobranchs collected in different seasons to verify the visual estimates of density (opacity or translucency: Wilson et al. 1983) used by Ishiyama (1951), Richards et al. (1963), Tanaka et al. (1978), Tanaka and Mizue (1979), and Waring (1984) to evalute growth-zone formation in elasmobranchs. In this paper, we evaluate the usefulness of the microanalytical technique in age determination and verification in two species of elasmobranchs.

To evaluate the microanalytical technique, we chose the tropical gray reef shark Carcharhinus amblyrhynchos from the Northwestern Hawaiian Islands and the more temperate common thresher shark Alopias vulpinus from California. Vertebrae were chosen both for whole-centrum and centrum-edge analyses. Verte-

brae for whole-centrum analytical scans were chosen from one
large and one small specimen of each species. Vertebrae from two
specimens of the common thresher shark were chosen from each of
two oceanographic periods (Bolin and Abbott 1963), one
representing warm water (hereafter called summer) and the other
representing cold water (hereafter called winter-spring), to
evaluate seasonal depositional patterns exhibited on the centrum
edge. These vertebrae had been previously aged using bands in
longitudinal sections (gray reef) and x-radiographs (common
thresher; Cailliet et al. 1983b).

  The longitudinally sectioned centra (Cailliet et al. 1983a)
of both species were embedded in epoxy resin with one side free
to form a 1-inch diameter disk. The surface of the disks and the
sample were then highly polished to prevent analytical errors and
diffraction of x-rays. The specimen disks and standards (made of
apatite) were then coated with carbon to further dampen diffraction of resultant x-rays and to increase electron conductance.
Calcium and phosphorus concentrations, expressed as weight-
fraction percentages, were analyzed using an electron microprobe.
The operating voltage was 15 kV, the beam current 3.25 mA, and
the absorbed current 15 nA. For readings the beam was focused on
a 5 m^2 area, and counts were for 30 seconds for each element
every 25 microns across one half of the centrum edge from the
outer edge to the center. Scans were initiated at the outer edge
to standardize the starting location, which could then be
conveniently compared with the counts.

  Whole centrum scans for calcium and phosphorus levels across
the sections of each species were compared with band patterns and
counts from the centra which had been prepared and aged using
standard techniques (Cailliet et al. 1983a, 1983b).

  Peaks in calcium and phosphorus from whole centrum scans of
gray reef sharks were in perfect agreement with the numbers and
spacing of the bands seen using light microscopy of sections
(Figs. 1-2). A distinct embryonic growth band, followed by a
peak in calcium and phosphorus associated with birth and first
feeding, was evident in both scans. There were three reduced
(valleys) and four elevated (peaks) regions of both elements
(Fig. 1) in the small gray reef shark, estimated to be three
years old from band counts. The large gray reef shark provided
similar concordance, with 12 bands being evident both in the
sectioned vertebra and in the calcium and phosphorus scan (Fig.
2). Small disruptive zones were observed in the interior portion
of the centrum of both specimens and are interpreted as variable
deposition during gestation. The first peak in calcium and
phosphorus is interpreted as the first summer's growth after
birth, with subsequent peaks indicative of subsequent summers.
The reductions in calcium and phosphorus are interpreted as
winter growth.

  The outer edge exhibited a peak in concentrations of both
elements in both specimens of gray reef shark, which were collected during the summer months of June and August (Figs. 1 and

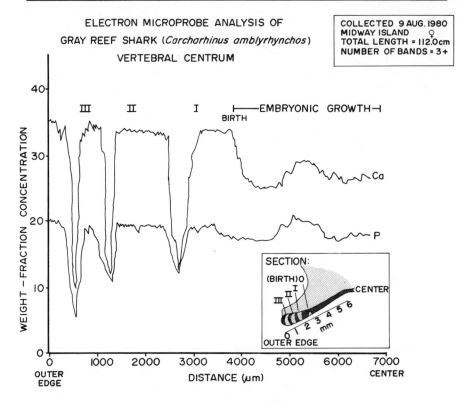

Figure 1. Electron microprobe analysis of calcium and phosphorus weight-fraction concentrations across the surface
of a sectioned vertebral centrum from a 112 cm (TL) gray
reef shark. Inset diagrams the band pattern discerned
from an adjacent centrum, with the darker portions representing the more opaque zones.

2). This is interpreted as the time during which the opaque
summer band was being deposited. The agreement of the microprobe
analyses with band counts and the elemental peaks during the
summer months lends credibility to age estimates made from
vertebral centra.

Other, more subtle, events were also detected by electron
microprobe analysis of gray reef shark centra. Elemental concentrations demonstrated a double peak during the second year of
growth in the larger specimen (Fig. 2), a feature not found in
the other areas of the sectioned vertebrae. This condition must
relate to a life-history event independent of annual periodic or
metabolic events, such as a temporarily decreased food supply, a
temperature change, a metabolic pause of some nature, or an
injury-related phenomenon.

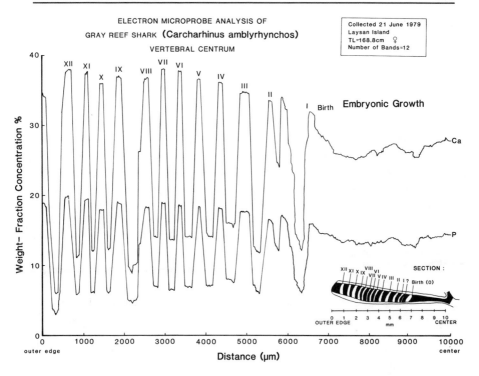

Figure 2. Electron microprobe analysis of calcium and phosphorus weight-fraction concentrations across the surface
of a sectioned vertebral centrum from a 168.8 cm (TL)
female gray reef shark. Inset diagrams the band pattern
discerned from an adjacent centrum, with the darker portions representing the more opaque zones.

Similarly, analysis of the two common thresher shark vertebrae demonstrated a strong concordance between light microscopy
and electron microprobe results (Figs. 3 and 4). The age estimate was two years in the smaller common thresher shark, and the
elemental concentrations indicated two peaks after birth (Fig.
3). Again, embryonic growth was represented by a zone with small
disruptions, followed by a burst in calcium and phosphorus after
birth. One difference between this species and the gray reef
shark is the higher variability exhibited within seasons by the
common thresher, which may be indicative of short-term changes in
the temperate environment occupied by this species, a fast-
moving, pelagic fish. The larger specimen exhibited smoother
elemental concentrations, and the number of bands and their
spacing agree with those obtained from a light microscope
analysis of an x-ray of a vertebra from this specimen.
Both of the common thresher sharks that were collected 27

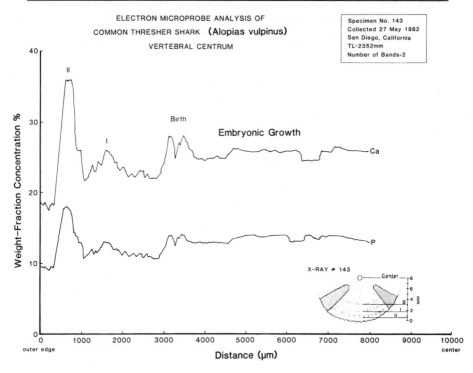

Figure 3.   Electron microprobe analysis of calcium and phosphorus weight-fraction concentrations across the surface
of a sectioned vertebral centrum from a 2352 mm (TL)
common thresher shark.   Inset diagrams the band pattern
discerned from an x-ray from an adjacent centrum, with
the darker portions representing the more opaque zones.

May 1982 (typically the last cold water month off California:
Bolin and Abbott 1963) had valleys in their calcium and phosphorus concentrations (Figs. 3 and 4).   As does the gray reef
shark, this evidence indicates this temperate species deposits
its opaque bands during the summer and its translucent bands
during the winter.

To further substantiate the temporal periodicity of band
formation in vertebral centra of the common thresher shark, the
four electron microprobe analyses of centrum edges indicated that
more calcium and phosphorus were being deposited during the summer than in the winter-spring period (Fig. 5).   The two summer-
caught thresher sharks exhibited peaks in their elemental concentrations at the centrum's outer edge, while the two winter-spring
specimens exhibited valleys.   It is presumed that the differences
in the sizes of specimens used did not influence these results.

These data suggest that the electron microprobe analysis of
calcium and phosphorus, although expensive and time-consuming, is

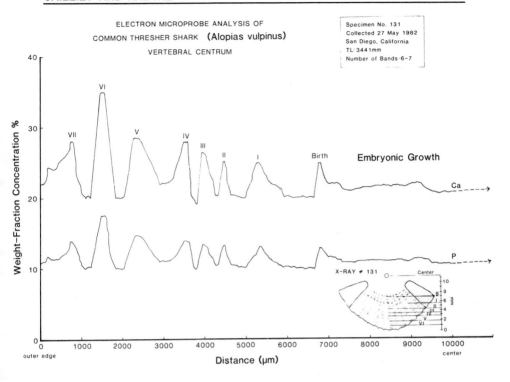

Figure 4. Electron microprobe analysis of calcium and phosphorus weight-fraction concentrations across the surface
of a sectioned vertebral centrum from a 3441 mm (TL)
common thresher shark. Inset diagrams the band pattern
discerned from an x-ray from an adjacent centrum, with
the darker portions representing the more opaque zones.

a promising tool for age verification in elasmobranch fishes.
The centrum-edge microanalysis for calcium and phosphorus required considerably less time, effort and expense than did the
whole-centrum scan. Controlled laboratory rearing and maintenance programs, coupled with further microanalyses of vertebral
centra from all size/age classes and seasons, should help to
explain the minor deflections observed in elemental
concentrations.

ACKNOWLEDGMENTS
        Bruce Welden and Lisa Natanson helped process and age the
common thresher shark vertebrae; James Parrish provided us with
the gray reef shark vertebrae; and Lynn McMasters drafted the
figures. Several anonymous reviewers provided valuable and
constructive criticisms. This research was sponsored in part by
NOAA, National Sea Grant College Program, Department of Commerce,
under Project nos. R/F-57 and R/F-84 to G.M. Cailliet through the

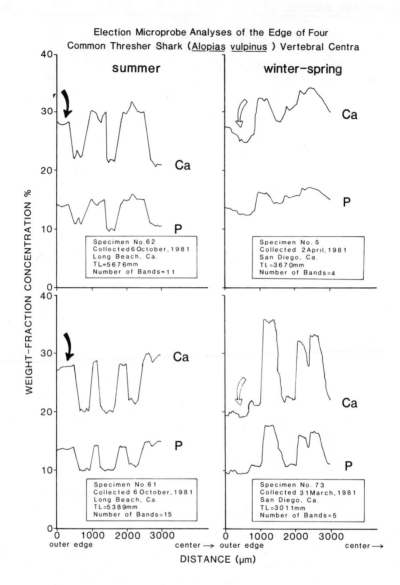

Figure 5. Electron microprobe analysis of calcium and phosphorus weight-fraction concentrations across the outer 3
mm of the centra of 2 summer-caught (warm water period)
and 2 winter-spring-caught (cold water period) common
thresher sharks off southern California in 1981. Black
arrows indicate peaks in both calcium and phosphorus at
the outer edge of the centra from the two summer specimens, while open arrows indicate valleys in these
concentrations for the winter-spring specimens.

California Sea Grant College Program, and under a cooperative
grant to R.L. Radtke, through the University of Hawaii Sea Grant
Program. The U.S. Government is authorized to reproduce and
distribute for governmental purposes.

REFERENCES
APPLEGATE, S. P. 1967. A survey of shark hardparts. Pages
37–67 in P. W. Gilbert, R. G. Mathewson and D. P. Rall,
editors. Sharks, skates and rays. Johns Hopkins Press,
Maryland, U.S.A.
BEAMISH, R. J. and G. A. MCFARLANE. 1983. The forgotten
requirements for age validation in fisheries biology.
Transactions of the American Fisheries Society
112:735–743.
BOLIN, R. L. and D. P. ABBOTT. 1963. Studies on the marine
climate and phytoplankton of the central coastal area of
California, 1954–1960. California Cooperative Oceanic
Fisheries Investigations, Reports, 9:23–45.
BROTHERS, E. B. 1983. Summary of round table discussions on age
validation. Pages 35–44 in E. D. Prince and L. M. Pulos,
editors. Proceedings of the international workshop on age
determination of oceanic pelagic fishes: tunas, billfishes,
and sharks. National Oceanographic and Atmospheric
Administration, Technical Report, National Marine Fisheries
Service 8.
CAILLIET, G. M., L. K. MARTIN, D. KUSHER, P. WOLF, and B. A.
WELDEN. 1983a. Techniques for enhancing vertebral bands in
age estimation of California elasmobranchs. Pages 157–165
in E. D. Prince and L. M. Pulos, editors. Proceedings of
the international workshop on age determination of oceanic
pelagic fishes: tunas, billfishes, and sharks. National
Oceanographic and Atmospheric Administration, Technical
Report, National Marine Fisheries Service 8.
CAILLIET, G. M., L. K. MARTIN, J. T. HARVEY, D. KUSHER, and B. A.
WELDEN. 1983b. Preliminary studies on the age and growth
of blue, Prionace glauca, common thresher, Alopias vulpinus,
and shortfin mako, Isurus oxyrinchus, sharks from California
waters. Pages 179–188 in E. D. Prince and L. M. Pulos,
editors. Proceedings of the international workshop on age
determination of oceanic pelagic fishes: tunas, billfishes,
and sharks. National Oceanographic and Atmospheric
Administration, Technical Report, National Marine Fisheries
Service 8.
CAILLIET, G. M., R. L. RADTKE, and B. A. WELDEN. 1986. Elasmobranch age determination and verification: a review. Pages
000–000 in T. Uyeno, R. Arai, T. Taniuchi and K. Matsuura,
editors. Indo-Pacific fish biology: Proceedings of the
second international conference on Indo-Pacific fishes.
Ichthyological Society of Japan, Tokyo.
CASSELMAN, J. M. 1983. Age and growth assessment of fish from

their calcified tissue--techniques and tools. Pages 1-17 in
E. D. Prince and L. M. Pulos, editors. Proceedings of the
international workshop on age determination of oceanic
pelagic fishes: tunas, billfishes, and sharks. National
Oceanographic and Atmospheric Administration, Technical
Report, National Marine Fisheries Service 8.

FRANCIS, M. P. 1981. Von Bertalanffy growth rates in species of
Mustelus (Elasmobranchii: Triakidae). Copeia 1981:189-192.

HOLDEN, M. J. 1974. Problems in the rational exploitation of
elasmobranch populations and some suggested solutions.
Pages 117-137 in F. R. Harden-Jones, editors. Sea fisheries
research research. John Wiley and Sons, New York, U.S.A.

HOLDEN, M. J. 1977. Elasmobranchs. Pages 187-214 in J. A.
Gulland, editor. Fish population dynamics. John Wiley and
Sons, New York, U.S.A.

ISHIYAMA, R. 1951. Studies on the rays and skates belonging to
the family Rajidae found in Japan and adjacent regions. 2.
On the age determination of Japanese black skate, Raja fusca
Garman. Bulletin of the Japanese Society of Scientific
Fisheries 16:112-118.

JEARLD, A., JR. 1983. Age determination. Pages 301-324 in L.
A. Nielsen and D. L. Johnson, editors, Fisheries techniques.
American Fisheries Society, Bethesda, Maryland.

JONES, B. C. and G. H. GEEN. 1977. Age determination of an
elasmobranch (Squalus acanthias) by x-ray spectrometry.
Journal of the Fisheries Research Board of Canada 34:44-48.

LA MARCA, M. J. 1966. A simple technique for demonstrating
calcified annuli in the vertebrae of large elasmobranchs.
Copeia 1966:351-352.

MOSS, S. A. 1972. Tooth replacement and body growth rates in
the smooth dogfish Mustelus canis (Mitchell). Copeia 1972:
808-811.

RICHARDS, S. W., D. MERRIMAN, and L. H. CALHOUN. 1963. Studies
on the marine resources of southern New England. IX. The
biology of the little skate Raja erinacea Mitchell.
Bulletin of the Bingham Oceanographic Collection 18:4-67.

RIDEWOOD, W. G. 1921. On the calcification of the vertebral
centra in sharks and rays. Philosophical Transactions of
the Royal Society, London, Series B 210:311-407.

TANAKA, S., C.-T. CHEN, and K. MIZUE. 1978. Studies on sharks.
XVI. Age and growth of Eiraku shark, Galeorhinus japonicus
(Muller et Henle). Bulletin of the Faculty of Fisheries,
Nagasaki University 45:19-28.

TANAKA, S. and K. MIZUE. 1979. Studies on sharks. XV. Age and
growth of the Japanese dogfish, Mustelus manazo, Bleeker, in
the East China Sea. Bulletin of the Japanese Society of
Fisheries 45:43-50.

URIST, M. R. 1961. Calcium and phosphorus in the blood and
skeleton of the elasmobranchs. Endocrinology 69:778-801.

WARING, G. T. 1984. Age, growth and mortality of the little
skate off the northeast coast of the United States.

Transactions of the American Fisheries Society 113:314–321.
WELDEN, B. A. 1986. Comparison of radiometric with vertebral band age estimates in four California elasmobranchs. In R. C. Summerfelt and G. E. Hall, editors. Age and growth of fish. Iowa State University Press, Ames, Iowa.
WILSON, C. A., E. B. BROTHERS, J. M. CASSELMAN, C. L. SMITH, and A. WILD. 1983. Glossary. Pages 207–208 in E. D. Prince and L. M. Pulos, editors. Proceedings of the international workshop on age determination of oceanic pelagic fishes: tunas, billfishes, and sharks, National Oceanographic and Atmospheric Administration, Technical Report, National Marine Fisheries Service 8.

Analysis of length-frequency distributions

PETER D. M. MACDONALD
DEPARTMENT OF MATHEMATICS AND STATISTICS
McMASTER UNIVERSITY
HAMILTON, ONTARIO L8S 4K1
CANADA

ABSTRACT
    With appropriate statistical analyses, length-frequency distributions give information about population structure and the growth of age-groups. However, the more overlapped the component length-at-age distributions are, the more important it is to incorporate additional information, whether from extra sampling (aged subsamples or a time sequence of length-frequency distributions) or in the form of biologically reasonable constraints imposed on the parameters being estimated. This paper reviews recent work and demonstrates an improved version of the Macdonald and Pitcher (1979) computer program which gives consistent results under a variety of constraints and assumptions.

INFORMATION IN LENGTH-FREQUENCY DISTRIBUTIONS

The Shape of a Length-Frequency Distribution

    The shape of a length-frequency distribution is the result of recruitment, growth, mortality, and sampling bias, obscured by variation in recruitment dates and individual growth-rates. Length-frequency analysis is most useful when recruitment is annual (or at least seasonal) and growth is sufficiently fast relative to potential sources of variation so there is little overlap between sizes of individuals in successive age-groups. In this case, size alone can identify with some certainty to which age-group a fish belongs, and basic statistics such as mean length and relative abundance can be computed for each group. While it may then still be necessary to adjust for sampling bias, if sampling is at all related to size or age, that is a separate problem not considered in this paper.

    In a typical application, considerable variation is present, and the separate components may not even show up as distinct modes

The Age and Growth of Fish, edited by Robert C. Summerfelt and Gordon E. Hall © 1987 The Iowa State University Press, Ames, Iowa 50010.

on the length-frequency distribution. This paper considers what can be done to recover information about the mean lengths and relative abundances of the component age-groups in the sample.

Extra Information from Extra Sampling

Aged subsamples. Hosmer (1973) showed that estimation of the relevant parameters of each age-group, such as mean length and relative abundance, is greatly facilitated when separate samples are available from one or more of the component groups. He outlined the computations for the case of just two normally distributed components, but they could be extended to more general situations. A reliable way to sort samples by age-group is to draw a subsample at random from the overall length-frequency distribution and determine the age of each individual in that subsample. Unfortunately, in many applications the population will consist of strong, well-defined young age-groups with relatively small numbers of older fish; the random subsample will consist mainly of young fish and hence not give enough information about the older age-groups (Fournier 1983).

Length-bias may be introduced by any attempt to select fish from specific age-groups instead of from the whole population. Historical length-at-age data could be used instead, but may be invalid due to changes in recruitment, age-specific mortality or growth patterns in the intervening time.

Age-at-length data, obtained by ageing all individuals in a subsample stratified by length, is a more promising approach for two important reasons: the problem of length bias in the sample is avoided; it is possible to put extra sampling effort into the longer lengths where, typically, overlap between age-groups is greatest.

Fournier (1983) presented a computerized method for fitting any combination of length-frequency data, length-at-age data, and age-at-length data simultaneously. In contrast, Macdonald and Pitcher (1979) proposed less formal use of an aged subsample: to determine the number of component age-groups in the length-frequency distribution; to obtain rough estimates of the parameters of the component distributions; and to provide guidelines for finding the most biologically meaningful fit to the overall mixed distribution.

Time sequences of length-frequency distributions. A time sequence of length-frequency distributions may allow separation of component age-groups which are otherwise obscured by overlap in the length-frequency distribution at one time, so there is a better chance of following the evolution in time of each cohort (year-class). Pauly and David (1981) analyse such time sequences by assuming that the mode of each age-group lies on a von Bertalanffy growth curve. They have developed a microcomputer program to trace the best-fitting set of growth curves through the modes of successive length-frequency distributions. Pitcher and Macdonald (1973) analyzed length-frequency distributions for minnows sampled every three or four weeks for 2 1/2 years. They showed that the mean length followed a modified form of von Bertalanffy growth curve with seasonally fluctuating growth rate.

The relative proportion for a given cohort can also be estimated at successive times; it can be used to estimate the mortality rate over each time interval, but only if the total population size or a measure of catch-per-unit effort is known at each time, and if adjustment can be made for age bias in sampling. In practice, there does not appear to be any satisfactory method of measuring such bias.

METHODS OF ANALYSIS

Modes, Probability Plots or Areas?

When component age-groups are well separated they will show up as distinct modes in the length-frequency distribution. The components may then be isolated by visual inspection of modes (Petersen 1891), or by a more sophisticated scoring of modes (Pauly and David 1981; Shepherd 1986). A probability plot is a graph of cumulative area from left to right, and the valleys between the modes will show up as inflection points. Manual probability-plot methods of Cassie (1954), Bhattacharya (1967), and others were for many years the most commonly-used methods of decomposing fisheries length-frequency data into component age-groups.

Modes are unreliable when there is substantial overlap between components, as the number of modes may be much less than the number of components (Figure 1). Theoretical normal distributions (Figure 1) are unlike sample data (Figures 2 and 3), which show a much more irregular appearance that tends to obscure modes or generate spurious modes which have no relation to component age-groups. The histogram in Figure 2 displays a sample of 200 observations generated by computer from the distribution shown in Figure 1D. The modes of the histogram are, clearly, poor indicators of the modes of the underlying theoretical distribution in this example. Thus, any method based on modes or on the visual interpretation of probability plots will be subjective and unreliable, unless the components are as well separated as, for example, those in Figure 1C.

Efficient statistical methods, such as maximum likelihood, work by matching areas under the fitted distribution to areas under the sample length-frequency histogram. Since areas under the component curves correspond to relative proportions, it is evident that maximum likelihood will give efficient estimates of the proportions. It can also be shown that it gives theoretically efficient estimates of all the parameters (Rao 1965). McNew and Summerfelt (1978) and Macdonald and Pitcher (1979) reported that maximum likelihood works well in practical applications.

The Importance of Constraints

Inspection of length-frequency distributions, such as those in Figure 1F or Figure 3, should suffice to illustrate the impos-

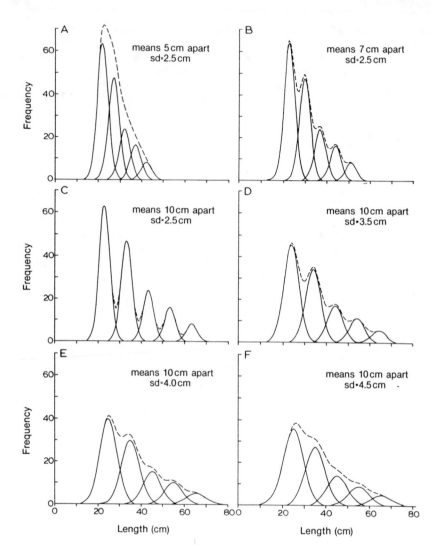

Figure 1. Effects of small changes in the means, standard deviations, and relative sizes of the component age-groups on the appearance of modes in a size-frequency plot. Dashed curves represent the frequency plot for an infinite population, or one which would be obtained from large samples collected within very fine size intervals. Individual components are drawn with solid lines. Proportions of population in each component are, from left to right, 0.4, 0.3, 0.15, 0.10, 0.05, which remain the same in each of A-F; (A, Unimodal; B, 2 clear modes and 2 indistinct modes; C, 5 clear modes; D, 2 clear modes and 2 indistinct modes; E, Only 2 modes; F, Unimodal.). (Reprinted from Journal of the Fisheries Research Board of Canada, with permission.)

374

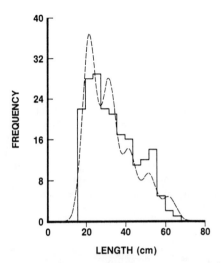

Figure 2. A histogram (solid line) from 200 computer-generated observations compared with the theoretical distribution (dashed line) from which the data were generated.

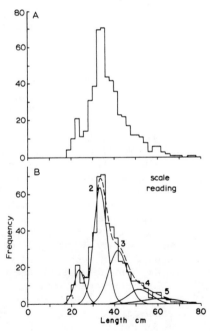

Figure 3. Length frequency histogram and age-groups determined by scale reading for 523 pike from Heming Lake, Manitoba, Canada, August and September 1965. (Reprinted from Journal of the Fisheries Research Board of Canada, with permission.)

sibility of a "black box" computer program, which inputs the overall length-frequency distribution and returns with the number of
components and the proportion, mean, and standard deviation of
each one. There is simply too much overlap. One can find a great
many fits which are mathematically reasonable, in the sense of
reproducing the overall shape of the length-frequency distribution, but not all of which are biologically reasonable. In
general, it is necessary to assume that the number of component
age-groups is known; if necessary, this can be achieved by lumping
the oldest age-groups into a single group (Macdonald and Green
1985). Macdonald and Pitcher (1979) then proceed by imposing
constraints on the parameters, which alleviates the problem by
reducing the number of parameters being estimated. Choosing biologically reasonable constraints forces the fit to be biologically
reasonable. The most recent version of their program (Macdonald
and Green 1985) allows for any of the following constraints:
specified parameters held fixed; component means all equal, equally spaced, or on a von Bertalanffy growth curve; component standard deviations all equal; component coefficients of variation all
equal or all equal to a fixed value.

Computational Methods for Maximum Likelihood Estimation

There are essentially three computational methods available:
direct search, quasi-Newton methods and the EM algorithm. Direct-
search algorithms use intelligent trial-and-error to adjust
initial guesses at the parameter values successively until the fit
is optimized. They are safe in that they will never converge or
diverge to a fit that is worse than the initial guess, but they
are expensive in terms of computer time and very slow to converge
to a final optimum. Macdonald and Pitcher (1979) chose the
Nelder-Mead algorithm (Nelder and Mead 1965) as one option in
their program, because it is reliable and easily portable between
computers; it is also simple to understand and admits constraints
of range, by using penalty functions, to keep the search away from
inadmissible parameter values. Efficient direct-search algorithms
are available in modern program libraries.

Quasi-Newton methods such as scoring (Macdonald and Pitcher
1979) will converge very quickly to a local maximum of the likelihood function if the initial parameter values are sufficiently
close to that solution. Direct search methods can be used to get
them sufficiently close. An array of derivatives must be computed, but these can be used to give standard errors for the estimates. Constraints are handled by appropriate manipulation of the
array of derivatives. Disadvantages are that the iterations may
fail to converge if the initial values are not sufficiently close
to the solution or if the parameters are not subjected to enough
constraints. Also, the iterations may lead to unacceptable parameter values such as negative values for the proportions. With
practice these are not serious problems, but they tend to confound
the novice user. Clark (1981) has used quadratic programming to
avoid negative proportions.

The EM algorithm has been proposed as the most natural way to calculate maximum likelihood estimates from a distribution mixture, such as a length-frequency distribution (Dempster et al. 1977; Everitt and Hand 1981; Agha and Ibrahim 1984), despite its being much slower to converge than quasi-Newton methods when the initial values are close to the best fit. In experience, when the length-frequency distribution is more or less unimodal the unconstrained EM algorithm will, for certain initial values, attempt to fit the data by a single component, regardless of the number of components actually present.

The data of Figure 3 were analyzed by a FORTRAN 77 version of Agha and Ibrahim's (1984) program. The results were not unacceptable but multiple solutions were found; even after 800 or 1000 iterations different runs from different initial values agreed to only two significant digits when approaching the same solution. This is a reflection of the flatness of the likelihood surface and is best alleviated by constraints on the parameters. Hathaway (1983) presented an EM algorithm for fitting a mixture of normal component distributions subject to weak constraints on the component standard deviations. It would be worthwhile to investigate ways of imposing other, stronger constraints with the EM algorithm.

Growth Curve Assumptions

Several authors have constrained the mean lengths of successive age-groups to lie along a growth curve, usually the von Bertalanffy curve (Schnute and Fournier 1980; Pauly and David 1981; Fournier 1983; Fournier and Breen 1983; Breen and Fournier 1984; Shepherd 1986). The idea is attractive in cases where the first three or more age-groups show as clear modes but the older age-groups are more overlapped because of a reduced growth rate. There may be enough information from the first three age-groups to fit a growth curve, even if the mean length at every older age cannot be fitted. Shepherd (1986) points out that often the objective of a length-frequency analysis is to determine the growth characteristics of the population, so this approach will accomplish that directly.

The disadvantage is that any growth curve is a mathematical idealization which might not apply very well to a particular cohort. One of the more-complicated growth curves could be used for a better fit, allowing for a more flexible shape (Richards 1959) or for seasonal growth (Pitcher and Macdonald 1973). However, such complicated curves will seldom be justified. Even with the simplest von Bertalanffy growth curve there are three parameters L_∞, k, and t_0, and these will not be well estimated, unless there are at least three well-defined components in the length-frequency distribution. A "banana-shaped" goodness-of-fit surface, with many combinations of L_∞ and k giving equally good fits, will result if there are not three or more well-defined modes. In this case, it is best to report the region of acceptable fits rather than a single value for each estimate (Shepherd 1986). Schnute and Fournier (1980) pointed out that estimates of the usual von Bertalanffy parameters L_∞, k, and t_0 are too sensitive to small errors in the data and suggested an appropriate reparameterization.

Furthermore, while it may often be reasonable to assume that the mean length in a given cohort follows some growth curve fairly well, it is less reasonable to assume that the mean lengths of successive age-groups will always lie along the same growth curve because of variation between recruitment classes, in recruitment dates and in growth rates. Thus if the growth curve model does apply, the analysis will be much more powerful for having made use of it. If the model is in doubt, the data might be fitted with and without the growth curve constraint; if removing the constraint does not significantly improve the fit then the growth curve analysis may be accepted.

Mortality Assumption

Fournier (1983) and Fournier and Breen (1983) assumed that successive age-groups in a single length-frequency distribution satisfy the relation $\pi_i = \pi_{i-1}\exp(-Z)$, where π_i is the proportion of the population attributed to age class, i, π_{i-1} is the proportion for the previous age class, and Z is the instantaneous rate of total mortality. Breen and Fournier (1984) point out that this model will apply only to unfished populations, populations fished for a long time at a constant rate, or virgin populations fished for a very short time. It should also be noted that total mortality must be age-independent, the initial size of each age-group must be the same (i.e., recruitment must not vary from year to year), and the sampling of the length-frequency distribution must be free of age bias. When these conditions hold, the mortality assumption constraint can be used to reduce the set of proportions to a single parameter Z, and hence provide an estimate of total mortality. As with growth-curve assumptions, fitting a length-frequency distribution with and without a mortality assumption may help to decide whether that assumption is acceptable.

EXAMPLES OF LENGTH-FREQUENCY ANALYSIS USING PROGRAM MIX

This computer program is documented by Macdonald and Green (1985). The data on northern pike, Esox lucius, from Heming Lake, Manitoba, Canada, August and September, 1965, analyzed in Macdonald and Pitcher (1979), are re-analyzed here. Eighteen different analyses are given, by taking all possible combinations of three different distributional assumptions (normal, lognormal, gamma), three different constraints on the component means (all free, on a growth curve, equally spaced) and two different constraints on the component standard deviations (coefficients of variation all equal; standard deviations all equal). The histogram for the data is shown in Figure 3 and the results are displayed in Tables 1 to 6. The first column of each table gives the true values of the parameters as determined by ageing all 523

Table 1. Parameters for the five age-groups of the northern pike data, estimated from the length-frequency distribution, with component coefficients of variation constrained to be equal. The χ^2 tests show goodness-of-fit and standard errors appear with estimates. First column values were obtained by sorting the fish into age-groups after ageing each fish by scale reading.

Parameter	By Scale Reading	Normal	Lognormal	Gamma
π_1	0.105 ± 0.013	0.092 ± 0.014	0.100 ± 0.015	0.097 ± 0.015
π_2	0.465 ± 0.022	0.467 ± 0.071	0.519 ± 0.120	0.495 ± 0.100
π_3	0.298 ± 0.020	0.259 ± 0.055	0.227 ± 0.077	0.241 ± 0.068
π_4	0.090 ± 0.013	0.128 ± 0.040	0.107 ± 0.081	0.117 ± 0.060
π_5	0.042 ± 0.009	0.054 ± 0.022	0.048 ± 0.048	0.050 ± 0.033
μ_1	23.33 ± 0.33	22.75 ± 0.43	23.07 ± 0.46	22.95 ± 0.44
μ_2	33.09 ± 0.19	32.97 ± 0.58	33.61 ± 0.97	33.33 ± 0.80
μ_3	41.27 ± 0.34	39.78 ± 1.69	41.10 ± 4.20	40.46 ± 2.91
μ_4	51.24 ± 0.74	48.63 ± 2.43	49.88 ± 6.99	40.29 ± 4.38
μ_5	61.32 ± 1.51	60.13 ± 2.27	60.47 ± 4.86	60.37 ± 3.43
σ_1	2.44 ± 0.23	2.17 ± 0.18	2.37 ± 0.28	2.28 ± 0.23
σ_2	3.00 ± 0.14	3.15	3.46	3.31
σ_3	4.27 ± 0.24	3.80	4.23	4.02
σ_4	5.08 ± 0.53	4.64	5.13	4.89
σ_5	7.07 ± 1.09	5.74	6.22	5.99
		$\chi^2 = 11.29$	$\chi^2 = 11.95$	$\chi^2 = 11.73$
		14 df	14 df	14 df
		P = 0.66	P = 0.61	P = 0.63

Table 2. Parameters for the five age-groups of the northern pike data, estimated from the length-frequency distribution, with the means constrained to lie on a growth curve and component coefficients of variation constrained to be equal; $t_1 - t_0$ is the time (in years) between hypothetical "zero length" and the exact age of age-group 1.

Parameter	By Scale Reading	Normal	Lognormal	Gamma
π_1	0.105 ± 0.013	0.092 ± 0.016	0.101 ± 0.016	0.098 ± 0.015
π_2	0.465 ± 0.022	0.570 ± 0.036	0.581 ± 0.043	0.576 ± 0.040
π_3	0.298 ± 0.020	0.232 ± 0.036	0.216 ± 0.044	0.223 ± 0.040
π_4	0.090 ± 0.013	0.045 ± 0.037	0.046 ± 0.043	0.045 ± 0.040
π_5	0.042 ± 0.009	0.061 ± 0.033	0.056 ± 0.042	0.058 ± 0.038
μ_1	23.33 ± 0.33	22.85 ± 0.52	23.19 ± 0.47	23.08 ± 0.48
μ_2	33.09 ± 0.19	33.76 ± 0.38	34.12 ± 0.47	33.99 ± 0.43
μ_3	41.27 ± 0.34	43.24 ± 0.73	43.63 ± 0.99	43.48 ± 0.87
μ_4	51.24 ± 0.74	51.48	51.89	51.74
μ_5	61.32 ± 1.51	58.63	59.07	58.92
σ_1	2.44 ± 0.23	2.44 ± 0.16	2.53 ± 0.19	2.48 ± 0.18
σ_2	3.00 ± 0.14	3.61	3.72	3.65
σ_3	4.27 ± 0.24	4.62	4.76	4.67
σ_4	5.08 ± 0.53	5.50	5.66	5.56
σ_5	7.07 ± 1.09	6.27	6.44	6.34
		$\chi^2 = 14.36$	$\chi^2 = 12.46$	$\chi^2 = 12.84$
		16 df	16 df	16 df
		P = 0.57	P = 0.71	P = 0.68
L_∞		106.003	106.840	107.015
$t_1 - t_0$		1.726	1.747	1.744
k		0.1407	0.1401	0.1393

Table 3. Parameters for the five age-groups of the northern pike data, estimated from the length-frequency distribution, with the means constrained to be equally spaced (linear growth) and component coefficients of variation constrained to be equal.

Parameter	By Scale Reading	Normal	Lognormal	Gamma
π_1	0.105 ± 0.013	0.093 ± 0.017	0.103 ± 0.016	0.100 ± 0.016
π_2	0.465 ± 0.022	0.606 ± 0.034	0.622 ± 0.037	0.615 ± 0.035
π_3	0.298 ± 0.020	0.222 ± 0.028	0.202 ± 0.031	0.210 ± 0.029
π_4	0.090 ± 0.013	0.069 ± 0.020	0.066 ± 0.022	0.067 ± 0.021
π_5	0.042 ± 0.009	0.010 ± 0.010	0.007 ± 0.010	0.008 ± 0.010
μ_1	23.33 ± 0.33	23.06 ± 0.55	23.34 ± 0.49	23.24 ± 0.51
μ_2	33.09 ± 0.19	34.02 ± 0.36	34.49 ± 0.43	34.32 ± 0.40
μ_3	41.27 ± 0.34	44.98	45.63	45.40
μ_4	51.24 ± 0.74	55.95	56.78	56.48
μ_5	61.32 ± 1.51	66.91	67.93	67.56
σ_1	2.44 ± 0.23	2.57 ± 0.17	2.66 ± 0.19	2.61 ± 0.18
σ_2	3.00 ± 0.14	3.79	3.94	3.85
σ_3	4.27 ± 0.24	5.01	5.21	5.09
σ_4	5.08 ± 0.53	6.23	6.48	6.34
σ_5	7.07 ± 1.09	7.45	7.75	7.58
		$\chi^2 = 16.79$ 17 df $P = 0.47$	$\chi^2 = 13.60$ 17 df $P = 0.70$	$\chi^2 = 14.41$ 17 df $P = 0.64$

Table 4. Same as Table 1, but constraining the component standard deviations, rather than the coefficients of variation, to be equal.

Parameter	By Scale Reading	Normal	Lognormal	Gamma
π_1	0.105 ± 0.013	0.115 ± 0.019	0.126 ± 0.019	0.123 ± 0.019
π_2	0.465 ± 0.022	0.533 ± 0.036	0.552 ± 0.042	0.545 ± 0.039
π_3	0.298 ± 0.020	0.214 ± 0.031	0.194 ± 0.036	0.201 ± 0.034
π_4	0.090 ± 0.013	0.095 ± 0.018	0.089 ± 0.021	0.091 ± 0.020
π_5	0.042 ± 0.009	0.042 ± 0.010	0.040 ± 0.011	0.041 ± 0.010
μ_1	23.33 ± 0.33	23.80 ± 0.65	24.30 ± 0.60	24.15 ± 0.61
μ_2	33.09 ± 0.19	33.78 ± 0.38	34.21 ± 0.45	34.06 ± 0.42
μ_3	41.27 ± 0.34	42.02 ± 0.84	42.67 ± 1.14	42.45 ± 1.01
μ_4	51.24 ± 0.74	51.16 ± 0.93	51.66 ± 1.22	51.49 ± 1.10
μ_5	61.32 ± 1.51	62.15 ± 0.93	62.38 ± 1.12	62.30 ± 1.04
σ_1	2.44 ± 0.23	3.25 ± 0.22	3.43 ± 0.28	3.34 ± 0.25
σ_2	3.00 ± 0.14	3.25	3.43	3.34
σ_3	4.27 ± 0.24	3.25	3.43	3.34
σ_4	5.08 ± 0.53	3.25	3.43	3.34
σ_5	7.07 ± 1.09	3.25	3.43	3.34
		$\chi^2 = 17.47$ 14 df $P = 0.23$	$\chi^2 = 14.38$ 14 df $P = 0.42$	$\chi^2 = 15.27$ 14 df $P = 0.36$

Table 5. Same as Table 2, but constraining the component standard deviations, rather than the coefficients of variation, to be equal.

Parameter	By Scale Reading	Normal	Lognormal	Gamma
π_1	0.105 ± 0.013	0.120 ± 0.022	0.130 ± 0.021	0.127 ± 0.021
π_2	0.465 ± 0.022	0.573 ± 0.032	0.584 ± 0.034	0.580 ± 0.033
π_3	0.298 ± 0.020	0.198 ± 0.025	0.183 ± 0.028	0.188 ± 0.027
π_4	0.090 ± 0.013	0.072 ± 0.015	0.069 ± 0.016	0.070 ± 0.015
π_5	0.042 ± 0.009	0.036 ± 0.012	0.034 ± 0.012	0.035 ± 0.012
μ_1	23.33 ± 0.33	24.29 ± 0.70	24.64 ± 0.60	24.53 ± 0.62
μ_2	33.09 ± 0.19	34.13 ± 0.37	34.53 ± 0.42	34.40 ± 0.39
μ_3	41.27 ± 0.34	43.75 ± 0.54	44.16 ± 0.61	44.02 ± 0.58
μ_4	51.24 ± 0.74	53.14	53.52	53.40
μ_5	61.32 ± 1.51	62.31	62.64	62.53
σ_1	2.44 ± 0.23	3.50 ± 0.24	3.65 ± 0.29	3.57 ± 0.26
σ_2	3.00 ± 0.14	3.50	3.65	3.57
σ_3	4.27 ± 0.24	3.50	3.65	3.57
σ_4	5.08 ± 0.53	3.50	3.65	3.57
σ_5	7.07 ± 1.09	3.50	3.65	3.57
		$\chi^2 = 21.01$ 16 df $P = 0.18$	$\chi^2 = 16.04$ 16 df $P = 0.45$	$\chi^2 = 17.44$ 16 df $P = 0.36$
L_∞		444.231	388.998	410.793
$t_1 - t_0$		2.369	2.376	2.378
k		0.02373	0.02753	0.02589

Table 6. Same as Table 3, but constraining the component standard deviations, rather than the coefficients of variation, to be equal.

Parameter	By Scale Reading	Normal	Lognormal	Gamma
π_1	0.105 ± 0.013	0.122 ± 0.022	0.132 ± 0.021	0.129 ± 0.021
π_2	0.465 ± 0.022	0.565 ± 0.031	0.578 ± 0.033	0.572 ± 0.032
π_3	0.298 ± 0.020	0.203 ± 0.025	0.187 ± 0.027	0.193 ± 0.026
π_4	0.090 ± 0.013	0.075 ± 0.014	0.072 ± 0.015	0.073 ± 0.015
π_5	0.042 ± 0.009	0.035 ± 0.011	0.032 ± 0.011	0.033 ± 0.011
μ_1	23.33 ± 0.33	24.51 ± 0.54	24.84 ± 0.53	24.74 ± 0.53
μ_2	33.09 ± 0.19	34.04 ± 0.38	34.44 ± 0.41	34.31 ± 0.39
μ_3	41.27 ± 0.34	43.57	44.04	43.87
μ_4	51.24 ± 0.74	53.10	53.64	53.44
μ_5	61.32 ± 1.51	62.63	63.24	63.01
σ_1	2.44 ± 0.23	3.48 ± 0.24	3.64 ± 0.29	3.56 ± 0.26
σ_2	3.00 ± 0.14	3.48	3.64	3.56
σ_3	4.27 ± 0.24	3.48	3.64	3.56
σ_4	5.08 ± 0.53	3.48	3.64	3.56
σ_5	7.07 ± 1.09	3.48	3.64	3.56
		$\chi^2 = 21.32$ 17 df $P = 0.21$	$\chi^2 = 16.48$ 17 df $P = 0.49$	$\chi^2 = 17.84$ 17 df $P = 0.40$

fish by scale reading. For the accuracy that would be required
for biological or management purposes the results are consistent
with each other and with the scale-reading results, with the
exception that the growth-curve fit with standard deviations
constrained to be equal (Table 5) gives quite misleading values
for L_∞ and k. According to the goodness-of-fit χ^2 values given,
all eighteen fits are acceptable, although those assuming a
constant coefficient of variation are better than those assuming
equal standard deviations.

DISCUSSION
     The northern pike data of the previous section were not
difficult to analyse because exactly five age-groups were known to
be present. In some applications there might be two or three
reasonably clear components on the left of the length-frequency
distribution, and then a right tail with an indeterminate number
of age-groups, perhaps ten or more, heavily overlapped because of
reduced growth rates in the older fish. Obviously, there is no
way that information in the older groups can come from the
length-frequency distribution alone without strong assumptions
about population structure. Length-stratified sampling is
probably the best way to get additional information about the
older age-groups. In the absence of such additional information,
the interval boundaries of the length-frequency distribution
should be set so that most of the older fish are lumped together
into the rightmost interval and treated as a single component, as
only the parameters of the younger age groups can be estimated
reliably.
     If a set of data can be analysed under different models,
assumptions, or constraints, then confidence in the results should
be commensurate with the consistency of the different analyses;
if they are consistent, the data speak for themselves. There is
some justification for picking the most parsimonious of the
models, since simplicity of interpretation increases as the number
of parameters decrease. The simplest case is, of course, a
population structure described by the three parameters of a von
Bertalanffy growth curve and a constant total force of mortality,
sampled without age or length bias. If a population cannot be
described in such simple terms, then more parameters will have to
be estimated from field data, and application of the results to
management of the stock will be correspondingly more complicated.
That is the case when recruitment, growth, and mortality vary from
one cohort to the next.
     With the increased power and availability of microcomputers,
more comprehensive analyses and more statistically efficient methods are becoming available for routine application. Correspondingly, users will be dealing with computer packages, the workings
of which are beyond their competence to verify. It is important
that such packages meet high standards of convenience,
reliability, and accuracy. For the present, there is also the

requirement that a package be adapted to as many as possible of
the microcomputers currently in use.

Two useful areas for future work will be to develop likeli-
hood-based methods for analysing time sequences of length-
frequency distributions and to develop procedures for determining
how much length-stratified sampling is needed to supplement a
given length-frequency distribution.

ACKNOWLEDGMENTS

Daniel Pauly, David Fournier, Jon Schnute, and John Shepherd
read the first draft of this paper and made suggestions which
substantially improved it. This work owes much to Tony Pitcher
for his continued encouragement. He drew Figures 1 and 3, which
are reproduced, with permission, from the Journal of the Fisheries
Research Board of Canada.

REFERENCES
AGHA, M., and M. T. IBRAHIM. 1984. Algorithm AS 203. Maximum
    likelihood estimation of mixtures of distributions. Applied
    Statistics 33:327-332.
BHATTACHARYA, C. G. 1967. A simple method of resolution of a dis-
    tribution into Gaussian components. Biometrics 23:115-135.
BREEN, P. A., and D. A. FOURNIER. 1984. A user's guide to
    estimating total mortality rates from length frequency data
    with the method of Fournier & Breen. Canadian Technical
    Reports of Fisheries and Aquatic Sciences No. 1239.
CASSIE, R. M. 1954. Some used of probability paper in the
    analysis of size frequency distributions. Australian Journal
    of Marine and Freshwater Research 5:513-522.
CLARK, W. G. 1981. Restricted least-squares estimates of age com-
    position from length composition. Canadian Journal of
    Fisheries and Aquatic Sciences 38:297-307.
DEMPSTER, A. P., N. M. LAIRD, and D. B. RUBIN. 1977. Maximum
    likelihood from incomplete data via the EM algorithm (with
    discussion). Journal of the Royal Statistical Society,
    Series B 39:1-38.
EVERITT, B. S., and D. J. HAND. 1981. Finite mixture
    distributions. Chapman and Hall, London, England.
FOURNIER, D. A. 1983. Use of length and age data for estimating
    the age structure of a collection of fish. Pages 206-208 in
    W. G. Doubleday and D. Rivard, editors. Sampling commercial
    catches of marine fish and invertebrates. Canadian Special
    Publications of Fisheries and Aquatic Sciences 66.
FOURNIER, D. A. and P. A. BREEN. 1983. Estimation of abalone
    mortality rates with growth analysis. Transactions of the
    American Fisheries Society 112:403-411.
HATHAWAY, R. J. 1983. Constrained maximum-likelihood estimation
    for normal mixtures. Pages 263-267 in J. E. Gentle, editor.

Computer science and statistics: The interface. North-Holland, Amsterdam, The Netherlands.

HOSMER, D. W. 1973. A comparison of iterative maximum likelihood estimates of the parameters of a mixture of two normal distributions under three different types of sample. Biometrics 29:761-770.

MACDONALD, P. D. M. and P. E. J. GREEN. 1985. User's guide to program MIX: An interactive program for fitting mixtures of distributions. Ichthus Data Systems, Hamilton, Ontario, Canada.

MACDONALD, P. D. M., and T. J. PITCHER. 1979. Age-groups from size-frequency data: a versatile and efficient method of analyzing distribution mixtures. Journal of the Fisheries Research Board of Canada 36:987-1001.

MCNEW, R. W., and R. C. SUMMERFELT. 1978. Evaluation of a maximum-likelihood estimator for analysis of length-frequency distributions. Transactions of the American Fisheries Society 107:730-736.

NELDER, J. A., and R. MEAD. 1965. A simplex method for function minimization. Computer Journal 7:308-313.

PAULY, D., and N. DAVID. 1981. ELEFAN I, a BASIC program for the objective extraction of growth parameters from length-frequency data. Meeresforschung 28(4): 205-211.

PETERSEN, C. G. 1891. Eine Methode zur Bestimmung des Alters und Wuchses der Fische. Mitteilungen. Deutscher Seefischerei-Verein 11:226-235.

PITCHER, T. J., and P. D. M. MACDONALD. 1973. Two models for seasonal growth in fishes. Journal of Applied Ecology 10:599-606.

RAO, C. R. 1965. Linear statistical inference and its applications. Wiley, New York, New York, U.S.A.

RICHARDS, F. J. 1959. A flexible growth function for empirical use. Journal of Experimental Botany 10:290-300.

SCHNUTE, J., and D. FOURNIER. 1980. A new approach to length-frequency analysis: growth structure. Canadian Journal of Fisheries and Aquatic Sciences 37: 1337-1351.

SHEPHERD, J. G. 1986. A weakly parametric method for the analysis of length composition data. In press in D. Pauly and G. R. Morgan, editors. Length-based methods in fisheries research. ICLARM conference proceedings, International Center for Living Aquatic Resources Management, Manila, Philippines.

An inexpensive microcomputer-based system for making rapid and precise counts and measurements of zonations in video displayed skeletal structures of fish

MICHAEL F. McGOWAN,[1,2] **ERIC D. PRINCE, AND DENNIS W. LEE**
U.S. DEPARTMENT OF COMMERCE
NATIONAL OCEANIC AND ATMOSPHERIC ADMINISTRATION
SOUTHEAST FISHERIES CENTER, MIAMI LABORATORY
75 VIRGINIA BEACH DRIVE
MIAMI, FL 33149-1099

ABSTRACT
       A personal computer and a commercially available digitizing board were combined with a video camera fitted to a microscope for analyzing zonations in skeletal structures of fish. The system provides the cost-effective capability of making rapid and precise counts and measurements of zonations displayed on a video monitor at a resolution of 256 times 256 pixels times 256 brightness levels. An inexperienced reader did as well as a more experienced one, with no decrease in precision of counts within and between readers; total processing time (i.e., counting and measuring zonations, compiling and editing data sets) was less when using the computer system compared to conventional stereoscope and ocular micrometer procedures and data transcription errors associated with standard techniques were eliminated. An elementary image processing algorithm was used to demonstrate the microcomputer-based system's potential for automated detection of zonations. Individual components of the modular system can be upgraded to improve performance.

[1]Present Address: C.I.M.A.S., Rosenstiel School of Marine and Atmospheric Science, University of Miami, 4600 Rickenbacker Causeway, Miami, FL 33149-1099.
[2]Supported in part by National Oceanic and Atmospheric Administration under Cooperative Agreement #NA 84-WC-H-06098 with C.I.M.A.S., University of Miami.

The preferred skeletal structures used to determine the age
of fishes differ among species and sometimes between workers on
the same species (Prince and Pulos 1983). Interpretations of zonations vary and validation of the periodicity of increments and
the time of year (or day) of their formation are subjects of
active research and sometimes contentious dispute (Brothers 1983).
However, there are at least two points of general agreement among
fishery scientists concerning age determination. First, knowledge
of the age structure of fish populations is important for managing
fisheries; second, the work of collecting, collating, editing, and
analyzing the data for ageing studies is time consuming and tedious, can be subjective, and may be imprecise or inaccurate between
methods or between workers using the same method. Therefore, no
matter which skeletal structure, method, or interpretation may be
favored, a rapid, objective counting and measuring tool should be
welcomed by all. The recent decreases in cost, miniaturization of
electronic components, and increases in the capability of microcomputer products have put sophisticated image-processing hardware
within reach of most research agencies. In this paper, we describe a visual analysis system modified for counting and measuring zonations in video-displayed skeletal structures of fish.
This system is composed of a microcomputer and digitizing board
combined with a video camera fitted to a microscope. The precision (repeatability) of counts and total time for data acquisition
and analysis are compared between the computer-based system and
conventional methods consisting of a microscope and ocular micrometer.

METHODS
     We used dorsal spine sections from Atlantic blue marlin
Makaira nigricans to compare the speed and precision in counting
zonations and processing the data between conventional microscope
methods and our visual analysis system. We did not compare zonation measurements between methods because discrepancies in counts
among and between readers inherently affected measurement results
and we could find no practical way of partitioning these effects.
However, we decided that comparing mean counts, precision of
counts, and processing time was adequate to assess the overall
performance of our visual analysis system.
     Dorsal spines were sectioned following procedures of Prince
et al. (1984). Cross sections of the fifth dorsal spine were
selected from 21 blue marlin ranging in size from 183.0-333.0 cm
lower jaw fork length (LJFL) and 51.3-429.6 kg round weight.
Each cross section was examined using transmitted light (6.0-
7.5X) and the translucent zonations were measured and counted.
Three separate replicate counts of the 21 samples were made by 2
readers using each of the two analysis methods described below.
The order in which the sections were read was randomized.

Conventional Microscope Method (CMM)

Dorsal spine sections were examined by each reader through a stereoscope at 7.5X. The measurements of zonations and section radius were made with the eyepiece ocular micrometer and recorded by the reader in a data log along with counts of zonations for each specimen. Supplemental biological data (LJFL, weight, sex, and date of capture) were added to the data logs after the counts and measurements were made to prevent possible biases. The data for the 3 replicate counts and measurements were entered into a data entry terminal. After correcting key punch errors, the data were transferred to a main frame computer system for analysis. The time spent on each phase of the procedure was recorded.

Computer-Based Method (CBM)

The computer system is based on a microcomputer with 256 k bytes memory. Two additional circuit boards provide an interface to a video camera, digitizing ability, and enhanced video display capability. A standard video monitor displays the image from the stereoscope through a video camera and a second standard monitor is the computer terminal display and can display a digitized version of the video image (Fig. 1). The resolution of the computer digitizer board is 256 times 256 pixel units, horizontal times vertical. This means that a single dot is about 1 mm square on the display monitor. The digitizer measures 256 levels of brightness. The video controller board is only capable of displaying the digitized brightness values as 16 colors or 16 grey levels, depending on the monitor used, but the full 8 bits of information per pixel (256 intensities) are available for image processing. The video camera and video display monitor are both connected to the digitizer board which fits inside the computer (Fig. 1). The video camera image is the input to the digitizer board in the computer. The output of the digitizer board is the same video image plus an intensified point of light, called a cursor, which can be moved around on the video image using four arrow keys on the computer keyboard. The x-y coordinates of the cursor and the brightness of the original video image under the cursor are available to the computer for further processing.

The basic acquisition of counts and measurements of zonations proceeds as follows: (1) display the skeletal structure through the stereoscope and video camera onto the video screen; (2) move the cursor to the focus of the structure; (3) press the keyboard key which has been programmed to record the x-y coordinates of the cursor; (4) move the cursor along a selected counting structure. The array of x-y coordinates can be calibrated and processed, stored on diskette, or transmitted via telephone to a mainframe computer for further processing and analysis. We combined these primary functions with an interactive program in BASIC language which prompts the user for collection information and provides menus of choices for repeating the reading, doing a replicate reading of the same structure, saving data or not, and editing the data again before storing it.

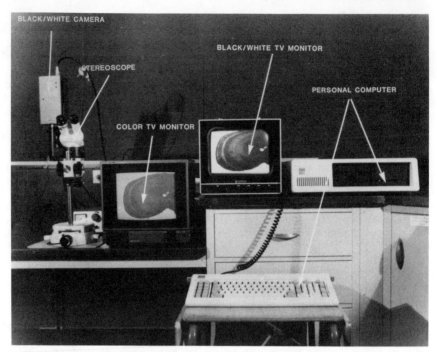

Figure 1. Hardware components of the computer-based visual
analysis system (from left to right): black/white camera,
stereoscope, color TV monitor, black/ white TV monitor,
and personal computer. The computer also includes 2 additional circuit boards (not shown); digitizing board and
video controller board.

We evaluated the system by comparing it to conventional
microscope methods which required using a microscope with an ocular micrometer, hand recording the data, and having the data key-
punched for processing. The variables compared between methods
were mean count of zonations, average percent error (APE), which
is a measure of precision (Beamish and Fournier 1981); index of
precision (D), which is a modification of APE (Chang 1982); and
total time for reading and processing the data. These variables
were compared between two readers; one reader was experienced in
both the conventional microscope method and the new computer-based
method (reader 1) and the other reader was inexperienced in both
methods (reader 2).

Data Analysis

The non-parametric sign test (Sokal and Rohlf 1969) was used
to test the null hypotheses of no difference of mean counts and
index of precision (D) between methods and between readers within
methods.

RESULTS

There was a significant difference (P < 0.01) in mean counts of zonations and D on dorsal spine sections using the conventional microscope method (CMM) between readers 1 and 2 (Table 1). For the most part, reader 2 had higher average counts than reader 1. In addition, reader 1 was able to make counts more consistently as evidenced by the smaller values for mean APE and mean D (0.020 and 0.008, respectively) compared to reader 2 (mean APE = 0.090 and mean D = 0.084, Table 2).

Table 1. Comparison of mean counts of zonations and index of precision (D) on dorsal spine sections of blue marlin for the conventional microscope method (CMM), computer-based method (CBM), reader 1 (R_1) and reader 2 (R_2) using the Sign test. Significant difference (s.d.) and no significant difference (n.d.) are indicated for each comparison. Significance level was set at P < 0.01.

Null Hypothesis	Result
Conventional Microscope Method (CMM)	
Mean Count (R_1) = Mean Count (R_2)	s.d.
$D_{(R_1)} = D_{(R_2)}$	s.d.
Computer-Based Method (CBM)	
Mean Count (R_1) = Mean Count (R_2)	n.d.
$D_{(R_1)} = D_{(R_2)}$	n.d.
Reader 1 (R_1)	
Mean Count (CMM) = Mean Count (CBM)	n.d.
D(CMM) = D (CBM)	n.d.
Reader 2 (R_2)	
Mean Count (CMM) = Mean Count (CBM)	s.d.
D (CMM) = D(CBM)	n.d.

Table 2. Mean average percent error (APE) and mean index of precision (D) for counts of zonations from dorsal spine sections of blue marlin by reader 1 and reader 2 using conventional microscope and computer based methods of analysis.

Conventional Microscope Method	
Reader 1	**Reader 2**
Mean APE = 0.02	Mean APE = 0.090
Mean D = 0.008	Mean D = 0.084
Computer-Based Method	
Mean APE = 0.060	Mean APE = 0.083
Mean D = 0.048	Mean D = 0.063

There was no detectable difference (P > 0.01) in mean counts of zonations on dorsal sections between readers 1 and 2 using the computer-based method (CBM, Table 1). The precision of these counts between readers 1 and 2 were also very close (reader 1 mean APE = 0.060, mean D = 0.048; reader 2 mean APE = 0.083, mean

D = 0.063). The two readers both had mean APE values of less than 10% (range = 0.020 to 0.090) for all methods. Reader 1 was a little more precise in counting zonations using the CMM (mean APE = 0.020) compared to the CBM (mean APE = 0.060). In contrast, reader 2 was slightly more precise using the CBM (mean APE = 0.083) compared to the CMM (mean APE = 0.090).

Reader 1 had considerably more experience using both methods of analysis and consistently less trouble in relocating his position on the spine section counting path. Reader 2 had difficulty relocating his position on the counting path using the CMM, especially for larger/older fish.

Comparisons between mean counts and D of the two methods were also examined for each reader (Table 1). The mean counts of zonations by reader 1 using the CMM were not significantly different ($P > 0.01$) than those tabulated for the CBM. Conversely, there was a significant difference ($P < 0.01$) in mean counts by reader 2 using the CBM compared to the CMM (Table 1). Reader 1 generally had higher mean APE and mean D values for the CBM compared to those from the CMM (Table 2). The mean APE and mean D values for reader 2, however, were lower for the CBM than for the CMM (Table 2).

Total data processing time using the CBM (76 minutes) took 61% as long as using the CMM (124 minutes, Fig. 2A). Most of the saving in processing time was due to eliminating the separate data entry step which was part of the CMM. Readings of the skeletal structures and checking the data (excluding data entry) took about the same time for both methods, although the CBM was slightly faster. However, if the delay in key-punch scheduling (turnaround time) is included (Fig. 2B), the CBM was 50 times faster than the CMM.

DISCUSSION

These results showed a 40% savings in time with no loss of precision when the computer-based system was compared to the conventional microscope method. Both methods had mean APE values less than 10%. This level of precision was reported by Powers (1983) as acceptable for stock assessment, and was within the range of APE measurements published for other species (Beamish and Fournier 1981; Cayre and Diouf 1983). The time savings of the CBM was due to eliminating a separate step for data entry which is part of the CMM. Clerical errors and editing time were also reduced by eliminating this step.

Although we did not present zonation measurement comparisons, we believe the CBM has several other advantages over the CMM. For example, the reader does not have to reposition the skeletal part within the range of the ocular micrometer, does not have to look away to record data, and does not have to interpolate between micrometer divisions when using the CBM. All of these advantages facilitate accurate and precise measurements, as well as counts. In addition, our system is parfocal, i.e., the focal

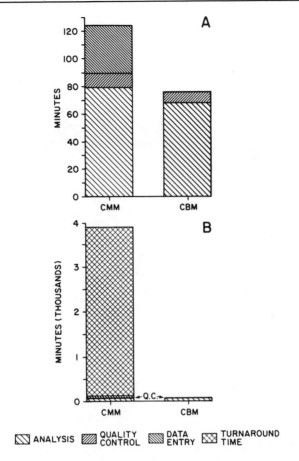

Figure 2. Processing time required for using conventional
  microscope method (CMM) and our computer-based method
  (CBM) to analyze skeletal structures for age determination. Processing time (minutes) for analysis, data entry,
  and quality control are given in (A); and processing time
  (minutes) for analysis, data entry, quality control, and
  turnaround time are given in (B).

planes of the microscope and attached TV camera are the same. The
reader is able to alternately view the microscope image and the
video-displayed image without refocusing. This is helpful in
reading structures with indistinct zonations, and may decrease the
sample rejection rate without giving up the advantages of the CBM.
    Because the digitizer reads brightness as well as position
of the cursor, the system has potential for automatic reading of
skeletal structures. To demonstrate this potential, we incorporated a subroutine in our program to graphically display the

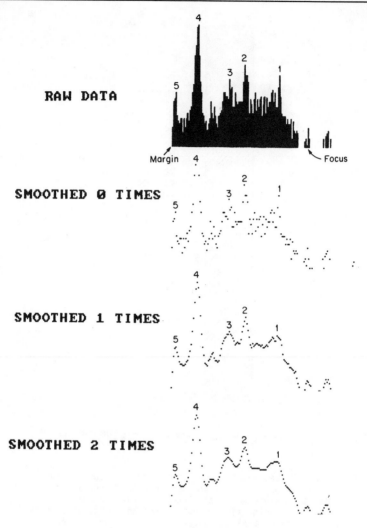

Figure 3. Computer analysis of the brightness levels of
light and dark zonations along a transect of the video
displayed dorsal spine section from Figure 4. Raw data
are plotted in the histogram (top) and an algorithm was
used to smooth the curves (below) to remove "noise" and
better isolate the alternating light and dark zonations.

brightness of each point along a specified counting path (Fig. 3,
top two plots). Numerical processing of these data (three point
moving average, LaFara 1973) removed random variation and revealed
the peaks of brightness (translucent zones in this case) more
clearly (Fig. 3, bottom two plots). We are investigating other

numerical techniques, such as spectral analysis and polynomial curve-fitting, which would permit automatic counts and peak-to-peak measurements by the CBM. However, this function would depend on criteria which are programmed into the computer for distinguishing true year marks. Therefore, validation of automatic age-ing techniques will be just as critical as it is for conventional methods.

The digitizer is capable of digitizing the entire image of the skeletal part, not just the points along the counting path. This image may be saved on floppy disk, printed, or enhanced by suitable image-processing techniques prior to reading by the technician or computer program. An enhanced contrast version of the video image in Figure 4A is shown in Figure 4B. Our microcomputer display is only capable of showing 16 gray levels so the enhanced image appears coarse relative to the original image. However, the full 256 brightness levels are available to image-processing software.

When our system was designed in 1983, a digitizing board with the 256 times 256 resolution and 256 brightness levels was available for $440, including software. A functionally similar interface custom built for National Marine Fisheries Service's Southwest Fisheries Center (Methot 1981) cost $5000. Digitizing systems including 8 bit microcomputers were available from major microscope manufacturers for about $20,000. A system equivalent to ours with black and white video camera and monitor, including the 16 bit microcomputer, could be purchased for less than $5000 in 1983, but prices can be expected to decrease.

In addition to age and growth data acquisition, the system can be used for morphometric studies (Fink 1985), and contains a powerful microcomputer for statistical analysis and word processing. We plan to improve our system by including back-calculation and growth model fitting in the program, by upgrading the hardware to improve resolution, and by automating the counting and measuring process. A complete list of the brand names and model numbers of the various components of this system can be obtained from the authors.

REFERENCES

BEAMISH, R. J., and D. A. FOURNIER. 1981. A method for comparing the precision of a set of age determination. Canadian Journal of Fisheries and Aquatic Sciences 38:982-983.

BROTHERS, E. B. 1983. Summary of round table discussions on age validation. Pages 35-44 in E.D. Prince and L.M. Pulos, editors. Proceedings of the International Workshop on Age Determination of Oceanic Pelagic Fishes: Tunas, Billfishes, and Sharks. National Marine Fisheries Service, National Oceanic and Atmospheric Administration Technical Report NMFS8.

CAYRE, P. M., and T. DIOUF. 1983. Estimating age and growth of little tunny, Euthynnus alletteratus, off the coast of Senegal, using dorsal fin spine sections. Pages 99-104 in E.D.

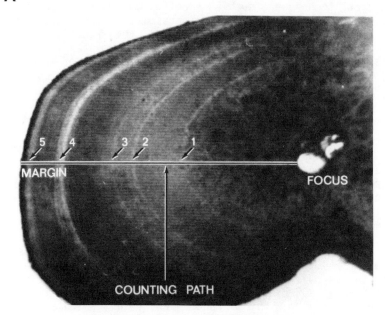

Figure 4. Microscope image (A) and video enhanced image (B) of a dorsal spine section from a blue marlin (218 cm LJFL, 100.0 kg). The focus, counting path, and margin of the structure are shown and five zonations can be clearly distinguished.

Prince and L.M. Pulos, editors. Proceedings of the International Workshop on Age Determination of Oceanic Pelagic Fishes: Tunas, Billfishes, and Sharks. National Marine Fisheries Service, National Oceanic and Atmospheric Administration Technical Report NMFS8.

CHANG, W. Y. B. 1982. A statistical method for evaluating the reproducibility of age determination. Canadian Journal of Fisheries and Aquatic Sciences 39:1208-1210.

FINK, W. L. 1985. A video digitizer system for shape analysis. Sixty Fifth Annual Meeting of American Society of Ichthyologists and Herpetologists. University of Tennessee, Knoxville, Tennessee, 9-14 June 1985. (Abstract) 62-63.

FRIE, R. V. 1982. Measurements of fish scales and back-calculation of body lengths using a digitizing pad and microcomputer. Fisheries 7(5):5-8.

LAFARA, R. L. 1973. Computer methods for science and engineering. Hayden Book Company, Inc., Rochelle Park, New Jersey. USA.

METHOT, R., Jr. 1981. Growth rates and age distributions of larval and juvenile northern anchovy, Engraulis mordax, with inferences on larval survival. Ph.D. Thesis, Scripps Institute of Oceanography, La Jolla, California, USA.

POWERS, J. E. 1983. Some statistical characteristics of ageing data and their ramifications in population analysis of oceanic pelagic fishes. Pages 19-24 in E.D. Prince and L.M. Pulos, editors. Proceedings of the International Workshop on Age Determination of Oceanic Pelagic Fishes: Tunas, Billfishes, and Sharks. National Marine Fisheries Service, National Oceanic and Atmospheric Administration Technical Report NMFS8.

PRINCE, E. D., D. W. Lee, C. A. Wilson, and J. M. Dean. 1984. Progress in estimating age of blue marlin, Makaira nigricans, and white marlin, Tetrapterus albidus, from the western Atlantic Ocean, Caribbean Sea, and Gulf of Mexico. International Commission for the Conservation of Atlantic Tunas, Collective Volume of Scientific Papers, Madrid, Spain. Standing Committee for Research and Statistics-1983, 20(2): 435-447.

PRINCE, E. D., and L. M. Pulos, editors. 1983. Proceedings of the International Workshop on Age Determination of Oceanic Pelagic Fishes: Tunas, Billfishes, and Sharks. National Marine Fisheries Service, National Oceanic and Atmospheric Administration Technical Report NMFS8.

SOKOL, R. R., and F. J. Rohlf. 1969. Biometry. Wm. Freeman and Company, San Francisco, California, USA.

Computer-assisted age and growth pattern recognition of fish scales using a digitizing tablet

GREGG J. SMALL[1] AND
GEORGE HIRSCHHORN
NATIONAL MARINE FISHERIES SERVICE
NORTHWEST AND ALASKA FISHERIES CENTER
SEATTLE, WA 98115-6349

ABSTRACT

A new method of computer-assisted age determination using fish scale growth patterns has been developed which involves tracing annular rings on a digitizing tablet coupled to a personal computer. The program is capable of interactively computing von Bertalanffy growth parameters from selected rings and provides the investigator with the necessary information to decide if the age and ring pattern is annual. Preliminary study has focused on Pacific cod Gadus macrocephalus and walleye pollock Theragra chalcogramma, but several other marine and freshwater species are also being examined to determine the utility of this approach.

A computerized digitizing method using parameters derived from the von Bertalanffy growth equation (von Bertalanffy 1951) was developed to identify growth patterns on individual scales of Pacific cod Gadus macrocephalus. This system is a practical means of aging scales and quantifying the growth pattern for each individual fish using areas of annular rings. In addition, an objective criterion is used to aid in the final choice of annular ring location. Traditional aging techniques are not able to draw on both measurements and growth criteria during the aging process.

The development of objective criteria for aging a fish species is an important goal in the formulation of aging methodology; and is used to increase the accuracy of age determination. In practice, agreement between replicate age

[1]Present address: Natural Resources Consultants, 4055 21st Avenue, West, Seattle, Washington 98199

The Age and Growth of Fish, edited by Robert C. Summerfelt and Gordon E. Hall © 1987 The Iowa State University Press, Ames, Iowa 50010.

determinations is an important aspect of age and growth
validation, although agreement between readings does not
necessarily imply that ages are properly determined (Carlander
1974; Beamish and Fournier 1981).

Several previous attempts have been made to develop
criteria to age Pacific cod scales. Kennedy (1970) used
marginal increments and measurements of the last two annuli to
define the age of a scale. Later attempts by Westrheim and Shaw
(1982) to use this method on cod scales from other areas in
British Columbia were unsuccessful, and they recommended the
length-frequency partitioning method of Foucher and Fournier
(1982). A different approach to aging was that of Frost and
Kipling (1959) with pike <u>Esox</u> <u>lucius</u>; they measured radius
length at age in scales and opercular bones, both relatively
flat structures, in order to estimate lengths at age in
individuals, and fitting these by the von Bertalanffy growth
function. Results of the method were considered satisfactory
but too time consuming for general use.

Recent computer hardware and software developments permit
measurement of growth in a practical and efficient manner. For
example, digitizing tablets, coupled with a microprocessor or
large mainframe computer, have made it possible to measure
growth precisely along any chosen axis of a scale (Frie 1982)
and, at the same time, to calculate the areas within annular
rings on any flat image in one simple procedure. Interactive
computer processing of scales has allowed the on-screen display
of numeric and graphic information enabling rapid material
processing and data editing. The purpose of this paper is to
demonstrate an application of this new technology in aging
Pacific cod.

MATERIALS AND METHODS

The digitizing program in use at the Northwest and Alaska
Fisheries Center in Seattle, Washington was designed for use on
a mainframe computer and a microcomputer. Both used a 11" X 11"
digitizing tablet (Fig. 1) for input using FORTRAN 77 software
written at the Northwest and Alaska Fisheries Center. The
executable and source code for both programs as well as
equipment specifications are available upon request.

This system was developed using scales from Pacific cod
from the Bering Sea and Gulf of Alaska (Fig. 2). During the
summers of 1979, 1981, and 1982, scales were systematically
collected from the left side of the fish below the second
dorsal fin, immediately above the lateral line, and a vertical
series from the area between the lateral line and second dorsal
fin (Fig. 3a). After removal, each scale was attached to a
serially numbered gummed card (40 locations per card). These
numbers corresponded to the area on the fish where the scale was
taken. A plastic impression was made of each scale, enlarged
(X 58) and photographed with a microfiche reader/printer.

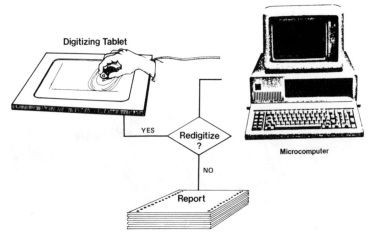

Figure 1. Illustration of equipment used in conjunction with
the digitizing program.

Regenerated or damaged scales were eliminated prior to
photographing.

Annular rings were marked on each photographic enlargement
by an experienced age reader and positioned on the digitizing
tablet for both radius and area measurement. For digitizing

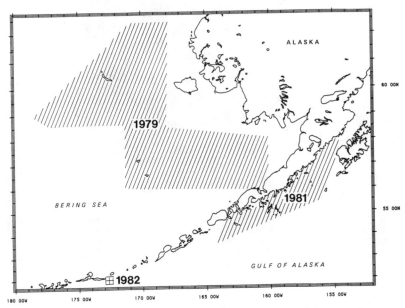

Figure 2. North Pacific and Bering Sea waters with year of
collection indicated for areas sampled.

3a

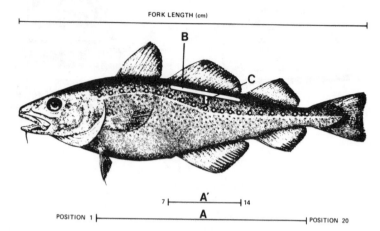

3b

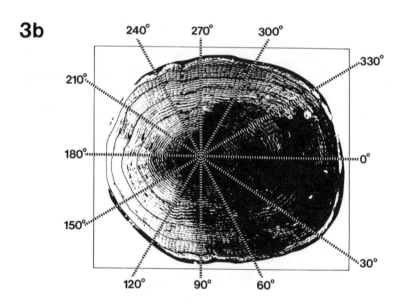

Figure 3. Scale-sampling positions and scale radius
directions obtained from Pacific cod. Scale locations A
and A' were along lateral line. B and C as shown (3a).
Direction of 12 scale radii generated and retained for
analysis during complete tracing of each annulus (3b).

purposes, annular rings were identified in the same manner used
in standard production aging practice (Westrheim and Shaw
1982). The photography and marking process was slower than
having scales projected directly onto the digitizing tablet, but
the photographs were permanent records, invaluable for later
reference.

Collection and specimen data, such as sex, length, and
weight for each scale, were entered from the keyboard. Editing
traps within the program permitted error detection before
proceeding to the next step. The program guides the user
through the operation of locating each annulus, either along
either a chosen radius or tracing the perimeter of each annular
ring. Both of these measuring routines were developed to allow
flexibility in reading scales.

Operation of the Digitizing Tablet

In the one-radius technique, the scale focus is located
with the pen stylus or cross-hair cursor and the button is
depressed at the point of intersection of each annulus and the
outer margin along the chosen radius line. The digitizer
measures the distances between annular rings and the focus along
any one axis. In contrast, complete digitizing begins by
holding the cursor at the intersection of the longest axis and
the smallest annular ring and tracing its perimeter in a
clockwise manner; each successive annular ring is then traced as
well as the scale margin. The complete digitizing technique has
the dual advantage of measuring the radii and accumulated areas
simultaneously as each marked annulus is being traced.

In both techniques, linear distances from the focus and
square root of the area measurements are stored in the
computer's memory for later calculations. The square root of
areas is used in all calculations to enable comparison of both
radii and area measurements in one dimension. The cod scales
used in this study were nearly symmetrical and used the full
capability of the program by taking radius measurements spaced
at 30 degree intervals (Fig. 3b). In addition to radius
measurements, up to four scale half-areas and a complete scale
area are also selectable. The full area is the total enclosed
by the annular ring. The half-areas correspond to dorsal and
ventral, or anterior and posterior halves of the scale. In the
scales of some species, only portions of scales may be usable.
In these cases, the programs permit users to specify alternative
partial areas and radii for analysis.

The digitizing software calculated areas using X, Y
coordinates in a continuous sweep method. As the annular ring
or outer margin is traced, two X,Y coordinate points/second are
sent from the digitizing tablet and processed by the program.
The output rate of points/second is a user-defined option, and
it determines the resolution of the output. Because the
points/second are constant, varying the speed of tracing
increases or decreases the distance between points and affects

the area approximations. Increasing the points/second rate
option also increases the data storage requirement. For
example, increasing the rate from 2 to 20 points/second will
result in a tenfold increase in data storage, assuming the size
of the ring and the speed at which the area is traced remains
constant. High transmission rates may cause the data storage
allotted by our programs to be exceeded. A circle with a 10 cm
diameter can be digitized in 15 seconds with approximately 200
X,Y pairs generated.

Analysis of Growth Pattern

The analysis of the growth pattern is the most difficult
aspect of the digitizing process. The following examples of the
procedures are intended to clarify the steps necessary to accept
a ring pattern.

After all rings have been traced, the data is automatically
standardized to the scale size of the last ring (100 percent)
and fitted to the von Bertalanffy growth equation (Fabens
1965). The outer margin measurement, although stored, is not
used in these calculations. Estimates of asymptotic length
(L_∞), curvature (e^k), intercept (t_0), standard deviation of the
residuals from fit (sigma), and standardized ring measurements
are displayed. Summary displays generated from a four-year-old
scale are the simplest and the easiest to understand (Fig. 4).
Because the ring sizes are standardized, (size of the last
annular ring equal to 100 percent) the value of L_∞ displayed is
117.8, a percentage. This indicates an estimate of 17.8 percent
additional growth beyond size at age 4 for this scale. The
parameter estimates of e^k and t_0 in this example fall within the
acceptable range of 0 to 1.0. Under the von Bertalanffy
hypothesis, estimates of both t_0 and e^k are expected to assume
values between 0 and 1.0. The standard deviation of residuals
from fit, along with acceptable t_0 and e^k parameter estimates
and biological information, indicate a correctly aged scale.
For example, Pacific cod in this study typically had standard
deviations of residuals from fit under 3.0. However,
exceedances of this value from single radii were double those
from full areas (20 - 26% vs 12%). A standard deviation above
3.0 from areas was an indication to the age reader that the
marked annuli should be re-examined.

Selected-radius results and a combination of area and radii
measurements from the complete-scale digitizing routine contain
different amounts of data, but use a similar format. These
values can be used to establish the growth pattern by examining
the parameter estimates generated from either the radius
measurement technique only or, more reliably, from a combination
of area and radius sets. Fish older than 4 years permit
estimation from age subsets, lagged by one year. Subsets of
rings 1 to 4 (1-4), rings 2 through 5 (2-5), and the complete
set of 1-n displayed at the terminal allow the operator to check
for uniformity of growth parameters between the time

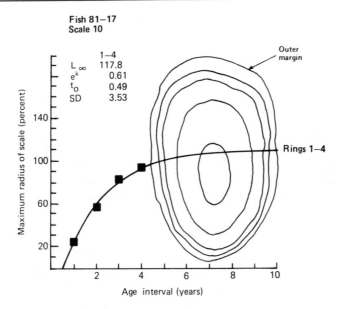

Fish 81—17
Scale 10

Figure 4. Example of terminal display after digitizing annulus
intersections along one radius on a 4-year-old scale.
Observed ring radii are shown in dark squares.

intervals. This is based on the assumption that growth
parameters do not change during the life of an individual.
Comparison of the three growth periods displayed should yield
similar results under the von Bertalanffy hypothesis. In
addition to absolute ring sizes, resultant von Bertalanffy
parameters are saved from each radius or area digitized.

If a non-annular ring is mistakenly included in a fit as
the first annular ring, an entirely different set of parameter
values will be generated than those obtained from a correctly
aged scale(Fig. 5a,b). With a misidentified ring (dashed ring
in Fig. 5a), parameter values from a fit of the first four rings
do not meet the parameter constraints defined previously because
of a negative intercept (t_0 = -.93). Parameter estimates for the
entire age range (1-5 in this example) are also unacceptable
because of the inclusion of a non-annular ring corresponding to
age one. The second subset (rings 2-5) does not include the
incorrect ring and parameter estimates of t_0 and e^k are within
the defined range (Fig. 5b).

Upon elimination of the incorrect ring and redefinition of
the first annulus, the parameters from all subsets of the
correctly aged scale are similar. This indicates that the
pattern of growth with the first annular ring closer to the
origin of the scale is the best choice (Fig. 5b). In a similar
manner, the addition of an extra ring in other locations, or the
exclusion of a valid annular ring in the marking of a scale,
will yield a "poor" fit, e.g., one or more of the parameters
estimated falling outside of the established range.

When a problem scale is encountered the age reader can
remove, add, or retrace any questionable annular rings after
examining the parameter estimates. After making any change to
the original scale, the new pattern is fitted to the von
Bertalanffy equation without having to re-digitize all "correct"
rings again. In such cases, obvious age determination problems,
and the location of rings which cause unreasonable results, are
identified and corrected at the most opportune moment with the
least amount of time expended.

Following the completion of the aging and digitizing of
scales from a particular fish, parameter values and the standard
deviation of residuals from fit are summarized and displayed for
all scales. One useful comparison was a plot of the estimate k
versus L_∞ (Fig. 6). When viewing such an example on the screen,
the sample can be immediately re-examined and a common age for
all of the other scales can be established. An outlying value
from a specific scale can be identified based on frequency of
occurrence of parameter estimates, rather than by reference to a
predetermined rigid range of scale sizes at age.

At the successful completion of scale measurements,
absolute measurements from area and length routines and the
calculated von Bertalanffy parameters are labeled with
collection identification information and stored on computer
disks.

RESULTS
    Square root of area of each ring was used, rather than
total area, to allow comparisons between linear measurement and
areas within single dimensional units. We found that

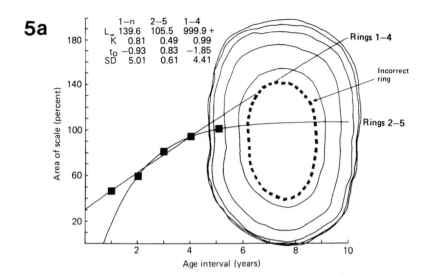

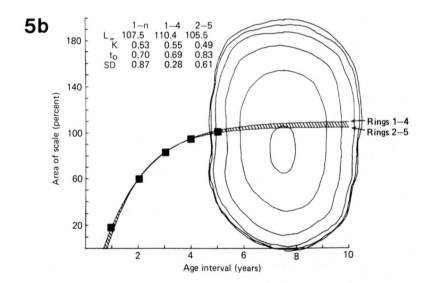

Figure 5. Example of observed ring areas (dark squares) and
graphic representation of von Bertalanffy growth curves:
a) for incorrectly aged 5-year-old scale. b) correctly
positioned rings for 5-year-old scale.

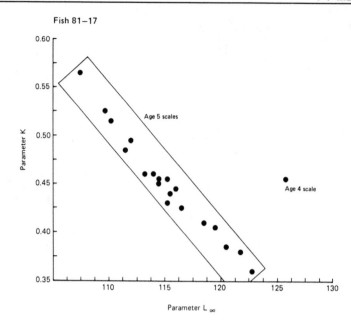

Figure 6. Plot of estimates of k on L_∞ for 20 Pacific cod
    scales from a single specimen (81-17). Scales included by
    rectangle were correctly aged. Age 4 scale was considered
    an outlier.

proportionate change in square root of areas from one ring to
the next was not different among scales taken from different
body positions. Absolute scale shape varied within a fish, but
the growth patterns of numerous scales from an individual fish
were similar (Fig. 7).

Measurement of growth patterns along a single axis was a
much faster technique than tracing complete sets of annuli. It
produced information on approximate positions of annuli, or,
with some species, it was the only method required to derive
growth-history information.

Including the preparation steps, it was possible to mark
and digitize 30 to 40 scales in an eight-hour period by the
quick-pass method. Once the photographs were marked, the
digitizing step proceeded rapidly.

DISCUSSION
    Von Bertalanffy parameter values, which are used to
determine the age of a scale, are obtained solely from the
digitizer hardware and software. The procedure is expected to
be most helpful in determining true annual rings for fish four
years or older and in eliminating problems associated with

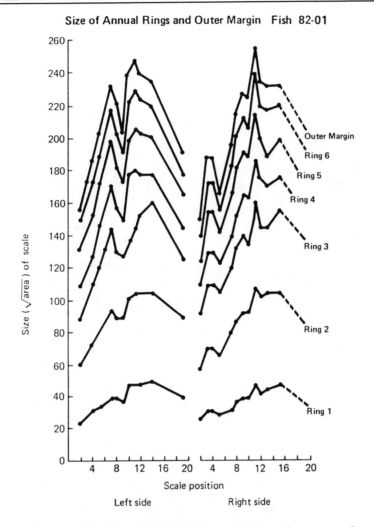

Size of Annual Rings and Outer Margin   Fish 82-01

Figure 7. Plot of proportionate change in areas from one scale
to the next among scales taken from different body
positions (specimen 82-01). Dashed line represents
measurements on damaged scale from position 19 on right
side of scale.

selection of structures other than scales for aging purposes.
Four years is also the minimum age which will produce an error
term in addition to parameter estimates. The criterion of scale
age determination requires that the parameters t_0 and e^k
generated from each scale must adhere to the biologically based
constraint of falling between 0 and 1.0. Secondarily,
comparisons are made using the standard deviation of the

residuals from fit in conjunction with ring sizes from several
scales of an individual fish.

We have found that parameter values calculated using
standardized measurements from correctly marked scales of a
Pacific cod scale generally fall into a narrow range. This
range may vary in other species, or in age structures other than
scales. In generating parametric criteria to help in aging from
scales (or other hard parts) the kind of growth record utilized
here has some distinctive features. E.g., the "map" being
measured is a reflection of each individual's entire growth
history from the date of scale formation. Using this kind of
data base amounts to reduced dependence on less direct criteria
such as correspondence of the length of the aged fish at time of
capture, to historic or other length means of fish whose growth
patterns (hence expected size at age) may or may not have
resembled that of the individual aged. This uncertainty is not
resolved by resorting to other standards of size at age, e.g.
those invoking fish-length relations to scale length. A close
relation can be expected to hold through most of life; however,
the relation of such size-at-age estimates or their means or
modes, to that of a single fish remains unclear. Given these
alternatives, the comparative self sufficiency of scale patterns
as growth descriptors may recommend them as a less ambiguous
basis for distinguishing between true and false annuli.

Parameter estimates from the von Bertalanffy growth
equation (or any other model) used to evaluate a particular
growth pattern are processed rapidly, once annular rings and an
individual's growth pattern have been identified. This record
is generated and displayed in nearly simultaneous fashion by the
interaction of the age reader with the digitizing equipment.
Experience in identifying annuli speeds up the digitizing
process and reduces variability. The majority of fish scales
digitized in this study were done over a six-month period.
After a six-month break, two fish in a 1982 sample were
digitized using techniques learned in the first year of data
collection. The standard deviation of residuals from fit
decreased as experience in this new method increased.

Quality of individual scales photographed or projected is
also very important, and uniform results depend upon
readability. Normally, a problem scale can be re-examined using
information determined from adjoining scales or, in the extreme
case, by re-digitizing the material.

The presence of non-annular rings in the data set, or the
exclusion of an annulus, is inferred from differences in the
parameters determined for the three growth periods. Given a
general knowledge of ring patterns on scales, an age reader can
readily identify a mis-marked scale and, from the breakdown
provided, localize the problem area. When more information is
available for particular species, the computer routine itself
may be able to flag and identify annuli according to predefined
biological criteria.

We found that different scales from the same fish may have a different number of annuli, some higher but, more often, lower than the modal age reading. By comparing parameter values and relative ring-size information, the sample can be brought to a uniform age. Examination of a scale ring pattern indicated that, although absolute scale sizes, and therefore annular ring sizes, varied over the body surface of a fish, growth parameter estimates showed little variation in certain regions (Hirschhorn and Small 1986). Uniformity of growth patterns from scales on the same individual was used in the 1982 sample (Pacific cod 82-1) to locate the initially missed rings in under-aged scales and bring the sample to a uniform age.

In the development study, parameter estimates obtained using the one-scale radius procedure were found to yield more variable estimates than those generated by measuring the area inside each annular ring (Hirschhorn and Small 1986). The usefulness of the one-radius technique lies in its ability to rapidly measure the growth pattern, and in its use as a guide to locate annular-ring positions before complete tracing is undertaken. The growth information obtained by tracing entire annular rings is more time consuming than digitizing one radius, but it is a more comprehensive technique.

In addition to cod scales, the digitizing process has been extensively tested on scales from walleye pollock Theragra chalcogramma and sablefish Anoplopoma fimbria with satisfactory results. In addition, small samples of scales from haddock Melanogrammus aeglefinus, dace Rhinichthys sp., chub Semotilus sp., green sunfish Lepomis cyanellus, dover sole Microstomus pacificus, and hake Merluccius gayii have been aged with good results.

ACKNOWLEDGMENTS
We would like to acknowledge the extensive technical assistance provided to us by Frank M. Kikuchi in the early development stages of this work, and to acknowledge Kenneth L. Davis, M.P. Bailey, James P. Hughes, and Alan C. Lindsay who provided valuable steps in software development. Jane A. Small and Betty J.Goetz aided us in criticizing the manuscript.

REFERENCES
BEAMISH, R. J. AND D. A. FOURNIER. 1981. A method for comparing the precision of a set of age determinations. Canadian Journal of Fisheries and Aquatic Sciences 38:982-983.
CARLANDER, K. J. 1974. Difficulties in ageing fish in relation to inland fisheries management. Pages 200-205 in T. B. Bagenal, editor. Ageing of fish. Unwin Brothers, Ltd., Surrey, England.

FABENS, A. J. 1965. Properties and fitting of the von Bertalanffy growth curve. Growth 29:265–289.

FOUCHER, R. P., AND D. FOURNIER. 1982. Derivation of Pacific cod age composition using length–frequency analysis. North American Journal of Fisheries Management 2:276–284.

FRIE, R. V. 1982. Measurement of fish scales and back-calculation of body lengths using a digitizing pad and microcomputer. Fisheries 7(6):5–8.

FROST, W. E., AND C. KIPLING. 1959. The determination of age and growth of pike (Esox lucius) from scales and opercular bones. Journal of the International Council for Exploration of the Sea 24:321–341.

HIRSCHHORN, G., AND G. J. SMALL. 1986. Variability in growth parameter estimates from scales of single specimens of Pacific cod based on radius and area measurements. In R. C. Summerfelt and G.E. Hall, editors. Age and growth of fish. Iowa State University Press, Ames, Iowa, USA.

KENNEDY, W. A. 1970. An examination of criteria for determining the age of Pacific cod (Gadus macrocephalus) from otoliths. Fisheries Research Board of Canada 27:915–922.

VON BERTALANFFY, L. 1951. Theoretische Biologie, Vol. 2. A. G. Francke, Berne, Switzerland.

WESTRHEIM, S. J., AND W. SHAW. 1982. Progress report on validating age determination methods for Pacific cod (Gadus macrocephalus). Canadian Manuscript Report of Fisheries and Aquatic Sciences No. 1670. Department of Fisheries and Oceans, Fisheries Research Branch, Pacific Biological Station. Nanaimo, British Columbia V9R 5K6.

VI
Larval and juvenile fish

Daily increments on the otoliths of larval American shad and their potential use in population dynamics studies

THOMAS F. SAVOY AND
VICTOR A. CRECCO
CONNECTICUT DEPARTMENT OF ENVIRONMENTAL PROTECTION
MARINE FISHERIES OFFICE
P.O. BOX 248
WATERFORD, CT 06385

ABSTRACT

Two groups of known-age American shad larvae _Alosa sapidissima_ were reared at 15 and 18 C to determine the precision and accuracy of otolith increment counts, the age at which the first otolith increment is deposited, the relationship between otolith half-length and larval growth, and the accuracy of using otolith age estimates to place wild American shad larvae from the Connecticut River into various hatching intervals. There was a significant linear relationship between the number of otolith increments and the true age of the 15 and 18 C groups. The slope of the overall mean regression for the 18 C group did not differ significantly from 1.0, indicating that the number of increments reflect the true ages of these larvae. The slope for the slow-growing 15 C group fell below 1.0, mainly due to underaging of the older larvae. The onset of increment deposition occurred before yolksac absorption for both groups. The precision of repeated age estimates for the 18 C larvae was high and independent of age, whereas the level of precision for the 15 C group was low among the newly-hatched larvae. The otolith half-length and body-length relationships were significant for the 18 C and wild larvae, but not the 15 C group. Underaging of the 15 C larvae was due to problems in differentiating between their tightly spaced increments. A hatching interval of 5 days (±2 days) satisfied the tradeoff among accuracy, sample size and the number of hatching intervals. When the 317 wild larvae were separated into nine consecutive 5-day hatching intervals, cohort-specific larval growth was related to short term changes in water temperature.

Knowledge of the true ages of larval and juvenile fishes can provide important early life history information with which to enhance our understanding of recruitment success. The relative dis-

The Age and Growth of Fish, edited by Robert C. Summerfelt and Gordon E. Hall © 1987 The Iowa State University Press, Ames, Iowa 50010.

tance between the increments can be used to back-calculate larval growth rates (Methot 1981; Radtke and Scherer 1982). Knowing the collection dates of the larvae, the number of daily or primary increments can be used to establish their hatching dates, spawning intensity and duration (Graham et al. 1984). Additionally, cohort-specific changes in larval growth and mortality can be related to fluctuations in biotic and abiotic conditions (Methot 1983; Graham et al. 1984; Crecco and Savoy 1985).

The extent to which otolith increments can be used to accurately estimate age, growth and mortality of larval and juvenile fishes depends on the assumptions that the otolith increments are deposited daily (Pannella 1971), or at some constant rate (Lough et al. 1982), that the date of initial otolith increment formation is known (Taubert and Coble 1977; Wilson and Larkin 1980; Volk et al. 1984), and that there is a functional relationship between the growth of the otoliths and the larvae (Laroche et al. 1980). Although growth increments on the otoliths are assumed to be deposited daily (Pannella 1974), only recently have known-age fish been used to assess the accuracy of otolith age determinations (Geffen 1982; Campana 1983). Moreover, few studies have used known-age fish to evaluate the reliability of fixing larval birthdates and spawning distributions (Miller and Storck 1984; Rice et al. 1985), and to our knowledge, no studies have examined the precision of repeated age estimates from the same larvae.

We used two groups of known-age American shad Alosa sapidissima larvae to determine the precision and accuracy of otolith increment counts, the age at which the initial increment is deposited on the otoliths, the feasibility of placing wild larvae into birthdate intervals ranging from 1 to 11 days, and the otolith half-length and body-length relationship. Also, the results from the known-age larvae were used to fix the hatching intervals and assess the growth rates of American shad larvae from the Connecticut River in 1984.

METHODS

Data Source

One group of 69 larvae, ranging in age from 1–20 days and total length from 8.0 to 19.0 mm, was received from the Lamar National Fish Hatchery, Lamar Pennsylvania (Table 1). Eggs came from the Pamunkey River, Virginia; the Delaware River, Delaware; and the Columbia River, Oregon. The Columbia and Pamunkey River larvae were reared in 0.7 m x 0.8 m x 0.6 m rectangular tanks at densities of less than 100/L and less than 50/L, respectively. The Delaware larvae were raised in 1.2 m diameter x 0.6 m deep circular tanks at densities of 50 and 100/L. The 69 larvae were fed brine shrimp Artemia salina and reared in water temperatures of 18 ± 1 C, with a mean pH of 8.0 (Ronald Howey, Director of Research pers. comm.) under a 12:12 light-dark photoperiod. These larvae are hereafter referred to as the 18 C group.

Table 1. Relationship between true age and mean total
length of two groups of American shad larvae reared
at temperatures of 15 and 18 C; SD is the standard
deviation of the mean length.

	15 C			18 C		
Age	Sample Size	Mean Length (mm)	SD	Sample Size	Mean Length (mm)	SD
(days)						
1				9	8.0	-
2	8	9.8	0.38			
3	11	10.1	0.35			
4	11	10.2	0.34			
5	11	10.0	0.39			
6	6	10.3	0.41	12	11.4	0.43
7	10	10.4	0.41			
8				10	11.4	0.59
11	10	10.6	0.52	6	10.1	0.22
12	10	10.7	0.42	13	12.3	1.15
13	10	11.3	0.48			
14	10	10.9	0.57			
15	10	10.8	0.49	10	13.5	2.35
16	12	10.9	0.31			
17	10	10.9	0.41			
18	13	10.7	0.52			
19	11	11.2	0.40			
20	10	11.2	0.67	9	17.4	0.80
21	10	11.0	0.50			
22	10	11.7	0.71			
23	10	11.8	1.01			
25	10	11.3	0.63			
26	9	11.9	1.53			
27	10	11.1	0.72			

Another group of 218 larvae, ranging in age from 2-27 days
and in total length from 9.0 to 13.0 mm, were obtained from the
Pennsylvania Fish Commission's Van Dyke Fish Hatchery, Bellefonte,
Pennsylvania (Table 1). These larvae, which originated from
adults captured in the Pamunkey River, Virginia, were reared in a
circular 1100 L, 1.5 m diameter 0.8 m depth fiberglass tank and
fed brine shrimp under a 12:12 light-dark regime (Wiggins et al.
1984). The larvae were reared in 13.3 to 18.9 C water with a mean
of 15.6 C and pH of 6.5. These larvae are referred to as the 15 C
group. Both sets of known-age larvae were shipped from the hatcheries in buffered alcohol (80% ETOH buffered to pH 7.0 with
$CaCO_3$).
A total of 42,500 larvae were collected from the Connecticut
River, from June through mid-July, 1984, of which 380 were selected randomly in the field and frozen.

Otolith Preparation
Larvae were measured to total length (mm), placed on a glass
slide in a drop of distilled water and the otoliths teased from
the saccullar chamber under a binocular dissecting microscope
(10 - 70X). Only the sagittae were used in this study; they are

the largest of the three pairs in American shad, and thus the most consistently and easily extracted. The otoliths were allowed to air dry, covered with a drop of permanent mounting media (Histoclad) and a glass coverslip, and assigned a random number.

Both sets of known-age larvae were aged under transmitted light at 430 and 1000X by four independent observers. Increments were counted according to the criteria of Barkman (1978) and without prior knowledge of the length or developmental stage of the larva, thus ensuring that age determinations were based solely on the number of otolith increments. Otolith half-length, the distance from the center to the widest edge, was recorded in micrometer units.

Otoliths from wild larvae were processed similarly to hatchery-reared larvae except that station, sampling date and total length were recorded. Increment counts were made and the half-length and distance from the focus to the 4, 5, 6, 7, 8, 14, 21, and 28 increments were measured. The corresponding body lengths (mm) were backcalculated by the Lea method (Ricker 1975) with a constant y-axis intercept of 8.0 mm (Maxfield 1953). We used linear regression to examine the otolith-body growth relationship for the larvae, and compared it to the otolith-body growth relationship for hatchery-reared fish. Mean increment widths were calculated by age and compared to the age-specific increment widths of the 15 and 18 C groups.

Daily Increment Validation

We conducted simple linear regression (SAS 1982) on both sets of known-age larvae, using the true age as the (X) variable and estimated age as the dependent (Y) variable. Regression analyses were conducted separately on the age estimates of the four observers, and on the overall mean age estimates. If the increments on the otoliths were daily, there should be a significant ($P <$ 0.05) linear regression between the true and estimated ages, with the slope approaching 1.0. Additionally, if the first increment was deposited at hatching, the y-axis intercept should not differ significantly from zero. We used paired t-tests to determine the statistical significance of the slope and intercept values.

We estimated the coefficients of variation (CV), the standard deviation/mean, based on the four age estimates of each otolith, to examine the degree of repeatability of otolith age estimates. The CV value for each otolith age was regressed against the true age in both data sets. Low CV values ($< 25\%$), and a weak correlation between the CV and true age, would indicate that the precision of repeated age estimates was high and independent of larval age.

Hatching Intervals

Assuming that wild larvae can be aged with at least the same degree of accuracy as hatchery-reared larvae, the difference between the true and estimated ages can be used to measure the error in establishing various hatching intervals for wild larvae.

We used the otolith age estimates of known-age larvae to evaluate the percentage accuracy of assigning a larva to an interval which included its true hatching date using 1, 3, 5, 7, 9, and 11-day intervals, corresponding to the estimated age plus or minus 1, 2, 3, 4, and 5 days, respectively. These narrow intervals are required for American shad in the Connecticut River, where larval hatching is usually confined to a 40-day period from June through early July (Cave 1978; Crecco et al. 1983). The hatching intervals within which a larva was placed correctly depended on the difference between its estimated and true age. For example, if 12 otolith increments were counted for a 14-day old larva, this would represent a 2-day error, allowing the correct placement of this larva into the ±2-day and all wider intervals.

The hatching interval used to examine spawning intensity and duration, growth and mortality should satisfy the trade-off among accuracy, sample size and the number of intervals. Specifically, the hatching interval should be wide enough to correctly assign the larvae, require realistic sample sizes (N < 500) for a given level of error, and yet be narrow enough to obtain the maximum number of intervals.

The total sample sizes (N) needed to place the larvae into each interval with a 20% error were determined by:

$$N = [(t_{0.05}/0.20)^2/(1 - P/P)]C$$

for a binomial distribution (Elliot 1971), where (P) was the percentage of larvae placed within each successive interval, and (C) was the number of 1, 3, 5, 7, 9, and 11-day intervals that were established within 40 days. The use of the binomial model to determine (N) was justified given that the estimated age either fell inside or outside of each interval.

Larval-Otolith Growth Relationship

A significant linear relationship between the growth of the otoliths and larvae is needed to correctly back-calculate length-at-age using the regression method (Carlander 1981). The mean increment widths (x half-length - 30 / true age) for each age were estimated for both groups of larvae (30 microns is the mean radius of the core), and statistically compared by t-test analysis. We then conducted linear regression analyses between the total length of the larvae and otolith half-length. A lack-of-fit test (Sokal and Rolf 1969) to a linear regression was performed by partitioning the unexplained sums of squares into the mean square error unexplained (MSE_U) and mean square error pure error (MSE_{PE}). If the F-statistic for MSE_U/MSE_{PE} exceeded the tabulated F value at r-2, n-2 degrees of freedom, where r is the number of age groups and n is the sample size, then a nonlinear model would provide a better fit to the data.

Age-specific instantaneous growth rates (mm/day) were estimated from mean length-at-age data with:

$$G_t = LN(L_{t+1}) - LN(L_t)/dt$$

where G_t = the mean instantaneous growth rate of larvae from the
ages t to t+1; L_{t+1} = mean total length of larvae at each age t+1;
L_t = mean total length of larvae at age t; dt = the number of days
between L_t and L_{t+1}.

Age-Specific Growth

Age-specific instantaneous growth rates (mm/d) for wild larvae were estimated based on back-calculated length-at-age data
from 4-7 days, and for each seven-day interval thereafter (ie. 7-
14, 14-21, 21-28) by the method of Ricker (1975). Larvae were
partitioned into these four groups to examine whether growth rates
differed over time and among ages. Mean age-specific growth rates
(G_{it}) and standard deviations (SG_{it}) of all larvae within a cohort
were estimated with the above growth equation. We also compared
the larval growth rates of the 15 and 18 C groups to the growth
rates of all wild larvae combined.

Temperature and Growth Relationship

Linear and nonlinear regression methods were used to relate
the growth rates among cohorts to short-term changes in river
temperature (C). Mean temperatures were calculated for all dates
inclusive within each of the intervals (4-7 d, 7-14 d, 14-21 d,
and 21-28 d) to determine whether growth rates among ages and cohorts differed with water temperatures. We considered the median
date (day 3) within each cohort as a reasonable starting point
from which to estimate temperature conditions throughout larval
development. The hatching intensity (the number of larvae
assigned to each cohort) was examined against the corresponding
mean water temperature measured within each 5-day period. The
hatching intensity was also related to the cumulative number of
days (degree days) in which the daily water temperature exceeded
15 C starting on May 15, 1984. Water temperatures were recorded
daily in 1984 by the U.S. Geological Survey (U.S.G.S. Water Year
Reports 1984) at the Thompsonville, Connecticut (river km 89)
gauging station.

RESULTS

There was close agreement between the number of increments
and the true ages of the known-age larvae for both data sets
(Table 2). The slopes of both grouped and ungrouped regressions
in the 18 C group did not differ significantly from 1.0 (Figure
1). By contrast, the slopes of the four age estimates for the 15
C group were significantly less than 1.0, which was attributed
mainly to the underaging of older larvae (> 20 days)(Figure 2).

The y-axis intercepts from three of the four linear regression equations in the 18 C group did not differ significantly from
0, with the remaining intercept not being significantly different
from 1.0. Moreover, 17 of 18 one-day-old larvae were correctly

Table 2. Linear regression equations of estimated (Y) versus true age (X) of known-age American shad larvae reared at temperatures of 15 and 18 C.

Observer	Regression	r^2
	18 C Group	
1	Y = 1.0633 X + 1.2971	0.80
2	Y = 1.0924 X + 0.7165	0.82
3	Y = 0.9948 X − 0.1274	0.89
4	Y = 0.9196 X + 1.7987	0.88
MEAN	Y = 1.0166 X + 0.9522	0.90
	15 C Group	
1	Y = 0.7699 X + 2.5324	0.80
2	Y = 0.7311 X + 2.4270	0.80
3	Y = 0.9113 X − 1.5493	0.76
4	Y = 0.9296 X + 1.1528	0.85
MEAN	Y = 0.8358 X + 1.1458	0.88

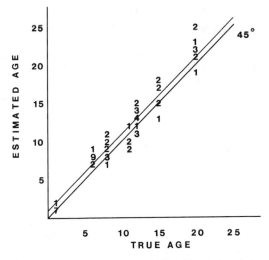

Figure 1. Relationship between the estimated and true ages (days) for the 18 C group of American shad larvae. (Numbers represent numbers of observations).

aged from the otoliths. Conversely, the four y-axis intercepts of the 15 C group differed significantly from 0, two of the four intercepts differed from 1.0, but none differed significantly from 2.0.

No significant linear relationship existed (r = -0.055) between CV and true age in the 18 C group (Figure 3). These CV values were generally less than 25%, indicating that the variability of repeated otolith age estimates was low and independent of larval age. By contrast, a significant inverse relationship (r = -0.59, p < 0.001) was evident between CV and true age in the 15 C

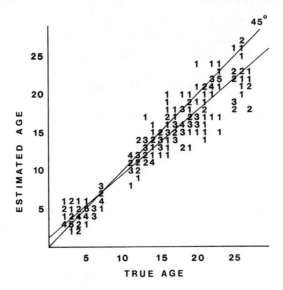

Figure 2. Relationship between the estimated and true ages (days) for the 15 C group of American shad larvae. (Numbers represent numbers of observations).

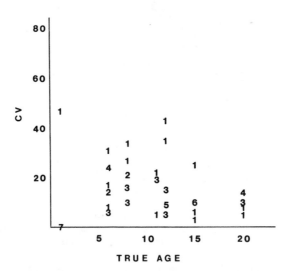

Figure 3. Relationship between the precision (CV values) of repeated age estimates and the true ages (days) for the 18 C group of American shad larvae.

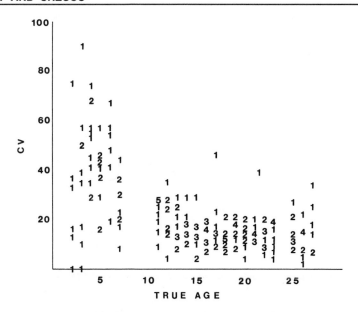

Figure 4. Relationship between the precision (CV values) of
repeated age estimates and the true ages (days) for the 15
C group of American shad larvae.

group (Figure 4). This was due mainly to low precision of repeated otolith age estimates among newly-hatched larvae (ages 2-5
days). When the newly-hatched larvae were removed from the regression, the CV values were unrelated (r = 0.10) to the true age.

Hatching Interval
        The percentage accuracy of assigning larvae into successive
hatching intervals was always higher for the 18 C than the 15 C
group (Table 3). Whereas less than 22% of the larvae from both
groups were placed correctly into 1-day hatching intervals, the
percentage accuracy rose rapidly with increasing interval length
to 99% for a ± 5-day interval. When the larvae from both groups
were assigned to the 1-day interval, the required sample sizes
were too large (N > 1200) and the percentage accuracy too low (P =
51-53%) for practical use in field studies. By contrast, the sample sizes and percentage accuracy for the ± 4 and ± 5-day intervals
were reasonable, but only four such intervals were created within
a 40-day hatching period, thereby allowing only two degrees of
freedom in regression analysis. The ±2 and ±3-day hatching intervals best satisfied the trade-off among accuracy, sample size and
number of intervals.

Larval Otolith Growth Relationship
        Larval lengths were linearly related to otolith half-lengths

Table 3. Sample size, percentage accuracy and
         number of age intervals per interval length
         for American shad larvae reared at 18 and
         15 C.

Interval Length (days)	Sample Size/ Interval	Percent Accuracy	No. of Age Intervals	Total Sample Size (N)
		18 C Group		
1	376	21	40.0	15040
3	96	51	13.3	1248
5	20	83	8.0	160
7	5	95	5.7	30
9	3	97	4.4	12
11		100	3.6	
		15 C Group		
1	456	18	40.0	18240
3	113	53	13.3	1469
5	45	69	8.0	360
7	19	84	5.7	114
9	10	90	4.4	40
11	4	96	3.6	12

for the 15 and 18 C groups (Figures 5 & 6). The linear model explained 80% of the variation in the 18 C group, showing that the
relative distance between increments can be used to accurately
back-calculate larval lengths with the regression method. By contrast, only 58% of the variation in length for the 15 C group was
ascribed to otolith half-length. This was because half-length
continued to increase with age whereas growth in length ceased after 14 days. When total length of the 15 C group was regressed
against both otolith half-length and true age, the r^2 rose to
0.90.

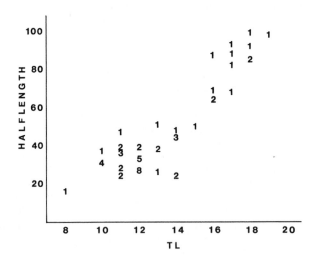

Figure 5. Relationship between sagittal half-lengths (m.u.)
          and larval lengths (mm) for the 18 C group of American
          shad larvae.

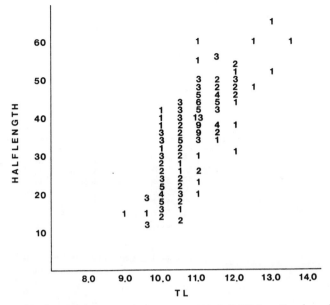

Figure 6. Relationship between sagittal half-lengths (m.u.) and larval lengths (mm) for the 15 C group of American shad larvae.

The slope of the otolith-body growth relationship for the 18 C group was significantly greater (t = 10.6, P < 0.001) than that of the 15 C group. Similarly, the age-specific increment widths of the 18 C group were consistently wider than those of the 15 C group (Table 4). The mean increment widths of the 15 C group demonstrated a decreasing trend with age, whereas no pattern was discernable among the 18 C group. The faster growth rates and more widely spaced increments of the 18 C group were primary reasons why their otolith increment counts were more accurate and precise than those for the 15 C group.

Application to Wild Larvae
    We applied the otolith-aging technique to American shad larvae collected in the Connecticut River and compared their growth rates to those of the hatchery-reared larvae. Of the 380 larvae collected, 317 were aged and backcalculated.
    Having shown that the ± 2 and ± 3-day intervals were the most reasonable, we used the estimated age and date of capture to assign the 317 larvae to one of nine consecutive ±2-day intervals, hereafter referred to as a 5-day cohort. Larval hatching extended from May 25 to July 8, 1984, with most larvae hatching after mid-June when water temperatures exceeded 20 C (Table 5). These results are in accord with previous field studies (Marcy 1976; Cave 1978; Crecco et al. 1983). The intensity of larval hatching among

Table 4. Age-specific changes in the average increment widths (microns) of wild and hatchery-reared American shad larvae.

Age (days)	18 C N	18 C X	18 C SD	15 C N	15 C X	15 C SD	Wild Larvae N	Wild Larvae X	Wild Larvae SD
2				7	5.47	0.49			
3				10	4.63	0.18			
4				7	5.27	0.83	3	13.50	1.69
5				11	5.39	0.63			
6	11	18.21	2.25	6	5.42	0.87	12	13.96	1.97
7				10	5.33	0.43	11	15.07	3.11
8	9	18.09	1.28				12	17.14	2.99
9							7	17.67	2.78
10	4	12.89	1.32				11	18.57	4.05
11				9	6.68	3.22	12	17.24	3.57
12	11	12.21	3.89	10	5.31	0.51	16	16.53	4.03
13				10	5.42	0.46	14	19.21	4.21
14				9	5.02	0.55	20	19.31	3.97
15	10	14.65	3.99	7	5.07	0.53	22	20.24	4.30
16				11	5.06	0.41	18	22.81	4.06
17				10	4.81	0.40	18	24.15	4.07
18				12	4.51	0.52	11	21.45	4.08
19				11	4.54	0.40	5	16.44	3.12
20	9	20.67	1.35	9	3.51	0.83	12	25.71	5.59
21				8	4.00	0.36	21	23.09	4.55
22				9	3.58	0.44	8	21.13	3.11
23				7	4.66	0.70	18	24.20	5.16
24							8	24.61	4.37
25				7	4.34	0.61	12	22.10	4.50

Table 5. Hatching intensity, mean water temperature and degree days for wild larval American shad, 1984.

Cohort	Hatching intensity	Water temperature	Degree days
May 27	7	17.6	8
June 1	17	12.3	8
June 6	60	14.5	10
June 11	42	17.9	15
June 16	48	20.2	20
June 21	89	21.6	25
June 26	44	22.3	30
July 1	8	23.4	35
July 6	2	23.5	40

all cohorts exhibited no trend with mean water temperature. However, when intensity was related to degree days, a parabolic relationship was evident ($r = 0.79$). Frequency counts of larvae may not reflect true hatching intensity, as no age-specific mortality corrections were made (Miller and Stork 1984).

There was a strong linear relationship between the growth of the otolith and the larvae (r^2 of 0.89). This supports the validity of both the regression and Lea methods of backcalculating the lengths of wild shad larvae. The total length-otolith size relationship for the wild larvae was slightly higher than the fit for the 18 C group (Figure 5), but much stronger than the r^2 of 0.58

for the 15 C group (Figure 6). Mean increment widths of the wild larvae increased slowly with age (Table 4) and were of similar magnitude to the 18 C larvae, but considerably higher than the increment widths of the 15 C larvae.

Mean instantaneous growth rates (G_t) were variable among age groups and cohorts (Table 6). The G_t values for young larvae (4-

Table 6. Age-specific changes in instantaneous growth (G) rates of larval American shad among 5-day cohorts, and water temperatures (T), 1984.

COHORT		AGE-GROUP			
		4-7	7-14	14-21	21-28
May 27	G	0.0231	0.0255	0.0329	0.0298
	SD	0.0065	0.0051	0.0047	0.0093
	N	7	7	7	3
	T	12.0	14.8	19.3	21.3
June 1	G	0.0272	0.0283	0.0340	0.0275
	SD	C.0079	0.0066	0.0072	0.0022
	N	17	15	11	2
	T	14.9	18.3	20.8	22.4
June 6	G	0.0317	0.0347	0.0329	0.0250
	SD	0.0080	0.0070	C.0074	0.0153
	N	53	35	16	2
	T	18.3	20.3	22.0	23.3
June 11	G	0.0354	0.0402	0.0317	0.0291
	SD	C.0095	0.0081	0.0065	0.0095
	N	39	29	13	9
	T	20.1	21.4	22.8	23.3
June 16	G	0.0410	0.0444	0.0318	0.0258
	SD	0.0081	0.0079	0.0060	0.0079
	N	45	34	17	10
	T	21.9	22.6	23.6	22.2
June 21	G	0.0422	0.0433	0.0338	0.0277
	SD	0.0053	0.0070	C.0062	0.0041
	N	87	49	31	5
	T	22.1	23.3	22.8	23.3
June 26	G	0.0406	0.0444	0.0329	0.0285
	SD	C.0055	0.0062	0.0054	-
	N	44	40	19	1
	T	23.4	23.1	22.9	23.9
July 1	G	0.0379	0.0398	-	-
	SD	0.0080	0.0094	-	-
	N	8	8	0	0
	T	23.4	22.3	-	-
July 6	G	0.0348	-	-	-
	SD	0.0094	-	-	-
	N	2	0	0	0
	T	21.9	-	-	-
ALL COHORTS					
	G	0.0376	0.0401	0.0329	0.0275
	SD	C.0088	0.0089	0.0062	0.0077

7, 7-14 d) were lowest among the early cohorts, rose to a maximum during late June, and declined slightly thereafter. The G_t values for older larvae exhibited no apparent trend among cohorts. The age-specific growth rates of the wild larvae were similar to the growth rates of the 18 C group, but were two orders of magnitude higher than the growth rates of the 15 C group (Table 7). Even the slow-growing May 25 cohort grew four to five times faster than the 15 C larvae, further demonstrating the atypical growth rates of this group.

Table 7. Age-specific changes in instantaneous growth (G) rates of wild and hatchery-reared larval American shad.

GROUP	AGE INTERVALS			
	4-7	7-14	14-21	21-28
15	0.0028	0.0029	0.000057	0.0026
18a	0.0300	0.0100	0.0200	
WILD	0.0376	0.0401	0.0329	0.0275

a Age intervals from 1-6 days, 8-15 and 15-20

Significant linear relationships existed between the G_t values among the young larvae and water temperatures (4-7 d, r = +0.93, 7-14 d, r = +0.96), whereas no relationship was evident among the older larvae (14-21 d, r = 0.20, 21-28 d, r = 0.26). Growth rates of young larvae were lowest when water temperatures were below 18 C and highest when temperatures exceeded 20 C (Table 6). Small fluctuations in water temperature after mid-June were the most likely reason for the lack of fit between temperature and growth among the older larvae.

DISCUSSION

This study demonstrated that otolith increments for the 18 C larvae were deposited daily and thus reflect their true age. Given that the otolith increments of wild larvae were as distinct as those of the 18 C group, increment counts for wild American shad larvae should accurately estimate their true ages. Other researchers (Laroche et al. 1980; Lough et al. 1982; Campana 1984) have also noted that the otolith increments of wild larvae are usually clearer and easier to count than the increments on laboratory-reared larvae.

These conclusions are consistent with the results of other clupeid studies including Atlantic herring Clupea harengus (Townsend and Graham 1981), and gizzard shad Dorosoma cepedianum (Davis et al. 1985). Although Geffen (1982) reported that the periodicity of otolith increments on Atlantic herring larvae were more related to larval growth rates than age, Campana and Neilson (1985) argued that the increment widths of the slow-growing larvae may have fallen below the resolving power of the light microscope, thereby accounting for the underestimation of age.

The number of otolith increments for the 15 C group were linearly related to the true age, however increment counts fell below one per day. The tendency here to underage the older larvae was likely a problem in differentiating between the tightly spaced increments. The mean increment widths for the 15 C group declined steadily with age, suggesting that the outermost increments for the older larvae may have fallen below the resolving power of the light microscope. Although it has been suggested that otolith increment deposition may cease after larvae stop growing (Taubert and Coble 1977; Marshall and Parker 1982; Geffen 1982), this is unlikely in our case, given that the otoliths of the 15 C group grew continuously despite a lack of larval growth. In addition, the significant linear relationship between increment counts and true ages would not be expected if otolith increment deposition stopped for older larvae.

The low growth rates and narrow increment widths among the 15 C larvae may be attributed to either low rearing temperatures or pH. A pH of 6.5 is unlikely to have caused stunting of these larvae, as wild larvae in the Connecticut River grew well at pH levels of 6-7.0 (U.S.G.S. Water Year Reports 1985). That low rearing temperatures adversely affected larval growth was supported by the wider otolith increments on the 18 C larvae and the significant positive correlations between water temperature and growth rates among the wild larvae. These results indicate that both otolith and larval growth are linked to water temperature, as was reported for juvenile sockeye salmon <u>Oncorhynchus</u> <u>nerka</u> (Marshall and Parker 1982). Other factors influencing the width of otolith increments that we have not investigated include: feeding periodicity (Struhsaker and Uchiyama 1976; Pannella 1980; Tanaka et al. 1981; Neilson and Geen 1982), rearing conditions (Laroche et al. 1980; Geffen 1982) and light-dark cycles (Taubert and Coble 1977; Radtke and Dean 1982).

The problem of aging the slow-growing 15 C group is unlikely to occur for wild American shad larvae, since they are rarely encountered at temperatures below 17 C in northern rivers (Walburg and Nichols 1967; Cave 1978; Crecco et al. 1983). Moreover, it is unlikely that such stunted larvae would survive very long in the wild due to their high vulnerability to predation and starvation (Lasker 1975; Hunter 1976).

The onset of otolith increment deposition for American shad occurs shortly after hatching. Since yolk-sac absorption for American shad larvae begins three to five days after hatching (Wiggins et al. 1984), this indicates that the first otolith increment of American shad larvae is deposited prior to external feeding. The positive y-axis intercept noted for the estimated and true age regression for the 15 C group was likely due to the systematic underaging of the older larvae and/or the presence of subdaily increments. Subdaily increments have been documented on the otoliths of other species (Pannella 1974; Taubert and Coble 1977; Wilson and Larkin 1980; Campana and Neilson 1982); they would account for the low precision of our age estimates among the

newly-hatched 15 C larvae. Our conclusion that initial increment
deposition occurs before yolk absorption is consistent with results for sockeye salmon (Marshall and Parker 1982), haddock
Melanogrammus aeglefinus and cod Gadus morhua (Bolz and Lough
1983). Other studies have reported that the first ring is deposited after exogenous feeding for the Atlantic herring (Geffen 1982;
Lough et al. 1982), English sole Parophrys vetulus (Laroche et al.
1980), and turbot Scopthalmus maximus (Geffen 1982).

    Otolith increment counts have been used to establish the
hatching dates of wild larvae (Methot 1983; Miller and Storck
1984; Crecco and Savoy 1985). This study demonstrates that the
practice of assigning wild American shad larvae to 5 and 7-day
hatching intervals will allow a more precise determination of
hatching intensity and periodicity, and a more complete examination of how short-term changes in larval growth and mortality relate to climatic fluctuations, predation, and food availability
(Crecco and Savoy 1985). Because year-class strength of many
fishes, including American shad (Crecco et al. 1983), is established during the larval period, a method of generating multiple
hatching intervals per year may help resolve the long standing
controversy of whether biotic or abiotic factors most influence
the stock recruitment relationship (Ricker 1954; Cushing 1982).

ACKNOWLEDGMENTS
    We thank Linda Gunn Alexander, Deborah Pacileo, David
Witherell and Rita Lorenzetti Langan for their help in the laboratory and field sampling.

REFERENCES
BARKMAN, R. C. 1978. The use of otolith growth rings to age
        young Atlantic silversides, Menidia menidia.
        Transactions of the American Fisheries Society 107:790-792.
BOLZ, G. R., AND R. G. LOUGH. 1983. Growth of larval Atlantic
        cod, Gadus morhua, and haddock, Melanogrammus aeglefinus, on
        Georges Bank, Spring 1981. United States National Marine
        Fisheries Service Fishery Bulletin 81:827-836.
CAMPANA, S. E. 1983. Feeding periodicity and the production of
        daily growth increments in the otoliths of steelhead trout
        (Salmo gairdneri) and starry flounder
        (Platichthys stellatus). Canadian Journal of Zoology
        61:1591-1597.
CAMPANA, S. E. 1984. Microstructural growth patterns in the
        otoliths of larval and juvenile starry flounder, Platichthys
        stellatus. Canadian Journal of Zoology 62:1507-1512.
CAMPANA, S. E., AND J. D. NEILSON. 1982. Daily growth increments
        in otoliths of starry flounder, Platichthys stellatus, and
        the influence of some environmental variables in
        their production. Canadian Journal of Fisheries and
        Aquatic Sciences 39:937-942.
CAMPANA, S. E., AND J. D. NEILSON. 1985. Microstructure of fish

otoliths. Canadian Journal of Fisheries and Aquatic Sciences 42:937-942.

CARLANDER, K. D. 1981. Caution on the use of the regression method of back-calculating lengths from scale measurements. Fisheries 6:2-4.

CAVE, J. R. 1978. American shad (Alosa sapidissima) larval distribution, relative abundance and movement in the Holyoke Pool, Connecticut River, Massachusetts. Masters thesis, University of Massachusetts, Amherst Massachusetts, USA.

CRECCO, V., T. SAVOY, AND L. GUNN. 1983. Daily mortality rates of larval and juvenile American shad (Alosa sapidissima) in the Connecticut River; an assessment of the "critical period" hypothesis. Canadian Journal of Fisheries and Aquatic Sciences 40:1719-1728.

CRECCO, V. A., AND T. F. SAVOY. 1985. Effects of biotic and abiotic factors on the growth and survival of young American shad in the Connecticut River. Canadian Journal of Fisheries and Aquatic Sciences 42:1640-1648.

CUSHING, D. H. 1982. Climate and Fisheries. Academic Press, New York, New York, USA.

DAVIS, R. D., T. W. STORCK, AND S. J. MILLER. 1985. Daily growth increments in the otoliths of young-of-the-year gizzard shad. Transactions of the American Fisheries Society 114:304-306.

ELLIOT, J. M. 1971. Statistical analysis of samples of benthic invertebrates. Freshwater Biological Association Scientific Publication 25.

GEFFEN, A. J. 1982. Otolith ring deposition in relation to growth rate in herring (Clupea harengus) and turbot (Scopthalmus maximus) larvae. Marine Biology 71:317-326.

GRAHAM, J. J., B. J. JOULE, C. L. CROSBY, AND D. W. TOWNSEND. 1984. Characteristics of the Atlantic herring (Clupea harengus L.) spawning population along the Maine coast, inferred from larval studies. Journal Northwest Atlantic Fisheries Science 5:131-142.

HUNTER, J. R. (ed). 1976. Report on a colloquium on larval fish mortality studies and their relation to fishery research, January 1975. NOAA Technical Report NMFS Circular 395, 5 p.

LAROCHE, J. L., S. L. RICHARDSON, AND A. A. ROSENBERG. 1980. Age and growth of a pleuronectid, Parophrys vetulus, during the pelagic larval period in Oregon coastal waters. United States National Marine Fisheries Service Fishery Bulletin 80:93-104.

LASKER, R. 1975. Field criteria for survival of anchovy larvae: the relation between inshore chlorophyll maximum layers and successful first feeding. United States National Marine Fisheries Service Fishery Bulletin 73:453-462.

LEGGETT, W. C. 1976. The American shad (Alosa sapidissima), with special reference to its migration and population dynamics in the Connecticut River. Pages 169-225 in D. Merriman and L. M. Thorpe, editors. The Connecticut River ecological study: the impact of a nuclear power plant. American Fisheries Society Monograph I.

LOUGH, R. G., M. PENNINGTON, G. R. BOLTZ, AND A. A. ROSENBURG.

1982. Age and growth of larval Atlantic herring, Clupea
harengus L., in the Gulf of Maine- Georges Bank region based
on otolith growth increments. United States National Marine
Fisheries Service Fishery Bulletin 80:187-199.

MARCY, B. C. JR. 1976. Early life history studies of American
shad in the lower Connecticut River and the effects of the
Connecticut Yankee plant. Pages 141-148 in D. Merriman and
L.M. Thorpe editors. The Connecticut River ecological study:
the impact of a nuclear power plant. American Fisheries
Society Monograph I.

MARSHALL, S. L., AND S. S. PARKER. 1982. Pattern identification
in the microstructure of sockeye salmon (Oncorhynchus nerka)
otoliths. Canadian Journal of Fisheries and Aquatic Sciences
39:542- 547.

MAXFIELD, G. H. 1953. The food habits of hatchery-produced pond-
cultured shad (Alosa sapidissima) reared to a length of two
inches. Chesapeake Biological Laboratories 98, Solomons,
Maryland, USA.

METHOT, R. D. JR. 1981. Spatial covariation of daily growth
rates of larval northern anchovy, Engraulis mordax, and the
Northern Lampfish, Stenobranchius leucopsarus. Rapports et
Proces-verbaux des Reunions, Conseil International pour
l'Exploration de la Mer 178:424-431.

METHOT, R. D. JR. 1983. Seasonal variation in survival of larval
northern anchovy, Engraulis mordax, estimated from the age
distribution of juveniles. United States National Marine
Fisheries Service Fishery Bulletin 81:741-750.

MILLER, S. J., AND T. STORCK. 1984. Temporal spawning
distribution of largemouth bass and young-of-year growth,
determined from daily growth rings. Transactions of the
American Fisheries Society 113:571-578.

NEILSON, J. D., AND G. H. GEEN. 1982. Otoliths of chinook salmon
(Oncorhynchus tshawytscha): daily growth increments and
factors influencing their production. Canadian Journal of
Fisheries and Aquatic Sciences 39:1340-1347.

PANNELLA, G. 1971. Fish otoliths: daily growth layers and
periodical patterns. Science 173:1124-1127.

PANNELLA, G. 1974. Otolith growth patterns: an aid in age
determination in temperate and tropical fishes. Pages 28-39
in T. B. Bagenal, editor. Ageing of fish. Unwin Brothers,
Old Woking, Surrey, England.

PANNELLA, G. 1980. Growth patterns in fish sagittae. Pages 519-
560 In Skeletal growth of aquatic organisms. D. C. Rhoads
and R. A. Lutz, editors. Plenum Press, New York.

RADTKE, R. L., AND J. M. DEAN. 1982. Increment formation in the
otoliths of embryos, larvae, and juveniles of the mummichog,
Fundulus heteroclitus. United States National Marine
Fisheries Service Fishery Bulletin 80:201-215.

RADTKE, R. L., AND M. D. SCHERER. 1982. Daily growth of winter
flounder (Pseudopleuronectes americanus) larvae in
the Plymouth Harbor estuary. Pages 1-5 in Bryan, J. V.
Conner, and F. M. Truesdale, editors. Fifth annual larval

fish conference. Louisiana State University, Baton Rouge, Louisiana, USA.

RICE, J. A., L. B. CROWDER, AND F. P. BINKOWSKI. 1985. Evaluating otolith analysis for bloater Coregonus hoyi: Do otoliths ring true? Transactions of the American Fisheries Society 114:532-539.

RICKER, W. E. 1954. Stock and recruitment. Journal of the Fishery Research Board of Canada 11:559-586.

RICKER, W. E. 1975. Handbook of computations for biological statistics of fish populations. Fisheries Research Board of Canada Bulletin 191.

SAS (Statistical Analysis System). 1982. SAS user's guide: statistics. SAS Institute, Cary, NC, USA.

SOKAL, R. R., AND F. J. ROLF. 1969. Biometry. W. H. Freeman and Company, San Francisco, USA.

STRUHSAKER, P., AND J. H. UCHIYAMA. 1976. Age and growth of the nehu, Stolephorus purpureus (Pisces: Engraulidae), from the Hawaiian Islands as indicated by daily growth increments of sagittae. United States National Marine Fisheries Service Fishery Bulletin 74:9-17.

TANAKA, K., Y. MUGIYA, AND J. YAMADA. 1981. Effects of photoperiod and feeding on daily growth patterns in otoliths of juvenile Tilapia nilotica. United States National Marine Fisheries Service Fishery Bulletin 79:459-466.

TAUBERT, B. D., AND D. W. COBLE, 1977. Daily rings in otoliths of three species of Lepomis and Tilapia mossambica. Journal of the Fishery Research Board of Canada 34:332-340.

TOWNSEND, D. W., AND J. J. GRAHAM. 1981. Growth and age structure of larval Atlantic herring, in the Sheepscot River estuary, Maine, as determined by daily growth increments in otoliths. United States National Marine Fisheries Service Fishery Bulletin 79:123-130.

UNITED STATES GEOLOGICAL SURVEY. 1984. Water resources data, Connecticut, Water Year Reports.

VOLK, E. C., R. C. WISSMAR, C. A. SIMENSTAD, AND D. M. EGGERS. 1984. Relationship between otolith microstructure and the growth of juvenile chum salmon (Oncorhynchus keta) under different prey rations. Canadian Journal of Fisheries and Aquatic Sciences 41:126-133.

WALBURG, C. H., AND P. R. NICHOLS. 1967. Biology and management of the American shad and status of the fisheries. Atlantic coast of the United States, 1960. U.S. Fish and Wildlife Service Special Scientific Report 550.

WIGGINS, T. A., T. R. BENDER, JR., V. A. MUDRAK, AND J. A. COLL. 1984. The development, feeding, growth, and survival of cultured American shad through the transition from endogenous to exogenous nutrition. Pennsylvania Fish Commission. Benner Springs Fish Research Station, Bellefonte, Pennsylvania.

WILSON, K. H., AND P. A. LARKIN. 1980. Daily growth rings in the otoliths of juvenile sockeye salmon (Oncorhynchus nerka). Canadian Journal of Fisheries and Aquatic Sciences 37:1495-1498.

Estimation of age and growth of larval Atlantic herring as inferred from examination of daily growth increments of otoliths

SHOUKRI N. MESSIEH, DAVID S. MOORE,[1] AND PETER RUBEC

FISHERIES RESEARCH BRANCH
DEPARTMENT OF FISHERIES AND OCEANS
MONCTON, NEW BRUNSWICK E1C 9B6
CANADA

ABSTRACT

Estimates of age and growth of autumn-spawned herring Clupea harengus larvae in the southern Gulf of St. Lawrence, Cape Breton area were obtained from examination of daily growth increments in otoliths. A check was clearly seen in the center of the sagittae with a mean width of 20.3 μm. The age of larvae at the time of check formation was estimated at 15-17 days, which agrees with the transitional period from yolk sac absorption to active feeding. Outside the nuclear check, growth increments were observed at regular intervals. The number of these increments increased with the size of otolith, ranging from 13 for a larva of 14 mm to 150 for one of 36 mm. Growth estimates based on otolith increments were in the same range as those of laboratory-reared larvae, indicating a daily periodicity of otolith increments after the nuclear check formation.

The growth rate of larval herring was estimated by fitting von Bertalanffy and Gompertz growth curves. Both curves fitted the empirical data, but the latter gave higher estimates for the early part of larval growth. The asymptotic length estimates were 39.8 mm and 39.4 mm for the two curves respectively, which are comparable to the observed metamorphic lengths of herring larvae. Based on the fitted von Bertalanffy curve, the growth rates of herring larvae increased from 0.28 mm/day to 0.38 mm/day at 20 days of age and declined to 0.05 mm/day at the end of winter, shortly before metamorphosis.

[1] Present Address: Invertebrate Division, Department of Fisheries and Oceans, Halifax, NS, B3J 2S7

The Age and Growth of Fish, edited by Robert C. Summerfelt and Gordon E. Hall © 1987 The Iowa State University Press, Ames, Iowa 50010.

Techniques have been developed for age estimation of larvae
of several freshwater and marine fish from daily growth
increments deposited on otoliths (Pannella 1971; Brothers
et al. 1976; Radtke and Waiwood 1980; Campana and Neilson 1982;
Bolz and Lough 1983).   However, such studies on larval Atlantic
herring Clupea harengus on either side of the Atlantic are
limited, with only one published report on northeast Atlantic
herring in Scottish waters (Geffen 1982); in the northwest
Atlantic, Townsend and Graham (1981) studied the growth rate and
age structure of Atlantic herring larvae in the Sheepscot River
estuary, Maine, and Lough et al. (1982) conducted a similar study
in the Gulf of Maine.   The present study was undertaken to
estimate age and growth of autumn-spawned herring larvae in the
southern Gulf of St. Lawrence and to compare those estimates with
those from length-frequency distributions of field-collected and
labratory-reared larvae.

MATERIALS AND METHODS

Samples of autumn-spawned herring larvae were collected from
the southern Gulf of St. Lawrence off Cape Breton in October
1980, November 1980, and February 1981.   Freshly caught larvae
were placed in plastic Petri dishes and frozen for future otolith
analysis.   Measurements of frozen larvae were made to the nearest
0.5 mm and a shrinkage factor of 20% was applied (Radtke and
Waiwood 1980; Theilacker 1980).   A total of 75 larvae was
examined to establish the relationship between larval length and
otolith diameter.   However, otoliths from 52 larvae ranging in
total length from 14.0 mm to 36.0 mm were successfully extracted
and used for examination of otolith microstructure.

In the laboratory, larvae were thawed and the sagittae
extracted with a fine glass needle as viewed through a binocular
microscope.   The sagittae were then placed on a microscope slide
and covered with a coverslip.   The numbers of growth increments
were counted under a compound photo-microscope (1000X, oil
immersion) equipped with a video system; resolving power was in
the range of 0.2 - 0.5 µm.

Growth increments were counted independently by 2 readers,
and in case of disagreement another count was made until
agreement was reached.   Disagreement between the 2 readers was
less than 20% in the first count, and in all cases differences
did not exceed 2 growth increments.   Measurements of the diameter
of each sagitta , to the nearest micrometer (µm), were made along
the longest axis.

Growth rates of field-collected samples were compared with
length-frequency distributions of laboratory-reared larvae.
Herring eggs were artificially fertilized from ripe and running
fish; the fertilized eggs were kept in tanks at constant
temperature (10 C), which is similar to the average temperature
found when recently-hatched larvae were collected in field.   In
the laboratory, the photoperiod was set at 16 hours of light and
8 hours of darkness (16:8 LD); the care and feeding of larvae
were by techniques described elsewhere (Messieh et al. 1981).

A total of 254 of laboratory-reared, known-aged larvae (hatching until 55 days old) were measured and their length-frequency distribution used for growth rate estimates. The otoliths of these larvae, however, were not suitable for increment study because they had been preserved in formalin. Measurements were made for fresh larvae before preservation; hence no adjustment for shrinkage was made.

Growth of larvae were expressed by the Gompertz and von Bertalanffy growth models, respectively, as follows:

$$L_t = L_\infty \ (b)^{c^t}$$

$$\text{and} \quad L_t = L_\infty \ (1-e^{-K(t-t_0)})$$

RESULTS

Otolith Morphology and Growth

Sagittae from the early developmental stage of herring larvae are spherical in shape (Fig. 1). As the larvae develop (about 20 mm), the otoliths grow faster along the antero-posterior axis and assume an oval shape. At metamorphosis (35 - 40 mm), the rostrum develops as a protrusion at the anterior margin of the otolith. The development and morphology of larval otoliths from Bay of Fundy larval herring were described by Messieh (1975).

A nuclear check around the center of the sagitta was observed in all specimens examined (Fig. 1). The nucleus had a mean diameter of 20.3 ± 1.3 μm which did not change significantly with the age of the larvae. Within the nucleus, growth increments were indistinct; 1 or 2 and sometimes up to 5 faint increments were observed within the nucleus (Fig. 1). Lough et. al. (1982), using a similar magnification, observed only 1 increment within the otolith nucleus of field collected larval herring from the Gulf of Maine, and no increments in otoliths of the laboratory-reared larvae.

Outside the nucleus, growth increments of regularly spaced dark and light bands were observed. The number of these increments increased with the size of otoliths, ranging from 13 increments for a larva of 14.0 mm T.L. to 150 increments for one of 36.2 mm. These increments are presumably daily growth increments (Pannella 1974, Townsend and Graham 1981, Lough et al. 1982).

Relationship Between Larval Length and Otolith Diameter

Otolith diameters plotted against length of larval herring showed a linear relationship. Least squares parameters for the equation were as follows:

$$y = -54.6 + 6.9x \qquad r = 0.98 \ \ldots\ldots\ldots (1)$$

where y = otolith diameter in μm
       x = larval length in mm

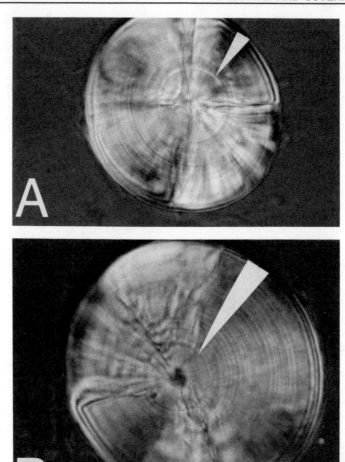

Figure 1. Sagittae of herring larvae enlarged 850x:
A = larva 16 mm total length, otolith diameter
60.5 μm with 21 growth increments; B = larva 18mm total
length, otolith diameter 82.8 μm with 29 increments; the
nuclear check is at the tip of the pointer.

Substituting in equation (1) and assuming linearity right
through the origin, an otolith diameter of 20.3 μm corresponds to
a larval length of 10.9 mm.

Larval Growth
A plot of larval length against age, as determined from the
number of growth increments of field-collected larvae, showed a

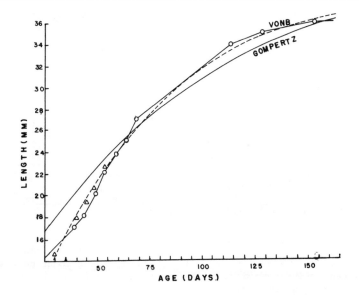

Figure 2. Growth of Gulf of St. Lawrence fall-spawned
herring larvae up to metamorphosis, fitted to von
Bertalanffy (VONB) and Gompertz (GOMP) growth curves.
○Mean length-at-age (days from hatch) from
field-collected larvae; △ Mean length-at-age (days from
hatch) for laboratory-reared larvae.

curvilinear relationship (Fig. 2). In the early part of growth
(about 50 growth increments), the increase in size by age was
fastest. The rate of acceleration decreased gradually from day
50 to 125, until it reached an asymptotic length at about day
150. A von Bertalanffy growth curve and a Gompertz growth curve
were fitted to the data. Mean length at age and the fitted
growth curve (Figure 2) showed that both models fit the empirical
data; the maximum estimated lengths were 39.8mm and 39.4 mm
respectively. However, the von Bertalanffy curve provided a
better fit of the early part (<25mm) of the observed larval
lengths. Growth parameters estimated from von Bertalanffy curve
fitting are the following:

L_∞ = 39.8 mm; K = 0.016; t_0 = -12.48

Growth parameters estimated from Gompertz curve fitting were:

L_∞ =39.4mm; b = 0.50; c = 0.92

Based on these growth parameters, the predicted length-at-
age and growth rates of larval herring of the size range examined
were calculated (Table 1).Assuming a daily frequency of increment
formation, larvae at 150 d age-group (mean number of daily growth
increments excluding the nucleus) were estimated to be 36.0 mm in
total length. Growth rates were estimated from the von
Bertalanffy equation at 0.38 mm/d for age group 20, and decreased
steadily with age until they reached 0.05 mm/d for age-group

Table 1. Mean length-at-age[a] of field-collected herring larvae. Predicted values were estimated by fitting the von Bertalanffy and Gompertz growth curves to observed data.

NUMBER OF INCREMENTS ON OTOLITHS	OBSERVED	LENGTH OF LARVAE (mm) PREDICTED	
		Bertalanffy	Gompertz
15	15	14.1	
20	16	16.1	19.7
25	17	17.9	20.8
30	18	19.5	21.9
35	20	21.1	23.0
40	22	22.5	24.0
45	24	23.9	25.1
50	25	25.1	26.0
75	30	29.3	36.0
100	34	33.2	33.0
125	35	35.5	35.1
150	36	36.8	36.0

[a]Number of growth increments excluding the nucleus.

150. The average growth rate for all ages examined was 0.17 mm/d. Based on Gompertz growth equation, the growth rate was estimated at 0.39 mm/d for age group 35, and decreased to 0.14 mm/d for age group 150.

Comparison with Laboratory-Reared Larvae

Length-frequency distributions of laboratory-reared herring larvae showed a size range of 5.5 mm-22.5 mm, from hatching to day 55. These larvae were reared from artificially fertilized eggs in tanks at 10C (Messieh et al. 1981). Hatching occurred after 10.1 d of incubation with larvae ranging from 5.5-6.0 mm in length; absorption of the yolk sac occurred at 4-7 days after hatching.

In the first 20 days, the growth rate of larvae was slow, ranging from 0.24 mm/d in the 1st week to 0.33 mm/d in the 3rd week. At the age of 20-30 d, the growth rate dropped from 0.33 mm/d to 0.27 mm/d; after d 30, the growth rate steadily declined to 0.20 mm/d at day 55. These growth rate estimates reasonably agree with those based on the otolith increment counts for field-collected larvae (Table 2).

DISCUSSION

Otolith microstructure examination has found increasing application in recent years. In a comprehensive review of the microstructure of fish otoliths, Campana and Neilson (1985) discussed present applications and their assumptions. They presented a hypothesis of increment formation which implies the presence of an endogenous, circadian rhythm entrained by photoperiod, but susceptible to modification by other cyclic environmental variables.

Otoliths of herring larvae in the present study appeared to have daily growth increments deposited at regular intervals as dark and light bands which could be used to estimate their age.

Table 2. Comparison of growth rates (mm/d) of field-collected and laboratory-reared herring larvae in the Gulf of St. Lawrence and Gulf of Maine.

AGE (days)	Gulf of St. Lawrence			Gulf of Maine LOUGH et al. (1982)
	LAB-REARED	FIELD-COLLECTED		
		Bertalanffy	Gompertz	
5	0.24			0.27
10	0.22			0.29
15	0.25	0.28		
20	0.33	0.38		0.30
25		0.35		
30	0.27	0.33		0.29
35		0.30	0.39	
40	0.25	0.27	0.23	0.27
45	0.23	0.27	0.22	
50	0.23	0.23	0.22	0.23
55	0.20	0.23	0.20	
60		0.20	0.20	0.19
85		0.16	0.18	0.12
110		0.15	0.18	0.07
135		0.09	0.16	0.04
160		0.05	0.14	0.02

In the center of the otolith, a clearly defined nucleus embracing vague increments was shown. Between 1 and 5 increments inside the nucleus could be seen in some specimens. The appearance of these increments was different from regular daily growth increments, and the spacing between them was irregular, which suggests that they might not be deposited at daily intervals.

From the regression of otolith diameter size on larval length, the nucleus diameter (20.3 μm) was estimated to correspond to a larval length of 10.9 mm, assuming linearity right through the origin. Based on the growth of laboratory-reared larvae, the age of larvae at the completion of nucleus formation was estimated as 15-17 days, which coincided with the time of yolk sac absorption.

The estimated age at the time of nuclear check is not much different from the estimates of Lough et al. (1982) of 17.79 days based on the Gompertz growth model for Jeffreys Ledge herring. From laboratory-reared larvae, the same authors found that the 1st increment was laid 4.5 days after hatching; the 2nd increment was delayed 7.5 days from the initial increment, and the 3rd increment was observed for the first time on a larva 16 days after hatch. While Lough et al. (1982) did not describe the otolith nucleus 'core', nor did they calculate larval length at time of nucleus completion, their description of the area of interrupted slow growth for the first 3 increments appears to correspond to the nucleus observed in this study.

The growth estimates in this study are comparable with estimates derived from analysis of length-frequency distributions of larval herring reported earlier. Messieh and Kohler (1972) reported a growth rate of 0.27 mm/d in the early part of larval growth, decreasing to 0.09 mm/d near winter for the Gulf of St. Lawrence herring larvae; Das (1968) estimated growth rates of Bay of Fundy herring larvae as 0.29 mm/d in early September, 0.21 mm/d in October, and 0.14 mm/d over the winter; Lambert (1984) estimated a growth rate of 0.2 mm/d in September-November, decreasing to 0.05 mm/d in December-March for larval herring in St. Mary's Bay, southwest of Nova Scotia; for the Gulf of Maine, larval herring growth was estimated at 0.13 mm/d (Graham and Davis 1971) and 0.17 mm/d (Boyar et al. 1973); for St. Margaret's Bay, Nova Scotia, larval herring growth was estimated at 0.13 mm/d from October through June (Ware and Henricksen 1978).

The agreement between the growth-rate estimates derived from otolith increment counts and those from length-frequency distribution analyses for both laboratory-reared and field-collected larvae, provides supporting evidence for validation of daily growth increments in larval herring otoliths (Lough et al. 1982), and for other fish species (Campana and Neilson 1982; Bolz and Lough 1983; Campana 1984).

Our estimates are in good agreement with Lough et al. (1982) estimates except that the maximum theoretical length was lower (30.9 mm) than that reported here (39.4-39.8 mm). Based on the Gompertz growth model (Lough et al. 1982), the average growth rate of larval herring in the Gulf of Maine was 0.20 mm/d for the first 150 days; growth rates increased from 0.25 mm/d shortly after hatching to 0.30 mm/d at 20 days, and decreased thereafter to 0.03 mm/d at 150 days. Approaching the asymptote, the growth rate was 0.01 mm/d.

In the Gulf of St. Lawrence, the values for maximum asymptotic length of 39.4-39.8 mm agree well with the metamorphic length (35-40 mm) from field observations (Messieh, unpublished data) and with those from theoretical studies (Messieh 1969; Ware and Henriksen 1978). It appears that the estimate of 30.9 mm of Lough et al. (1982) is an under-estimate since, in the Gulf of Maine, Graham et al. (1972) and Boyar et al. (1973) found herring larvae of 40.0-40.5 mm length before metamorphosis. Similarly, Saila and Lough (1981) estimated the maximum size of herring larvae at 42.5 mm in the same area.

We conclude that otolith growth increments are suitable for estimating age and growth of larval herring in the Gulf of St. Lawrence during the fall-winter period. For practical purposes, this is the period when herring larvae are usually caught in the ichthyoplankton research surveys and for which estimates of growth and mortality rates are required. The Gompertz and von Bertalanffy growth models adequately described the larval growth up to metamorphosis, but the latter gave a better fit, particularly for the early part of larval growth.

ACKNOWLEDGMENTS
    We thank Dr. Stephen Campana and Dr. John Neilson for their review of an earlier version of this manuscript.

REFERENCES

BOLZ, G.R., and R.G. LOUGH. 1983. Growth of larval Atlantic cod
Gadus morhua and haddock Melanogrammus aeglefinus on
Georges Bank, Spring 1981. United States National Marine
Fisheries Service Fishery Bulletin 81(4): 827-836.

BOYAR, H.C., R.R. MARAK, F.E. PERKINS, and R.A. CLIFFORD. 1973.
Seasonal distribution and growth of larval herring Clupea
harengus in the Georges Bank-Gulf of Maine area from 1962 to
1970. Journal du Conseil, Conseil International pour
l'Exploration de la Mer, 35: 36-51.

BROTHERS, E.B., C.P. MATHEWS, and R. LASKER. 1976. Daily growth
increments in otoliths from larval and adult fishes. United
States National Fisheries Service Fishery Bulletin, 74: 1-8.

CAMPANA, S.E. 1984. Microstructural growth patterns in the
otoliths of larval and juvenile starry flounder, Platichthys
stellatus. Canadian Journal of Zoology, 62: 1507 - 1512.

CAMPANA, S.E., and J.D. NEILSON. 1982. Daily growth increments
in otoliths of starry flounder Platichthys stellatus and the
influence of some environmental variables in their
production. Canadian Journal of Fisheries and Aquatic
Sciences, 39: 937-942.

CAMPANA, S.E., and J.D. NEILSON. 1985. Microstructure of fish
otoliths. Canadian Journal of Fisheries Aquatic Sciences,
42:1014-1032.

DAS, N. 1968. Spawning, distribution, survival, and growth of
larval herring Clupea harengus in relation to hydrographic
conditions in the Bay of Fundy. Fisheries Research Board of
Canada Technical Report, 88: 1-129

GEFFEN, A.J. 1982. Otolith ring deposition in relation to growth
rate in herring Clupea harengus and turbot Scophthalmus
maximus larvae. Marine Biology, 71: 317-326.

GRAHAM, J.J., S.B. CHENOWETH, and C.W. DAVIS. 1972. Abundance,
distribution, movement and lengths of larval herring along the
western Coast of the Gulf of Maine. United States National
Marine Fisheries Service Fishery Bulletin, 70: 307-321.

LAMBERT, T.C. 1984. Larval cohort succession in herring Clupea
harengus and capelin Mallotus villogus Canadian Journal of
Fisheries and Aquatic Sciences, 41: 1552-1564

LOUGH, R.G., M. PENNINGTON, G.R. BOLZ, and A. ROSENBERG. 1982.
Age and growth of Georges Bank larval herring Clupea harengus
in the Gulf of Maine area based on otolith growth increments.
United States National Fisheries Services Fishery Bulletin,
80(2): 187-199.

MESSIEH, S.N. 1969. Similarity of otolith nuclei in spring and
autumn-spawning Atlantic herring in the southern Gulf of St.
Lawrence. Journal of the Fisheries Research Board of Canada,
26: 1889-1898.

MESSIEH, S.N. 1975. Growth of the otoliths of young herring in
the Bay of Fundy. Transaction of the American Fisheries
Society, 104(4): 770-772.

MESSIEH, S.N., and A.C. KOHLER. 1972. Distribution, relative abundance and growth of larval herring Clupea harengus in the southern Gulf of St. Lawrence. Fisheries Research Board of Canada Technical Report, 318: 1-31.

MESSIEH, S.N., D. WILDISH, and R.H. PETERSON. 1981. Possible impact of dredge disposal on the Miramichi Bay herring fishery. Canadian Technical Report of Fisheries and Aquatic Sciences, 1008: 1-33.

PANNELLA, G. 1971. Fish otoliths: daily growth layers and periodical patterns. Science 173: 1124-1127.

RADTKE, R.L., and K.G. WAIWOOD. 1980. Otolith formation and body shrinkage due to fixation in larval cod Gadus morhua. Canadian Technical Report of Fisheries and Aquatic Sciences, 929: 1-10.

SAILA, S.B., and R.G. LOUGH. 1981. Mortality and growth estimation from size data - an application to some Atlantic herring larvae. Rapports et Procès-Verbaux dex Réunions Conseil Permanent International pour l'Exploration de la Mer, 178: 7-14.

SAMEOTO, D.D. 1971. The distribution of herring Clupea harengus larvae along the southern coast of Nova Scotia with some observations on the ecology of herring larvae and the biomass of macrozooplankton on the Scotian Shelf. Fisheries Research Board of Canada Technical Report, 252: 1-72.

THEILACKER, G. H. 1980. Changes in body measurements of larval northern anchovy Engraulis Mordax and other fishes due to handling and preservation. United States National Marine Fisheries Service Fishery Bulletin, 78: 685-692.

TOWNSEND, D.W., and J.J. GRAHAM. 1981. Growth and age structure of larval Atlantic herring, Clupea harengus in the Sheepscot River estuary, Maine, as determined by daily growth increments in otoliths. United States National Marine Fisheries Service Fishery Bulletin, 79: 123-130.

WARE, D.M., and B.L. HENRICKSEN. 1978. On the dynamics and structure of the southern Gulf of St. Lawrence herring stocks. Fisheries Marine Service Technical Report, 800: 1-83.

Occurrence of daily growth increments in otoliths of juvenile Atlantic menhaden

LAWRENCE F. SIMONEAUX AND
STANLEY M. WARLEN
NATIONAL MARINE FISHERIES SERVICE
NATIONAL OCEANIC AND ATMOSPHERIC ADMINISTRATION
SOUTHEAST FISHERIES CENTER, BEAUFORT LABORATORY
BEAUFORT, NC 28516

ABSTRACT
    Laboratory and field-held juvenile Atlantic menhaden
Brevoortia tyrannus were injected with oxytetracycline (OTC) to
determine the rate of increment formation on their otoliths.
Concentrations of 0.013 to 0.020 mg OTC/g wet weight of fish
produced fluorescent marks on the otoliths of 78% of all fish and
caused no mortalities. The mean counts between the OTC mark and
the otolith margin in fish from one laboratory and two field
experiments closely approximated the number of elapsed days, and
the daily increment formation rate in each experiment approached
1.0. Regression analysis of the pooled data set indicated that
average increment formation rate was not significantly different
from one increment per day, although individual counts were quite
variable and, in some instances, quite different from the expected
1.0 per day (range 0.64 to 1.46).

    Accurate determination of age in larval and juvenile fishes is
an essential step in determining growth and mortality rates,
investigating population dynamics, and establishing periods when
various morphological and biological changes occur. Since
Pannella (1971, 1974) first suggested that fishes can form daily
growth increments on their otoliths many investigators have used
daily increments to age young-of-the-year fishes (Brothers et al.
1976; Taubert and Coble 1977; Wilson and Larkin 1980; Warlen
1982). Because of differences among species, this aging technique
must be validated for each species studied (Beamish and McFarlane
1983).
    There are several methods of validation. One is to compare
the number of increments deposited on the otoliths to the known
age of laboratory-reared fish (Brothers et al. 1976). A second
method is to chemically mark the otoliths, hold the fish for a

The Age and Growth of Fish, edited by Robert C. Summerfelt and Gordon E. Hall © 1987 The Iowa
State University Press, Ames, Iowa 50010.

specified time, and then compare the number of increments
deposited between the mark and the otolith edge to the elapsed
time in days (Wild and Foreman 1981).

Tetracycline in several forms has been used to chemically mark
scales, vertebrae, and otoliths of various species of fishes. In
such studies, a mark was induced either by injection of
tetracycline (Jensen and Cumming 1967; Wild and Foreman 1980;
Campana 1983), by addition of tetracycline to the diet (Choate
1964; Weber and Ridgway 1967), or by total immersion of the fish
in water containing the drug (Campana and Neilson 1982; Hettler
1984).

The purpose of this study was to validate the frequency of
increment formation in otoliths of juvenile Atlantic menhaden
Brevoortia tyrannus with the use of oxytetracycline (OTC) and the
procedures of Wild and Foreman (1980).

MATERIALS AND METHODS

Juvenile Atlantic menhaden were collected with a 1.5-m cast
net from estuarine creeks located near Beaufort, North Carolina.
A laboratory experiment (20 July - 1 August 1983) was designed to
determine the OTC dosage required to produce a distinguishable
mark on the otoliths of menhaden, to establish a method for
handling and injecting the fish that would minimize mortality, and
to determine the frequency of increment formation on their
otoliths. Subsequent complementary field experiments were
conducted to determine if the frequency of increment formation was
daily in fish confined in a natural area.

Laboratory Experiment

Juvenile menhaden, 67-92 mm fork length (FL), were held in the
laboratory for 24 h, and then randomly divided into four treatment
groups of 10 fish each. All fish were injected intramuscularly
with 0.1 ml OTC (Terramycin, Pfizer Inc.) diluted with 1% aqueous
sodium chloride. Dosages for the groups were: 25, 50, 100, and
200 µg OTC/0.1 ml of solution administered without regard to the
weight of individual fish. The highest dosage resulted in
concentrations of 13 µg/g wet weight for the largest fish (16g) to
20 µg/g wet weight for the smallest fish (10g). Members of each
treatment group were fin-clipped to aid in identification at
recovery. The marked fish were held in a 950 l circular holding
tank that received ambient running seawater from the Newport River
estuary adjacent to the laboratory. Mean water temperature over
the holding period was 28.4°C. Fish were fed ad-libitum each day
using a commercial size-00 salmon starter feed. After 11 days,
all fish were removed and preserved in 95% ethanol. The otoliths
were removed and processed for determination of mark deposition
and increment reading.

Field Experiments

After determining the correct dosage required to provide a
distinguishable mark, we conducted two field experiments. In the

first (26 August - 9 September 1983), we injected 360 fin-clipped
fish with 0.1 ml of 200 µg OTC/0.1 ml of solution and released
them into a holding area adjacent to the capture site. This area
was at the head of a small estuarine creek which was blocked with
a 3.1-m deep, 6.4-mm mesh net stretched across the width and
anchored on each bank. The enclosure had an average depth of 1.5
m and a surface area of approximately 650 m^2. After 14 days, the
holding area was repeatedly seined and a total of 48 fin-clipped
juveniles (x=73 mm, range = 63-88 mm FL) were recovered. These
fish were an average of 6 mm larger than a sample of the fish at
the time of marking. The size difference in attributed to growth
during the experiment. This experiment terminated on day 18 when
the block net was stolen.

The design of the second field experiment (August 1984)
replicated that of the 1983 experiment. On 7 August 1984, we
captured 182 juvenile Atlantic menhaden in the headwaters of the
same creek used in 1983. The fish were fin-clipped, injected with
OTC as before and released into the same kind of enclosure as
previously described. The holding area was seined after periods
of 7, 13, and 20 days. The first seining netted 10 (72-98 mm FL)
fin-clipped juveniles, while the second produced 12 (70-95 mm FL).
No fin-clipped juveniles were collected on day 20 and the
experiment was terminated. Water temperature in both experiments
was between 28 and 32°C. Recaptured fish in both experiments were
preserved immediately in 95% ethanol and brought to the
laboratory, where otoliths were removed and prepared for
analysis.

Otolith Preparation and Reading

Otoliths (sagittae) were removed, cleaned in distilled water,
mounted on a clear glass microscope slide, and embedded - proximal
side down - in clear acrylic adhesive. No grinding or sectioning
was performed.

The OTC mark on the otoliths was detected by use of incident
ultraviolet light from a 200-W mercury lamp attached to a compound
microscope and the position of the fluorescent mark was noted with
an ocular micrometer. Then the otoliths were viewed at either 400
or 630X magnification under transmitted white light and counts of
concentric increments were made from the edge of the OTC mark to
the edge of the otolith. Otolith increments were as defined by
Tanaka et al. (1981) and each was a light, thick incremental band
and a dark, thin discontinuous band.

Increment counts were made by the same person at least twice
on all otoliths. Mounted otoliths from all of the experiments
were pooled, labelled with numbers chosen at random, and then
presented to the reader. After all the otoliths were read once,
the process was repeated after re-labelling the slides with a new
sequence of randomly chosen numbers. The results of both readings
or counts were compared and a mean count calculated. Individual
fish were eliminated from the data set when mean increment counts
differed from each other by more than 20%. Only two fish (1983

field experiment) were thus eliminated. The incremental counts from the remaining otoliths were then compared to the known elapsed time.

RESULTS

All 40 of the laboratory fish survived injection and handling and 32 (80%) were found to have a distinguishable OTC mark (Fig. 1) on their otoliths that fluoresced under UV light. However, a higher percentage of the otoliths receiving the two highest doses

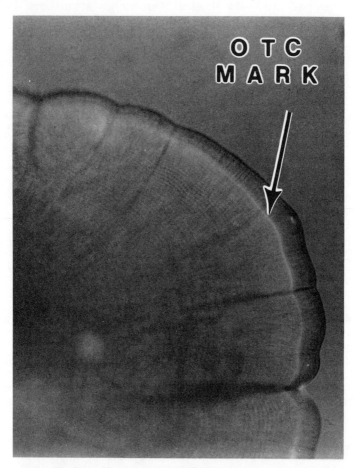

Figure 1. Light microscope photograph of juvenile Atlantic menhaden otolith (X 400) with transmitted white light and UV light. The fluorescent OTC mark is most distinct on the posterior edge of the otolith. Growth increments appear as pairs of incremental (light) and discontinuous (dark) bands.

Table 1. Mean fork length and number of laboratory-held
juvenile Atlantic menhaden marked with oxytetracycline
at four concentrations per 0.1 ml solution.

Treatment group (µg OTC/fish)	Number of fish	Mean fork length (mm)	Number with OTC mark
25	10	78.1	6
50	10	78.1	8
100	10	85.1	10
200	10	80.0	9

(100 and 200 µg OTC/fish) were marked (Table 1) and those marks
were more distinct. Seventy of the 540 field-injected fish were
recaptured and, of these, 55 (78.6%) had a distinguishable OTC
mark on their otoliths. We have no explanation for the absence of
a detectable OTC mark in some of the injected fish in either
laboratory or field experiments, except that we did observe
freshly injected OTC solution leaking out of the needle wound in
some fish.

There was consistency between the pairs of otolith readings
(Table 2); 80% of the first and second readings differed by $\leq$ one
increment and 93% by < two rings. Only in the 1983 field
experiment did some differ by three increments.

Table 2. Differences in two readings of the number of otolith
increments from the OTC mark to the otolith margin in juvenile Atlantic menhaden.

Experiment	Number of fish	Elapsed days	Difference between counts			
			0	1	2	3
Laboratory	32	11	15	14	3	0
Field - 1983	36	14	8	14	8	6
Field - 1984	10	7	4	6	0	0
Field - 1984	8	13	3	5	0	0
TOTAL	86		30	39	11	6

Laboratory Experiment

In the laboratory experiment, mean individual fish increment
counts varied between 9.5 and 11.5, and most otoliths (72%) had an
increment count within one standard deviation of the mean. These
fish held for 11 days had an overall mean count of 10.63
increments (s.d. = 0.48, n = 32) and a mean daily deposition rate
of 0.966 increments per day (s.d. = 0.04, n = 32). Although these
results are very close to the expected values and suggest that one
increment is formed each day they were significantly different
(t-test, P < 0.05) from both the expected number of days (11) and
the expected rate (1.0 per day).

Field Experiments

The incremental mean counts for the otoliths from 36 fish from
the 1983 field experiment was 13.43 (s.d. = 1.77) for 14 elapsed
days and the daily increment formation rate of 0.959 (s.d. =

0.1269) was slightly less than 1.0. The counts ranged from 9 to 20.5 increments; however, 92% were within one standard deviation of the mean. These results were not significantly different from either the expected number of increments (14) or the expected daily rate of 1.0 (t-test, P > 0.05).

Two samples were obtained in the 1984 field experiment. Ten fish sampled 7d after OTC marking had a mean increment value of 6.35 (s.d. = 0.34) and a daily increment formation rate of 0.907 (s.d. = 0.048). These estimates were significantly different (t-test, P < 0.05) from both the expected number of increments (7) and a daily increment formation rate. All the increment mean counts were between 6 and 7. If there was error associated with reading increments along the otolith margin, or if the OTC mark was deposited over more than one day, the number of increments could have been consistently under-estimated. If there was such a systematic bias, it would be relatively less in fish kept longer in the same experiment. Eight juveniles kept for 13 d in the field enclosure in the 1984 field experiment, did not have a mean increment count (x = 12.88, s.d. = 0.52) or daily increment formation rate (x = 0.990, s.d. = 0.04) significantly different from the expected values of 13 elapsed days and an increment formation rate of 1.0 (t-test, P > 0.05). Counts for any fish did not differ from the elapsed number of days by more than one increment. A sample of fish taken at an even longer elapsed time also would have been useful; however, no additional marked fish were collected beyond 13 d.

Regression analysis of the pooled data indicates that increments form at the rate of approximately one per day (Fig. 2). The regression coefficient is not significantly different (t-test, P > 0.05) from 1.0.

DISCUSSION AND CONCLUSIONS

A concentration of $\geq$ 0.01 mg OTC/g wet weight is sufficient to produce a detectable mark on the otoliths of most juvenile Atlantic menhaden. Z. C. Villavicencios-Rios, a visiting investigator from Lima, Peru at the Beaufort Laboratory in 1976, recommended a concentration of 0.05 mg OTC/g wet weight to mark scales, vertebrae, and otoliths of similar size juvenile Atlantic menhaden. Ralston and Miyamoto (1983) used a similar dosage (0.03 mg OTC/g wet weight) to mark the otoliths of Hawaiian snapper Pristipomoides filamentosus.

The average number of increments in each of the four data sets was slightly below the number of days after marking. This may have been due to the apparent delay before first increment deposition because of handling and injection, or to a traumatic reaction to the drug. Closely spaced increments may not have been resolved, since the resolution of the light microscope is about 0.2 µm (Campana and Neilson, 1985). Also, if we did not adequately detect increments along the periphery of the otolith, the number of growth increments could have been underestimated by

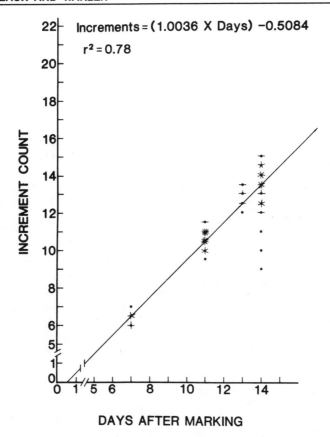

Figure 2.   Relationship between increment count and the
    number of days from OTC marking of the otolith to
    sacrifice of fish. Overlapping points are described by the
    "Sunflower" symbols of Chambers et al. (1983). (Key to
    figures: • = 1 observation; ⊷ = 2 observations;⊤ =
    3 observations, etc.).

one increment. The only instance of overestimating daily
increment formation rates (Fig. 2) is probably attributable to
erroneous counting of sub-daily increments.

In three of four cases the daily formation rates were greater
than 0.95 increments per day, which is close to a formation rate
for all data of 1.0 per day. This suggests that the otolith
growth increments on otoliths of juvenile Atlantic menhaden can be
used to estimate their age.

ACKNOWLEDGMENTS
   The authors appreciate review of the manuscript by William
Hettler, Alex Chester and David Peters. Chester also provided
statistical advice, and Peters suggested the problem and
approach. Reference to trade names in this article does not imply
endorsement by the National Marine Fisheries Service, NOAA.

REFERENCES
BEAMISH, R.J., and G.A. MCFARLANE. 1983. The forgotten require-
    ment for age validation in fisheries biology. Transactions of
    the American Fisheries Society 112:735-743.
BROTHERS, E.B., C.P. MATHEWS, and R. LASKER. 1976. Daily growth
    increments in otoliths from larval and adult fishes. Fishery
    Bulletin, U.S. 74:1-8.
CAMPANA, S.E. 1983. Feeding periodicity and the production of
    daily growth increments in otoliths of steelhead trout (Salmo
    gairdneri) and starry flounder (Platichthys stellatus).
    Canadian Journal of Zoology 61:1591-1597.
CAMPANA, S.E., and J.D. NEILSON. 1982. Daily growth increments in
    otoliths of starry flounder (Platichthys stellatus) and the
    influence of some environmental variables in their production.
    Canadian Journal of Fisheries and Aquatic Sciences 39:937-942.
CAMPANA, S.E., and J.D. NEILSON. 1985. Microstructure of fish
    otoliths. Canadian Journal of Fisheries and Aquatic Sciences
    42:1014-1032.
CHAMBERS, J.M., W.S. CLEVELAND, B. KLEINER, and P.A. TUKEY. 1983.
    Graphical methods for data analysis. Duxbury Press, Boston,
    Massachusetts, USA.
CHOATE, J. 1964. Use of tetracycline drugs to mark advanced fry
    and fingerling brook trout (Salvelinus fontinalis). Trans-
    actions of the American Fisheries Society 93:309-311.
HETTLER, W.F. 1984. Marking otoliths by immersion of marine fish
    larvae in tetracycline. Transactions of the American Fisheries
    Society 113:370-373.
JENSEN, A.C., and K.B. CUMMING. 1967. Use of lead compounds and
    tetracycline to mark scales and otoliths of marine fishes.
    Progressive Fish-Culturist 29:166-167.
PANNELLA, G. 1971. Fish otoliths: daily growth layers and
    periodical patterns. Science 173:1124-1127.
PANNELLA, G. 1974. Otolith growth patterns: an aid in age deter-
    mination in temperate and tropical fishes. Pages 28-39 in T.B.
    Bagenal, editor. The ageing of fish. Unwin Brothers, Surrey,
    England.
RALSTON, S., and G.T. MIYAMOTO. 1983. Analyzing the width of
    daily otolith increments to age the Hawaiian snapper
    Pristipomoides filamentosus. Fishery Bulletin, U.S. 81:
    523-535.
TANAKA, R., Y. MUGIYA, and J. YAMADA. 1981. Effects of photoperiod
    and feeding on daily growth patterns in otoliths of juvenile
    Tilapia nilotica. Fishery Bulletin, U.S. 79:459-466.

TAUBERT, B.D., and D.W. COBLE. 1977. Daily rings in otoliths of three species of Lepomis and Tilapia mossambica. Journal of the Fisheries Research Board of Canada 34:332-340.

WARLEN, S.M. 1982. Age and growth of larvae and spawning times of Atlantic croaker in North Carolina. Proceedings of the Annual Conference of the Southeastern Association of Fish and Wildlife Agencies 34:204-214.

WEBER, D., and G.J. RIDGWAY. 1967. Marking Pacific salmon with tetracycline antibiotics. Journal of the Fisheries Research Board of Canada 24:849-865.

WILD, A., and T.J. FOREMAN. 1980. The relationship between otolith increments and time for yellowfin and skipjack tuna marked with tetracycline. Inter-American Tropical Tuna Commission Bulletin 17:509-560.

WILSON, K.H., and P.A. LARKIN. 1980. Daily growth rings in the otoliths of juvenile sockeye salmon (Oncorhynchus nerka). Canadian Journal of Fisheries and Aquatic Sciences 37:1495-1498.

Size and growth of juvenile chinook salmon back-calculated from otolith growth increments

MICHAEL J. BRADFORD AND
GLEN H. GEEN
DEPARTMENT OF BIOLOGICAL SCIENCES
SIMON FRASER UNIVERSITY
BURNABY, BRITISH COLUMBIA V5A 1S6
CANADA

ABSTRACT
     The reliability of otolith daily growth increments in esti-
mating growth rates and size-at-age of juvenile chinook salmon
Oncorhynchus tshawytscha was determined using laboratory-reared
fish. Growth rates and lengths backcalculated from increment
width data were compared to actual values. Back-calculated mean
lengths were not significantly different from observed mean
lengths in 8 of 9 comparisons made at intervals of 10-20 days
over a 120-day experiment. No correlation was found between
observed growth rates over 7-15 day intervals and those calculated from otolith increment widths of individually marked fish; the
relationship was significant over a 51-day interval. There was
less variability in back-calculated compared to observed growth
rates. While approximate estimates of growth and back-calculated
size can be derived from increment widths, the conservative
nature of otolith growth compared to fish growth may preclude
very detailed analyses.

INTRODUCTION
     The recent discovery that otolith growth increments of many
teleosts are deposited on a daily basis has provided a method of
determining the age and growth of larval and juvenile fish in
more detail than was previously possible (Pannella 1980; Campana
and Neilson 1985). Increment counts can be used to age fish, and
increment widths have the potential to provide estimates of daily
growth. While the increment deposition rate has been confirmed
as daily in some species, the reliability of estimates of short-
term growth rates or back calculations of size-at-age based on
otolith microstructure data have rarely been tested. If

increment widths are to provide information about growth rates
during the early life of fish, a close coupling of fish growth
and otolith growth will be required. To date, increment widths
have only been related to average growth rates for groups of fish
over longer-term experiments (>40 days, Volk et al. 1984; Neilson
and Geen 1985). Few studies have assessed the accuracy of size
backcalculations, although Wilson and Larkin (1982) reported a
15% error in the back-calculated weights of individual juvenile
sockeye salmon Oncorhynchus nerka.

Here we report experiments which enabled us to examine the
accuracy and precision of lengths backcalculated from otolith
increment measurements of juvenile chinook salmon O. tschawytscha. In addition, we compared the observed growth rate of
fish to estimates derived from otolith increment widths to
determine if these microstructure features accurately reflected
growth over shorter time intervals.

METHODS AND MATERIALS

Chinook salmon fry (mean fork length 60 mm) from the
1982 run to Campbell River, British Columbia, were transported
from the Quinsam Hatchery to the Canada Department of Fisheries
and Oceans West Vancouver Laboratory on March 25, 1983. They
were placed in 400-liter tanks supplied throughout the
experiments with 10-11 C aerated well water. A constant ration
of 8% of dry body weight of Oregon Moist Pellets was provided
once per day. After 5 days fry were divided into two groups:

Treatment 1: 180 fish kept for 120 days. Approximately 40 fish
             were sacrificed on days 30, 60, 90, and 120.
Treatment 2: 40 fish held for 67 days, and sacrificed at the end
             of the experiment. Fish in this group were
             individually marked on day 1 with wire brands cooled
             in liquid nitrogen.

On day 1, and at approximately 14-day intervals thereafter,
all fish from both groups were anaesthetized with 100 mg/liter
MS 222 and their fork lengths measured.

Left sagittal otoliths were removed from the sacrificed fish
and prepared as described by Neilson and Geen (1981). Otolith
radius and increment widths were measured along a standard
radius, 90° from the long axis of the otolith on the pararostral
side. Otolith radius was measured from the center of the nuclear
area to the outer edge of the otolith, using a compound microscope equipped with an ocular micrometer. Photomicrographs were
taken along the standard radius. The resulting negatives were
projected on a screen and the location of the increments were
marked on a paper strip. Increment widths were then measured
from the edge of the otolith inward on a digitizing tablet as
described by Neilson and Geen (1982).

We did not test the rate of increment formation in these
experiments because we did not have a time marker on the

otolith. We assumed increments were formed daily, based on the
experiments of Neilson and Geen (1982, 1985), who found an
average of one increment produced per day for juvenile chinook
salmon reared under similar conditions.

A predictive relationship of the form $Y = a + bX^2$ was fitted
to the fork length and total otolith radius data from fish in
Treatments 1 and 2. To backcalculate fish length, the otolith
radius on the measurement dates was calculated by summing
increment widths from the edge of the otoliths inward to the
increments formed on previous measurement dates; that sum was
then subtracted from the total radius to give the radius of the
otoliths on the measurement dates. For each fish, back-calculated mean length was estimated at each measurement date by
inserting the otolith radius into the otolith radius-fish size
regression equation. For Treatment 1 back-calculated mean fish
lengths were averaged and compared to the mean lengths of all
fish in the tank on each measurement date.

The marking of individual fish in Treatment 2 allowed the
comparison of growth rates and lengths backcalculated from
increment measurements to observed lengths and growth rates of
individual fish. Back-calculated and observed lengths of
individual fish were compared with paired t-tests; observed
growth rates between measurement dates were calculated from the
measured lengths. Daily growth in length was estimated from
otolith measurements as the product of the derivative of the
otolith radius-fish length relationship and increment width:
$G = 2bRI$, where G is fish growth as mm/day, b is the coefficient
of the fish length-otolith radius regression, R is the otolith
radius at which the increment measurement was made, and I is the
width of the increment. Daily growth estimates were averaged
over the intervals between measurement dates and compared with
observed growth rates using correlation analysis.

RESULTS

Although the curvilinear relationship used to fit the
otolith radius-fish length data was not significantly better than
a simple linear regression (F test, $P > 0.1$), there appeared to
be a slight curvilinear trend in the data (Fig. 1). Data from
the 4 samples of Treatment 1 and the single sample of Treatment 2
were pooled, as the range of fish lengths in any single sample
was insufficient to form a regression.

The number of increments measured from otoliths of the four
Treatment 1 fish samples was greater than the number of days
between samples, so that otolith radius data were available for
more than one sample on 7 of 9 measurement dates. Because the
samples were drawn randomly from the tank we expected that the
mean radii of the samples would not be different from each other
on the measurement dates. Significant differences might indicate
that errors in reading or interpretation of the increments had
occurred. No significant differences existed in any of these

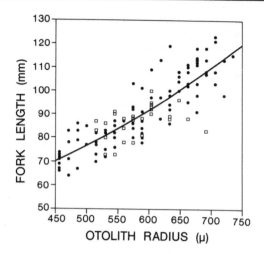

Figure 1. Relationship between fish fork length and otolith
radius for all chinook salmon used in the experiments.
Equation fitted was $Y = 42.2 + 0.000138X^2$, N = 141, r = 0.87.
● Treatment 1 fish, □ Treatment 2 fish.

comparisons (t-tests, all P > 0.1); consequently data from all
samples were pooled to backcalculate mean fish length (Fig. 2).
      Back-calculated lengths were not significantly different
from the observed mean lengths on 8 of 9 occasions in Treatment 1
(Table 1). Back-calculated length was different than observed
length at the end of the experiment when there was a rapid
increase in fish growth. The back-calculated size-at-time curve
was less variable than the actual size curve (Fig. 2).
      The mean difference between back-calculated and observed
fork lengths of individual fish in Treatment 2 were not significantly different on any of the 6 measurement dates (Table 2, all
P > 0.1). The difference between back-calculated and observed
lengths of individual fish in Treatment 2 ranged from –8 to +19
mm, with an overall standard deviation of 5.64 mm. The 95%

Table 1. Back-calculated and observed mean fork lengths (mm) of
juvenile chinook salmon in Treatment 1. Back-calculated data
are from all available otoliths, observed sizes are from all
fish in the treatment. Significant difference between observed and back-calculated lengths indicated as * (P < 0.05).

	Back-calculated			Observed		
Day	N	Length	SE	N	Length	SE
0	22	61.8	1.15	83	64.9	0.54
15	35	66.8	0.85	149	67.9	0.45
30	36	71.5	0.85	149	71.5	0.49
45	24	78.0	1.59	120	76.8	0.59
60	37	82.8	1.18	118	80.3	0.64
81	37	90.1	1.17	69	89.1	0.88
91	38	93.4	1.15	60	94.9	0.94
105	24	98.2	1.18	33	98.9	1.16
120	24	103.9	1.15	33	109.8	2.28*

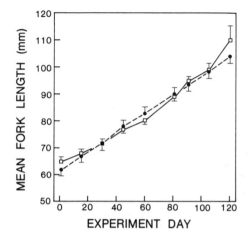

Figure 2. Back-calculated (●) and observed (□) mean fork
    lengths of laboratory-reared juvenile chinook salmon in
    Treatment 1. Back-calculated lengths are from data pooled
    from the 4 samples taken over the experiment. Error bars
    indicate ±2 SE.

confidence bounds (±2 SD) of 11.3 mm on the back-calculated
length of individual fish represents a 9-19% error over the size
range of fish used in this experiment.

Growth rates calculated from increment widths and observed
growth rates of individual fish were not significantly correlated
in 4 of 5 intervals of Treatment 2 (Table 3), although the r

Table 2. Back-calculated and observed mean fork lengths (mm) of
    juvenile chinook salmon in Treatment 2. BCL: back-calculated
    length, OL: observed length, SD: standard deviation of the
    difference between back-calculated and observed lengths of
    individual fish.

Day	N	BCL	OL	Difference	SD
0	6	67.2	66.1	1.07	6.21
15	14	72.8	71.8	1.04	6.36
30	25	77.0	75.4	1.63	4.17
45	30	81.7	80.0	1.72	5.37
60	30	86.2	82.8	3.43	5.87
67	30	88.3	88.3	-0.04	5.58

Table 3. Correlations between observed and increment width
    derived growth in length of individual juvenile chinook salmon
    in Treatment 2. Last column is the range of observed growth
    values. Significant correlation indicated as * (P < 0.05).

Interval	Days	N	r	Range
1	0-14	12	0.15	0.00-0.33
2	15-29	25	0.21	0.07-0.67
3	30-44	30	0.16	0.13-0.60
4	45-59	30	0.45*	0.21-0.71
5	60-67	30	-0.13	0.10-0.86

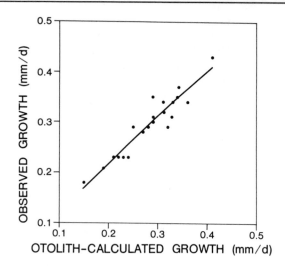

Figure 3. Observed and back-calculated growth rates of
individual chinook salmon in Treatment 2. Data are averaged
over the last 51 days of the experiment. Equation of the
line: Y = 0.97X - 0.003, r = 0.94.

values were positive in 4 of 5 intervals. Fish growth rates in
these correlations ranged from 0 to 0.86 mm/day.

We examined the relationship between mean increment width
and fish growth of individual fish over a longer time interval by
comparing the calculated rates against the observed values
averaged over the last 51 days of the experiment. The regression
was significant (P < 0.001, n = 24) with the slope of 0.93 not
significantly different from 1.0, the expected slope if increment
width accurately reflected fish growth (Fig. 3).

For each interval of Treatment 2, the variability in otolith-derived growth rates among individual fish was much less
than was observed for actual rates (coefficient of variation:
16-20% vs. 29-55%), indicating that otolith growth was not as
variable as body growth.

DISCUSSION
Mean fish lengths backcalculated from otolith increment
widths were generally accurate when compared to observed lengths.
In Treatment 1, a significant difference occurred because otolith
growth did not follow an unexplained increase in fish growth.
Back-calculated mean fish lengths had 95% confidence limits 2-5%
of mean length.

The 9-19% errors in the back-calculated lengths of individual fish were similar to the 15% error reported by Wilson and
Larkin (1982) for backcalculations of juvenile sockeye salmon

weight. The magnitude of these errors may limit the utility of
size back-calculations of individual fish.

Otolith-derived growth rates did not exhibit the same range
of fluctuations as the observed growth rates. Although considerable variability exists between the widths of individual increments (e.g., Volk et al. 1984), growth rates derived from
averaged increment widths were less variable than observed fish
growth rates. Increases or decreases in fish growth rate were
not necessarily coupled with corresponding changes in otolith
growth. Thus size-at-time curves backcalculated from increment
widths may not indicate the actual variation in fish growth.

Fish growth and otolith growth were not strongly correlated
over the short sampling intervals of Treatment 2 (Table 3).
Brothers (1981) suggested that the existence of an otolith size-
fish size relationship is not sufficient to conclude that
increment widths will closely reflect fish growth. The uncoup-
ling of otolith growth and fish growth has been well documented
for fish receiving little or no ration (Marshall and Parker 1982;
Volk et al. 1984). In these cases, otolith growth continued
after fish growth halted. Even for the well-fed fish used in our
experiments otolith growth and fish growth were not closely
correlated for individual fish (Table 3). The scatter in the
otolith radius-fish length relationship reflects the considerable
variation that can occur in fish growth relative to otolith
growth.

Over the long term, however, the relationship between
otolith size and fish size implies that increment width must be
related to fish growth. This was confirmed by the significant
regression of actual fish growth on otolith-derived growth of
individual fish for the final 51 days of Treatment 2. Significant correlations of fish growth rates and otolith increment
widths have been found for groups of fish reared under different
feeding regimes in the 40-60 day experiments of Volk et al.
(1984) and Neilson and Geen (1985).

Our results suggest that for individual chinook salmon
increment widths will have to be averaged over at least 30-40
days for growth rate estimates to be accurate. When averaged
over >50 days, it may be possible to resolve differences in
growth between individuals of as little as 0.1 mm/day (Fig. 3).

The closeness of the coupling between fish growth and
otolith growth will be species and fish-size specific. Smaller
fish than those used in these experiments may have otoliths that
respond more rapidly to changes in fish growth. In addition, the
variability in the otolith size-fish size relationship is species
specific (e.g., Fig. 1; Taubert and Coble 1977; Laroche et al.
1982) which would affect the reliability of growth estimates
derived from increment widths.

Detailed examination of growth of juvenile fish based on
individual otoliths or a short time-series of increments may have
minimal value given the conservative nature of otolith growth in

relation to fish growth. More accurate estimates of back-calculated size and growth rates may be possible over longer time
intervals. The experiments reported here emphasize the need to
carefully assess the reliability of fish size and growth
estimates backcalculated from otolith increments in controlled
experiments before field studies are undertaken with the object
of reconstructing size and growth-rate histories.

ACKNOWLEDGMENTS
    We thank Dr. C. D. Levings and the staff of the West
Vancouver Laboratory for the use of their facilities and their
invaluable assistance. Collette O'Reilly prepared the otoliths,
S. Meester provided statistical advice and Dr. J. D. Neilson
reviewed the manuscript. Financial support for this work was
provided by the Canada Department of Fisheries and Oceans and the
Natural Sciences and Engineering Research Council.

REFERENCES
BROTHERS, E. B. 1981. What can otolith microstructure tell us
    about daily and subdaily events in the early life history of
    fish? Rapports et Process-Verbaux Reunions, Conseil
    International Pour l'Exploration de la Mer 178: 393-394.
CAMPANA, S. E. and J. D. NEILSON. 1985. The microstructure of
    fish otoliths. Canadian Journal of Fisheries and Aquatic
    Sciences 42: 1014-1032.
LAROCHE, J. L., S. L. RICHARDSON and A. A. ROSENBERG. 1982. Age
    and growth of a pleuronectid, Parophrys vectulus, during the
    pelagic larval period in Oregon coastal waters. United
    States National Marine Fisheries Service. Fishery Bulletin
    80: 93-104.
MARSHALL, S. L. and S. S. PARKER. 1982. Pattern identification
    in the microstructure of sockeye salmon (Oncorhynchus nerka)
    otoliths. Canadian Journal of Fisheries and Aquatic
    Sciences 39: 542-547.
NEILSON, J. D. and G. H. GEEN. 1981. Method for preparing
    otoliths for microstructure examination. Progressive Fish-
    Culturist 43: 90-91.
NEILSON, J. D. and G. H. GEEN. 1982. Otoliths of chinook salmon
    (Oncorhynchus tshawytscha): daily growth increments and
    factors influencing their production. Canadian Journal of
    Fisheries and Aquatic Sciences 39: 1340-1347.
NEILSON, J. D. and G. H. GEEN. 1985. Effects of feeding regimes
    and diel temperature cycles on otolith increment formation
    in juvenile chinook salmon (Oncorhynchus tshawytscha).
    United States National Marine Fisheries Service. Fishery
    Bulletin 83: 91-101.
PANNELLA, G. 1980. Growth patterns in fish sagittae. Pages
    519-560 in D. C. RHOADS and R. A. LUTZ, editors. Skeletal

growth of aquatic organisms: Biological records of environmental change. Plenum Press, New York, New York, U.S.A.

TAUBERT, B. D. and D. W. COBLE. 1977. Daily rings in otoliths
of three species of Lepomis and Tilapia mossambica. Journal
of the Fisheries Research Board of Canada 34: 332-340.

VOLK, E. C., R. C. WISSMAR, C. A. SIMENSTAD and D. M. EGGERS.
1984. Relationship between otolith microstructure and the
growth of juvenile chum salmon (Oncorhynchus keta) under
different prey rations. Canadian Journal of Fisheries and
Aquatic Sciences 41: 126-133.

WILSON, K. H. and P. A. LARKIN. 1982. Relationship between
thickness of daily growth increments in sagittae and change
in body weight of sockeye salmon (Oncorhynchus nerka) fry.
Canadian Journal of Fisheries and Aquatic Sciences 39:
1335-1339.

Environmental regulation of growth patterns in juvenile Atlantic salmon

JOHN E. THORPE

FRESHWATER FISHERIES LABORATORY
PITLOCHRY PH16 5LB
SCOTLAND

ABSTRACT
    Wild Atlantic salmon Salmo salar vary widely in growth rates
during their freshwater phase, both within and among stocks.
Over the past 12 years, salmon have been reared in sibling groups
at Almondbank, Scotland, in radial-flow tanks designed to minimise
agonistic behaviour among these territorial fish.   Genetic
differences in growth capacity have been found between sibling
groups, particularly in the expression of bimodal length-frequency
distributions.   This   developed   regularly   during   the   first
growing season, such that the upper modal group smolted in the
following spring, while the lower group required a further year in
freshwater.   The proportion of a population which formed the
upper modal group could be modified environmentally, ultimately
through the relative opportunity to feed: populations experiencing
local crowding in the presence of cover grew less rapidly than
dispersed populations lacking cover; those experiencing relatively
long periods of light grew more rapidly than others experiencing
shorter periods.
    The developmental switch determining which modal group an
individual would enter was tripped in mid- to late summer.
The choice that a fish made was governed apparently by individual
sensitivity to stimulation of the neuroendocrine system over this
restricted period.   First, the members of the upper modal group
were derived from throughout the original unimodal population; and
secondly, the tripping of the switch could be delayed by setting
the photoperiod regime that the population experienced out of
phase with ambient conditions.   The switch did not reflect
incipient sexual maturation, as populations sterilised by
cobalt-60 gamma irradiation did not differ in bimodal growth
patterns from controls.   Estimation of individual food intake,
using diets labelled with iron powder and serial X-rays, showed
that bimodality arose due to growth arrest in those individuals

which ultimately formed the lower modal group. As this arrest occurred at a time when conditions were favourable for growth, it is postulated that bimodality occurs as the consequence of environmental control of the neuroendocrine regulators of appetite and feeding behaviour, and not directly those of growth.

Throughout its geographical range Atlantic salmon Salmo salar varies greatly in freshwater growth rate both within and among stocks (Power 1981; Saunders 1981; Thorpe and Mitchell 1981). Within stocks, this variation in growth leads to a range of ages at which the juveniles smolt and migrate to sea, and this population characteristic has been correlated with population age-structure and density (Egglishaw and Shackley 1980; Buck and Hay 1984). A series of experiments were conducted over the past 12 years at the Almondbank Smolt-Rearing Station, Scotland, to examine the nature of this developmental variation more closely. The first experiments revealed a bimodal segregation within the length-frequency distribution among sibling populations during their first growing season (Simpson and Thorpe 1976; Thorpe 1977). Similar growth patterns have since been reported by Knutsson and Grav (1976), Bailey et al. (1980), Saunders et al. (1982), and Bagliniere and Maisse (1985). The Almondbank experiments were designed to explore the genetic and environmental regulation of these patterns of salmon growth, and to evaluate factors which might determine growth rate and age at smolting. The results of these experiments are summarised here.

THE RADIAL-FLOW TANK

Typically, Atlantic salmon fry occupy individual feeding territories on the bed of shallow streams, which they defend aggressively against conspecifics of a similar size (Kalleberg 1958; Keenleyside and Yamamoto 1962). To evaluate factors regulating growth, a standard tank was needed which would provide the essential elements of territory for all its occupants. Wankowski and Thorpe (1979) showed experimentally that the intensity of agonistic behaviour between territorial contestants was correlated negatively with the number of channels through which they could obtain food. So, in a tank, if there were as many equal opportunities to feed as there were fish, minimal energy would be wasted in aggressive activities. Since salmon territory consists principally of a site from which to intercept drifting food items, a tank was designed in which water carrying food particles entered at the centre and travelled radially outwards to a circular peripheral drain (full description in Thorpe 1981). Fish arranged themselves in a ring, heads toward the source of food.

The mean coefficient of variation of body length in the populations after 9 months feeding was significantly less at 0.055 ($P < 0.05$) than that of 0.072 in sibling populations grown in conventional tangential-flow tanks with food dispensed from the surface (Thorpe and Wankowski 1979). That is, fish reared in radial tanks showed more uniformity of growth response than those reared in tangential-flow tanks. All the experimental data discussed in this paper were obtained using these radial-flow tanks, and relate to sibling populations of 0+ salmon, reared from individual pairings of wild River Almond parent fish.

BIMODALITY OF LENGTH DISTRIBUTION

Initial experiments revealed an immediate analytical problem (Fig. 1): length-frequency distributions of all populations in 52

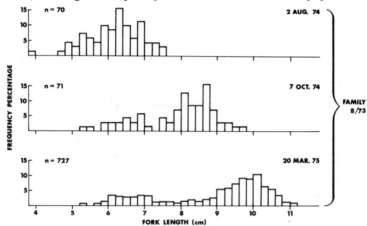

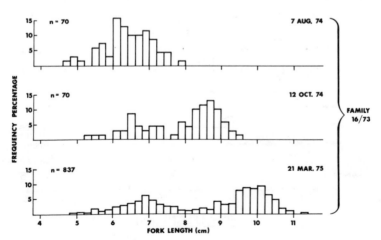

Figure 1. Development of bimodality in length frequency distribution of two groups of sibling juvenile Atlantic salmon Salmo salar (from Thorpe 1977).

tanks were not normal at the end of their first growing season in October (Thorpe 1977). From repeated monitoring of these populations throughout subsequent months, length frequency distributions became clearly bimodal. The upper modal group became completely distinct from the lower by March, due to continued growth during the winter (Fig. 2). Fish of the upper modal group smolted, while fish of the lower group required a further year in freshwater (Simpson and Thorpe 1976). From a set of auxiliary experiments (Thorpe 1977) it was established that the separation of the modes was not due to differential growth of the two sexes, nor to maturation, nor to behavioural dominance of the upper over the lower group. In a 4 * 4 factorial

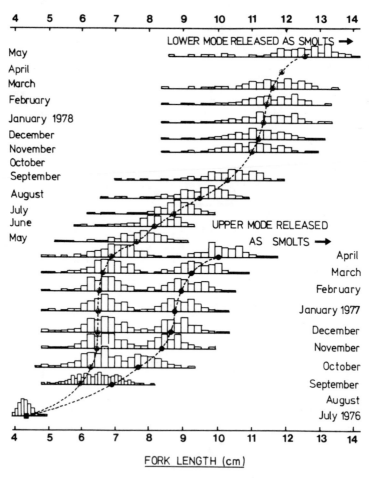

Figure 2. Monthly length frequency distributions in a sibling salmon population (modified from Thorpe et al. 1980).

experiment, statistically significant differences of growth performance were found between 2 families (P < 0.001), between populations reared with and without overhead cover (P < 0.01), and between populations reared under two different water flow regimes (P < 0.02) (Thorpe 1977).

From a 4 * 4 half-sib factorial experiment (Thorpe and Morgan 1978) it was found that only the female parent influenced the mean size of her progeny at the end of their first growing season: larger females of a given age produced larger progeny. This was subsequently confirmed from progeny groups from 65 pairings (Thorpe et al. 1984). However, both parents, particularly the male, had significant effects on developmental rate of progeny as measured by the proportion of the population in the upper modal group (P < 0.01). In subsequent experiments (Thorpe and Morgan 1980; Thorpe et al. 1983), it was shown that parents which matured early (i.e. at a younger total age), and thus had demonstrated relatively rapid developmental rates themselves, produced progeny which also developed relatively rapidly, in growth and smolting as well as in maturation.

In more detailed experiments (Thorpe and Wankowski 1979) that varied the type of cover available to the fishes, the upper modal group contained 44% of the population when 30 cm wide annular covers were provided. This was significantly greater (P < 0.02) than the 22.5% in this group when the covers were only 15 cm wide. Characteristically, the fish held stations with their heads immediately below the inner edge of these cover rings. Maximal proportions of the population in the upper modal group were found when the water velocity across the tank floor at the head of the fish was maintained at 5 body lengths per second throughout the growth season.

It was concluded that while growth capacity was evidently influenced genetically, ultimate performance was modified environmentally, through the relative opportunity for individuals to obtain food.

TIME OF DIVERGENCE OF MODAL LENGTH GROUPS

Eight sibling groups, in 25 populations, were monitored at monthly intervals to determine the time at which divergence into the two modal length groups occurred (Thorpe et al. 1980). Mean fish length, proportion in the upper modal group, and time of separation between modal groups varied between sibling groups, the separation becoming evident between late September and December. However, backward extrapolation of points on the modal growth curves suggested that the developmental switch which resulted in the segregation into two modal groups must have been operated in mid- to late summer (Fig. 2). Thorpe et al. (1982) established that the changes of RNA:DNA ratios of salmon epaxial muscle between sample dates served as a valid index of specific growth rate over the same intervals. This method showed promise as a useful index of short-term growth changes. However, a problem arose when Villarreal (1983) applied this method to sibling populations, in an attempt to define the precise time at which the segregating switch was tripped. The nucleic-acid ratios correlated well with specific growth rates up to the end of June, and then again after September, when the separate modal groups were distinguishable in the length-frequency distribution; in the intervening period, no correlations were evident. Such a situation would arise if the developmental switch operated independently of fish size, such that future, upper-modal group members would be drawn at random from within the entire size distribution in July (Fig. 3).

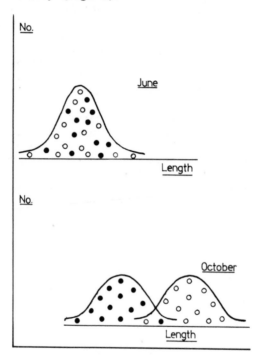

Figure 3. Model of the random segregation of individual
sibling Atlantic salmon, from a unimodal length frequu-
ency distribution in June to a bimodal one in October.

ENVIRONMENTAL FACTORS INFLUENCING THE SWITCH
    Villarreal (1983) proposed that photoperiod synchronised an endogenous rhythm, which was genetically determined and which timed the moment when the switch became operable. By rearing separate sibling populations under a range of photoperiods set to run at 3, 6, and 9 months out of phase with ambient conditions, Thorpe (in press) showed that delaying photoperiod did delay the time when the separation into two modal groups occurred (Fig. 4).

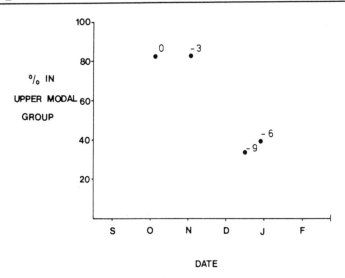

Figure 4. Time and percentage level at which bimodal segregation of growth groups became stable in sibling
salmon populations reared under controlled photoperiod
regimes 0, 3, 6, and 9 months out of phase with the
natural light regime (from Thorpe, in press).

It was suggested that the fish were aware of their growth rate,
through their rate of acquisition of surplus energy, and the
hormone kinetics associated with its storage. Provided that
this rate was above a genetically determined level at the critical
time when the fish were susceptible to photoperiodic stimulation
of the switch, then growth would be maintained.

NATURE OF THE DEVELOPMENTAL CHANGE
Using iron-labelled diets and continuous feeding Higgins
(1985) and Higgins & Talbot (1985) estimated the daily food intake
of freely feeding fish in bimodally distributed populations, by a
series of X-rayed samples. They showed that upper-modal group
fish took significantly larger meals (dry weight of food as
percentage of dry body weight) than did lower-modal group
individuals (P < 0.001), and fed for a longer proportion of each
24 hour period. The upper group showed higher food-conversion
efficiencies, and increased in caloric content throughout the
measurement period, June to May. The lower group showed lower
conversion efficiencies and did not increase in caloric content
after mid-September, at which time their daily intake had fallen
to a maintenance level only (Fig. 5). It appears, therefore,
that the alternative developmental pathways consisted of
maintaining or curtailing growth; that the switch operated at a

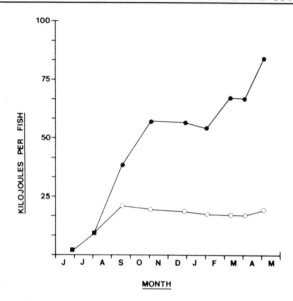

Figure 5. Changes in caloric content of juvenile salmon
    during the development of the bimodal length-frequency
    distribution.   Upper curve: upper modal group; lower
    curve: lower modal group (from Higgins and Talbot
    1985).

time   (July/August) when temperatures (10-14°C) were favourable for
growth;   and   that it operated despite plentiful food.   Hence,
the   bimodal   length-distribution   arose through growth arrest of a
part   of   the population, rather than an increase in growth-rate of
the other part.
       An   earlier   hypothesis   postulated   that the modal separation
represented   the   first   expression   of   sexual   maturation   in the
population   (Thorpe   et   al.   1982).   Later this was shown to be
false,   when   populations sterilised by exposure to cobalt-60 gamma
radiation   showed   statistically   identical   growth   patterns   with
fertile sibling control populations (Villarreal and Thorpe 1985).

DISCUSSION
       The   crucial   developmental change, determining whether or not
a   young   Atlantic salmon will continue to develop towards smolting
in   the spring of the following year, appears to occur in the weeks
shortly   after midsummer, and this is synchronised by photoperiod.
The   choice   that a fish made was governed apparently by individual
sensitivity   to   stimulation of the neuroendocrine system over this
restricted   period.   Present evidence implies that absolute size
by   itself   does   not   determine   the   outcome,   and that if growth
arrest   occurs,   it does so independently of the fish's position in

the size distribution. This conclusion is supported by actual growth data from populations of individually marked Atlantic salmon (Kristinsson et. al. 1985), and of masu salmon Oncorhynchus masou (T. Hirata, Laboratory of Embryology and Genetics, University of Hokkaido, Hakodate, Japan, pers. comm.). In both these cases, the fish were marked before the populations segregated, and subsequent identification of those individuals which became the upper modal group showed that they had been derived from the entire range of the original unimodal distribution of length.

Photoperiod has been shown to have profound effects on growth of Atlantic salmon at other stages of development (Saunders and Henderson 1970; Komourdjian et al. 1976). However, those experiments were concerned primarily with the effect of the direction of change of daylength, as was the earlier work of Swift (1961) on Salmo trutta, and Gross et al. (1965) on Lepomis cyanellus. In the present studies, as in those of Eriksson and Lundqvist (1982), and Lundqvist (1983), on smolting and maturation, photoperiod change serves as a zeitgeber to define a precise season at which a train of neuroendocrine events can be induced, which either permit or prevent sustained growth. As those events took place at a time when temperature and food availability were favourable for growth, it is postulated that regulation must act on appetite and feeding behaviour, and not on growth itself, which is dependent on adequate intake. Studies of the endocrine control of appetite and feeding are now needed.

The occurrence of growth arrest, under environmental conditions which would appear to favour continued feeding and growth, is puzzling. However, both Gibson (1978) and Rimmer et al.(1983) noted that young salmon sought cover in the bottom gravel when temperatures fell below 10 °C in autumn, which coincides approximately with the time and temperature conditions of arrest in the hatchery fish. Two kinds of advantage result from this behaviour: first, reduced activity and metabolic rate, and consequent low food requirement, confer energetic advantage over winter; and second, in males, a variable proportion of the arrested population will mature the following year in freshwater, while their growth-maintaining siblings will not do so until returning from the sea 12 months after that. The population genetic consequences of this phenomenon remain to be explored.

ACKNOWLEDGMENTS
I thank my colleagues at the Pitlochry laboratory, R. I. G. Morgan, C. Talbot, D. Pretswell, C. A. Villarreal, J. W. J. Wankowski, P.J. Higgins, I. M. Irvine, and A. McQueen, and at the Almondbank hatchery, M. S. Miles, D. S. Keay, T. J. W. Fletcher, and J. S. Muir, without whose assistance the work reviewed above would not have been possible.

REFERENCES

BAGLINIERE, J.-L., AND G. MAISSE. 1985. Precocious maturation and smoltification in wild Atlantic salmon in the Armorican Massif, France. Aquaculture 45:249-263.

BAILEY, J. K., R. L. SAUNDERS AND M. I. BUZETA. 1980. Influence of parental smolt age and sea age on growth and smolting of hatchery-reared Atlantic salmon (Salmo salar). Canadian Journal of Fisheries and Aquatic Sciences 37:1379-1386.

BUCK, R. G. J., AND D. W. HAY. 1984. The relation between stock size and progeny of Atlantic salmon, Salmo salar L., in a Scottish stream. Journal of Fish Biology 24:1-11.

EGGLISHAW, H. J., AND P. E. SHACKLEY. 1980. Survival and growth of salmon, Salmo salar (L.), planted in a Scottish stream. Journal of Fish Biology 16:565-584.

ERIKSSON, L-O., AND H. LUNDQVIST. 1982. Circannual rhythms and photoperiod regulation of growth and smolting in Baltic salmon (Salmo salar L.). Aquaculture 28:113-121.

GIBSON, R. J. 1978. The behaviour of juvenile Atlantic salmon (Salmo salar) and brook trout (Salvelinus fontinalis) with regard to temperature and water velocity. Transactions of the American Fisheries Society 107:703-712.

GROSS, W. L., P. O. FROMM AND E. W. ROELOFS. 1965. Influence of photoperiod on growth in green sunfish, Lepomis cyanellus (Rafinesque). Transactions of the American Fisheries Society 92:401-408.

HIGGINS, P. J. 1985. Metabolic differences between Atlantic salmon (Salmo salar) parr and smolts. Aquaculture 45: 33-53.

HIGGINS, P. J., AND C. TALBOT. 1985. Feeding and growth in juvenile Atlantic salmon (Salmo salar). Pages 000-000 in C. B. Cowey, A. M. Mackie and J. G. Bell, editors: Nutrition and Feeding in Fish. Academic Press, London.

KALLEBERG, H. 1958. Observations in a stream tank of territoriality and competition in juvenile salmon and trout (Salmo salar L. and S.trutta L.). Report of the Institute of Freshwater Research Drottningholm 39:55-98.

KEENLEYSIDE, M. H. A., AND F. T. YAMAMOTO. 1962. Territorial behaviour of juvenile Atlantic salmon. Behaviour 19:139-169.

KNUTSSON, S., AND T. GRAV. 1976. Seawater adaptation in Atlantic salmon (Salmo salar L.) at different experimental temperatures and photoperiods. Aquaculture 8:169-187.

KOMOURDJIAN, M. P., R. L. SAUNDERS AND J. C. FENWICK. 1976. Evidence for the role of growth hormone as a part of a 'light-pituitary axis' in growth and smoltification of Atlantic salmon (Salmo salar). Canadian Journal of Zoology 54:544-551.

KRISTINSSON, J., R. L. SAUNDERS AND A. J. WIGGS. 1985. Growth dynamics during the development of bimodal length frequency distribution of juvenile Atlantic salmon (Salmo salar L.). Aquaculture 45:1-20.

LUNDQVIST, H. 1983. Precocious sexual maturation and smolting in Baltic salmon (Salmo salar): photoperiodic synchronisation and adaptive significance of annual biological cycles. Ph.D.Thesis, University of Umea, Umea, Sweden.

POWER, G. 1981. Stock characteristics and catches of Atlantic salmon (Salmo salar) in Quebec, and Newfoundland and Labrador in relation to environmental variables. Canadian Journal of Fisheries and Aquatic Sciences 38:1601-1611.

RIMMER, D. M., U. PAIM AND R. L. SAUNDERS. 1983. Autumnal habitat shift of juvenile Atlantic salmon (Salmo salar) in a small river. Canadian Journal of Fisheries and Aquatic Sciences 40:671-680.

SAUNDERS, R. L. 1981. Atlantic salmon (Salmo salar) stocks and management implications in the Canadian Atlantic Provinces and New England, USA. Canadian Journal of Fisheries and Aquatic Sciences 38:1612-1625.

SAUNDERS, R. L., AND E. B. HENDERSON. 1970. Influence of photoperiod on smolt development and growth in Atlantic salmon (Salmo salar). Journal of the Fisheries Research Board of Canada 27:1295-1311.

SAUNDERS, R. L., E. B. HENDERSON AND B. D. GLEBE. 1982. Precocious sexual maturation and smoltification in male Atlantic salmon (Salmo salar L.). Aquaculture 28:211-229.

SIMPSON, T. H., AND J. E. THORPE. 1976. Growth bimodality in the Atlantic salmon. International Council for the Exploration of the Sea C.M.1976/M:22. 7pp.

SWIFT, D. R. 1961. The annual growth rate cycle in brown trout (Salmo trutta Linn.) and its cause. Journal of Experimental Biology 38:595-604.

THORPE, J. E. 1977. Bimodal distribution of length of juvenile Atlantic salmon (Salmo salar L.) under artificial rearing conditions. Journal of Fish Biology 11:175-184.

THORPE, J. E. 1981. Rearing salmonids in freshwater. Pages 325-344 in A.D.Hawkins, editor. Aquarium Systems. Academic Press Inc., London.

THORPE, J. E. 1986. Age at first maturity in Atlantic salmon Salmo salar L.: freshwater period influences and conflicts with smolting. Canadian Journal of Fisheries and Aquatic Sciences 43.

THORPE, J. E., AND K. A. MITCHELL. 1981. Stocks of Atlantic salmon (Salmo salar) in Britain and Ireland: discreteness and current management. Canadian Journal of Fisheries and Aquatic Sciences 38:1576-1590.

THORPE, J. E., AND R. I. G. MORGAN. 1978. Parental influence on growth rate, smolting rate and survival in hatchery-reared juvenile Atlantic salmon, Salmo salar L. Journal of Fish Biology 13:549-556.

THORPE, J. E., AND R. I. G. MORGAN. 1980. Growth-rate and smolting-rate of progeny of male Atlantic salmon parr, Salmo salar L. Journal of Fish Biology 17:451-460.

THORPE, J. E., AND J. W. J. WANKOWSKI. 1979. Feed presentation and food particle size for juvenile Atlantic salmon, <u>Salmo</u> <u>salar</u> L. Pages 501–513 <u>in</u> J. E. Halver & K. Tiews, editors Finfish Nutrition and Fishfeed Technology: 2 volumes. Heenemann, Berlin.

THORPE, J. E., M. S. MILES AND D. S. KEAY. 1984. Developmental rate, fecundity and egg size in Atlantic salmon, <u>Salmo</u> <u>salar</u> L. Aquaculture 43:289–305.

THORPE, J. E., C. TALBOT AND C. VILLARREAL. 1982. Bimodality of growth and smolting in Atlantic salmon, <u>Salmo</u> <u>salar</u> L. Aquaculture 28:123–132.

THORPE, J. E., R. I. G. MORGAN, E. M. OTTAWAY AND M. S. MILES. 1980. Time of divergence of growth groups between potential 1+ and 2+ smolts among sibling Atlantic salmon. Journal of Fish Biology 17:13–21.

THORPE, J. E., R. I. G. MORGAN, C. TALBOT AND M. S. MILES. 1983. Inheritance of developmental rates in Atlantic salmon, <u>Salmo</u> <u>salar</u> L. Aquaculture 33:119–128.

VILLARREAL, C. A. 1983. The role of light and endocrine factors in the development of bimodality of growth in the juvenile Atlantic salmon (<u>Salmo</u> <u>salar</u> L.). Ph.D.Thesis, University of Stirling, Stirling, Scotland.

VILLARREAL, C. A., AND J. E. THORPE. 1985. Gonadal growth and bimodality of length frequency distribution in juvenile Atlantic salmon (<u>Salmo</u> <u>salar</u>). Aquaculture 45:265–288.

WANKOWSKI, J. W. J., AND J. E. THORPE. 1979. Spatial distribution and feeding in Atlantic salmon, <u>Salmo</u> <u>salar</u> L., juveniles. Journal of Fish Biology 14:239–247.

Use of daily otolith rings to interpret development of length distributions of young largemouth bass

J. JEFFERY ISELY[1] AND
RICHARD L. NOBLE[1]
DEPARTMENT OF WILDLIFE AND FISHERIES SCIENCES
TEXAS A&M UNIVERSITY
COLLEGE STATION, TX 77843

ABSTRACT
    The length distributions of seven populations of large-
mouth bass Micropterus salmoides in small impoundments in Texas
were investigated through the use of daily otolith rings. Daily
rings were recognizable by microscopic observation as long as
growth continued, up to 184 days in one population. The difference in age between the oldest and youngest fish within a sample, reflecting the length of the spawning period, ranged from
15 - 68 days. Comparison of samples of fish of similar mean
ages indicated that those populations with short spawning periods subsequently exhibited less variation in total lengths than
populations with longer spawning periods. Length distributions
were typically characterized by a significant correlation
between length and age. This relationship suggests that
extended spawning seasons contribute to the development of
skewed and bimodal length distributions of young largemouth bass
cohorts.

    Laboratory and field studies have verified the daily for-
mation of otolith rings in largemouth bass Micropterus salmoides
less than 150 days of age (Schmidt and Fabrizio 1980; Miller and
Storck 1984; Isely 1984). Daily otolith rings appear to be
deposited for the length of the growing season on otoliths of
largemouth bass; and they have been used to verify annulus formation (Taubert and Tranquilli 1982), and to back-calculate
spawning dates (Miller and Storck 1984).
    Bimodal, or strongly skewed length-frequency distributions
commonly observed by the time of first annulus formation in

    [1]Present address: Department of Zoology, North Carolina State
University, Raleigh, NC 27695

largemouth bass have been attributed to extended spawning periods, growth advantages of larger fish, or a combination of these
factors (Aggus and Elliot 1975; Summerfelt 1975; Shelton et al.
1979; Timmons et al. 1980). Recently, Miller and Storck (1984)
used age estimates from daily otolith rings to evaluate size
variation in young largemouth bass; they observed that temporal
changes in food availability resulted in differential growth of
fish spawned at different times.

This paper further evaluates the relationship between
length and age, and between growth and age within cohorts, and
examines the relationship between length and age distributions
of largemouth bass in several pond populations.

MATERIALS AND METHODS

Young largemouth bass were collected from seven small
impoundments in Texas by seining, electrofishing or rotenone
application. Ponds were privately owned and ranged in size from
0.3-9.0 hectares in surface area. Single samples were taken
from each population between June 24 and September 6, 1983.
Sample size was subject to availability of fish and varied
between 31 and 113.

All bass were returned to the laboratory for processing.
Fish were measured in total length to the nearest millimeter,
and sagittal otoliths were removed. Sections of the otoliths
were prepared according to the methods of Taubert and Coble
(1977), except that otoliths were not polished. Instead, a
small amount of immersion oil was placed on each otolith to
improve optical quality (Isely 1984). Consecutive ring counts
by the senior author were made until a pair of counts agreed
within an arbitrary two rings. This was usually achieved within
the first two counts and rarely was it necessary to count the
otolith rings more than three times. When counts were not identical, the larger of the two counts was chosen. Occasionally,
otoliths were destroyed in the grinding process and were discarded. When necessary, the second otolith was used. Length-
frequency and age-frequency distributions were developed for
each population. Average daily growth rate for each fish was
calculated as total length divided by number of rings.

RESULTS

Total length of largemouth bass from the seven Texas farm
ponds ranged from a mean of 39 mm in the PMS New Pond sample to
83 mm in the Redhead Pond sample (Table 1). Variance in total
length was highest in July and September samples. Age ranged
from a mean of 71 days in the PMS New Pond sample to 145 days in
the Redhead Pond sample and was related to sampling date. Mean
daily growth rate varied from 0.54 mm/day in the PMS Old Pond
sample to 0.62 mm/day in Isbell Pond. Despite differences in

Table 1. A summary of sampling data, and daily growth rate (DGR) for young largemouth bass from Texas small impoundments.

Location	Date	N	Age (range)	Variance	Age (range)	Variance	DGR
Lower Tranquility	June 23	48	82 (82-72)	89.8	50 (35-68)	61.5	0.60
Upper Tranquility	June 23	55	75 (56-89)	57.5	45 (34-81)	36.3	0.60
PMS New	June 24	31	71 (52-84)	108.4	39 (26-49)	27.6	0.56
PMS Old	June 24	41	77 (69-84)	11.7	41 (34-54)	31.1	0.54
Isbell	July 25	75	79 (55-123)	140.9	48 (36-93)	118.2	0.62
Redhead	Sept. 2	50	145 (128-184)	102.5	83 (57-139)	198.2	0.58
Gazebo	Sept. 6	113	128 (116-141)	34.3	78 (61-136)	93.3	0.61

age of fish and environment, growth rates were similar among populations.

The difference in age between the oldest and youngest fish within a sample has been used to estimate length of spawning period for largemouth bass (Miller and Storck 1984). The range in age within a sample varied from 15 days in the PMS Old Pond to 68 days in Isbell pond. A comparison of variance among samples of similar mean age indicated that variance in total length increased with variance in age.

A significant positive correlation between total length and age occurred for all samples except the PMS New Pond sample (Table 2). Although the relationship between length and age was significant in most cases, correlation coefficients were low, indicating that other factors are also important in length determination.

The relationship between growth rate and age was less consistent. While a positive correlation was also observed between daily growth rate and age in the Redhead Pond sample, a negative correlation between daily growth rate and age was observed in the PMS New Pond sample (Table 2.).

Table 2. Correlation coefficients between length and age, and daily growth rate (DGR) and age, for young largemouth bass from small impoundments in Texas.

Location	Length X Age	DGR X Age
Gazebo	0.301*	-0.080
Isbell	0.782*	0.079
PMS New	0.241	-0.680*
PMS Old	0.558*	0.264
Redhead	0.680*	0.282*
L. Tranquility	0.602*	-0.051
U. Tranquility	0.624*	0.009

* significant at the a = 0.05 level.

The significant correlation between age and total length in most cases suggests that length- and age-frequency distributions also should be related. Length-frequency distributions showed variation typical of first-year bass populations. Distributions at the time of sampling varied from normal in the PMS New Pond sample, to bimodal in the Isbell Pond sample (Figure 1). Age-frequency distributions varied from normal in the PMS Old Pond sample to highly skewed in the Redhead Pond sample. Sample size in most cases was insufficient to test for differences between length and age distributions. However, skewness and kurtosis calculated for each sample were consistent between length and age for all samples except the Gazebo Pond sample (Table 3).

Table 3. Skewness and kurtosis values for length and age distributions of young largemouth bass from Texas small impoundments.

Location	Length		Age	
	Skewness	Kurtosis	Skewness	Kurtosis
Gazebo	-0.03	-0.43	2.17	10.10
Isbell	0.87	2.25	2.78	7.92
PMS New	-0.61	-0.86	-0.08	-0.31
PMS Old	0.16	-0.16	0.77	-0.13
Redhead	1.88	5.63	1.65	4.06
L. Tranquility	1.43	1.68	0.56	0.10
U. Tranquility	-0.96	0.55	2.26	7.81

DISCUSSION

The range and distribution of ages varied considerably between samples, independent of sampling date, suggesting that spawning periods varied among populations. Our comparison of samples of fish of similar mean age indicated that those populations with short spawning seasons, as indicated by age, subsequently exhibited less variation in length than populations with longer spawning periods. Further analysis showed that length distributions were typically characterized by a significant correlation between length and age. These results indicate that length may be in part determined by individual age, that is, individual time of hatching, within a cohort.

A comparison of age-frequency with length-frequency distributions indicated the latter were similar to the frequency distributions of successful bass spawning activity. In general, normal or skewed distributions in spawning activity, as indicated by age-frequency distributions, resulted in normal or skewed length distributions respectively, and bimodal age distributions resulted in bimodal length distributions. The relationship between bass spawning activity and the resulting length-frequency distribution of young-of-the-year largemouth bass observed in this study supports earlier findings of Summerfelt (1975) on the production of bimodal length-frequency distributions following disjunctive spawning.

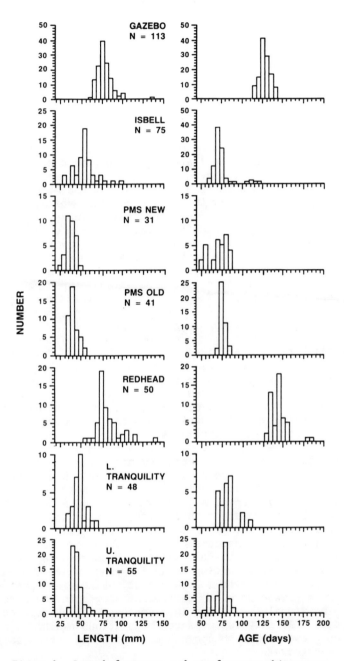

Figure 1. Length-frequency and age-frequency histograms for largemouth bass from Texas small impoundments.

Comparison of fish data from more than one population often required the assumption that the fish were the same age. Differences in size then were attributed to differences in growth rate. Daily otolith rings provide an opportunity to control for differences in spawning dates when evaluating growth, by facilitating an estimate for daily growth rate. Although size varied among populations sampled during the same period, growth rates were similar. Further, daily otolith rings provide a basis for comparison of fish within a population. The significant positive correlation between age and daily growth rate in the Redhead Pond sample indicates that, in some cases, older fish grow faster than younger fish, which amplifies age-related size differences, such as reported by Miller and Storck (1984). However, it is also likely that a decrease in available prey with size results in reduced growth rates of larger individuals, thereby allowing younger fish to obtain comparable lengths. Decreased growth of larger, older individuals may then mask age effects. The negative correlation between age and daily growth rate in the PMS New Pond sample indicates that older fish grew slower than younger fish in that population, dampening age-caused size differences. The lack of a significant relationship between length and age in this sample, therefore, was a result of age-specific differences in growth rate.

The age structure within a population, predicted from daily otolith rings, explained the development of fall length distributions in young largemouth bass. The correlation between length and age suggests that extended spawning periods contribute to the development of skewed or bimodal length distributions. However, compressed distributions in length may not necessarily represent short spawning seasons. Spawning date differences result in an initial size distribution proportional to successful spawning activity, and each individual within the population is acted upon by variables affecting growth. The initial variability in age provides the variation in size necessary for the differential action of many of these variables. Although genetic differences in growth rate, as well as other variables, may act independent of size, much of the variation in length distributions of young largemouth bass can be associated with variation in the age distribution of the population.

Differential growth generally has been reported to favor larger individuals, resulting in an increase in age-caused size differences. Our results indicate that, while age-dependent increases in growth do occur, differential growth may also favor younger individuals resulting in a decrease in age-related size differences. As only two of seven populations exhibited age dependent growth, this phenomenon may not be as common as previously suggested.

AKNOWLEDGMENTS
This study was conducted as part of Texas Agricultural Experiment Station project S-6206, with additional support provided by the Tom Slick Graduate Research Fellowship program. Appreciation is extended to Drs. William H. Neill, William E. Grant, and James A. Matis for their contributions to the development of this manuscript and to Alan E. Rudd for his frequent field assistance.

REFERENCES
AGGUS, L.R., AND G.V. ELLIOT. 1975. Effects of cover and food on year-class strength of largemouth bass. pages 317-322 in R.H. Stroud and H. Clepper, editors. Black Bass Biology and Management, Sport Fishing Institute, Washington, District of Columbia, USA.

ISELY, J.J. 1984. Relative age and first-year growth of largemouth bass. Doctoral dissertation. Texas A&M University, College Station, Texas, USA.

MILLER, S.J., AND T. STORCK. 1984. Temporal spawning distribution of largemouth bass and young-of-year growth, determined from daily otolith rings. Transactions of the American Fisheries Society 113:571-578.

SCHMIDT, R.E., AND M.C. FABRIZIO. 1980. Daily growth rings on otoliths for aging young-of-the-year largemouth bass from wild populations. Progressive Fish Culturist 42:78-80.

SHELTON, W.L., W.D. DAVIES, T.A. KING, AND J. TIMMONS. 1979. Variation in the growth of the initial year class of largemouth bass in West Point Reservoir, Alabama and Georgia. Transactions of the American Fisheries Society 108:142-149.

SUMMERFELT, R.C. 1975. Relationship between weather and year-class strength of largemouth bass. pages 166-174. in R.H. Stroud and H. Clepper, editors. Black Bass Biology and Management. Sport Fishing Institute, Washington, District of Columbia, USA.

TAUBERT, B.C., AND D.W. COBLE. 1977. Daily rings in otoliths of three species of Lepomis and Tilapia mossambica. Journal of the Fisheries Research Board of Canada 34:332-340.

TAUBERT, B.C., AND J.A. TRANQUILLI. 1982. Verification of formation of annuli in otoliths of largemouth bass. Transactions of the American Fisheries Society 111:531-534.

TIMMONS, T.J., W.L. SHELTON, AND W.D. DAVIES. 1980. Differential growth of largemouth bass in West Point Reservoir, Alabama and Georgia. Transactions of the American Fisheries Society 109:176-186.

Otolith aging of young-of-year smallmouth bass

ROBERT J. GRAHAM[1] AND
DONALD J. ORTH
DEPARTMENT OF FISHERIES AND WILDLIFE SCIENCES
VIRGINIA POLYTECHNIC INSTITUTE AND STATE UNIVERSITY
BLACKSBURG, VA 24061

ABSTRACT
    Otolith aging of young-of-year smallmouth bass Micropterus
dolomieui was evaluated using laboratory and field collections.
Increments in the sagittae of laboratory-reared smallmouth bass fry
were formed daily with the first increment formed the day of
swim-up. Wild young-of-year smallmouth bass could be aged using
sagittal rings if collected before water temperatures fell to
approximately 10 C. Instantaneous growth rates and variability in
growth rates of wild young-of-year smallmouth bass, determined by
counts of sagittal increments, decreased with increasing age. von
Bertalanffy growth parameters developed from data on length at
daily age had narrower confidence limits than those developed using
length at collection time data.

    Concentric increments (rings) observed in whole mounts or
sections of sagittae of several species of centrarchids have been
interpreted to represent daily growth (Taubert and Coble 1977;
Schmidt and Fabrizio 1980; Miller and Storck 1982). However, the
rate of sagittal ring formation may vary by species (Geffen 1982),
and the accuracy of age estimates obtained from sagittae may vary
with fish age (Miller and Storck 1982). Furthermore, the time of
first daily ring formation must be known to obtain precise age
estimates (Brothers et al. 1976). If discernable daily rings occur
regularly in the sagittae of young-of-year fish and other
assumptions of the technique are met, otolith aging should prove a
useful tool for determining precise growth rates. The objectives
of this study were to (1) verify the occurrence of daily rings in
sagittae of young-of-year smallmouth bass Micropterus dolomieui,

[1]Present address: North Anna Environmental Laboratory, Virginia
Power, Mineral, Virginia 23117

The Age and Growth of Fish, edited by Robert C. Summerfelt and Gordon E. Hall © 1987 The Iowa
State University Press, Ames, Iowa 50010.

(2) determine when the first ring formed, (3) determine whether sagittal rings are reliable age indicators of wild young-of-year smallmouth bass, and, (4) assess the applicability of otolith aging to growth studies of wild young-of-year smallmouth bass.

METHODS

A laboratory population of larval smallmouth bass was started from smallmouth bass eggs obtained from a nest in the South Fork of the Roanoke River, Virginia, on 26 May 1982. In the laboratory, the eggs were maintained under a 16-hour light by 8-hour dark daily cycle, at a temperature range of 20-22 C. Eggs hatched 3 days after collection, and larvae reached swim-up 7 days later. Larvae were raised within a temperature range of 17-23 C and fed live Artemia nauplii to satiation at 0800 and 2000 hours each day. Beginning at swim-up, 10 or 11 larvae were sampled every other day between 2000 and 2200 hours and frozen in water until the sagittae could be prepared for aging. Sampling was terminated 14 days after swim-up due to a mass mortality.

Sagittae were embedded in thermoplastic cement on microscope slides to view the sagittal plane. Slides were labeled with an identification number only to reduce bias when counting rings. Ring counts (Pannella 1971) were made in the posterior field of sagittae at 400-1000x magnification. Rings in both sagittae from each fish were counted three times each; repeat counts were not consecutive. The mean of three ring counts per sagitta was rounded to an integer and used in analyses.

The relation between mean ring count and known age (days after swim-up) was determined using Theil-Sen linear regression procedures (Hollander and Wolfe 1973). Theil-Sen procedures were used because unequal sample sizes among different sample days might invalidate the assumption of bivariate normality. The null hypothesis that the slope of this regression was equal to one, as would be expected if rings formed daily, was tested by Kendall's test for bivariate association. Confidence intervals on the slope and y-intercept were based on Hodges-Lehman estimators.

Beginning in May 1982, young-of-year smallmouth bass were collected periodically from 2 sites in the New River drainage in Virginia to determine when ring formation ceased to be one per day. Collections were continued into January 1983 to ensure that young-of-year with very low growth rates were represented. Fish collected between May and October were used to assess the applicability of otolith aging for growth studies. Descriptions of study sites and collection methods were detailed in Graham (1984).

Sagittae of fish less than 15 mm total length (TL) were prepared as described for laboratory-reared fish. Sagittae of fish longer than 15 mm TL were mounted on glass slides in thermoplastic cement with their convex side facing the slide. The concave side of sagittae were ground with number 600 wet sandpaper until a flat plane was achieved between sagitta and slide, as Miller and Storck

(1982) did for their sagittal sections. Sagittae were then turned, and the convex side was ground until the nucleus was clearly exposed. Sagittae were turned again and rings were counted. Three ring counts were made of one sagitta from each fish collected. The mean number of sagittal rings in the right and left sagittae of fifteen randomly selected fish from each of four 10-mm size categories (40-80 mm) were compared by Wilcoxon signed-ranks tests. Counts of right and left sagittae did not differ within any size group ($P < 0.05$); therefore, the use of only one sagittae from each wild fish did not introduce error in our analysis.

Two methods were used to determine when the rate of sagittal ring formation was less than one per day in wild fish. The first method consisted of plotting mean ring counts for sagittae against collection dates for each study site. These plots were then examined for departures from linearity. The second method involved taking notes during ring counts regarding the degree to which crowding of rings affected our ability to distinguish individual rings. Collection dates and study sites were matched with microscopic observations when counts were complete.

Growth of wild young-of-year smallmouth bass collected from the New River near Eggleston and Walker Creek was examined on a short-term and a seasonal basis. Instantaneous growth rates, G, (Ricker 1975) for the period from swim-up to collection were determined for individual smallmouth bass from total length at capture (l_t) and the mean number of rings counted in the sagitta

(age in days) of each fish collected. Length at time of swim-up was assumed to be 8.4 mm based on the mean total length of 17 swim-up fry collected from a nest in the New River near Eggleston on 30 June, 1982. Therefore, $G = \log_e (l_t/8.4)/t$, where l_t = length

(mm) at age t (days).

Length at age and length at collection time data were used to fit first year growth of smallmouth bass at each study site to von Bertalanffy's growth model (Ricker 1975) using nonlinear regression.

RESULTS AND DISCUSSION
      The sagittae of 82 fish (N=164) laboratory-reared fish were examined for validation of daily increment formation in smallmouth bass otoliths. Thirty-three sagittae were severely fragmented and 12 were considered unreadable; hence 119 sagittae were used in the analyses. Fragmentation was thought to be the result of a power failure during which samples had thawed and later refrozen. Sagittae were considered unreadable when the saccular membrane obscured the rings.

      The slope of the regression of mean ring counts on known ages of laboratory-reared smallmouth bass larvae was 0.970 (95% confidence interval 0.900-1.000) and did not differ significantly from 1.0 (P = 0.02), indicating that sagittal rings formed daily through 14 days post swim-up. However, mean ring counts of sagittae

from laboratory fish sacrificed on the same day generally exhibited
a wide range of values (Fig. 1). Variability in mean ring counts
may reflect variation in the rate of ring formation of individual
fish. If so, these individuals comprised a relatively small
proportion of the sample as evidenced by the small standard errors
about mean ring counts and a slope value not significantly different
from one. It is more likely that variation in ring counts resulted
from difficulty in distinguishing between daily and subdaily rings.

The sagittae of laboratory-reared smallmouth bass reared for
this study had faint, poorly defined rings in contrast to sagittae
from field-collected larval smallmouth bass collected from two
sites in the New River drainage. Radtke and Dean (1982) and Tsuji
and Aoyama (1982) also found that sagittae from field-collected
specimens exhibit more prominent rings than sagittae of the same
species from laboratory-reared fish. Further, difficulty in
distinguishing between subdaily and daily rings has been considered
a major source of counting error when aging laboratory-reared
larvae (Tsuji and Aoyama 1982). Therefore, in the present study,
most variation in mean ring counts of fish sacrificed on the same
day probably was due to erroneously assuming daily rings were
subdaily rings, or vice versa.

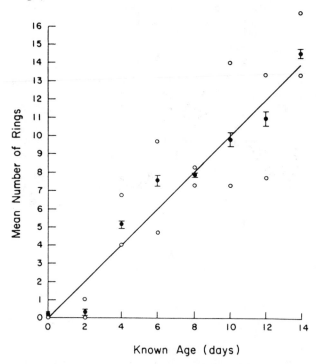

Figure 1. Relationship between mean number of sagittal rings and
known age of fish after swim-up. Bars represent one standard
error from the mean of any sampling day's counts and circles
denote the range in mean ring counts.

The y-intercept of the relationship between mean ring counts
and known ages (0.420; 95% confidence interval 0.350-0.500)
indicated that formation of the first daily ring occurred the day
of swim-up. This time of ring formation is consistent with
observations on other centrarchids. The first daily ring forms at
swim-up in pumpkinseed Lepomis gibbosus, green sunfish Lepomis
cyanellus (Taubert and Coble 1977), and largemouth bass Micropterus
salmoides (Miller and Storck 1982). These studies and our
observations support the hypothesis that initiation of daily
sagittal ring formation occurs at swim-up when centrarchids switch
to exogenous feeding.

The mean ring count of sagittae from 208 wild young-of-year
smallmouth bass increased rectilinearly with collection date before
November at both study sites (Fig. 2). Two linear trends were

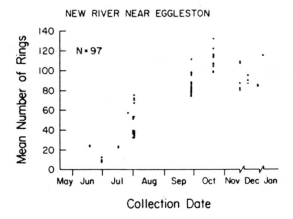

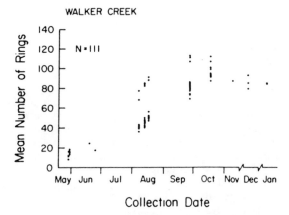

Figure 2. Mean number of rings counted on individual sagitta
    plotted against collection date of smallmouth bass from the New
    River near Eggleston and Walker Creek, Virginia.

apparent at each study site that were due to age differences among
young-of-year that had been spawned during different spawning
periods (Graham 1984). Beginning in November, a sharp departure
from linearity occurred (Fig. 2). November also was the time when
extensive crowding of rings in the posterior edge of sagittae began
to affect our ability to distinguish individual rings. Crowding
of rings was noted in sagittae of fish collected between late
October 1982 and January 1983, indicating that little or no otolith
growth occurred during this interval.

These observations agree with results of winter simulation
experiments by Taubert and Coble (1977) on green sunfish. They
found that ring formation apparently stopped when fish were exposed
to water temperatures around 10 C. The mean daily water
temperatures of the New River near Eggleston and Walker Creek during
the interval 16 October to 31 October were approximately 14 and 10
C, respectively. In the first two weeks of November the mean daily
water temperatures had dropped to approximately 11 and 8 C,
respectively. Since young-of-year smallmouth bass cease feeding
and become lethargic at temperatures around 10 C (Oliver et al.
1979), it appears that water temperature determines the time of year
when sagittal rings cease to be reliable indicators of age. Hence,
rings in the sagittae of smallmouth bass can be used to determine
the number of growing days prior to the first winter, as Taubert
and Tranquilli (1982) demonstrated for largemouth bass.

Only those fish collected prior to November were included in
growth analyses of field-collected young-of-year. Growth rates of
young-of-year smallmouth bass were inversely related to age (P <
0.01) (Fig. 3). Spearman's rank correlation coefficient for the
relationships between instantaneous growth rate and daily age for
the fish collected from the New River near Eggleston and Walker
Creek were -0.90 and -0.79, respectively.

Instantaneous growth rates of very young field-collected fish
(< 20 days past swim-up) were highly variable, particularly for the
Walker Creek collections (Fig 3). This may have been an artifact
of calculating instantaneous growth rates using an assumed constant
length at swim-up of 8.4 mm or aging error. Smallmouth bass total
lengths at swim-up have been reported to range between 8.0 mm
(Pflieger 1975) and 9.0 mm (Wallace 1972). Small differences in
the length at swim-up of individual fish would have had the greatest
effect on estimates of instantaneous growth rates calculated for
smaller fish. Similarly, aging error would have had the greatest
effect on calculated growth rates of smaller (younger) fish.
However, rings in the sagittae of wild young-of-year were very well
defined. Deviation about mean ring counts for individual fish never
exceeded 3 rings during the collection period May through July when
the youngest fish were captured. It is doubtful that these sources
of error can account for the wide range in estimates of
instantaneous growth rates of young fish. We conclude that the
variability of our calculated growth rates for very young
smallmouth bass was representative of actual environmental
conditions.

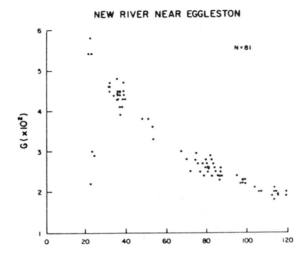

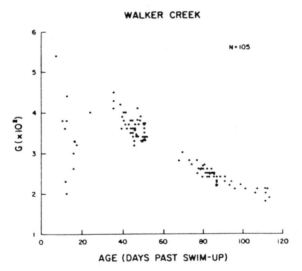

Figure 3.   Relationship between instantaneous growth rate (G) and
age of smallmouth bass from the New River near Eggleston and
Walker Creek, Virginia.

DeAngelis and Coutant (1979) observed that during the first
few weeks of life variance about the mean length of
laboratory-reared smallmouth bass increased through time. They
hypothesized that this relationship may be due to individual,
perhaps genetic, variability among fish.   In the dynamic

environments from which we collected young-of-year smallmouth bass, distributions of food items are generally patchy, unlike the laboratory environment. An environment where food availability is highly variable tends to augment any inherent variability in growth among individual fish and promote higher mortality among slower-growing individuals. Over time, this may reduce variability in growth rates of young-of-year.

The estimates of the von Bertalanffy growth parameters ($L\infty$, K, and t_o) determined from data on length at daily age and length at collection time did not differ greatly (Table 1). However, comparison of 95% confidence intervals on the parameter estimates indicates that use of daily age data increased the precision of these estimates over the traditional method of basing von Bertalanffy growth parameters on length at collection time data. Young-of-year fishes are particularly sensitive to environmental conditions, and growth rates reflect the environment in which fish are raised. Precise determinations of growth rates obtained from otolith-aging techniques can allow the effects of recent environmental changes to be evaluated and used to monitor the well-being of fish populations.

Table 1. Comparison of von Bertalanffy parameter estimates and 95% confidence limts for smallmouth bass from length at collection time (LC) versus length at daily age past swim-up (LA) data.

Method	$L\infty$ (mm)	K(day^{-1})	t_o (days)
	New River near Eggleston N = 81		
LC	92.5	0.014	167.1[a]
	(75.1, 109.9)	(0.007, 0.021)	(159.4, 174.8)
LA	98.2	0.016	3.4
	(85.7, 110.6)	(0.011, 0.021)	(−0.5, 7.3)
	Walker Creek N = 105		
LC	134.1	0.005	130.16[b]
	(57.4, 210.9)	(0.001, 0.009)	(117.6, 142.6)
LA	102.6	0.012	0.0
	(84.5, 120.7)	(0.008, 0.017)	(−4.3, 4.3)

[a] Julian date − June 16
[b] Julian date − May 10

ACKNOWLEDGMENTS

This study was supported by the National Park Service through Colorado State University. Steve Miller, Ted Storck, and Bruce Taubert reviewed portions of the manuscript and provided valuable suggestions. John Ney and Richard Neves made important comments during all phases of the study, and Douglas Austen and Paul Leonard assisted with collections.

REFERENCES

BROTHERS, E. B., C. P. MATHEWS, and R. LASKER. 1976. Daily growth increments in the otoliths from larval and adult fishes. Fisheries Bulletin 74:1-8.

DE ANGELIS, D. L., and C. C. COUTANT. 1979. Growth rates and size distributions of first-year smallmouth bass populations: some conclusions from experiments and a model. Transactions of the American Fisheries Society 108:137-141.

GEFFEN, A. J. 1982. Otolith ring deposition in relation to growth rate in herring (Clupea harengus) and turbot (Scophthalmus maximus) larvae. Marine Biology 71:317-326.

GRAHAM, R. J. 1984. Time and duration of spawning by smallmouth bass (Micropterus dolomieui) in the New River drainage. Master's thesis, Virginia Polytechnic Institute and State University, Blacksburg, Virginia, USA.

HOLLANDER, M., and D. A. WOLFE. 1973. Nonparametric statistical methods. John Wiley and Sons, New York, USA.

MILLER, S. J., and T. STORCK. 1982. Daily growth rings in otoliths of young-of-the-year largemouth bass. Transactions of the American Fisheries Society. 111:527-530.

OLIVER, J. D., G. F. HOLETON, and K. E. CHUA. 1979. Overwinter mortality of fingerling smallmouth bass in relation to their size, percent storage materials and environmental temperature. Transactions of the American Fisheries Society 108:130-136.

PANNELLA, G. 1971. Fish otoliths: daily growth layers and periodical patterns. Science 173:1124-1127.

PFLIEGER, W. L. 1966. Reproduction and survival of the smallmouth bass (Micropterus dolomieui) in a small Ozark stream. American Midland Naturalist 76:410-418.

RADTKE, R. L., and J. M. DEAN. 1982. Increment formation in the otoliths of embryos, larvae, and juveniles of the mummichog, Fundulus heteroclitus. Fisheries Bulletin 80(2):201-216.

RICKER, W. E. 1975. Computation and interpretation of biological statistics of fish populations. Fisheries Research Board of Canada Bulletin 191.

SCHMIDT, R. E., and M. C. FABRIZIO. 1980. Daily growth rings on otoliths for aging young-of-the-year largemouth bass from a wild population. Progressive Fish-Culturist 42:78-80.

TAUBERT, B. D., and D. W. COBLE. 1977. Daily rings in otoliths of three species of Lepomis and Tilapia mossambica. Journal of the Fisheries Research Board of Canada 34:332-340.

TAUBERT, B. D., and J. A. TRANQUILLI. 1982. Verification of the formation of annuli in otoliths of largemouth bass. Transactions of the American Fisheries Society 111:531-534.

TSUJI, S. and T. AOYAMA. 1982. Daily growth increments observed in otoliths of the larvae of Japanese Red Sea Bream Pagrus major (Temminck and Schlegel). Bulletin of the Japanese Society of Scientific Fisheries. 48(11):1559-1562.

WALLACE, C. R. 1972. Embryonic and larval development of the smallmouth bass at 23 C. Progressive Fish-Culturist 34:237-239.

VII
Applied studies

Application of information on age and growth of fish to fishery management[1]

DANIEL PAULY

INTERNATIONAL CENTER FOR
LIVING AQUATIC RESOURCES MANAGEMENT
MCC P.O. BOX 1501
MAKATI, METRO MANILA
PHILIPPINES

ABSTRACT
    Information on the age and growth of fish is usually related
to one or several of the following items: construction of
age/length keys for direct use in some assessment models (e.g.
Virtual Population Analysis); estimation of growth parameters,
used as inputs in some other models (e.g., yield-per-recruit);
comparative studies of growth performance in different stocks and
species (i.e., in studies of environmental factors affecting
growth); and studies of the anatomical and physiological
mechanisms underlying the growth performance in fish. The
relative importance of these aspects is discussed in the context
of fisheries management in the Northeast Atlantic, Kuwait and
Southeast Asia, which are considered as representative of three
major types of situations encountered worldwide in fishery
management.

    The collection and analysis of age and growth of fish
information is a time-consuming activity. It may be a
preoccupation of some fishery biologists. This Symposium is
certainly not the place to challenge the relevance of this
activity. What should be done here, however, is to ask ourselves
why knowing about the age and growth of fish is important, and
what are the uses to which such knowledge is being put in
different parts of the world .
    I have not undertaken a formal review; rather, I have chosen
a comparative approach, in which the points I want to make are
illustrated by the research (and/or routine assessments)
conducted in three specific areas of the world: Northeastern
Atlantic, Southeast Asia, and Kuwait.

[1]ICLARM Contribution No. 244.

The Age and Growth of Fish, edited by Robert C. Summerfelt and Gordon E. Hall © 1987 The Iowa
State University Press, Ames, Iowa 50010.

The first area, the Northeastern Atlantic, corresponds roughly to the jurisdiction of the International Council for the Exploration of the Sea (ICES). Scientists working there operate in developed countries on heavily exploited resources, about which a considerable amount of previous knowledge exists.

Southeast Asia, besides having the world's most diverse marine fauna, is characterized by extremely strong pressures on its natural resources and the lack of conservation of those resources. The fisheries differ in a number of important features from those of high latitudes. Among the differences important to fishery management are the short life-span of the bulk of the organisms (fish, squids, shrimps, etc.) exploited by the fisheries. Also, the resources are concentrated in shallow waters, with commercially exploitable concentrations usually occurring in waters less than 50 m deep within the coastal belt accessible to artisanal fishermen. Fisheries research in Southeast Asia is hampered by a number of factors, the key one probably being the shoestring budgets under which most of the laboratories operate. Factors hampering fishery management are, in the main, governments' inabilities to formulate a science-based fishery policy and to enforce regulations.

Kuwait is a small country with enough oil income to set up and maintain a first-rate research institution, the Kuwait Institute of Scientific Research (KISR); the Fisheries and Aquaculture Division of KISR is in charge of developing the marine resources and supplying government with management advice. The contrast in Kuwait of a sophisticated research institution and partly underexploited, under-investigated tropical resources as the object of its research resembles other situations, namely, those faced by the US National Marine Fisheries Service Laboratory in Hawaii, in charge of resources in the Central Pacific; the French ORSTOM laboratory in Vanuatu, in charge of South Pacific resources; and the CSIRO Marine Laboratory in Cronulla, Australia, in charge of the marine resources on the Northwest Shelf of Australia.

Four aspects of age and growth studies to be considered with respect to these geographic areas are:

1. aging, age-validation and the construction of age-length keys;
2. estimation of growth parameters using data obtained in (1), or other methods (particularly length-based methods);
3. comparative studies of fish growth performance;
4. studies of the anatomical and physiological mechanisms underlying the growth performance of fish.

Only marine fish and fisheries will be discussed.

FISH GROWTH STUDIES VS. FISHERIES MANAGEMENT

The Northeast Atlantic Area
    The largely international fisheries of the Northeastern
Atlantic area are managed mainly through national quotas (i.e.,
through allocation of annual TAC's = Total Allowable Catch)
within the ICES area, of which a major part overlaps with the
waters of the European Community (EC). TAC's are determined
through a formal and annually repeated sequence to:

-    Establish age-length keys and compilation of quarterly
     catch-at-length and catch-at-age data for all major
     exploited species. This work is usually performed in
     national laboratories;
-    Amalgamate several rational data sets into "total
     international catch-at-age" data for each major species and
     assessment of these, usually by a form of Virtual Population
     Analysis (VPA, see Pope 1972). This analysis is usually
     performed by international working groups, whose members are
     scientists drawn from the national laboratories of
     interested countries.
-    Review of the working groups' reports and formulation of
     management advice by the Advisory Committee on Fisheries
     Management (ACFM), a body of ICES whose members (usually
     fishery scientists) act as representatives of their
     countries. The annual ACFM report is then sent to the
     relevant management authority, usually a department of the
     EC administration in Brussels;
-    Political negotiations over TACs, etc.

    Almost all management advice and assessments in the ICES
area are based on annually updated catch-at-age data provided by
this elaborate and expensive research machinery. Some estimates
of the cost of the annual information updates are as high as 80%
of all research costs by major laboratories, which have about 20%
of their overall budget going to ageing and related studies
(see also Casselman et al. 1986).
    Important also is that ageing of very old individual fish in
the context of VPA-oriented research is not mandatory, because
one can create for them a so-called "plus-group" with
corresponding catch and biomass (Mesnil 1970; also Beamish and
McFarlane 1986).
    The fact that the major breakthroughs on the growth of
European fishes were achieved at the beginning of this century
(see bibliography in Mohr 1927, 1930, 1934) implies that fish
ageing has, in the ICES area, ceased to be a major area of
research; this contrasts with the Mediterranean area, where age
and growth studies are deemed important enough to justify special
emphasis by a major funding agency (The European Community, in
its guidelines for proposal submission, for the year 1984-1985).
    Major insights into the general area of fish growth have
been obtained in the ICES area, however, and these concern the

relationship between food intake (i.e., predation) and growth performance of individual year-classes. These studies, largely initiated by the landmark study by Daan (1973) and the work of the "Danish school" (Ursin 1967; Andersen and Ursin 1977; Helgason and Gislason 1979; and Sparre 1980), led to a strong awareness of the role played by major predators in exploited marine ecosystems and of the relationship between the food availability and the growth performance of the fishes of a given year-class. Unfortunately, the models used in the North Sea area required a large data base (e.g., over 10^5 stomach contents of North Sea fishes were analyzed in a recent ICES exercise). These models, therefore, do not seem applicable to other areas without empirical data on trophic dynamics. Additionally, most of the models used are based on ages instead of size, even when the latter (e.g., mesh-selection ogives) would be biologically more appropriate.

Although the Northeast Atlantic stocks are presently still managed as single-species stocks, the various multispecies assessments conducted in the North Sea area, have convinced most interested parties of the need to account for species interactions, and of the role of a variable food supply to the growth and maintenance of fish stocks. It can be expected that these insights will gradually percolate to other areas of the world, even if model-building does not evolve to the complex form it has taken in Western Europe.

Southeast Asia

While self-managing artisanal fisheries in Southeast Asia go back to time immemorial, commercial fisheries, particularly bottom trawl fisheries, are quite recent. They started on a grand scale, after relatively unsuccessful pre World War II attempts by the colonial powers, in the 50s in the Philippines, the 60s in Thailand, and the 70s in Indonesia and Malaysia. These new commercial fisheries, bursting as they did onto the national and international scene, faced governments without the institutional apparatus necessary for successfully managing them. As a consequence, major fisheries became overcapitalized a decade after they started (Pauly 1979).

Half-hearted attempts were undertaken by several governments to prevent trawlers from fishing too close inshore, particularly where they were depleting the resources upon which artisanal fishermen depended. These attempts largely failed, as did efforts to increase the mesh size used by these trawlers from the commonly used 20 mm stretched mesh. The only "success story" related to the management of the demersal trawl fisheries in Southeast Asia has been the total ban on trawling in Indonesian waters (Sarjono 1980). Here successful enforcement was put in the hands of the artisanal fishermen who demanded it.

This should not give the impression, however, that fishery biologists based in Southeast Asia have been idle. The life histories, mortality rates, and growth parameters of a number of commercial species have been elucidated or estimated (Simpson 1982; Ingles and Pauly 1984). Thus, it has become possible in principle to provide management advice based not only on the generally used surplus production models, but also on suitable modifications of analytical models such as the yield-per-recruit model of Beverton and Holt (1957). An example of this is the length-structured yield-per-recruit model of Beverton and Holt (1964), which requires estimates of $L\infty$, of the mean size at first capture (L_c), and of the ratio M/K (expressing natural mortality as related to growth). This model, which provides estimates of relative yield per recruit as a function of exploitation rate ($E = F/Z$), has been extended for use in a multispecies context by Sinoda et al. (1979).

A nother model which might turn out to be particularly useful for Southeast Asian stocks is that of Csirke and Caddy (1983), in which annual catch is plotted on corresponding estimates of total mortality (Z) or on the ratio Z/K. These latter quantities can be estimated through length-converted catch curves (sensu Pauly 1982, 1984), or methods proposed by Jones (1984), i.e., combining growth parameter estimates ($L\infty$ and K) with length-frequency samples.

It appears that estimates of growth parameters in commercial species will become increasingly important in Southeast Asia. Their increased availability would make possible the application of some of the more elaborate management schemes that have been proposed when fishery management institutions and regulatory mechanisms have emerged.

It is my opinion that daily rings, often proposed as a panacea for the ills of those trying to age tropical fish, will not replace length-frequency data as the basic source of age and growth information on Southeast Asian fish. Rather, daily rings will be used to corroborate inferences drawn from length-frequency data (Gjøsaeter and Sousa 1983). An even better approach is to simultaneously estimate growth parameters from length and age data, as is now possible using a method proposed by Morgan (1986).

Kuwait

An interesting intermediate stage between the large-scale, routine ageing activities in the North Atlantic area and the difficult stage in which fishery research finds itself in Southeast Asia is provided in Kuwait. There, management advice is actively sought by both government and industry, and a well-equipped and well-staffed research institute (KISR), is faced with stocks not previously investigated.

Studies on fish stocks in Kuwait began in 1980, and a capability for ageing fish was establishing in 1981. Work at first concentrated on "age-length keys." In 1982, fisheries scientists in Kuwait also began using the ELEFAN I program of Pauly and David (1981), a computer-based objective method for the estimation of the growth parameters $L\infty$ and K of the von

Bertalanffy growth function from length-frequency data. Also, length-converted catch curves were used for the estimation of total mortality and estimation of mean length at first capture using the ascending, left arm of the curve (Pauly 1984; Morgan 1985).

Of the various species investigated in this fashion, only two Pampus argenteus and Otolithes argenteus are actually short-lived enough for the unmodified ELEFAN I program to be directly applied to them. In these two cases, results were obtained from the length-based analysis which matched closely those obtained from an age-based analysis (Morgan, 1985; Mathews and Samuel 1985).

Encouraged by this, the KISR researchers then attempted to apply the length-based ELEFAN methods to longer-lived fishes (for which the method was not originally designed), notably to Epinephelus tauvina, Lutjanus coccineus (= L. malabaricus) and Acanthopagrus latus. The results of these studies, reported in a number of technical reports, were summarized by Mathews (1986). They suggest that approaches based solely on length data become increasingly more unreliable as the investigated fish become longer-lived. This observation is confirmed by recent Monte-Carlo simulation studies of ELEFAN I (Hampton and Majkowski 1986; Rosenberg and Beddington 1986), which show that growth parameters estimated by ELEFAN I become increasingly biased when the overlap between adjacent length-frequency distributions increases, as occurs in fish with a long life-span and/or a long recruitment season. This bias is in part self-compensatory, with the overestimation of L_∞ partly compensating for the underestimation of K, and the resulting growth curves being very similar to the original curves.

The optimum solution to this problem, however, is the use of additional information on the growth of the fish concerned, either in the form of growth increments, i.e., tagging data (Kirkwood 1983), or, preferably, age-length data pairs obtained by reading otoliths. However, as emphasized by Morgan (1986), the goal, when working with both age- and length-frequency data, should not be for these to be analyzed separately, but to combine the strength of age-based and length-based approaches such that a single set of precise and accurate growth parameter estimates is obtained through simultaneous analysis of both types of data.

Morgan (1986) recently developed a method to perform such combined analysis. The method, a modification of ELEFAN I soon to be made available as "ELEFAN IV," estimates a best set of growth parameters from length-frequency data, age-length pairs obtained from reading of hard parts, and growth-increment data resulting from tagging/marking experiments, or any combinations of the three data types.

Exercises conducted to date with this method (Morgan 1986, and pers. comm.) suggest that it will provide cost-effective estimation of growth parameters in long-lived fishes, because the age information obtained from otoliths (i.e., the information

that is usually the most costly to obtain) is not used to
determine the shape of a growth curve. Rather it is used to
select from the number of local optima on the response surface
generated by the analysis of the length-data the one that is most
compatible with the available age-length observations.

This precedure requires length-frequency data and only a few
age readings to achieve stability of growth-parameter estimates
(Fig. 1). This is a feature of particular interest in areas where
fish destined for ageing studies must be purchased from fishermen
(Morgan 1983), and in which the expertise for otolith reading must
be imported (Mathews 1986).

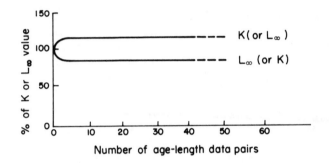

Fig. 1. Schematic representation of the effect of incorporating
        age-at-length information into an analysis of
        length-frequency data using ELEFAN V. Number of
        age-length data pairs in ordinate scale needed to
        stabilize growth parameter estimate stem from analysis of
        data on Lutjanus coccineus (G. R. Morgan, pers.
        comm.).

Growth studies of fish are being conducted in Kuwait for
stock assessment, and also to identify species suitable for
mariculture. In both Europe and Southeast Asia the screening of
species for large-scale culture is essentially completed, the
pros and cons of the various species, inclusive of their growth
performance under culture, being essentially known. Thus,
research on sea-going salmonids in Europe and on milkfish
Chanos chanos in Southeast Asia tends to be oriented toward
fine-tuning of long-established production systems, such as
genetic improvement of the cultivated animals (Gjedrem 1983), or
identification of constraints to intensive production (Chong et
al. 1984). In Kuwait, on the other hand, this stage has not been
reached, and species suitable for mariculture under Kuwaiti
conditions still need to be identified.

It can be expected that fish, which in the wild quickly
reach large size, should be anatomically and physiologically
equipped to perform well under aquaculture conditions (Pauly
1981). In the present Kuwaiti context, this implies that only

those fish for which previous age and growth studies have produced evidence of high growth rates should be screened for their suitability for mariculture. This procedure saves time and resources since it is cheaper and easier to age fish than to obtain live fry from the wild and perform growth experiments with them.

DISCUSSION

In temperate waters, where fish tend to be longer-lived than in the tropics, the use of large computers, the choice of analytic methods (mainly VPA), and the availability of detailed catch data for the major exploited species allow for fishery management to rely predominantly on catch-at-age information. Consequently, vast resources are invested in the annual reestimation of age-length keys. The research leading to ageing methods for the major exploited species was conducted in the first half of the century, and further major advances may not be expected in ageing methods per se; however, because of emphasis on multispecies modelling, further advances can be expected concerning the ecology of fish, particularly the relationships between food intake (i.e., predation) and growth, and hence on the density dependence of growth.

In tropical waters, the specific features of the marine resources and the lack of an adequate infrastructure in the fishery sector render detailed, species-specific investigations and stock assessments less relevant than in other areas. Rather, "gross" assessment methods have to be used, such as treating a species assemblage as if it consisted of a single species. This situation makes comparative studies of the growth performance of various species very useful, especially for the identification of groups of species similar enough for them to be lumped for assessments (Pauly 1979). However, this same situation and the overall scarcity of resources for research make major advances on growth of fish (including further advances based on daily rings in otoliths) unlikely to emerge in the near future from laboratories in the developing countries of the intertropical belt.

The combination of a well-equipped laboratory, with a well-trained staff and little-studied stocks seems to be most conducive to major advances concerning the growth of fish, especially as far as cost-effective, length-based approaches are concerned. Recent work conducted in Northern Australia by K. Sainsbury and associates (Sainsbury 1984, Sainsbury and Whitelaw 1984); in Hawaii by Wetherall et al. (1986); in the South Pacific area by Brouard and Grandperrin (1984); as well as work conducted in Kuwait and discussed above, illustrate this contention.

ACKNOWLEDGMENTS

I wish to thank Drs. Z. Shehadeh, G. Morgan, C. P. Mathews, KISR for their support during several stays in Kuwait. Thanks are also due to Drs. S. Saila, J. A. Gulland and the anonymous reviewers for their editorial suggestions.

REFERENCES

ANDERSEN, K. P., AND E. URSIN. 1977. A multispecies extension to the Beverton and Holt theory of fishing, with accounts of phosphorus circulation and primary production. Meddelelser fra Danmarks Fiskeri-og Havundersogelser (N.S.) 7:319-435.

BEAMISH, R. J., AND G. A. MCFARLANE. 1986. Current trends in age determination methodology. In R. C. Summerfelt and G. E. Hall, editors, Age and growth of fish, Iowa State University Press, Ames, Iowa, USA.

BEVERTON, R. J. H., AND S. J. HOLT. 1957. On the dynamics of exploited fish populations. Fisheries Investigations Series II, Vol. 19.

BEVERTON, R.J.H., AND S.J. HOLT. 1964. Tables of yield function of fishery management. Food and Agricultural Organization of the United Nations (FAO) Fisheries Technical Paper 38, Rome.

BROUARD, F., AND R. GRANDPERRIN. 1984. Les poissons profonds de la pente récifale externe à Vanuatu. Notes et Documents d'Océanographie No. 11, ORSTOM, Vanuatu.

CASSELMAN, J. M., M. POWELL, AND R. BIETTE. 1985. Age and growth determination in fisheries - a review of the science and technology as currently practised. International Symposium on Age and Growth of Fish, Des Moines, Iowa, June 9-12, 1985, Abstract 1-3:33-34.

CHONG, K. C., M. S. LIZARDO, Z. S. DELA CRUZ, C. V. GUERRERO, AND I. R. SMITH. 1984. Milkfish production dualism in the Philippines: a multidisciplinary perspective on continuous low yields and constraints to aquaculture development. International Center for Living Aquatic Resources Management Technical Reports 15, Manila.

CSIRKE, J., AND J. F. CADDY. 1983. Production modelling using mortality estimates. Canadian Journal of Fisheries and Aquatic Science 40:43-51.

DAAN, N. 1973. A quantitative analysis of the food intake of North Sea Cod, Gadus morhua. Netherlands Journal of Sea Research 6:479-517.

GJEDREM, T. 1983. Genetic variation in quantitative traits and selective breeding in fish and shellfish. Aquaculture 33:51-72.

GJØSAETER, J., AND M. I. SOUSA. 1983. Reproduction, age and growth of the Russel's scad, Decapterus russelii (Rüppel, 1828) (Carangidae) from Sofala Bank. (Maputo) Mozambique. Revista de Investigaciones Pesqueria 8:83-107.

HAMPTON, J., AND J. MAJKOWSKI. 1986. An examination of the

accuracy of the ELEFAN computer programs for length-based
stock assessments. In D. Pauly and G. Morgan, editors,
Length-based methods in fisheries research, International
Center for Living Aquatic Resources Management Conference
Proceedings, Manila, Philippines (in Press).

HELGASON, T., AND H. GISLASON. 1979. VPA-analysis with species
interactions due to predation. International Council for
the Exploration of the Sea, Council Meetings 1979/G:52.

INGLES, J., AND D. PAULY. 1984. An atlas of the growth,
mortality and recruitment of Philippine fishes.
International Center for Living Aquatic Resources Management
Technical Reports 13, Manila.

JONES, R. 1984. Assessing the effects of changes in
exploitation pattern using length composition data. Food
and Agricultural Organization of the United Nations (FAO)
Fisheries Technical Paper No. 256, Rome.

KIRKWOOD, G. P. 1983. Estimation of von Bertalanffy growth
curve parameters using both length-increment and age-length
data. Canadian Journal of Fisheries and Aquatic Science
40:1405-1411.

MATHEWS, C. P. 1986. Fisheries management in a tropical
country: the most appropriate balance of size and age/length
- related methods for practical assessments. In D. Pauly
and G. Morgan, editors, Length-based methods in fisheries
research, International Center for Living Aquatic Resources
Management Conference Proceedings, Manila, Philippines (in
Press).

MATHEWS, C. P. AND M. SAMUEL. 1985. Stock assessment
of Newaiby, Hamoor and Hamra in Kuwait. Pages 67-115 in
C. P. Mathews, editor, Proceedings of the 1984 Shrimp and
Fin Fisheries Management Workshop, Kuwait Institute for
Scientific Research, Salmiyah, Kuwait.

MESNIL, B. 1980. Théorie et pratique de l'analyse des
cohortes. Revue des Travaux de l'Institut des Pêches
Maritimes, 44:119-155.

MOHR, E. 1927. Bibliographie des Alters - und
Wachstumsbestimmung bei Fischen. Journal du Conseil
International pour l'Exploration de la Mer 2:236-258.

MOHR, E. 1930. Bibliographie des Alters-und Wachstumsbestimmung
bei Fischen II. Nachträge und Fortsetzung. Journal du
Conseil International pour l'Exploration de la Mer
5:88-100.

MOHR, E. 1934. Bibliographie des Alters - und
Wachstumsbestimmung bei Fischen III. Nachträge und
Fortsetzung. Journal du Conseil International pour
l'Exploration de la Mer 9:377-391.

MORGAN, G. R. 1983. Application of length-based stock
assessments to Kuwait's fish stocks. International Center
for Living Aquatic Resources Management Newsletter 6(4):3-4.

MORGAN, G. R. 1985. Stock assessment of pomfret (Pampus
argenteus) in Kuwait waters. Journal du Conseil

International pour l'Exploration de la Mer 42(1):3-10.

MORGAN, G. R. 1986. Incorporating age data into length-based stock assessment methods. In D. Pauly and G. Morgan, editors, Length-based methods in fisheries research, International Center for Living Aquatic Resources Management Conference Proceedings, Manila, Philippines (in Press).

PAULY, D. 1979. Theory and management of tropical multispecies stocks: a review, with emphasis on the Southeast Asian demersal fisheries. International Center for Living Aquatic Resources Management Studies and Reviews 1, Manila.

PAULY, D. 1981. The relationship between gill surface area and growth performance in fish: a generalization of von Bertalanffy's theory of growth. Meeresforschung/Reports on Marine Research 28:251-282.

PAULY, D. 1982. Studying single-species dynamics in a multispecies context. Pages 33-70 in D. Pauly and G. I. Murphy, editors, Theory and management of tropical fisheries, International Center for Living Aquatic Resources Management Conference Proceedings 9, Manila.

PAULY, D. 1984. Fish population dynamics in tropical waters: a manual for use with programmable calculators. International Center for Living Aquatic Resources Management Studies and Reviews 8, Manila.

PAULY, D., AND N. DAVID. 1981. ELEFAN I, a BASIC program for the objective extraction of growth parameters from length-frequency data. Meeresforschung/Reports on Marine Research 28:205-211.

POPE, J. G. 1972. An investigation of the accuracy of Virtual Population Analysis. International Commission for the Northwest Atlantic Fisheries Research Bulletin 9:65-74.

ROSENBERG, A. A., AND J. R. BEDDINGTON. 1986. Monte Carlo testing of two methods for estimating growth from length-frequency data, with general conditions for their applicability. In D. Pauly and G. Morgan, editors, Length-based methods in fisheries research, International Center for Living Aquatic Resources Management Conference Proceedings, Manila, Philippines (in Press).

SAINSBURY, K. J. 1984. Optimal mesh size for tropical multispecies trawl fisheries. Journal du Conseil International pour l'Exploration de la Mer 41:129-139.

SAINSBURY, K. J., AND A. W. WHITELAW. 1984. Biology of Peron's threadfin bream, Nemipterus peronii (Valenciennes) from the North West Shelf of Australia. Australian Journal of Marine and Freshwater Research 35:167-185.

SARJONO, I. 1980. Trawlers banned in Indonesia. International Center for Living Aquatic Resources Management Newsletter 3(4):3.

SIMPSON, A. C. 1982. A review of the data base on tropical multispecies stocks in the Southeast Asian Region. Pages 5-32 in D. Pauly and G. I. Murphy, editors, Theory and Management of Tropical Fisheries, International Center for

Living Aquatic Resources Management Conference Proceedings 9, Manila, Philippines.

SINODA, M., S. M. TAN, Y. WATANABE, AND Y. MEEMESKUL. 1979. A method for estimating the best cod-end mesh size in the South China Sea area. Bulletin of the Choshi Marine Laboratory, Chiba Univ. 11:65-80.

SPARRE, P. 1980. A goal function of fisheries (Legion analysis). International Council for the Exploration of the Sea, Council Meetings 1980/G:40.

URSIN, E. 1967. A mathematical model of some aspects of fish growth, respiration and mortality. Journal of the Fisheries Research Board of Canada 13:2355-2453.

WETHERALL, J. A., J. J. POLOVINA, AND S. RALSTON. 1986. Estimating growth and mortality in steady state fish stocks from length frequency data. In D. Pauly and G. Morgan, editors, Length-based methods in fisheries research, International Center for Living Aquatic Resources Management Conference Proceedings, Manila, Philippines (in Press).

Growth in length of Pacific halibut

DONALD A. McCAUGHRAN
INTERNATIONAL PACIFIC HALIBUT COMMISSION
P.O. BOX 95009
SEATTLE, WA 98145

ABSTRACT
    Mark-recapture data was used to estimate growth in length of Pacific halibut Hippoglossus stenolepis by sex, region, and year classes (1935-1965). Analysis showed that males and females form separate families of growth functions; females grow faster and attain a larger size than males. There was little difference in growth of both sexes from 1935 to 1965 and no relationship to stock biomass was found. Migrating halibut were shown to grow slightly faster than non-migratory individuals. Growth was greater in the central part of the range and diminished toward the ends.

    The growth of Pacific halibut Hippoglossus stenolepis was studied during the 1960s by Southward and Chapman (1965) and Southward (1967). Back-calculated lengths from unsexed individuals were used to estimate the parameters in a Von-Bertalanffy growth model. It had been noted earlier (IFC 1930) that female halibut tend to grow at a faster rate than males and attain a larger maximum size, and halibut from Hecate Strait grow faster than halibut from western Alaska. Hoag, Schmitt, and Hardman (1979) found an increasing trend in average weight of similar aged halibut in a comparison of average weight for periods 1920-1939, 1940-1959, and 1960-1977. The present study attempts to quantify these observed phenomena by examining growth of both sexes over time and area.

METHODS

Tag Recovery Data Set
    Since the mid-1920s, the International Pacific Halibut Commission (IPHC) has tagged 260,000 halibut over its total range

The Age and Growth of Fish, edited by Robert C. Summerfelt and Gordon E. Hall © 1987 The Iowa State University Press, Ames, Iowa 50010.

(Fig. 1) and has recovered approximately 32,000 tags. Most tag recoveries come from halibut vessels using longline gear and trawlers fishing for other species. Length and sex information is often not obtained from the fishermen returning tags. Longlines and trawls have different length selectivity curves (Myhre 1969); trawls tend to catch smaller fish than longlines. The trawling gear used to capture halibut for tagging by IPHC did not catch 1-year-olds and caught only the largest 2- and 3-year-old halibut. These halibut exhibited above average growth for their age and were not included in the analysis. To investigate possible bias from faster growing young halibut caught with longline gear, a review of the estimates of instantaneous fishing mortality rate (F) obtained from cohort analysis (Hoag and McNaughton 1978) was conducted. The results showed that F increases with age until age ten and was relatively stable thereafter. It was decided, therefore, to use only those data from halibut captured with longline gear, that were at least ten years of age when marked. Observations were deleted if time at liberty was less than one month. The data set with complete information on sex, length, and recovery region, with the biased data removed, contained approximately 3,500 observations.

To facilitate the study of growth changes over time and area the data were subdivided by regions (Fig. 1) and year classes. Five year classes were grouped together beginning with 1936-1940 and ending with 1961-1965. To determine the effect of migration on growth, the data were separated into two groups; one where fish were marked and recaptured in the same region and the other where recapturing took place at least one region removed from the tagging site.

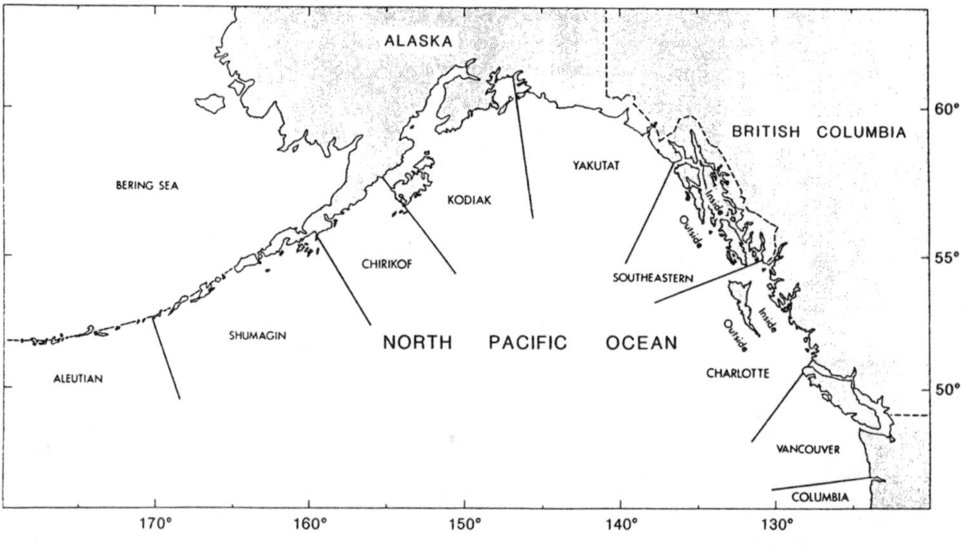

Figure 1. Map of Northeast Pacific with IPHC regions.

Growth Model

The growth model used in this analysis is a two parameter power function described in McCaughran 1981. The model is given by:

$L_j(t)$ = fork length of the jth halibut at age t

$L_j(t)$ ~ lognormal $(\mu(t), \alpha^2(t))$

Median $(L_j(t))$ = $e^{\mu(t)} = \alpha t^\beta$

where $\mu(t)$ and $\alpha^2(t)$ are the mean and variance of the underlying normal distribution, $L_j(t_m)$ = mark length, and $L_j(t_r)$ = recapture length.

Estimation

McCaughran (1981) derived an estimation procedure for the parameters $\ln \alpha$ and of the growth model using mark-recapture data. Minimum variance unbiased estimators for $\ln \alpha$ and β were constructed and the estimator $\hat{\alpha}t^{\hat{\beta}}$ is lognormally distributed with median αt^β. The median of the estimator is equal to the median of the distribution of length at age. This property is termed "median unbiased." The individual estimators are given by:

$$\hat{\beta}_j = \frac{\ln L_j(t_r) - \ln L_j(t_m)}{\ln t_r - \ln t_m}$$

$$\hat{\ln} \alpha_j = \ln L(t_m) - \hat{\beta}_j \ln t_m = \frac{\ln L_j(t_m)\ln t_r - \ln L_j(t_r)\ln t_m}{\ln t_r - \ln t_m}$$

and the population estimators by

$$\hat{\ln} \alpha = \frac{1}{n} \sum_{j=1}^{n} \ln \hat{\alpha}_j$$

$$\hat{\beta} = \frac{1}{n} \sum_{j=1}^{n} \hat{\beta}_j$$

with variances, $\text{Var}(\hat{\ln} \alpha) = c_1/n$ and $\text{Var}(\hat{\beta}) = c_2/n$ (c_1, c_2 are constants).

RESULTS

The growth model parameters $\ln \alpha$ and β were estimated for both males and females for each of the time area divisions for both migratory and non-migratory halibut. Figure 2 shows an example of a typical growth function and the data points used in estimation.

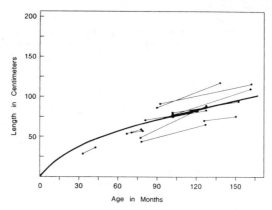

Figure 2. Typical male growth function showing data
points and estimated growth curve.

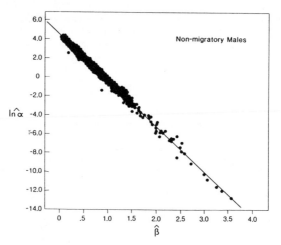

Figure 3. Relationship of the growth parameters $\hat{ln}\alpha_j$ and $\hat{\beta}_j$.

The relationship of $\hat{ln}\ \alpha$ to $\hat{\beta}$ was investigated by plotting
the pairs $(\hat{ln}\ \alpha_j, \hat{\beta}_j)$ (Fig. 3). The relationship proved linear,
with $ln\ \alpha_j = A_1 + A_2\ \hat{\beta}_j$ (A_1 and A_2 constants). The R^2 values
were 0.98 for females and 0.99 for males. The two relationships,
$ln\ \alpha = A_1 - A_2\beta$ and $L(t) = \alpha t^\beta$ imply a family of curves for
halibut of the form,

$$L(t) = K_1 \left(\frac{t}{K_2} \right)^\beta$$

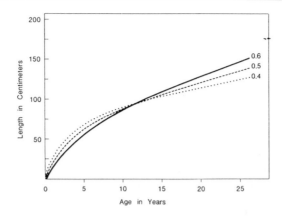

Figure 4. Examples of male growth curve family.

where $K_1 = e^{A_1}$ and $K_2 = e^{A_2}$. Hence, all growth curves pass through the common point (e^{A_1}, e^{A_2}). Figure 4 shows an example of the family of curves for male halibut. The estimates of e^{A_1} and e^{A_2} are given for both non-migratory and migratory males and females in Table 1.

Table 1. Estimates of the common length-age value (e^{A_1}, e^{A_2}) for the relationship $L = e^{A_1}(t/e^{A_2})^{\hat{\beta}}$ of male and female migratory and non-migratory halibut.

	Migratory		Non-Migratory	
	Males	Females	Males	Females
e^{A_1} (cms)	102.4	140.7	88.0	110.5
e^{A_2} (months)	155.1	190.3	127.0	144.6

Temporal and Regional Growth Rates

Since $\ln \alpha$ and β are linearly related, a test for differences in growth curves can be accomplished by testing for difference in β. The analysis of variance (ANOVA) is an appropriate test procedure since $\hat{\beta}_j$ are normal random variables. A two factor ANOVA was performed on the $\hat{\beta}_j$ using the year class groups and the regions as factors. To avoid missing cells in the analysis six 5-year birth periods 1936-1965 and 6 regions were used as factor levels. Columbia, Vancouver, Shumagin, and Aleutian regions were excluded from the analysis because of little or no data. The analysis was performed with data from non-migratory halibut so that the regional effect on growth could be determined.

The ANOVA results for males indicate that no interaction (P > 0.70) and no temporal effect (P > 0.07) were present but the main effect of regions was significant (P < 0.001). The results for females were slightly different. The interaction was significant (P = 0.01); there was no temporal effect (P > 0.10) and no regional effect (P > 0.08). The reason for the significant female interaction is the higher $\hat{\beta}_j$ values in all regions in the 1961-1965 cohorts except the Charlotte region. It was noted that the sample sizes in the groups showing higher $\hat{\beta}_j$ were small. Table 2 gives the cell and marginal mean $\hat{\beta}_j$ values for both males and females. A review of the age-length data from stock assessment surveys confirms the lack of significant growth change since the 1960s; however, large changes are observed in length at age when reviewed back to the 1920s. Unfortunately, the mark-recapture data have too few observations for analysis earlier than 1960.

Table 2. Estimates of the Parameter β and sample size (n) for regions andyear classes for non-migratory halibut.

Year classes	FEMALES						
	Regions						Mean
	Char	SEAK	Yaku	Kodi	Chir	BSea	
1936-1940	0.76	0.66	0.63	0.55	0.59	0.48	0.65
	(33)	(27)	(41)	(9)	(5)	(15)	(120)
1941-1945	0.77	0.66	0.63	1.06	0.56	0.63	0.74
	(178)	(60)	(59)	(34)	(10)	(27)	(368)
1946-1950	0.70	0.71	0.65	0.66	0.65	0.75	0.70
	(95)	(94)	(11)	(39)	(16)	(35)	(290)
1951-1955	0.67	0.60	0.57	0.72	0.69	0.70	0.68
	(134)	(26)	(4)	(56)	(39)	(28)	(287)
1956-1960	0.74	0.63	0.82	0.66	0.63	0.77	0.72
	(155)	(3)	(5)	(7)	(25)	(8)	(203)
1961-1965	0.65	0.92	0.85	0.89	0.89	0.84	0.71
	(120)	(11)	(13)	(5)	(5)	(7)	(161)
MEAN	0.72	0.69	0.66	0.77	0.66	0.69	0.71
	(715)	(221)	(133)	(150)	(100)	(120)	(1439)
	MALES						
1936-1940	0.56	0.51	0.64	0.54	1.06		0.58
	(37)	(12)	(17)	(1)	(1)		(68)
1941-1945	0.51	0.62	0.63	0.68	0.52		0.54
	(180)	(51)	(23)	(4)	(2)		(260)
1946-1950	0.46	0.65	0.65	0.86	0.38		0.55
	(115)	(83)	(8)	(8)	(1)		(215)
1951-1955	0.59	0.60	0.84	0.86	0.77		0.62
	(222)	(41)	(12)	(20)	(9)		(305)
1956-1960	0.60	0.58	0.62	0.55	0.71		0.60
	(243)	(3)	(11)	(5)	(5)		(267)
1961-1965	0.54	0.57	0.73	0.13	0.56		0.55
	(192)	(6)	(11)	(1)	(2)		(212)
MEAN	0.55	0.62	0.67	0.77	0.70		0.58
	(989)	(197)	(82)	(39)	(20)		(1327)

Migration Effects

Skud (1977) proposed that young halibut migrate in an eastward and southward direction to compensate for the north and westward drift of eggs and larvae. The mark-recapture data show extensive migrations, frequently as far as the Bering Sea to the coast of Oregon. A large amount of energy must be expended during such migrations. It was hypothesized, therefore, that migrating halibut should have lower growth rates than non-migrants.

The previous analysis did not show large differences between regions and age class groups, therefore, the data were combined over these categories to obtain larger sample sizes. The growth parameters (β) of migratory fish were compared to the growth parameters of non-migratory fish. The migratory females had significantly lower values of β (P > 0.01); the males were not significantly different. The estimates of β were 0.59 and 0.66 for migratory males and females and 0.59 and 0.70 for the non-migratory males and females. In addition, there were significantly different values of the parameters A_1 and A_2 for both migratory and non-migratory males and females (Table 1). The differences in the constants e^{A_1} and e^{A_2} and the growth parameter (β), which define the family of curves, show that migratory halibut grow faster and attain a larger size than non-migratory individuals. Growth curves for migratory and non-migratory males and females are shown in Figure 5.

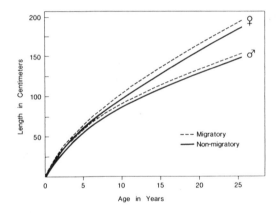

Figure 5. Growth curves for migratory and non-migratory males and females, all data combined.

DISCUSSION

Pacific halibut are long lived and may be subjected to differing population densities during their life span. Halibut under age 3 occupy shallow inshore areas and migrate to deeper water as they grow older; during their southeasterly migration, halibut often encounter varying population densities over different parts of the range. It is not clear, therefore, which densities might affect the growth of individual fish. Southward (1967) claimed that the average weight of all ages of halibut has increased since the 1950s. The data on length and weight at age collected on research cruises does not substantiate his claim. The weight at age from all regions has been similar for all year classes since 1950. Prior to that time the average weights were lower and the lowest weights for all ages occurred prior to 1920 when the stocks were very low.

The analysis of variance conducted on the growth parameter estimates $\hat{\beta}_j$ from the mark-recapture data indicate that no significant change in growth in length has occurred over the range of stock densities from 1935 to 1965. This result appears to be in contradiction to the weight-at-age data which shows lower weights for the period 1935 to 1950 than for the period 1950 to the present. The growth rates developed from the mark-recapture data and the weight-at-age estimates were not correlated with stock biomass (Fig. 6). These results indicate that other factors in the environment of halibut may be more important in influencing growth than the often reported density dependent phenomena.

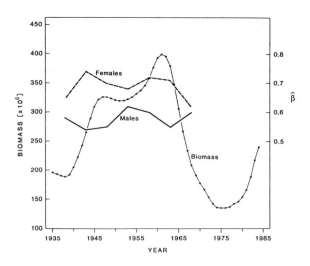

Figure 6. Estimates of the growth parameter β and
exploitable biomass from 1935 to 1985.

Halibut caught in the central part of the range (Kodiak, Chirikof) grow fastest and attain the largest size-at-age. The populations of fish from the Kodiak-Chirikof area are made up of unknown proportions of resident and immigrant fish. The immigrants are mostly young fish, and most of their growth takes place after becoming resident. It seems reasonable that conditions for growth are optimal in the center of the range and are less so at the ends.

The original intention of separating the recaptures into

migratory and non-migratory sets was to determine what loss in growth could be attributed to migration. It is somewhat surprising that the results indicate migrating halibut are faster growing than non-migrants. It appears that the energy requirement of migration is more than satisfied. Since most of the migration is completed before the onset of sexual maturity, the energy requirements of reproduction are not involved. It is a matter of speculation whether increased growth is genetically based or caused by increased encounters with food organisms during migration, or some other environmentally induced behavioral difference.

REFERENCES
HOAG, S. H. and R. J. McNAUGHTON. 1978. Abundance and fishing mortality of Pacific halibut, cohort analysis, 1935-1976. International Pacific Halibut Commission, Sci. Report No. 65.
HOAG, S. H., C. C. SCHMITT, and W.H. HARDMAN. 1979. Size, age, and frequency of male and female halibut: Setline Catches, 1925-1977. International Pacific Halibut Commission, Tech. Report No. 17.
INTERNATIONAL FISHERIES COMMISSION (IFC). 1930. Investigation of the International Fisheries Commission to December 1930 and their bearing on regulation of the Pacific halibut fishery. Int. Fish. Comm., Report No. 7.
McCAUGHRAN, D. A. 1981. Estimating growth parameters for Pacific halibut from mark-recapture data. Canadian Journal of Fisheries and Aquatic Sciences 38:394-398.
MYHRE, R. J. 1969. Gear selection and Pacific halibut. International Pacific Halibut Commission, Report No. 51.
SKUD, B. E. 1972. Drift, migration and intermingling of Pacific halibut stocks. International Pacific Halibut Commission, Scientific Report No. 63.
SOUTHWARD, G. M. and D. G. CHAPMAN. 1965. Utilization of Pacific halibut stocks: Study of Bertalanffy's growth equation. International Pacific Halibut Commission, Report No. 39.
SOUTHWARD, G. M. 1967. Growth of Pacific halibut. International Pacific Halibut Commission, Report No. 43.

Age determination of ocean bivalves

JOHN W. ROPES AND
AMBROSE JEARLD, JR.
NATIONAL MARINE FISHERIES SERVICE
NORTHEAST FISHERIES CENTER, WOODS HOLE LABORATORY
WOODS HOLE, MA 02543

ABSTRACT

Three commercially important ocean bivalve mollusks provide a varied, but fairly complete, methodological approach to age determinations of bivalves. Age determinations from examining the external valve surface and resilia of the Atlantic deep sea scallop Placopecten magellanicus, thin sections of the chondrophore of the Atlantic surf clam Spisula solidissima, and acetate peels of ocean quahogs Arctica islandica cover the range of methods applicable to population studies. Results support the hypothesis of an annual periodicity of age marks.

Bivalves accrete a relatively indestructible exoskeleton containing microstructural elements that are useful in determining age and growth. The elements may be recognized as fairly obvious macroscopic concentric bands or rings on the external valve surface to more subtle microscopic internal valve growth lines. Extremely fine daily and subdaily increments have been found for some species, such as the hard-shell clam Mercenaria mercenaria (Gordon and Carriker 1978). Considerable interspecific variation occurs in the valve microstructure of Bivalvia (Carter 1980). Age determinations at the macroscopic level for older individuals of a long-lived species are often impossible.

In bivalve population age research at the Woods Hole National Marine Fisheries Service Laboratory, the recognition of interspecific variability in growth lines denoting an annual periodicity has resulted in examinations of the external valve surface, radially-sectioned valves, and acetate peels from radially-sectioned valves. Accessory structures, such as resilia, chondrophores or hinge teeth, may be included or specifically prepared for examination, since the age record is often more complete in these structures than in the valve, and they constitute addition-

The Age and Growth of Fish, edited by Robert C. Summerfelt and Gordon E. Hall © 1987 The Iowa State University Press, Ames, Iowa 50010.

al confirming evidence for age determinations made from the valve.
Techniques for exposing the successive accretion of internal shell
structures have been known for some time by paleontologists and
biologists, but they have generally been avoided in routine fisheries research, because they are somewhat labor-intensive. However, the techniques have proved invaluable for accumulating accurate age data for two bivalve species.

The oceanic bivalves under investigation are the Atlantic
deep-sea scallop Placopecten magellanicus, the Atlantic surf clam
Spisula solidissima, and the ocean quahog Arctica islandica. After a brief historical review of age determinations of bivalves,
the method considered appropriate for each species is highlighted.
The adequacy of each method, alternate methods, and potential application of age estimates of other species are discussed, along
with validation of annuli.

HISTORICAL REVIEW

The literature on the basic structural components of bivalve
shells is voluminous. Investigations of bivalve shell structure
began in 1799, according to Schenck (1934). Important conclusions
of his review were that characteristic and definable structures
are produced and that traces of growth stages occur in the shells.
Weymouth (1923) reviewed 14 investigations, the earliest of which
was 1859, that suspected or used growth lines as age marks for
various bivalves, but he concluded that evidence for an annual periodicity of their formation was less than unanimous. Weymouth
(1923) identified suspected annuli in the valves of the Pismo clam
Tivela stultorum by exposing structural features in polished radial sections and thin-sections. Four general layers were evident,
the periostracum, an outer prismatic layer, an inner mother-of-
pearl layer, and a nacreous layer. The outer layer was interrupted by opaque lines alternating with translucent lines of growth
composed of modified groups of lamellae. These lines had continuity from a hinge tooth into the umbo and terminated at the external valve surface rings, which were later established as reliable
annuli. Weymouth's (1923) report has been a basic reference for
many later studies of bivalve age and growth.

Taylor et al. (1969, 1973) and Rhoads and Lutz (1980) have
greatly expanded our understanding of shell structure, mineralogy,
and growth of bivalves. These, and the work of Weymouth (1923),
form the basis for age determination of the three bivalves of concern to us at the NMFS Laboratory.

OPERATIONAL SPECIES METHODOLOGY

Atlantic Deep-Sea Scallop

Age is determined for sea scallops by the external valve
ring method developed for Pismo clams. A slight modification in
this technique is that the external surface of the left valve is

routinely examined; this species lives with the right valve opposed to the erosional effects of bottom substrata, which often
results in poorly defined rings on its dirty white surface (Merrill et al. 1966). Typically, dark rings suggestive of successive
growth increments interrupt shades of reddish-brown (rarely yellow
to lavendar) coloration on the valve. Microscopic inspection
of the valve surface reveals fine concentric and scalloped ridges,
termed circuli, that become more dense at each ring.

Age estimates of sea scallop valves may be confirmed after
disarticulating the paired valves and examining a part of the
hinge. Calcareous plates within the hinge form lateral linings of
a dark, V-shaped elastic resilium. The plates bond the resilium
to socket-like resilifers. Annuli appear on the resilium as dark
bands opposed to the lighter background coloration and are equal
in number and relative position to rings on the valves (Fig. 1);
constrictions on the lateral surfaces of the resilium also correspond to the annual rings. These internal age marks are sometimes
more prominent than the valve rings, due apparently to protection
from injury of the epithelial cells involved in their accretion.

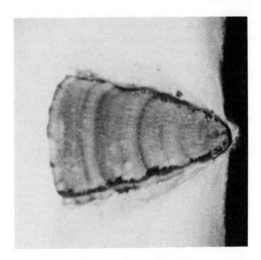

Figure 1. Annuli in the resilium of <u>Placopecten</u> <u>magellanicus</u>.

Atlantic Surf Clam
From 1910 to the mid-1970's, the few investigators of the
age of surf clams relied on external valve rings (Ropes 1980).
Ages to 17 years for individuals not larger than 163 mm in shell
length were reported, yet larger surf clams to a maximum of 226 mm
occur in the Middle Atlantic Bight population. The evidence for
accepting an annual periodicity of such age marks was marginally
qualified; investigators admitted to being unable to determine the
age of old clams with confidence, because the rings crowd together

at the valve margins or the earliest rings are eroded beyond detection. This method can effectively exclude age assessment for a substantial part of the population.

In recognition of deficiencies in the ring method, techniques were developed in 1975 by the NMFS for exposing internal growth lines in valves radially sectioned through the umbo to the ventral margin (Fig. 2). After polishing the cut edge, microscopic examination revealed dark, narrow growth lines terminating at the valve surface rings. These curved toward the umbo and to a thin middle layer, the myostracum, which separated inner from outer layers. The growth lines were clearly differentiated from broad, white growth increments in young clams, but, as expected, they crowded together at the valve margin in older clams. Nevertheless, their consistent dark coloration enhanced detection.

The radial-sectioning technique was a positive advance in the ability to determine age, but the hour or more preparation and examination procedures per specimen were excessive. A more rapid method was needed, since 1,500 or more surf clams are collected for age determination annually.

Radial sectioning routinely exposed a succession of growth lines in the robust surf clam chondrophore (Fig. 2). They appeared to be similar to the age marks in the valve, although they were more compressed in this smaller structure. Techniques have been developed to excise a portion of the chondrophore and umbo, glue it onto a slide, and produce sections about 0.25-mm thick (Ropes and O'Brien 1979). Preparation time was reduced to less than 30 min per specimen, but, more importantly, the small, flat section overcame the difficulties of holding a valve level and in focus during microscopic examination. The method greatly improved detection of annuli, which are translucent and in sharp contrast with opaque growth increments under transmitted light. Also, prints can be made easily by using the thin-section as a negative in a photographic enlarger (Fig. 3a). Measurements from the umbo

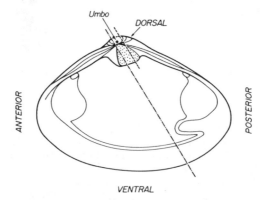

Figure 2. Sketch showing chondrophore (stippled area) excision of _Spisula_ _solidissima_ (Ropes and O'Brien 1979).

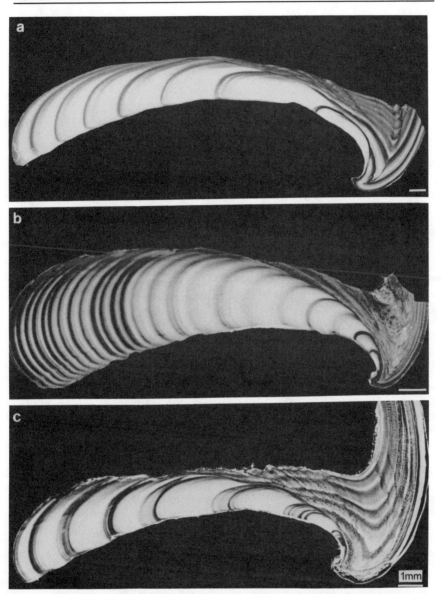

Figure 3. Thin-sectioned chondrophores of (a) <u>Spisula solidiss-ima</u>, (b) <u>S. polynyma</u>, and (c) <u>S. sachalinensis</u>. A millimeter scale is shown in (a).

to successive annuli in the chondrophore and valve have establish-ed a correspondence for their relative location and number (Fig. 4). A longevity of 37 years has been found for surf clams.

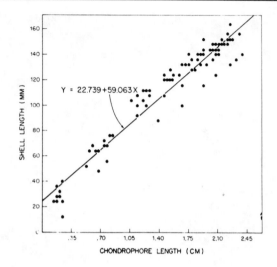

Figure 4. Plotted measurements of corresponding annuli in chondrophores and valves of <u>Spisula solidissima</u> (r = 0.97) (Ropes
and O'Brien 1979).

Ocean Quahog

The few age investigations of ocean quahogs conducted from
1929 to 1974 used the external valve ring method, even though evidence for an annual periodicity of their formation was poorly
founded (Murawski et al. 1982). Dark, spaced-apart rings often
contrast clearly with the light amber to mahogany periostracum of
small ocean quahogs; however, ring detection in the thick, uniformly black periostracum of large individuals, especially at the
valve margin, is generally considered to be impossible. Ages to
about 20 years were reported for individuals smaller than 60 mm in
shell length, but many ocean quahogs attain 100 mm in shell length,
and growth to a maximum of 139 mm is known for the species. Age
assessment of a part of the population was again excluded by the
method.

Age determination of ocean quahogs was significantly advanced by Thompson et al. (1980) and Jones (1980). They used the acetate peel method, which exposed internal age structural elements
for all sizes of clams. At our laboratory, left valves are radially sectioned from the umbo to ventral margin and include a hinge
tooth. This is followed by bleaching to remove the periostracum,
embedding the valves in an epoxy resin, polishing the cut edge to
a high luster and etching with 1% HCl for 1 min before sheet acetate is applied with acetone and allowed to dry. Then the sheet
is peeled off and compressed between slides for examination. Fine
annuli curve downward from exit locations at the outer valve surface to the myostracum and toward the umbo. Broad growth increments clearly separate annuli in small specimens and the exit lo-

cations of annuli occur at external rings; greatly compressed growth increments separate annuli with increasing age (Fig. 5).

The procedural steps of the peel method are time-consuming. Nevertheless, it has proved invaluable in gaining knowledge about the growth history of selected portions of the population and surpasses other known methods of determining age for ocean quahogs (Ropes 1984).

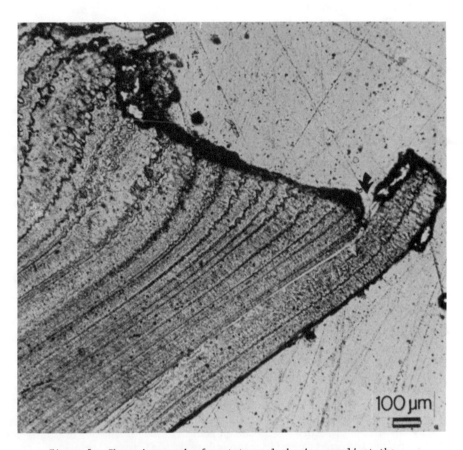

Figure 5. Photomicrograph of acetate peel showing annuli at the valve margin of a 100-yr-old, 100 mm in shell length <u>Arctica islandica</u> recovered two years after marking. Many annuli are shown and an arrow points to an interruption of growth and annulus formed at or soon after marking the clam. The flattened area was produced by the notching operation, which clearly separated shell growth and annuli formed before and after notching. Note that only one additional annulus was formed during the two years it was free in the environment, but that two increments of shell growth had been accreted (Ropes et al. 1984).

RESULTS AND DISCUSSION

The inadequacies of the external valve ring method for determining the age of longevous bivalves have been recognized for
some time. The general features of annual rings on the valves and
resilia of sea scallops are readily defined, but difficulties are
experienced in routine examinations of samples from the United
States resource (Merrill et al. 1966). Fortunately, the difficulties of determining age for sea scallops are not without manage-
able resolution, if procedural steps developed by Merrill et al.
(1966) are followed.

Unfortunately, thin-sectioned and peel preparations of sea
scallop shells are unsatisfactory for exposing age/growth lines.
Only portions of growth lines were detected at locations directly
beneath well-defined external rings. The shells of pectinids are
composed mostly of an outer, calcitic, foliated layer (Taylor et
al. 1969, 1973), and the irregular pattern of this microstructure
obscured detection of growth lines within the internal matrix.
Similar but weaker lines occurred throughout the layer between the
external rings, and growth lines were not clearly evident in other
layers of the shell or the resilifer. Preparations of the Iceland
scallop Chlamys islandica and Patagonian scallop C. patagonia,
were similarly disappointing. The poor resolution of growth lines
in the outer valve layer of molluscs having the foliated structure
has been recognized by Rhoads and Panella (1970). Thus, at present there is no better method for determining the age of sea scallops than examining valve rings and resilia.

The two-layered, aragonitic shells of mactrids have a thin
pallial myostracum separating outer and inner crossed-lamellar
layers (Taylor et al. 1969, 1973). More complex microstructural
elements have been found for S. solidissima by Jones (1980), and
those for growth increments were distinctly different from annuli.
Jones (1981) examined other than annual repeating layers in S.
solidissima shells, but these were variable in number and
aperiodic.

Properties of the growth-line microstructure of mactrids allow light transmission. Translucent growth lines occurred in
thin-section preparations of Stimpson's surf clam S. polynyma
(Fig. 3b) and the Japanese surf clam S. sachalinensis (Fig. 3c).
It is assumed that the translucent property of the growth lines
for these close relatives of S. solidissima is based on a similarity of microstructure, but the potential use of such lines for age
determination awaits validation of an annual periodicity.

Tag and/or mark-recapture experiments have been conducted
for sea scallops (Merrill et al. 1966), surf clams (Jones et al.
1978) and ocean quahogs (Murawski et al. 1982; Ropes et al. 1984).
The experiment for ocean quahogs was perhaps most daring. Almost
42,000 ocean quahogs were marked with a pair of parallel notches
at the ventral valve margin and released at a 53 m deep site off
Long Island, New York, in late July-early August 1978 (Murawski et
al. 1982; Ropes et al. 1984). Although only 1% have been recovered to date, they have shown that suspected annular lines are form-

ed annually (Fig. 5). Scanning electron microscope (SEM) examinations have revealed a distinctive microstructure for annuli and
growth increments accreted by ocean quahogs (Ropes et al. 1984).
Growth was characterized as slow, and evidence from the marking
experiment was instrumental in establishing a 225-year longevity
for the species.

Radiometric and stable isotope analyses support the hypothesis of an annual periodicity of annuli in the shells of ocean
quahogs and surf clams (Turekian et al. 1982; Jones et al. 1983).
These independent studies add confirmation to the results of mark/
recapture experiments. However, growth rates of sea scallops derived from isotope analyses were almost double estimates made from
external bands (Krantz et al. 1984). Isotope analyses of mark/
recaptured specimens would have substantially validated such results by fishery research standards. Even with such validation,
the method is very time-consuming and applicable for only a few
(2-3) specimens, conditions that make it inefficient for assessment of population age composition. Currently, then, the ring
method remains as the practical alternative for ageing scallops.

In conclusion, current population age studies of bivalves
have the advantage of examining internal age/growth phenomena, an
approach that is recommended for species exhibiting difficult-to-
interpret external rings, or ones that have a long life span. Examinations of microstructural elements and experiments to identify
an annual periodicity of age/growth phenomena greatly increases
the validity of such studies.

REFERENCES
CARTER, J. G. 1980. Chap. 2. Environmental and biological con-
    trols of bivalve shell mineralogy and microstructure. Pages
    69-113. In: D. C. Rhoads and R. A. Lutz (editors). Skele-
    tal growth of aquatic organisms. Plenum Press, New York.
GORDON, J., AND M. R. CARRIKER. 1978. Growth lines in a bivalve
    mollusk: Subdaily patterns and dissolution of the shell.
    Science 202(3):519-521.
JONES, D. S. 1980. Annual cycle of shell growth and reproduction
    in the bivalves Spisula solidissima and Arctica islandica.
    Doctoral dissertation. Princeton University, Princeton, New
    Jersey, USA.
JONES, D. S. 1981. Repeating layers in the molluscan shell are
    not always periodic. Journal of Paleontology 55:1076-1082.
JONES, D. S., I. THOMPSON, AND W. AMBROSE. 1978. Age and growth
    rate determinations for the Atlantic surf clam, Spisula
    solidissima (Bivalvia:Mactracea), based on internal growth
    lines in shell cross-sections. Marine Biology 47:63-70.
JONES, D. S., D. F. WILLIAMS, AND M. A. ARTHUR. 1983. Growth
    history and ecology of the Atlantic surf clam, Spisula soli-
    dissima (Dillwyn), as revealed by stable isotopes and annual
    shell increments. Journal of Experimental Marine Biology
    and Ecology 73:225-242.
KRANTZ, D. E., D. S. JONES, AND D. F. WILLIAMS. 1984. Growth
    rates of the sea scallop, Placopecten magellanicus, deter-

mined from the $^{18}O/^{16}O$ record in shell calcite. Biological
Bulletin 167:186-199.

MERRILL, A. S., J. A. POSGAY, AND F. E. NICHY. 1966. Annual
marks on the shell and ligament of sea scallop (Placopecten
magellanicus). Fishery Bulletin 65:299-311.

MURAWSKI, S. A., J. W. ROPES, AND F. M. SERCHUK. 1982. Growth of
the ocean quahog, Arctica islandica, in the Middle Atlantic
Bight. Fishery Bulletin 80:21-34.

RHOADS, D. C., AND G. PANELLA. 1970. The use of molluscan shell
growth patterns in ecology and paleoecology. Lethaia 3:
143-161.

RHOADS, D. C., AND R. A. LUTZ. 1980. Skeletal Growth of Aquatic
Organisms. Plenum Press, New York, 750 p.

ROPES, J. W. 1980. Biological and fisheries data on the Atlantic
surf clam, Spisula solidissima Dillwyn. U.S. Department of
Commerce, National Oceanic and Atmospheric Administration,
National Marine Fisheries Service, Technical Series Report
No. 24, 88 pp.

ROPES, J. W. 1984. Procedures for preparing acetate peels and
evidence validating the annual periodicity of growth lines
in the shells of ocean quahogs, Arctica islandica. Marine
Fisheries Review 46(2):27-35.

ROPES, J. W., AND L. O'BRIEN. 1979. A unique method for aging
surf clams. Bulletin of the American Malacological Union
1979:58-61.

ROPES, J. W., D. S. JONES, S. A. MURAWSKI, F. M. SERCHUK, AND A.
JEARLD, JR. 1984. Documentation of annual growth lines in
ocean quahogs, Arctica islandica Linné. Fishery Bulletin
82:1-19.

SCHENCK, H. G. 1934. Literature on the shell structure of pe-
lecypods. Bulletin de Museé d'Historie Naturelle de
Belgique 10:1-20.

TAYLOR, J. D., W. J. KENNEDY, AND A. HALL. 1969. The shell
structure and mineralogy of the Bivalvia: Introduction,
Nuclacea-Trigonacea. Bulletin of the British Museum (Natural History) Zoology, Supplement No. 3, 125 p.

TAYLOR, J. D., W. J. KENNEDY, AND A. HALL. 1973. The shell
structure and mineralogy of the Bivalvia. II. Lucinacea-
Clavagellacea: Conclusions. Bulletin British Museum (Natural History) Zoology 22:253-294.

THOMPSON, I., D. S. JONES, AND D. DREIBELBIS. 1980. Annual in-
ernal growth banding and life history of the ocean quahog,
Arctica islandica (Mollusca:Bivalvia). Marine Biology 57:
25-34.

TUREKIAN, K. K., J. K. COCHRAN, Y. NOZAKI, I. THOMPSON, AND D. S.
JONES. 1982. Determination of shell deposition rates of
Arctica islandica from the New York Bight using natural
^{228}Ra and ^{228}Th and bomb produced ^{14}C. Limnology and
Oceanography 27:737-741.

WEYMOUTH, F. W. 1923. The life history and growth of the Pismo
clam (Tivela stultorum Mawe). Fishery Bulletin of California No. 7, pp. 1-120.

Glossary

Accuracy - The proximity of an age estimate to the actual (absolute) value.

Age - A unit to express the passage of time measured in years, months, days, or other units.

Age-Group - A group of fish of a given *age (day, year)*. It is not synonymous with *year-class*.

Annual mark - Structural features (e.g., annuli) that correlate with a yearly event.

Annulus (plural, *annuli*) - A concentric mark on a structure that may permit its growth to be interpreted in terms of age. The optical appearance of these marks depends on the type of structure and the species and should be defined in terms of specific characteristics on the structure. For example, on scales it is a line or zone of discontinuity in the pattern of the circuli. On other hard structures the *annulus* may be defined as a continuous *translucent* (or less frequently *opaque) zone* that can be seen along the entire structure; or the *annulus* may be a ridge or a groove in or on the structure. A unit passage of time (i.e., 1 year) is not inherently implied unless specified. However, this term has traditionally been used to designate year marks even though dictionaries do not refer to time in the definition. Annulus does not apply to the microstructure of otoliths.

Band (rings, marks, zones) - Auxiliary descriptive terms used in interpreting the age of a fish from calcified structures.

Calendar age - The age of a fish based on calendar year rather than to the true date of hatch. In the northern hemisphere, January 1 is designated as the arbitrary birth date and in the southern hemisphere, July 1 is designated as that corresponding date. In the northern hemisphere *calendar age* 0 represents those fish that have not yet passed their first January 1, and fish of *calendar age 1 are those that have passed their first January 1.*

Check - An abrupt discontinuity in a zone or band. It is used in association with breaks in the configurations of the *circuli* on scales; however, it can also be used in reference to other hardparts where there is a incomplete zone or irregular spacing.

Circulus - A concentric bony ridge on the outer surface of a fish scale formed by the periodic addition of material. The *circulus* can be continuous or segmented by the scale radii.

Cohort - A group of fish that begins life at about the same time and is produced during a relatively discrete spawning event. It is difficult to apply this term to fish that are spawned

527

throughout the year. On the other hand it might not imply *year- class* when reproduction occurs monthly or with some other periodicity.

Core - The concentric area or areas of non-incremental growth surrounding the primordium or primordia of an otolith.

Focus (origin) - The hypothetical or real point of origin of the prepared or whole structure where the reader chooses to start a count or use as a reference point for measurement.

Increment - A general reference to the interval between like zones on a structure used for age-determination. The interval has an alternating bipartite composition and indicates a unit passage of time. Depending on the structure involved, the dimensions, chemistry, and period of formation of the two zones can vary widely. In otoliths, a daily or subdaily *increment* consists of two layers, or zones, dominated alternately by calcium carbonate and organic matrix.

Kernel - see *Nucleus*

Marginal increment (also referred to as edge) - The region beyond the last identifiable mark at the margin of a structure used for age determination. Quantitatively this increment is expressed in relative terms, i.e., as a fraction or proportion of the last complete growth increment. (See Casselman this proceedings for an example).

Nucleus, kernel - Collective terms used in the past which we consider ambiguous. Other terms listed in this glossary are preferred.

Opaque zone - A zone that inhibits the passage of light. Under transmitted light the opaque zone appears dark and the *translucent zone* appears bright. Under reflected light the *opaque zone* appears bright and the *translucent zone* appears dark.

Precision - A measure of the repeatability, or consistency, of counting concentric bands or other age-related features on the same skeletal structure. Provides a measure of the variability associated with an age estimate.

Primordium - A self-contained zone that represents the point (or points) of the initial growth of an otolith. Optically, it is a phase of growth within the core which is more opaque than the surrounding area of the core.

Radius - The linear distance from a focus to a specified point (may not conform to the mathematical definition). It is also an anatomical term referring to a line or fold scleritized and is in both the anterior and posterior fields of a scale.

Soft ray - A skeletal element in the fins of bony fishes, characterized by: (1) opened base; (2) distal radial element within open base; (3) right and left halves either separate or with a mid-sagittal suture; (4) branched distally; and (5) always segmented. Usually, criteria (4) and (5) hold true, however, there are exceptions.

Spiny ray (thin spine, spine) - A skeletal element in the fins of
     bony fishes, characterized by: (1) closed base; (2) distal
     radial always outside spine; (3) not separated into right
     and left halves; (4) unbranched distally; (5) unsegmented.
Translucent zone - A zone that allows the passage of light. The
     term "hyaline" has been used to define the same zone, but
     translucent is the preferred term.
Validation - The confirmation of the temporal meaning of an
     increment. It is used to determine the accuracy of an age
     determination. This term is frequently confused with
     verification, i.e., the repeatability of a numerical
     interpretation that may be independent of age. For example,
     if two readers agree on the number of zones present in a
     hardpart, or if two different age determination structures
     are interpreted as having the same number of zones, verifi-
     cation has been accomplished.
Year-class - Fish spawned or hatched in a given year. In the
     northern hemisphere, if spawning occurs in autumn and
     hatching in spring, the calendar year of hatching is
     commonly used to identify *year-class*.

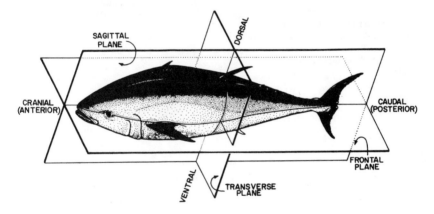

Figure 1. Axes and planes of an oceanic pelagic fish (tuna).

     Sections of structures under examination are to be described
under the following criteria based on the position of the hardpart
relative to organism being examined, Fig. 1. (Illustration by
J. C. Javech and reprinted from E. P. Prince and L. M. Pulos
1983. Proceedings of the International Workshop on Age Determination of Oceanic Pelagics Fishes: Tunas, Billfishes, and
Sharks. NOAA Tech. Report NMFS 8). A. Transverse (cross)
section. B. Longitudinal - horizontal = frontal vertical =
sagittal mid = a cut through the center, para = a cut that is off
center.

Glossary Committee

Dr. Charles A. Wilson, Chairman
Coastal Fisheries Institute
Center Center for Wetland Resources
Louisiana State University
Baton Rouge, La   70803-7503

Dr. R. J. Beamish
Dept. of Fisheries and Oceans
Fisheries Research Branch
Pacific Biological Station
Nanaimo, BC V9R 5K6
CANADA

Dr. Edward B. Brothers
3 Sunset West
Ithaca, NY 14850

Dr. Kenneth D. Carlander
Department of Animal Ecology
Iowa State University
Ames, IA   50011

Dr. John M. Casselman
Ontario Ministry of Natural Resources
Research Section, Fisheries Branch
Box 50
Maple, Ontario LOJ 1ED
CANADA

Dr. John M. Dean
Department of Biology and
    Marine Science Program
University of South Carolina
Columbia, SC 29208

Dr. Ambrose Jearld, Jr.
National Marine Fisheries Service
Northeast Fisheries Center
Woods Hole Laboratory
Woods Hole, MA 02543

Dr. Eric D. Prince
Southeast Fisheries
    Miami Laboratory
75 Virginia Beach Drive
Miami, Fl   33149-1099

Dr. Alex Wild
Inter-American Tropical
    Tuna Commission
Scripps Institute
La Jolla, CA 92092

Subject Index

Acetozolamide, 328
Africa, 103
African freshwater fishes, 86
Age
  accuracy of estimates, 203,
     241, 253, 427, 433, 483
  annual, 160, 203, 319, 359, 517
  comparison among methods, 24,
     203, 207, 241, 279, 301,
     359
  composition, 148, 253
  daily, 319, 413, 433, 517
  determination of, by use of
    bony structures, 159
    chemical microanalysis, 360
    cleithra, 25, 32
    computer assisted devices,
      385, 397
    electron microprobe analysis,
      359
    embryonic growth rates, 360
    fin rays, 23, 24, 25, 26, 30,
      115
    growth zone deposition, 360
    length-frequency analysis.
      See Length-frequency
    microanalytical technique,
      360
    opercle bones, 32, 115, 398
    otoliths, 26, 30, 115, 167,
      203, 223, 241, 253,
      267, 319, 413, 483
    radiometric dating, 301, 360
    scales, 3, 22, 115, 148, 177,
      203, 241, 397
    size-mode. See Length-
      frequency
    spines, 30, 32, 287, 359,
      360, 386
    tooth-replacement rates, 360
    vertebral centra, 32, 35,
      115, 301, 359
  estimates of, 1, 167, 287, 301,
    359, 433
  -frequency distribution, 476
  group
    composition, 6, 148
    effects on growth, 128
    growth of, 128, 371

  mean length of, 371, 372
  relative abundance of, 372
  separation of, 372
  -length keys, 149, 254, 495,
    496, 497, 499
  oldest recorded, 22, 26
  underestimates of, 8, 15, 22,
    140, 203, 241. See also
    Aging, errors in
  validation of, 3, 9, 16, 29,
    32, 148, 204, 223, 241,
    253, 279, 287, 301, 359,
    360, 398, 443, 496
Aging
  comparison of methods for, 24,
    203, 207, 241, 279, 301,
    359
  of elasmobranchs, 359
  errors in, 8, 22, 24, 28, 148,
    177, 189, 203, 211, 241,
    427
    effects of, on management
      decisions, 26, 28, 203.
      See also Stock assess-
      ment
  fish caught by anglers, 211,
    241
  methods, 301, 319
  techniques, 253
Alabama, 242
Alaska, 16, 507
Algorithm
  EM, 376
  Nelden-Mead, 376
  Quasi-Newton, 376
Alkaline phosphatase, 65
Allometric. See Growth,
  allometric
Amino acids
  extraction of, 345
  uptake of, 65
Anesthesia, 288, 454
Anglers, fish caught by, 211, 241
Annulus
  counting of, 320, 385, 399
  errors in recognition of. See
    Aging, errors in
  failure to form, 177, 179, 204
  first, 16, 161, 178, 183, 475

Taxonomic Index